Other monographs in the series, Major Problems in Clinical Pediatrics:

Altman and Schwartz: *Malignant Diseases of Infancy, Childhood and Adolescence*—1978

Avery and Fletcher: *The Lung and Its Disorders in the Newborn Infant*—Third Edition, 1974

Bell and McCormick: *Increased Intracranial Pressure in Children*—Second Edition, 1978

Bell and McCormick: *Neurologic Infections in Children*—1975

Brewer: *Juvenile Rheumatoid Arthritis*—1970

Cornblath and Schwartz: *Disorders of Carbohydrate Metabolism in Infancy*—Second Edition, 1976

Dubowitz: *Muscle Disorders in Childhood*—1978

Gryboski: *Gastrointestinal Problems in the Infant*—1975

Hanshaw and Dudgeon: *Viral Diseases of the Fetus and Newborn*—1978

Lubchenco: *The High Risk Infant*—1976

Markowitz and Gordis: *Rheumatic Fever*—Second Edition, 1972

Oski and Naiman: *Hematologic Problems in the Newborn*—Second Edition, 1972

Rowe and Mehrizi: *The Neonate with Congenital Heart Disease*—1968

Royer, et al: *Pediatric Nephrology*—1974

Scriver and Rosenberg: *Amino Acid Metabolism and Its Disorders*—1973

Smith: *Recognizable Patterns of Human Malformation*—Second Edition, 1976

Smith: *Growth and Its Disorders*—1977

Solomon and Esterly: *Neonatal Dermatology*—1973

Solomon, Esterly and Loeffel: *Adolescent Dermatology*—1978

Forthcoming Monographs

Belman and Kaplan: *Urologic Problems in Pediatrics*

Bluestone and Klein: *Otitis Media in Infants and Children*

Drash: *Juvenile Diabetes*

Glader: *Anemia in Children*

Griffin: *Children's Orthopaedics*

Solomon, Esterly, and Loeffel: *Adolescent Dermatology*

Wannamaker: *Streptococcal Infections*

DISORDERS OF CALCIUM AND PHOSPHATE METABOLISM IN CHILDHOOD AND ADOLESCENCE

by
Harold E. Harrison, M.D.
Professor of Pediatrics, Johns Hopkins University
School of Medicine, Baltimore

Helen C. Harrison, Ph.D.
Associate Professor of Pediatrics, Johns Hopkins University
School of Medicine, Baltimore

Volume XIX in the Series
MAJOR PROBLEMS IN
CLINICAL PEDIATRICS

ALEXANDER J. SCHAFFER
Consulting Editor
MILTON MARKOWITZ
Associate Consulting Editor

W. B. Saunders Company, Philadelphia, London, Toronto 1979

W. B. Saunders Company: West Washington Square
Philadelphia, PA 19105

1 St. Anne's Road
Eastbourne, East Sussex BN21 3UN, England

1 Goldthorne Avenue
Toronto, Ontario M8Z 5T9, Canada

Disorders of Calcium and Phosphate Metabolism
in Childhood and Adolescence

ISBN 0-7216-4537-2

 Made in the United States of America. Press of W. B. Saunders Company. Library of Congress catalog card number 78-64712.

Last digit is the print number: 9 8 7 6 5 4 3 2 1

To Our Children
Stephen, Richard and Ellen
and our grandchildren
Melissa and Rebekah

Foreword

One of the very first volumes I tried to corral for the Major Problems in Clinical Pediatrics series was one by the distinguished and delightful couple, Harold and Helen Harrison. I am ashamed to say that I continued to belabor them incessantly for 15 years. Finally, after Dr. Harold's retirement, so-called, they agreed to take on the task. It was worth waiting for.

Dr. Harold Harrison earned his B.S. and M.D. from Yale, served his residency and instructorship years at the New Haven Hospital, then moved on to Cornell for a few years. In 1945 he came to the Johns Hopkins where he has remained ever since, semi-retiring as an Emeritus Professor of Pediatrics in 1975.

Dr. Helen Harrison received her A.B. from Goucher College, her M.A. from Smith and her Ph.D. from Yale. After stops at Cornell and Yale, she too went to the Hopkins where she still serves as Associate Professor of Pediatrics. She works closely with Harold, and together they were granted the E. Mead Johnson Award of the American Academy of Pediatrics, the only such joint award of which I am aware.

I need not tell anyone in pediatrics of the Harrisons' productive work in their fields of chosen interest: the physiology of calcium and phosphorus, Vitamin D, parathyroid hormone and bone, renal and intestinal transport, and water and electrolyte balance. Their publications are almost beyond number. What some of you may not know is that Harold Harrison, over and above all of this research productivity, is one of the finest pediatricians I have ever come across.

ALEXANDER J. SCHAFFER, M.D.

Preface

We have been investigating the physiology of calcium and phosphate homeostasis for many years. In 1946 the late Dr. Edwards A. Park stimulated us to set up a laboratory for this purpose in The Harriet Lane Home of the Johns Hopkins Hospital. We were able to combine basic investigation into the physiological mechanisms governing calcium and phosphate transport with studies of patients with disorders of calcium and phosphate metabolism. These patients were studied at both the Baltimore City Hospital and the Johns Hopkins Hospital and we are indebted to our many colleagues who referred patients to us.

During the past two decades there have been major advances in our knowledge of the biochemistry and physiology of vitamin D, parathyroid hormone and other controlling factors in calcium and phosphate metabolism. Our more precise knowledge of physiological mechanisms has improved our understanding of clinical disorders of calcium and phosphate metabolism and given us more effective diagnostic techniques and methods of management. With the encouragement of Dr. Alexander J. Shaffer we prepared this monograph to coordinate present physiological knowledge with our patient experience in the expectation that this will be of help to physicians in their study and treatment of disorders of calcium and phosphate metabolism occurring during the period of growth and maturation.

The patient studies reported in this book required the cooperation and active participation of many residents and fellows at both the Baltimore City Hospital and the Johns Hopkins Hospital. It is not possible to list all, but we do wish to acknowledge the help of a number of colleagues who were particularly involved in patient investigations. Drs. Robert Klein and Buford L. Nichols collaborated with us in our early studies of hypoparathyroidism. Dr. Thomas P. Foley, Jr., and the late Dr. Frederick M. Kenney participated in the investigation of children with hypercalcemia. Drs. Jaakko Perheentupa, Fima Lifshitz and Russell Chesney provided valuable assistance in the investigation of patients with rickets. Dr. Laurence Finberg was a close collaborator in many areas and was of great help in the development of our methods of treatment of primary hypophosphatemic rickets.

During the course of years we visited a number of other departments of pediatrics and were able to see patients whose data are included in this book. We are indebted to Drs. Robert A. Blizzard, Allan Drash, Thomas P. Foley, Jr., Frederick Heldrich, Thomas Holcomb, Ann Johannson, John Rosen and Leo Stern in particular for providing details of certain of their patients and permitting us to include them.

Special thanks are owing to Dr. John Dorst for valuable help in selecting and interpreting many of the illustrative radiographs.

In our experimental studies in the laboratory we have been assisted by a number of fellows, medical students and technicians. We are particularly indebted to Mrs. Evelyn Bull for helpful technical assistance.

Mrs. Marian A. Nugent was of important help in the preparation of the manuscript and we thank the staff of the Harold E. Harrison Library at the Baltimore City Hospital for their assistance with the bibliography.

We also express our appreciation of the editorial contribution of Ms. Katherine Pitcoff, Associate Medical Editor, W. B. Saunders Company.

We have received support for our work from an NIH research grant, AM-00668, and an NIH contract 263-76-C-90.

HAROLD E. HARRISON, M.D.
HELEN C. HARRISON, PH.D.

Contents

Chapter One

BONE

Bone is a composite of a hard, crystalline mineral and a microfibrillar matrix into which the mineral is incorporated. The combination of hardness and tensile strength provided by these two components makes bone an excellent structural material for supporting the weight of the body and the pull of muscles. Bone is also a growing, living tissue which is remodeled and reorganized both with growth and in response to the forces acting upon it. There is, therefore, a cellular component of bone tissue. These cells not only function in the growth and reshaping of bone structure in response to external forces, but also act to regulate the flow of minerals into and out of bone; thus, the skeleton is not only a supporting organ but also a reservoir of calcium, magnesium and phosphate ions that can function to stabilize the concentrations of these ions in extracellular fluid.

THE BONE MINERAL AND MATRIX

The crystalline phase of bone has been classified by x-ray diffraction studies in the group of apatite minerals. Bone salt is regarded as a hydroxyapatite with the basic formula $Ca_{10}(PO_4)_6(OH)_2$, although some investigators have suggested that bone salt could be an octacalcium phosphate, $Ca_8[H_2(PO_4)_6\ 5H_2O]$. The composition of hydroxyapatite reflects in part the composition of the fluid phase from which the crystal has formed and which bathes the solid phase. A combination of surface adsorption, ionic exchange and substitution reactions accounts for most of the variable composition of bone hydroxyapatite. H_3O^+, Sr^{++} and Pb^{++} ions can replace Ca^{++} in the crystal lattice. F^- can replace OH^- on the surface and in the lattice, and bone in contact with F^--containing fluid will be converted to fluoroapatite. $CaCO_3$ forms an amorphous solid surrounding the crystal, and Na^+, Mg^{++} and citrate^{4-} are incorporated into the crystal surface. All the ions of extracellular fluid are found in the hydration shell, i.e., the layer of water surrounding the crystal phase. Bone crystals are exceedingly small and on electron microscopy appear as rods between 25 and 75 A in width, oriented in the same direction and forming chains up to 1000 A in length. Because of the small size of the crystals they have a tremendous surface area, 100 to 300 sq m per gm; from 25 to 50 per cent of their atoms are in the outermost layer, accounting for the ease with which atoms other than Ca^{++} or $PO_4^{\equiv}$ can enter the crystal surface and for the rapidity of exchange of ions between crystals and interstitial fluids.

The initial solid phase of bone mineral, however, is not crystalline; i.e., it does not have the long range order necessary to generate a well-defined x-ray diffraction pattern. It has been proposed that this initial phase which has a molar Ca/P ratio of 1.5 is amorphous calcium phosphate, $Ca_3(PO_4)_2$, which in contact with bone extracellular fluid spontaneously transforms to a crystalline apatite. The amorphous calcium phosphate, however, can be stabilized by Mg^{++} ion, pyrophosphates and nucleotides that inhibit its conversion to hydroxyapatite. The presence of these solutes in bone extracellular fluid may explain the presence of some amorphous calcium phosphate in old bone, but the highest proportion of the amorphous phase is found in the initial deposits of bone mineral. The composition of the mineral phase determines the mechanical properties of bone and the rates of exchange of ions between solid phase and extracellular fluid. Whether the composition of bone mineral differs in various metabolic bone diseases is still in question. The bone mineral makes up about 65 to 70 per cent of the dry weight of the bone, and the remaining 30 to 35 per cent consists of the organic matrix within which the mineral phase is formed.

The composition of bone matrix is about 96 per cent collagen, 3.5 per cent acid mucopolysaccharides and about 0.5 per cent noncollagen protein. The structure of the collagen fibril is probably the key to the organization of the bone mineral. This fibrous protein has a characteristic amino acid composition with an unusually large proportion of glycine, proline and hydroxyproline, and also a small fraction of hydroxylysine. The collagen macromolecule is thought to consist of three separate polypeptide chains, each of which is coiled in a left-handed helix, and the individual helical chains twist about a central common axis in a right-handed helix to form a coiled-coil structure. The orderly structure of the collagen macromolecule makes it essentially crystalline. The collagen fibril consists of rodlike collagen macromolecules polymerized both laterally and longitudinally, pointing in the same direction with ends of adjacent rods staggered with respect to one another by about one-fourth of their length. The structural regularity of the fibril results in a characteristic alteration of light and dark cross striations along the fibril when examined in the electron microscope with an axial repeat period of about 640 to 700 A (Fig. 1–1).

There are distinct differences among the collagens of various types of connective tissue depending on the primary structure of the polypeptide chain. Bone collagen is

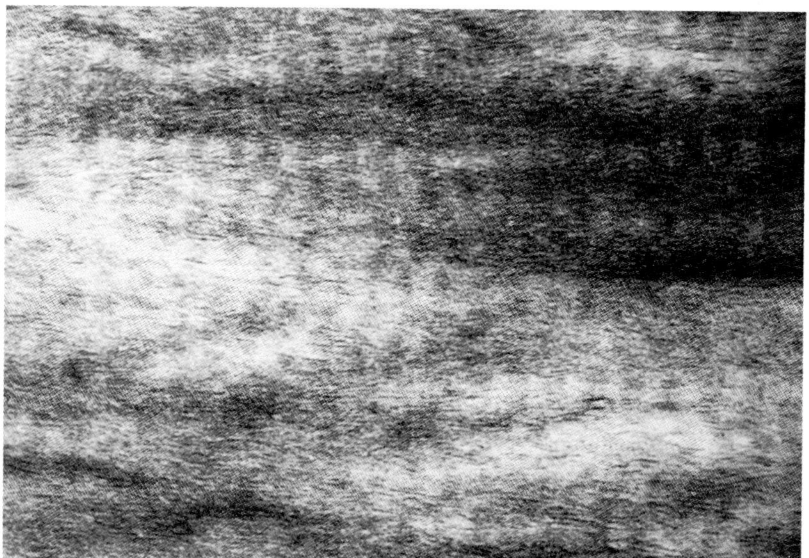

Figure 1–1. Electron micrograph of bone showing the 640-A repeating structure of collagen, which is accentuated in this instance by the apatite crystals deposited within and along the collagen fibrils. (From Glimcher, M. J.: Specificity of the molecular structure of organic matrices in mineralization. *In* Calcification in Biological Systems. American Association for the Advancement of Science, Washington, D.C., 1960. Reprinted with permission.)

classified as a type I collagen, similar to that of skin, tendon and fascia in its amino acid composition. It differs, nevertheless, from the type I collagens of these other tissues in physical properties that must depend on chemical modifications of the collagen after the polypeptide chains are formed, with resultant alterations in structure. There is evidence that bone collagen differs from that of noncalcifying tissues in the distribution, number and stability of the intra- and intermolecular cross-linkages, and these structural differences may be of importance in the calcifiability of bone collagen. It has also been suggested that one of the osteogenic disorders of bone, osteogenesis imperfecta, may result from a deficiency of type I collagen (see Chapter 7).

The collagen fibril forms the central element of connective tissue structure, an interlaced network of fibers in which the fibrils tend to be arranged in parallel bundles. The collagen is surrounded by a ground substance, the mucopolysaccharides. These are present in much greater quantities in cartilage than in bone. Much of the mucopolysaccharide is complexed with noncollagen protein, forming mucoprotein. The characteristic mucopolysaccharide of bone is chondroitin sulfate, which is a polyelectrolyte made up of approximately equimolar amounts of hexosamine, glucuronic acid and sulfate. Since chondroitin sulfate binds cations including calcium, it has been considered to play an essential role in the mechanism of bone crystal formation. Another theory of its action is that it functions as a cement substance linking collagen fibrils. Collagen and mucopolysaccharide together form the organic matrix of cartilage and bone, in which the mineral phase crystallizes during the process of calcification.

BONE CELLS

Although they make up only a small fraction of the total mass of mature bone, the bone cells, of course, are controlling factors not only in the original organization of bone but in its continuing function as a structural and metabolic unit: one that responds to physical stresses with remodeling of structure and to metabolic stress by uptake or release of calcium, magnesium, phosphate and sodium as well as such ions as fluoride, lead, radium and strontium that might enter the extracellular fluid. The osteoblasts are mononuclear cells more compact and thicker than the fibroblast. According to Pritchard the osteoblast is one stage of a "series of mutually transformable connective tissue cells resident in and around a bone within the confines of the fibrous periosteum, and including the osteocyte, osteoclast, marrow reticulum cell, flattened resting osteoblast, spindle cell and chondroblast as alternative forms." The active osteoblasts, like chondroblasts and fibroblasts, are collagen-producing cells. The presently held theory is that the collagen is secreted as tropocollagen consisting of rod-like molecules about 3000 A in length which then aggregate into the specific fibrillar pattern of collagen. This transformation probably occurs in the interstitial fluid at the surface of the cell. The most characteristic arrangement of osteoblasts is a one-cell-thick layer covering the surfaces of newly formed trabeculae of bone matrix. Histochemical studies have shown the presence of heavy concentrations of alkaline phosphatase in osteoblasts and chondroblasts. This enzyme is thought to be of importance in fibrogenesis but has also been considered to have a role in the mineralization of the organic matrix, as is discussed below.

The osteocyte is the cell of mineralized bone present in lacunae in the solid bone and possessing numerous fine branching processes. These flattened cells are evenly spaced and uniformly oriented with respect to the long and radial axis of the lamellar system they occupy. They arise from osteoblasts that have been surrounded by mineralized matrix. The osteocyte through its metabolic activity in some way maintains the integrity of the matrix, since osteocyte death results in eventual resorption of the surrounding matrix, with replacement by new living bone.

The osteoclast is the most striking cell in bone. Typically, it is a large, multinucleated cell most numerous in areas of physiologic or pathologic bone resorption. The origin of the osteoclasts is still in doubt, but it seems likely that they arise through coalescence of precursor mononuclear cells under conditions that stimulate bone resorption. They appear with great rapidity in the bone of animals injected with parathyroid extract. The specific role of the osteoclast in the dissolution of bone with removal of both mineral and organic matrix is now well accepted, and will be discussed in more detail under *Bone Resorption* later in this chapter.

CRYSTALLIZATION OF BONE MINERAL

The serum calcium concentration under normal conditions averages about 2.5 mM per liter (10 mg per dl), and there is little difference between the growing infant or child and the adult in this parameter. Approximately 50 to 55 per cent of this calcium is protein-bound so that the calcium content of the protein-low interstitial fluid is about 1.25 mM per liter (5 mg per dl). The serum phosphate concentration, unlike the calcium level, shows marked variations with age and growth rate. In the prematurely born infant the concentration of serum phosphate is 2.5 to 3.0 mM per liter (7.5 to 9.0 mg per dl), whereas in the adult the concentration is down to about 1.0 to 1.3 mM per liter (3 to 4 mg per dl). This range of serum phosphate levels at different ages is shown in more detail in Table 1–1. This almost threefold variation of serum phosphate concentrations at the extremes is sometimes forgotten in discussions of the calcification mechanism in terms of a solubility product. In contradistinction to calcium, little of the serum phosphate is bound in a nonfilterable form, so that the concentration of phosphate in interstitial fluid is approximately the same as that in serum. At the pH of extracellular fluid about four-fifths of the phosphate is present as bivalent $HPO_4^{=}$, about one-fifth as monovalent $H_2PO_4^-$; only a tiny fraction as trivalent $PO_4^{\equiv}$.

Table 1–1. Changes of Concentration of Phosphate in Serum with Age

Age	Serum P *mg/dl*	Serum P *mM/l*
Prematurely born infants	7.9	2.5
Full term infants – 1st week	6.1	2.0
1 to 10 years	4.6	1.5
Adult	3.5	1.1

The interstitial fluid may be considered as supersaturated with respect to hydroxyapatite but in a metastable state. Solutions in metastable equilibrium can persist indefinitely, but with a specific stimulus will form new, more stable phases. Nucleation is the process of formation of the initial fragments of a more stable phase from a metastable system. The stimulus for precipitation and crystallization of hydroxyapatitie from the calcium and phosphate ions of interstitial fluid could be local increase in concentration of one or both of these ions, change in pH or other alteration in the system. The crystallization of bone mineral is not random but occurs in close association with the collagen fibrils (Fig. 1–2). Glimcher et al demonstrated that "native type 640 A axial repeat," reconstituted collagen fibrils prepared from normally uncalcified tissues such as rat tail tendon, calf skin, guinea pig skin and fish swim bladder, were able to nucleate apatite crystals from metastable calcium phosphate solutions. Only the native-type collagen was able to induce crystal formation. Other types of fibrillar macromolecules or collagen in other states were ineffective. Once apatite nuclei are formed, growth of the crystals can occur with removal of calcium and phosphate from the surrounding fluid. The bone salt crystals do not grow to large size, however, and, as mentioned above, exchange reactions between the crystals and the surrounding solution phase are facilitated by the large surface area presented by the mass of minute crystals.

The concentration of inorganic phosphate is an important factor in the nucleation of hydroxyapatite by collagen. This is the basis of the empiric Ca × P product of Howland and Kramer. When this product (serum Ca × serum P expressed in mg per dl) is less

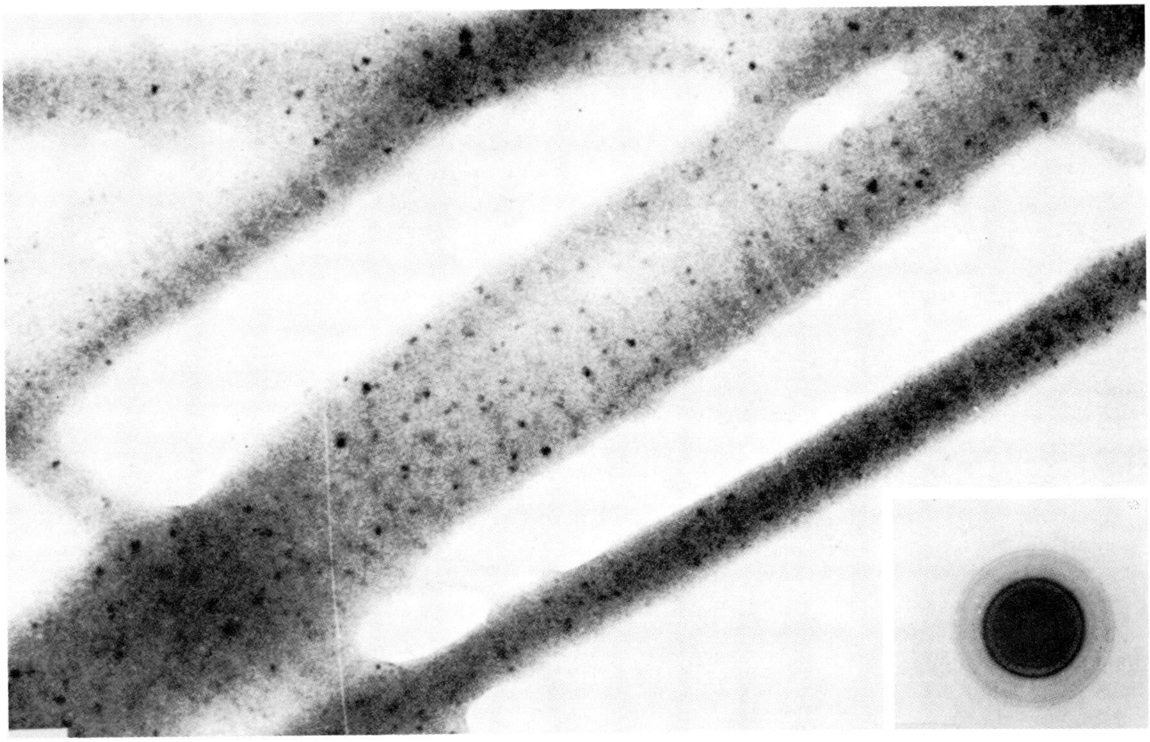

Figure 1–2. Electron micrograph of in vitro mineralization of reconstituted collagen fibrils. The periodic location of the small dense inorganic crystals is evident both longitudinally and laterally. (From Glimcher, M. J.: Specificity of the molecular structure of organic matrices in mineralization. *In* Calcification in Biological Systems. American Association for the Advancement of Science, Washington, D.C., 1960. Reprinted with permission.)

than 30, active rickets and osteomalacia will occur, i.e., failure of bone mineral to crystallize in organic matrix. At Ca × P products above 40, mineralization of the organic matrix takes place with healing of the rickets and osteomalacia. This relationship indicates that nucleation of bone mineral by crystalline collagen is directly related to the concentration of inorganic phosphate in the fluid phase: at high concentrations, crystals are deposited in collagen fibrils as they develop at the surface of the osteoblast, so that unmineralized collagen (osteoid) does not accumulate; at low concentrations, induction of crystallization of hydroxyapatite is delayed, and layers of unmineralized collagen accumulate at the surface of the bone trabeculae between the bone and the coating of osteoblasts. The critical concentration of inorganic phosphate is probably dependent on the rate of growth, and therefore presumably the rate at which new collagen is formed. In the young premature infant unmineralized collagen may accumulate despite inorganic phosphate concentrations of 1.6 mM per liter (5 mg per dl), whereas in the adult the collagen of remodeling bone will be completely mineralized at serum phosphate levels of 1.0 mM per liter (3 mg per dl).

Although mineralization of collagen occurs promptly as the organized fibril is laid down, the organic matrix is not at once completely mineralized. This can be seen best in the study of haversian systems, or osteons, cylinders of bone laid down in concentric lamellae about capillary blood vessels. They are formed in the remodeling of compact bone. The initial bone structure is penetrated by a blood vessel accompanied by osteoclasts that resorb the bone, and by osteoblasts that lay down new collagen in layers surrounding the blood vessel. The collagen close to the capillary is much less densely mineralized than that at the periphery of the osteon, and contains more water.

This difference in density can be shown by microradiogrpahy, i.e., x-ray photographs of thin sections of bone. The more hydrated bone takes up labeled phosphate or calcium much more rapidly than the bone in which the density of bone salt is greater, since in the completely mineralized bone water has been displaced by the solid bone salt and the rate of penetration of ions from the interstitital fluid is reduced. The same differential rate of uptake of ions from the interstitial fluid is found between the newly formed bone at the metaphyseal end of the shaft and the older and denser compact bone of the diaphysis.

The uptake by newly formed bone crystals of ions and organic solutes that are adsorbed on the surface of the crystal has been widely used in the study of bone growth. The radioactive isotopes of phosphorus, calcium and strontium have been used to label the growing and incompletely mineralized portion of bone. The deposition of these isotopes can be studied by autoradiographic methods in which a thin sliver of bone is placed against x-ray film and the patterns of radioactivity compared with microradiographic or histologic studies of adjacent sections of bone. The growth of bone can also be studied by administration of lead which is taken up at the sites of new crystal formation and can be detected by roentgenographic examination. Another method is the use of tetracycline, which has been shown by Milch et al to be complexed with bone mineral as it is deposited in growing areas of bone. The sites of deposition are detected by the tetracycline fluorescence in ultraviolet light. By these methods the rate of growth of bone and osteons in the remodeling of bone has been ascertained.

The theory of nucleation of hydroxyapatite by crystalline collagen from a metastable solution of calcium and phosphate ions, and the growth of these crystals by continued uptake of calcium and phosphate ions (depending on the concentration of these ions, particularly that of phosphate) is appealing but cannot be the complete story. Under in vitro conditions recrystallized collagen from nonbone sources is an effective nucleating system when placed in a solution with suitable concentrations of calcium and phosphate ions. Under in vivo conditions the collagen of tissues other than cartilage and bone rarely mineralizes, although such mineralization may occur in abnormal states. For example, the collagen of tendon may be mineralized, and the conversion of turkey leg tendon to bone has been extensively studied. There is some regulating factor, perhaps an inhibitor of mineralization, which prevents aberrant mineralization of nonskeletal tissues.

The role of the ground substance surrounding the collagen fibrils, the acid mucopolysaccharides, has been variously interpreted as a necessary factor in the initiation of mineralization or as an inhibitor of mineralization. The latter thesis has been put forth on the basis of the calcium-binding property of the mucopolysaccharides, which might make calcium ion unavailable for bone crystal formation. The role of metabolic activity of the cartilage cells and osteoblasts in the initiation of mineralization of cartilage and bone cartilage has been examined. In epiphyseal cartilage bone mineral deposition surrounds cells that have undergone certain changes in their microscopic structure. This can best be seen in the mineralization of rachitic cartilage, where the layer of bone salt first appears in conjunction with a layer of cells separated from the bony trabeculae rather than in the organic matrix surrounding the degenerated cartilage cells adjacent to the bone. The metabolic activity of the cell is presumably of importance in altering the state of the mucopolysaccharide at its surface. It is of interest that, in the developing chick, mineralization of the cartilage of a long bone such as the tibia begins in the cell layers of the hypertrophic zone beyond the flattened cells of the epiphyseal plate. The fibrous matrix becomes strongly honeycombed between the hypertrophic cartilage cells, and then mineral is deposited in the matrix at some distance from the cells. The content of hexosamine, which is a marker for mucopolysaccharide, drops precipitously in the hypertrophic cartilage when calcification has started, although collagen remains unchanged.

In addition to mediating the disappearance of polysaccharides in the matrix, the mature "hypertrophic" chondrocytes have another function in the initiation of precipitation of bone salt in the matrix surrounding these cells: namely, the extrusion of membrane-bound organelles, the matrix vesicles (Fig. 1–3). The local formation of calcium phosphate aggregates in the matrix is seen to occur within these membrane vesicles. Two theories of the function of these vesicles have been proposed:

(1) That these vesicles derive from intracellular membrane-bound organelles such as mitochondria or endoplasmic reticulum which can accumulate calcium and phosphate in concentrations sufficient for precipitation of the solid phase. Parts of these intracellular organelles are then extruded with their calcium phosphate aggregates surrounded by membrane, and these membrane vesicles act as nuclei for the growth of calcium phosphate precipitates in the matrix.

(2) That these membrane vesicles are formed from intracellular membranes, extruded from the cells, and accumulate calcium and phosphate from matrix fluid by ATP-dependent energy mechanisms catalyzed by enzymes in the membranes. When the concentrations of calcium and phosphate within the vesicle increase to the point of precipitation of calcium phosphate, solid aggregates form that rupture the membrane and are nuclei on which further growth of bone mineral occurs.

The initial step in the linear growth of tubular bones is calcification of the cartilage at the metaphyseal end of the bone forming the provisional zone of calcification. The cartilage cells in the epiphyseal cartilage are arrayed in rows aligned in the longi-

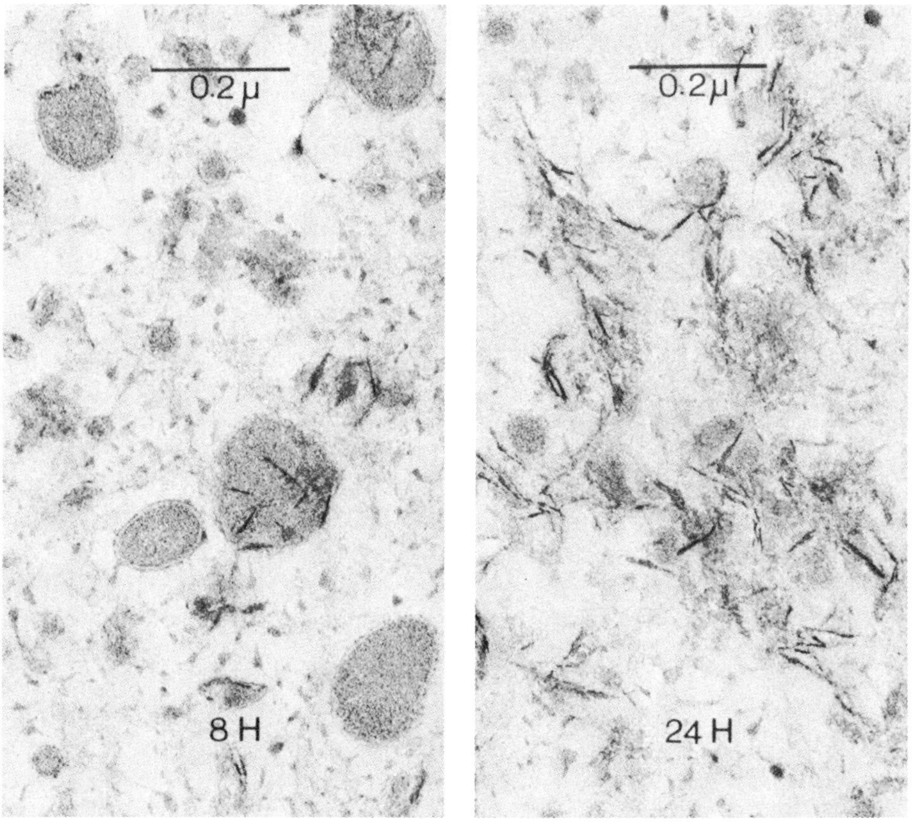

Figure 1–3. Matrix vesicles in hypertrophic zone of rachitic growth plate, 8 and 24 hours after intraperitoneal injection of rachitic rat with Na H_2 PO_4. The apatite accumulation within and at the surfaces of vesicles is similar to that seen when rachitic cartilage is incubated in vitro in a calcifying solution. Electron micrograph ×88,000. (From Anderson, H. C., and Sajdera, S. W. Reprinted from Federation Proceedings 35:148–153, 1976.)

tudinal axis of the bone. These chondrocytes, as mentioned, undergo physiologic and anatomic changes that are essential to the calcification process. The most proximal cells adjacent to the mineralizing front of bone undergo hypertrophy and degeneration and are invaded by capillaries from the bones that tunnel along the rows of cells. As the capillaries invade, calcium phosphate aggregates are formed in the matrix starting in the membrane-bound organelles derived from the chondrocytes, the membrane vesicles, which are dispersed in the matrix at the level of the hypertrophic cells that are being invaded by capillaries. It is probable that, in addition to the local concentration of calcium and phosphate within the vesicles, the calcification process, spreading from the vesicles as nuclei, requires the breakdown by cellular enzymes of the mucopolysaccharide protein complexes that act as inhibitors of mineralization by complexing calcium.

As the cartilage matrix mineralizes, the rows of degenerating chondrocytes are replaced by the invading capillaries which carry the bone cells, both cells that solubilize the calcified cartilage and osteoblasts that replace the solubilized cartilage by osteoid, i.e., new organized bone matrix, collagen and mucopolysaccharide. This osteoid is then mineralized by the same process involved in the calcification of cartilage. It is likely that matrix vesicles derived from osteoblasts are the nuclei for the spreading precipitation of calcium phosphate in the osteoid. Both chondrocytes and osteoblasts, therefore, have a multiple role in the mineralization process. They secrete the collagen which is organized to form the organic crystal structure on and in which bone mineral crystals are deposited, aligned with the collagen fibrils. They are the source of the membrane structures, the membrane vesicles, which are the sites of local concentration of calcium and phosphate and formation of the initial solid calcium phosphate phase that acts as a nucleus for further growth of the solid phase. They also secrete the enzymes that break down the inhibitors of bone mineral crystal formation and growth, the pyrophosphates, nucleotide triphosphates and the mucopolysaccharides.

The role of alkaline phosphatase in the mineralization process is also consistent with the matrix vesicle function. A high concentration of alkaline phosphatase is a characteristic feature of hypertrophic cartilage cells and osteoblasts. The original hypothesis of Robison that this enzyme functioned in the calcifying process by release of inorganic phosphate from an organic phosphoric ester, and initiated the precipitation of calcium phosphate by this increased concentration of inorganic phosphate, was not supported by evidence of a substrate in calcifying matrix which could be a source of this inorganic phosphate. Nevertheless, the view that this enzyme had a significant role in mineralization of cartilage and bone was buttressed by the observations in an inborn error of metabolism, hypophosphatasia. The content of alkaline phosphatase in tissues as well as plasma is markedly reduced in this genetically determined abnormality. The bones show defects of mineralization of epiphyseal cartilage and osteoid that have some resemblance to rickets, despite normal concentrations of calcium and phosphate in extracellular fluid. Alkaline phosphatase is normally concentrated in the membranes of chondrocytes and osteoblasts and is found in the matrix vesicles derived from cell membranes. This enzyme system has also pyrophosphatase and ATPase activity. It possibly functions in the transport of calcium and inorganic phosphate into the vesicle contents through an ATP-dependent system, and also hydrolyzes inorganic pyrophosphate that might be an inhibitor of bone crystal growth.

GROWTH OF CARTILAGE AND BONE

The growth of cartilage and of the bone that replaces cartilage in endochondral calcification depends on the same factors that control all cell proliferation and growth: growth hormone-mediated somatomedin;

the thyroid hormones; the adrenal androgens; insulin; and, during the pubertal growth spurt, the estrogen and androgen of gonadal origin. In addition there are specific stimuli that control bone density and bone organization, namely, the compressive and tensile forces which stress the bone and cause local cell proliferation and bone growth in response to these forces. The mechanisms of this reponse of bone cells to compression and bending are not completely known, but it has been suggested that bone mineral crystals are piezoelectric, i.e., generate an electrical potential when deformed by pressure, and that these electrical fields are of importance in stimulating local osteoblast function. Additionally there is some experimental evidence that osteoblasts react directly to compression by cyclic AMP production, which may stimulate cell growth and division. This characteristic response of bone cells to mechanical forces exerted upon them, whatever the mechanism, is naturally of importance in determining the shape and thickness of the bones so that they are best adapted to the pressures exerted upon them.

In addition to the growth factors already listed, bone growth is markedly dependent on the concentration of calcium and phosphate in body fluids, particularly phosphate, and consequently on vitamin D through its role in controlling calcium and phosphate concentrations (this role will be discussed in detail in Chapter 2). Although both calcium and phosphate are essential for bone mineral formation, the growth of bone is particularly dependent on the inorganic phosphate concentration in extracellular fluid. Skeletal growth can be normal in patients with hypoparathyroidism and hypocalcemia, but is invariably retarded in hypophosphatemic states whether due to vitamin D deficiency or to renal tubular disorders. In the latter situation extracellular calcium concentrations can be normal, emphasizing the significant role of phosphate in cartilage and bone growth. It is likely that this phosphate requirement is not only for bone mineral deposition but also for cell growth and proliferation, presumably through the essential role of organic phosphate in mediating cell energy production and the synthesis of essential cell components.

BONE RESORPTION

The dissolution of bone mineral proceeds concomitantly with the crystallization of new bone salt, and this removal of old bone and formation of new is the mechanism whereby the structure of bone is remodeled and change in form and size of bone already laid down occurs. As indicated above, the formation of new bone at the metaphysis is preceded by resorption of the calcified cartilage and reprecipitation of mineral in the new matrix laid down by the osteoblasts that line the invading capillaries. In the continual remodeling of bone, resorption of bone mineral is followed by new matrix formation and mineralization of this matrix. In addition the bone mineral provides a buffer store of calcium, phosphate and magnesium which is drawn upon to maintain physiologic concentrations of these ions in the extracellular fluid. The dissolution of bone mineral is not simply a physical solution of crystals by the permeating fluid phase. This process is probably slow, and despite the large surface area of the mass of bone crystals it is of only secondary importance in the regulation of the ionic concentrations in the fluid phase.

Bone salt dissolution is the result of metabolic activities of cells, and one cell species of particular importance in this function is the osteoclast. In areas of active bone resorption masses of these multinucleated giant cells are found. By means of time-lapse cinematography of bone growing in tissue culture, investigators have been able to visualize osteoclast function. The osteoclast is in very close contact with the resorbing bone surface. Its periphery is in constant motion and pushes pseudopods into every crevice formed in the bone structure as the bone is being dissolved. Large macrophages also apparently participate in the removal of the dissolved and fragmented bone.

In vitro studies have indicated that bone tissue in its metabolism produces organic acids, particularly lactate and citrate. A current theory of bone resorption is that the metabolic activity of osteoclasts, by production of CO_2 as well as organic acids, causes a reduction of pH at the surface of the cell which causes the solution of the bone mineral. The elaboration of citrate might also contribute to increasing the solubility of the crystalline bone salt. The resorption of bone mineral is accompanied by removal of the organic matrix of bone. This latter occurs simultaneously with the solution of the mineral phase, or at so brief an interval after that demineralized collagen fibers do not accumulate in areas of active bone resorption. In tissue culture preparations osteoclasts have vigorous proteolytic abilities, as demonstrated by liquefaction of the nutrient plasma clot upon which they are growing. Proteolytic enzymes apparently cannot attack collagen and mucoprotein until the bone mineral is removed.

The sequence of events in dissolution of bone by osteoclast activity is probably, therefore, solution of the bone crystals and enzymatic hydrolysis of the protein matrix with phagocytosis of fragments of destroyed bone by both osteoclasts and macrophages. Dissolution of bone also occurs as the result of the metabolic activity of the osteocytes, and during active bone resorption the space around the osteocytes is increased. Because of the large volume of osteocytes and their network of processes the activity of these cells is probably of major importance in the minute-to-minute regulation of Ca^{++} ion concentration in extracellular fluid (see Chapter 2, *Calcium Homeostasis*).

Balance Between Bone Mineral Formation and Solution

The processes of bone salt crystallization, sometimes termed *accretion*, and bone salt solution are going on together in the total bone structure. In growing bone accretion exceeds solution, so that total bone mineral mass increases parallel with formation of new matrix. In normal adult bone in a steady state the rates of the two processes are equal, so that the mass of bone mineral is unchanged although there is a constant turnover of the bone salt. Under conditions of progressive demineralization of bone the rate of solution is greater than the rate of accretion. It is important to know whether the imbalance in such states is due to an absolute increase of the rate of solution of bone crystal or a decrease of bone salt accretion with the rate of solution unchanged. This question has been asked particularly with regard to osteoporosis, whether spontaneous or secondary to excess of adrenocorticoids or immobilization.

When radioisotopes of elements which are part of the bone salt, such as calcium and phosphorus, or elements which can be incorporated into the bone crystal, such as strontium, became available, it was hoped that measurement of bone mineral formation and removal would be possible. This turned out to be a much more complex problem than had been anticipated. Administered radiocalcium and phosphorus are incorporated into the bone crystal by exchange with the stable calcium and phosphorus of the crystal. The uptake of isotopes measures the exchangeable mineral of the skeleton rather than the amount of new mineral deposited. The proportion of exchangeable calcium is much higher in young growing bone than in adult bone, as might be expected because of the greater water content of the young bone and the larger surface in contact with interstitial fluid.

Bauer, Carlsson, and Lindquist proposed a method for the treatment of data obtained following intravenous administration of radiocalcium which afforded a measure of the calcium accretion, i.e., the calcium added to the skeleton as new crystalline material. This does not mean the net addition of calcium to the skeleton, since resorption going on in the bone mass at the same time removes calcium. If the accretion and the total external balance of calcium are determined in the same experiment, the rate of accretion, the rate of resorption and the net increase of bone calcium can be determined.

Certain of the assumptions made in the calculations of Bauer et al, namely: that there are only two calcium pools in the body, exchangeable calcium and bone calcium; that exchange reaches equilibrium rapidly; and that over the short interval studied the radiocalcium removed from the skeleton by resorption can be neglected, have not been entirely accepted, and more complex systems of calcium kinetics have been developed. Although of great theoretic interest, the contribution of calcium kinetics to our understanding of disorders of skeletal mineralization has not been as great as might have been expected. One of the problems may be the extent of long-term exchange. Administered isotopic calcium exchanges rapidly with stable calcium on mineral crystal surfaces, and this plus the calcium in solution in body fluids constitutes the exchangeable calcium pool. In addition, isotopic calcium exchanges more slowly with stable calcium within the crystal, and this long-term exchange will be included with accretion, giving a falsely high value for the latter. In osteomalacia bone mineral accretion is almost certainly reduced but long-term exchanges may be more rapid, resulting in calcium kinetic measurements of increased rate of bone accretion in osteomalacia.

Because of restrictions on the use of radioisotopes in children, little is known of calcium kinetics in the growing child with or without disorders of mineralization. Data obtained in growing animals are of interest and may possibly be extended to children. For example, in rapidly growing rats the net increase in calcium is only about one-fourth of the calculated accretion, indicating that the rate of resorption is about 75 per cent that of the rate of accretion, even during active growth. This apparently wasteful type of growth is necessary to maintain the complex architecture of the bone during its expansion. The accretion rate is greater in growing animals than in adults, and in the adult animal it is increased in fractured bone. In vitamin D-deficient rats both the accretion rate and the resorption rate are less than comparable rates in vitamin D-treated animals, but the rate of accretion increases more than that of resorption following treatment, as would be expected from the known increase in bone mineral.

These studies do indicate that active bone resorption also requires the presence of vitamin D, which agrees with studies indicating the need for vitamin D for the action of parathyroid hormone on calcium homeostasis. Using a similar model, Heaney and Whedon concluded from studies of patients with osteoporosis that calcium accretion rates in this disorder are in the normal range. Since the total bone mass of these subjects is greatly reduced, they suggest that the rate of bone resorption is increased and that osteoporosis is the result of accelerated resorption of bone salt without a compensating increase in bone crystal accretion. In Paget's disease it has been found that flow of Ca^{45} into and out of bone are both markedly increased, which fits with the gross and histologic changes in bone structure.

Another method for studying the processes of addition and removal of bone mineral in the skeleton is the study of bone structure by microradiography. In this technique, sections of bone embedded in methyl methacrylate are cut or ground to a thickness of 100μ, and then contact microradiographs are made using a beam of x-rays from a small focal spot. In such a microradiograph the osteons in which new bone formation is going on show decreased density adjacent to the central blood vessel, and inactive or completely mineralized osteons can be differentiated. In addition the areas of bone resorption are detected as cavities with irregular walls which are not oriented with respect to a single haversian system or osteon. In this way the areas of active bone crystal formation and removal can be detected, and some estimation made of the proportion of each. Marked variations of the microradiographs are seen with age (Fig. 1–4). In young individuals the percentage of bone formation and resorption is high, which is compatible with a high turnover rate, whereas in the young adult there is little bone formation and resorption, and most of the osteons have a high mineral density. In later life there is an increase in the amount of resorption. The microradiographic appearance of osteoporotic bone from elderly persons is compatible with Heaney and Whedon's observation that bone resorption is increased in this disease.

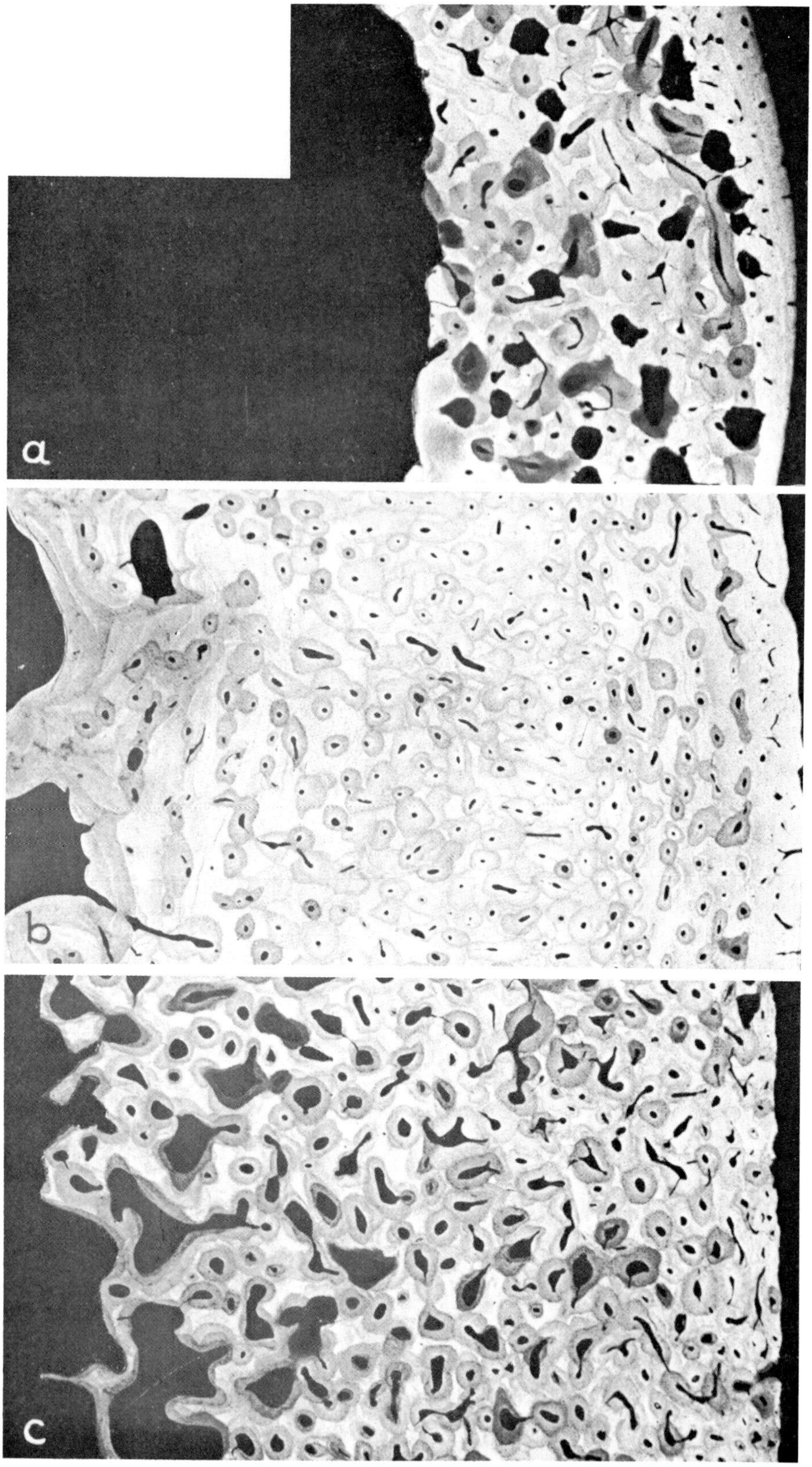

Figure 1–4. See legend on the opposite page.

CONTROL OF BONE RESORPTION

Bone resorption, as pointed out, has two functions: (1) the reshaping and remodeling of bone both during growth and in adult life; and (2) the release of calcium, phosphate and to a lesser extent magnesium and carbonate ions to maintain homeostasis of these ions in body fluids. The structural reorganization of bone is presumably under the control of local forces, since bone resorption and bone formation go hand in hand in the remodeling of bone. This has been mentioned already with reference to the role of local forces in determining the shape and density of bone.

The homeostatic role of bone resorption is controlled by the three major hormones of calcium and phosphate homeostasis: parathyroid hormone; vitamin D hormone; and calcitonin. The actions of these hormones will be discussed in Chapter 2.

METHODS FOR STUDYING RATES OF BONE RESORPTION

Calcium and phosphate balance studies and measurement of bone mineral density, whether by x-ray, the photon absorption technique or neutron activation, provide an estimate of the net change in skeletal mineral content which is the balance between bone mineral accretion and bone mineral resorption. It is sometimes of importance to determine whether alterations in this net balance are primarily the result of changes in rate of bone mineral accretion or bone mineral resorption, or both. The method of calcium kinetics as discussed above has been developed in an attempt to answer this need, but aside from theoretical problems in the assumptions on which this technique is based there are practical difficulties in its application. Because of the dependence on radiocalcium administration and also the problems involved in doing accurate calcium balances and obtaining numerous serum samples, the method is not suitable for children.

One of the methods of assessing the rate of bone resorption is the determination of the urine excretion of hydroxyproline-containing peptides. Hydroxyproline is a unique amino acid found only in collagen. The peptide chain of collagen as assembled contains proline, which is hydroxylated to hydroxyproline after it has been incorporated into the molecule. When bone mineral is resorbed the demineralized collagen matrix is quickly solubilized also by proteases with collagenase activity liberated by osteoclasts and osteocytes. The hydroxyproline-containing peptides liberated by the breakdown of collagen are excreted in the urine, and are thus a measure of the rate of bone matrix resorption and indirectly of bone mineral resorption, since matrix resorption occurs only after solubilization of the mineral phase. Not all of the hydroxyproline in urine, however, is derived from previously mineralized bone collagen. Hydroxyproline peptides derived from soluble collagen that has never been transformed into bone matrix and mineralized, or from collagen of soft tissue origin, are also excreted. These latter peptides are larger molecules than those derived from breakdown of demineralized bone collagen, and in the most precise studies can be separated so that the hydroxyproline which measures bone resorption can be determined. However, the determination and assessment of rate of bone resorption by hydroxyproline excretion in urine is essentially a research tool.

Figure 1–4. Microradiographs of the cortex of the femur from: (*a*) a 2-½ year old male; (b) a 17 year old male; and (c) a 77 year old female. The bone from the child shows a high percentage of both bone formation and resorption. In the bone of the young adult there is a low turnover rate with little bone formation or destruction, and most of the osteons show close to full mineralization. In the bone of the elderly person there is evidence of extensive bone resorption and little evidence of compensating increase in bone formation. The bone is porous and shows great variation in mineral density. (From Jowsey, J. Age changes in human bone. *In* Clinical Orthopedics, No. 17. J. B. Lippincott Co., 1960, p. 210. Reprinted with permission.)

REFERENCES

BONE MINERAL

Betts, F., Blumenthal, N. D., Posner, A. S. et al: Atomic structure of intracellular amorphous calcium phosphate deposits. Proc. Natl. Acad. Sci., USA *72*:2088–2090, 1975.

Pellegrino, E. D., and Biltz, R. M.: Bone carbonate and the calcium to phosphorus molar ratio. Nature *219*:1261–1262, 1968.

Posner, A. S.: Crystal chemistry of bone mineral. Physiol. Rev. *49*:760–792, 1969.

Posner, A. S., Eanes, E. D., Harper, R. A. et al: X-ray diffraction analysis of the effect of fluoride on human bone apatite. Arch. Oral Biol. *8*:549–570, 1963.

BONE MATRIX

Glimcher, M. J.: Molecular biology of mineralized tissues with particular reference to bone. Rev. Mod. Physics *31*:359–393, 1959.

Glimcher, M. J.: Specificity of the molecular structure of organic matrices in mineralization. In Calcification in Biological Systems. American Association for the Advancement of Science, Washington, D.C., 1960, p. 421.

BONE CELLS

Hancox, N.: The Osteoclast in the Biochemistry and Physiology of Bone. (Ed.: G. H. Bourne). Academic Press, New York, 1956, pp. 213–250.

Holtrop, M. E. and King, G. J.: The ultrastructure of the osteoclast and its functional implications. Clin. Orthop. *123*:177–196, 1977.

Pritchard, J. J.: The Osteoblast in the Biochemistry and Physiology of Bone (Ed.: G. H. Bourne). Academic Press, New York, 1956, pp. 179–211.

Rasmussen, H. and Bordier, P.: The cellular basis of metabolic bone disease. N. Engl. J. Med. *289*:25–32, 1973.

Wong, G. and Cohn, D. V.: Separation of parathyroid hormone and calcitonin-sensitive cells from non-responsive bone cells. Nature *252*:713–715, 1974.

CRYSTALLIZATION OF BONE MINERAL

Anderson, H. C.: Vesicles associated with calcification in the matrix of epiphyseal cartilage. J. Cell. Biol. *41*:59–72, 1969.

Anderson, H. C. and Sajdera, S. W.: Calcification of rachitic cartilage to study matrix vesicle formation. Fed. Proc. *35*:148–153, 1976.

Basset, C. A. L.: Biological significance of piezoelectricity. Calcif. Tissue Res. *1*:252–272, 1968.

Fraser, D.: Hypophosphatasia. Am. J. Med. *22*:730–746, 1957.

Posen, S.: Alkaline phosphatase. Ann. Inter. Med. *67*:183–203, 1967.

GROWTH AND REMODELING

Bauer, G. C. H., Carlsson, A. and Lindquist, B.: Bone salt metabolism in humans studied by means of radiocalcium. Acta Med. Scand. *158*:143, 1957.

Frost, H. M.: Bone dynamics in metabolic bone disease. J. Bone Joint Surg. *48-A*:1192–1203, 1966.

Glimcher, M. J. and Krane, S. M.: The organization and structure of bone and the mechanism of calcification. *In* A Treatise on Collagen II (Eds.: B. S. Gould and G. N. Ramachardra). Academic Press, London, 1968.

Heaney, R. P.: Evaluation and interpretation of calcium kinetic data in man. Clin. Orthop. *31*:153–183, 1963.

Heaney, R. P. and Whedon, G. D.: Radiocalcium studies of bone formation rate in human metabolic bone disease. J. Clin. Endocrinol. Metab. *18*:1246–1267, 1958.

Jowsey, J.: Age changes in human bone. Clin. Orthop. *17*:210–217, 1960.

Milch, R. A., Rall, D. P. and Tobie, J. E.: Fluorescence of tetracycline antibiotics in bone. J. Bone Joint Surg. *40-A*:897–910, 1958.

BONE RESORPTION

Raisz, L.: Physiologic and pharmacologic regulation of bone resorption. N. Engl. J. Med. *282*:909–916, 1970.

Raisz, L., Sandberg, A. L., Goodson, J. M. et al: Complement–dependent stimulation of prostaglandin synthesis and bone resorption. Science *185*:789–791, 1974.

Reynolds, J. J., Holick, M. F. and DeLuca, H. F.: The role of vitamin D analogues in bone resorption. Calcif. Tissue Res. *12*:295–301, 1973.

Vaes, G.: Lysosomes and the cellular physiology of bone resorption. *In Lysosomes in Biology and Pathology*, Vol. 1 (Eds.: J. T. Dingle and H. B. Fell). North Holland Publishing Company, Amsterdam, 1969.

BOOKS AND REVIEWS

Harris, W. H. and Heaney, R. P.: Skeletal Renewal and Metabolic Bone Disease. Little, Brown and Co., Boston, Mass., 1969.

Lutwak, L., Singer, F. R. and Urist, M. R.: Current concepts of bone metabolism. Ann. Intern. Med. *80*:630–644, 1974.

Rasmussen, H. and Bordier, P.: The Physiological and Cellular Basis of Metabolic Bone Disease. Williams and Wilkins, Baltimore, 1974, Chap. 2.

The Biochemistry and Physiology of Bone. 2nd Ed., Vol. 4 (Ed.: G. H. Bourne). Academic Press, New York, 1976.

Chapter Two

CALCIUM AND PHOSPHATE HOMEOSTASIS

The control of the concentrations of calcium and phosphate ions in extracellular fluids within narrow limits depends on four reactions: I, the net entrance of the ion into extracellular fluid from the intestine, which is the difference between the amount of ion absorbed from food sources and endogenous fecal loss, i.e., the quantity excreted into the intestine in the intestinal secretions and not reabsorbed; II, the balance between the flow of ions into bone mineral and their reentrance into extracellular fluid by solubilization of bone mineral; III, the balance between cell uptake of ion and the discharge of ions from intracellular contents; and IV, the excretion of the ions in urine, which represents the balance between the quantity filtered through the glomeruli and that retrieved by renal tubule reabsorption. Reactions I and II are the major determinants of calcium homeostasis; reactions II, III and IV are of particular importance in phosphate homeostasis, with IV being the dominant process. This relationship is diagrammed in Figures 2–1 and 2–2, which also show the role of the three major regulating hormones in these homeostatic mechanisms: the vitamin D hormone, 1,25-diOH vitamin D; parathyroid hormone; and calcitonin.

ESTIMATION OF EXTRACELLULAR CALCIUM

The physiologically important fraction of calcium in extracellular fluid, the concentration of which is closely controlled, is ionized calcium. The total calcium which is determined in samples of serum includes protein-bound un-ionized calcium, un-ionized calcium bound to diffusible molecules such as citrate and phosphate, and ionized calcium. The serum total calcium concentration varies only slightly with age, depending primarily on variations in serum protein concentration, and averages 10 mg/dl with a range between 8.8 and 10.7 mg/dl. The ionized calcium concentration does not vary appreciably with age and averages 4.5 mg/dl with a range of 4.0 to 5.6 mg/dl. Most of the difference between serum ionized calcium and total calcium is protein-bound calcium, only a small amount, possibly 0.5 mg/dl, being undissociated diffusible calcium. When serum phosphate is markedly increased, however, an appreciable fraction of calcium can be associated with phosphate, thus reducing ionized calcium. Most of the protein-bound calcium is complexed with the albumin fraction, so that variations in serum albumin

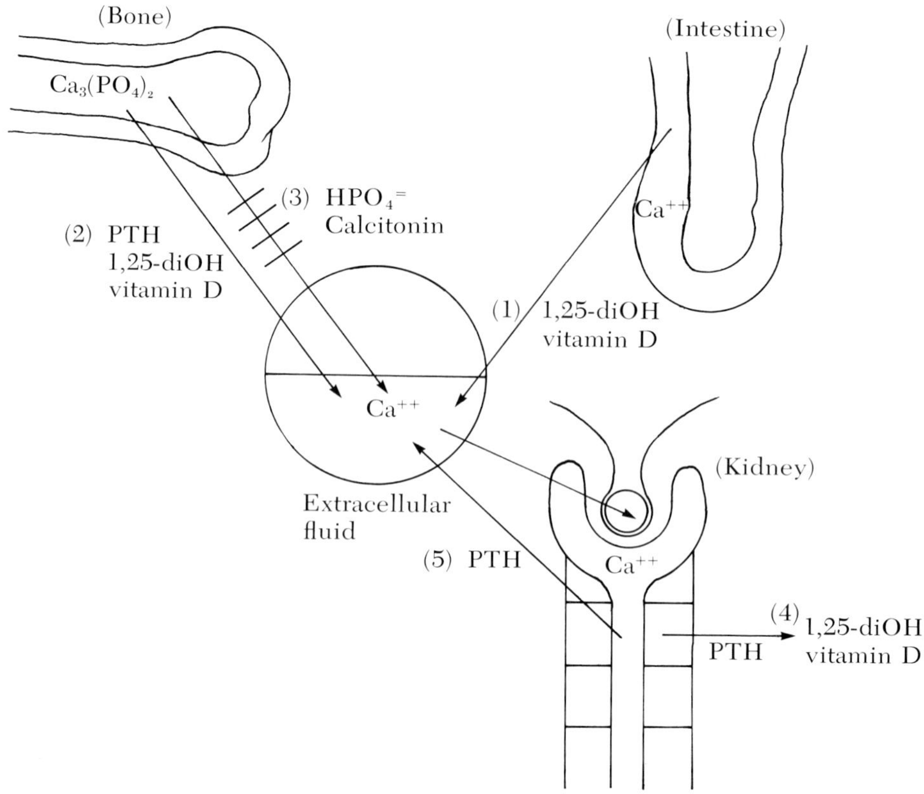

Figure 2–1. Major hormonal control of extracellular Ca^{++} concentration. (1) Intestinal transport of calcium dependent upon 1,25-diOH vitamin D; (2) Solubilization of bone mineral dependent on PTH and 1,25-diOH vitamin D; (3) Solubilization of bone mineral blocked by $HPO_4^{=}$ and calcitonin; (4) Renal conversion of 25-OH vitamin D to 1,25-diOH vitamin D dependent on PTH; (5) Renal tubular reabsorption of Ca^{++} increased by PTH.

concentration can cause marked variations in total serum calcium without change in serum ionized calcium concentration. The current method of determining the latter by means of a calcium ion electrode is reproducible and apparently accurate; it is not yet, however, generally available in diagnostic laboratories and has been primarily a research tool. For that reason estimations of ionized calcium concentration from total serum calcium and serum albumin concentrations are employed.

Such estimations were developed by the classic studies of McLean and Hastings, who developed a nomogram, given in Figure 2–3, for deriving calcium ion concentration from total serum calcium and total serum protein concentrations, assuming that the variations in total serum protein concentration reflect changes in the serum albumin fraction. This remains the best available method for obtaining an approximation of serum calcium ion concentration, but has been found to give discrepant values in individual cases when compared with results obtained by direct determination with the calcium ion electrode. This may reflect individual differences in binding of calcium by albumin or by other serum protein fractions. A major consequence of this discrepancy is the observation that determination of total serum calcium and serum protein concentrations may give a falsely high value for serum ionized calcium concentration in some infants with hypocalcemia in the newborn period, and conversely may not truly indicate an increased concentration of serum ionized calcium in patients with hyperparathyroidism. These special instances are discussed in the sections on hypocalcemic and hypercalcemic states (Chapters 3 and 4).

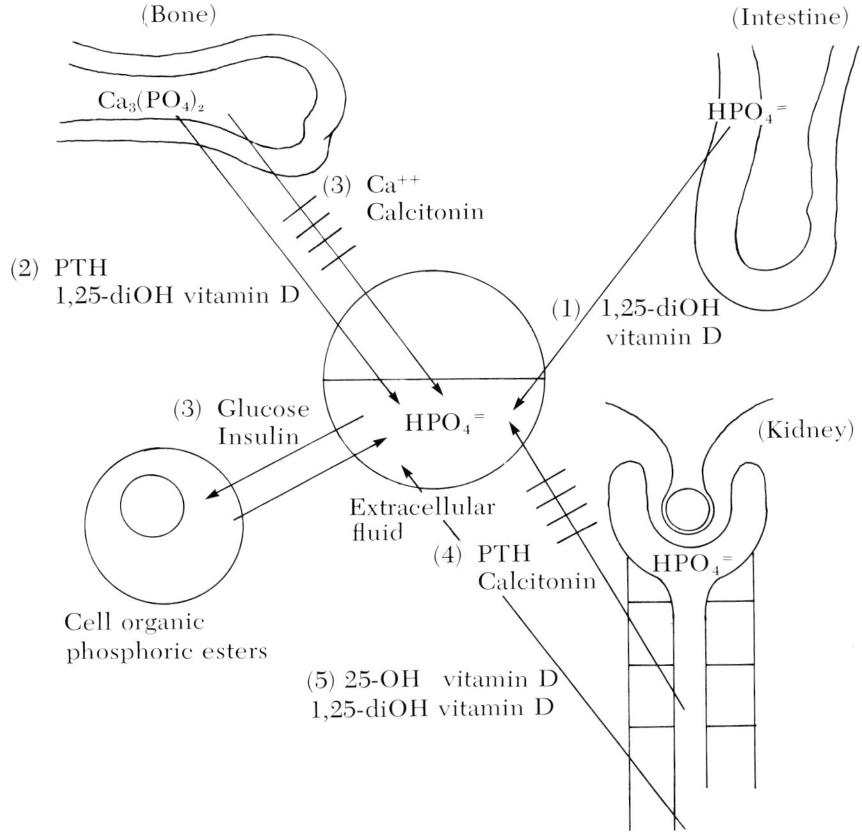

Figure 2–2. Major hormonal control of extracellular $HPO_4^=$ concentration. (1) Intestinal transport of phosphate enhanced by 1,25-diOH vitamin D; (2) Solubilization of bone mineral enhanced by PTH and 1,25-diOH vitamin D; (3) Flow of phosphate from bone to extracellular fluid blocked by Ca^{++} and calcitonin; (3) Uptake of phosphate by cell increased by glucose and insulin; (4) Tubular reabsorption of phosphate inhibited by PTH and calcitonin; (5) Tubular reabsorption of phosphate increased by vitamin D metabolites.

A rough but useful rule of thumb in deciding whether the hypocalcemia of a patient with hypoalbuminemia represents a disorder of calcium ion homeostasis or is simply a reduction of the protein-bound calcium fraction is as follows. For every gm/dl reduction of serum albumin the protein-bound calcium fraction is reduced by 0.8 mg/dl. For example, a patient with a serum albumin concentration of 1.5 gm/dl could have a total serum calcium concentration 2.0 mg/dl ($(4.0 - 1.5) \times 0.8$) below the normal concentration without reduction of serum ionized calcium concentration. For the best possible comparisons of serum calcium concentration it is important to draw blood samples without the loss of protein-free fluid from the plasma that occurs with prolonged venous stasis. Although the increased serum protein concentration of the serum of such concentrated blood samples can be used to correct the total serum calcium concentration, it has already been indicated that calcium binding by protein varies in individual instances so that such corrections may not be reliable.

EXTRACELLULAR PHOSPHATE

The serum phosphate is almost entirely inorganic phosphate which under normal physiologic conditions is freely diffusible, so that it measures the phosphate of extracellular fluid. When inorganic phosphate is

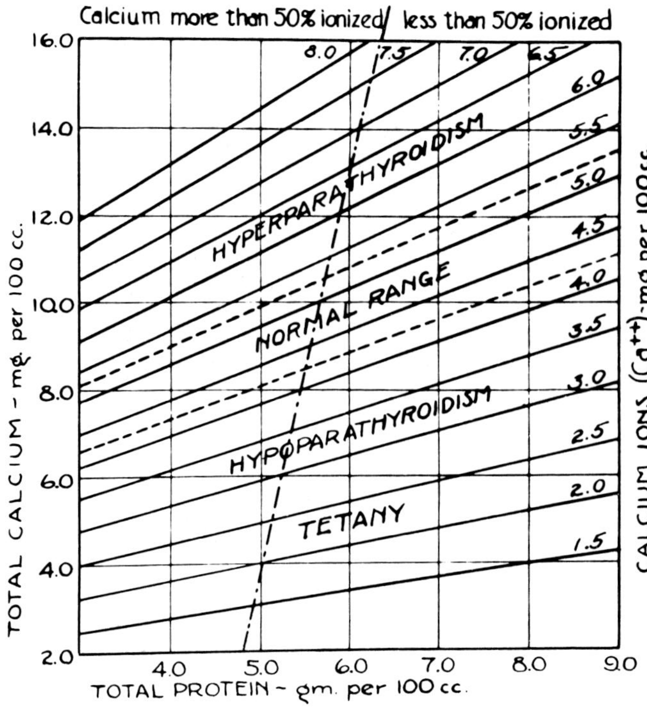

Figure 2–3. Calculation of Ca^{++} ion concentration by McLean-Hastings nomogram.

markedly increased, however, calcium phosphate associates are formed which are protein-bound so that a fraction of the serum phosphate becomes nondiffusible. These calcium phosphate aggregates are removed by macrophages and may be the nucleus for metastatic calcification in soft tissues. Intracellular phosphate concentrations are much greater than those of extracellular fluid but only a small proportion of intracellular phosphate is inorganic. Most of the phosphate within cells is present as organic phosphoric esters, phosphoproteins and phospholipids.

HORMONES OF CALCIUM AND PHOSPHATE HOMEOSTASIS

The three major hormones that control the concentrations of calcium and phosphate in body fluids and their flow into and out of bone mineral are the vitamin D hormone, parathyroid hormone and calcitonin. Cortisol, androgens and estrogens, growth hormone, prolactin and thyroid hormone influence the processes that control calcium and phosphate homeostasis, but these are secondary factors in comparison with the three primary hormones.

VITAMIN D HORMONE

Before vitamin D was discovered it was known that exposure to sunshine in some way prevented rickets, the bone-deforming disease of children, since this disease, so prevalent in children living in the northern latitudes, was not seen in warm sunny climates. When methods became available for determination of serum calcium and phosphate concentration in patients it was found that rickets was a disorder of calcium and phosphate homeostasis. In 1919 Huldschinsky reported that ultraviolet radiation produced by quartz lamps was effective in healing rickets, implying that the effective radiation of sunshine was that in the ultraviolet spectrum. At the same time Mellanby succeeded in producing an experimental model of rickets in puppies kept indoors and fed a low fat diet, and found that addition of cod liver oil or butter fat to the feed prevented rickets. He speculated that the active ingredient in these fats was vitamin A, which had already been discovered and was known to be present in cod liver oil and butter fat. However, McCollum and his colleagues disproved this. They produced nutritional rickets in laboratory rats and found that aerated cod liver oil that no longer had any vitamin A potency was still effective in preventing and healing experimental rickets. The new food factor was then called vitamin D.

The connection between ultraviolet radiation energy and vitamin D was solved separately by Alfred Hess and by Steenbock, who found that irradiation of foods devoid of vitamin D potency conferred vitamin D activity when tested in rachitic rats. They postulated that a provitamin D was present in the lipid fraction of the foodstuff which was activated to vitamin D by absorption of ultraviolet energy. With this clue the first provitamin D was found in food and identified as a plant sterol, ergosterol, present in large amounts in yeast. Subsequently, the provitamin D in the skin of man and other animals was identified as 7-dehydrocholesterol. These sterols absorb short wave ultraviolet radiation in the band between 280 and 305 mμ and in the process are converted to vitamin D by opening the B ring and formation of a secosterol (Figure 2–4). Only sterols with two double bonds in the B ring undergo this reaction, so that cholesterol which has only one double bond is not a provitamin D, whereas 7-dehydrocholesterol is. Ergo-calciferol or vitamin D_2, the product of irradiation of ergosterol, and cholecalciferol, vitamin D_3, the product of irradiation of 7-dehydrocholesterol, are equally effective in the rat and in man as measured by their potency in the healing of rickets. This is not true of the bird, which distinguishes between these two vitamins D, and in which cholecalciferol is ten times as effective as ergocalciferol. The only mammal in which there is known to be a similar distinction between the activity of these two forms of vitamin D is the New World monkey.

For many years vitamin D was considered to be the active agent in controlling calcium and phosphate homeostasis. However, when it became possible to prepare isotopically labeled vitamin D with high potency, and chromatographic methods became available for separation of compounds differing only slightly in their configuration, the metabolic transformation of vitamin D could be studied, and it became evident that vitamin D itself was not active until it was metabolized to an active form. The metabolic cycle of vitamin D is diagrammed in Figure 2–5 The first step is addition of an oxygen at C 25 in the side chain to form 25-OH vitamin D. This reaction occurs in the liver, and both a microsomal and a mitochondrial enzyme have been demonstrated 25-OH vitamin D has physiologic activity in both intestinal and bone systems but is not the active vitamin D hormone. 25-OH vitamin D released by the liver cells is bound in plasma to a specific binding protein that binds this molecule more avidly than precursor vitamin D, and is present in great excess so that it is highly unsaturated at physiologic concentrations of plasma 25-OH vitamin D.

A radioimmunoassay for 25-OH vitamin D has been developed based on the competitive binding principle, using this plasma protein or a 25-OH vitamin D binding protein in kidney cytosol. This determination has had clinical application in the study of vitamin D deficiency states and abnormalities of vitamin D metabolism (Chapter 5), and has been of major value in physiologic studies of vitamin D. Using this determination it has become evident that the formation of 25-OH vitamin D is only poorly controlled, although in vitro studies suggested that the 25-hydroxylase activity was inhibited by increased concentrations of the product, 25-OH vitamin D. The poor constraint is demonstrated by the marked variation in serum 25-OH vitamin D, depending on availability of the substrate. For example, there is marked seasonal variation with several-fold increase in the concentration of 25-OH vitamin D during periods of intensive exposure to sunshine (Table 2–1). The concentration of 25-OH vitamin D following exposure to sunshine or ultraviolet lamps is never high enough to produce the toxic manifestations of hypervitaminosis D. This may indicate some degree of constraint on 25-hydroxylation of vitamin D but also

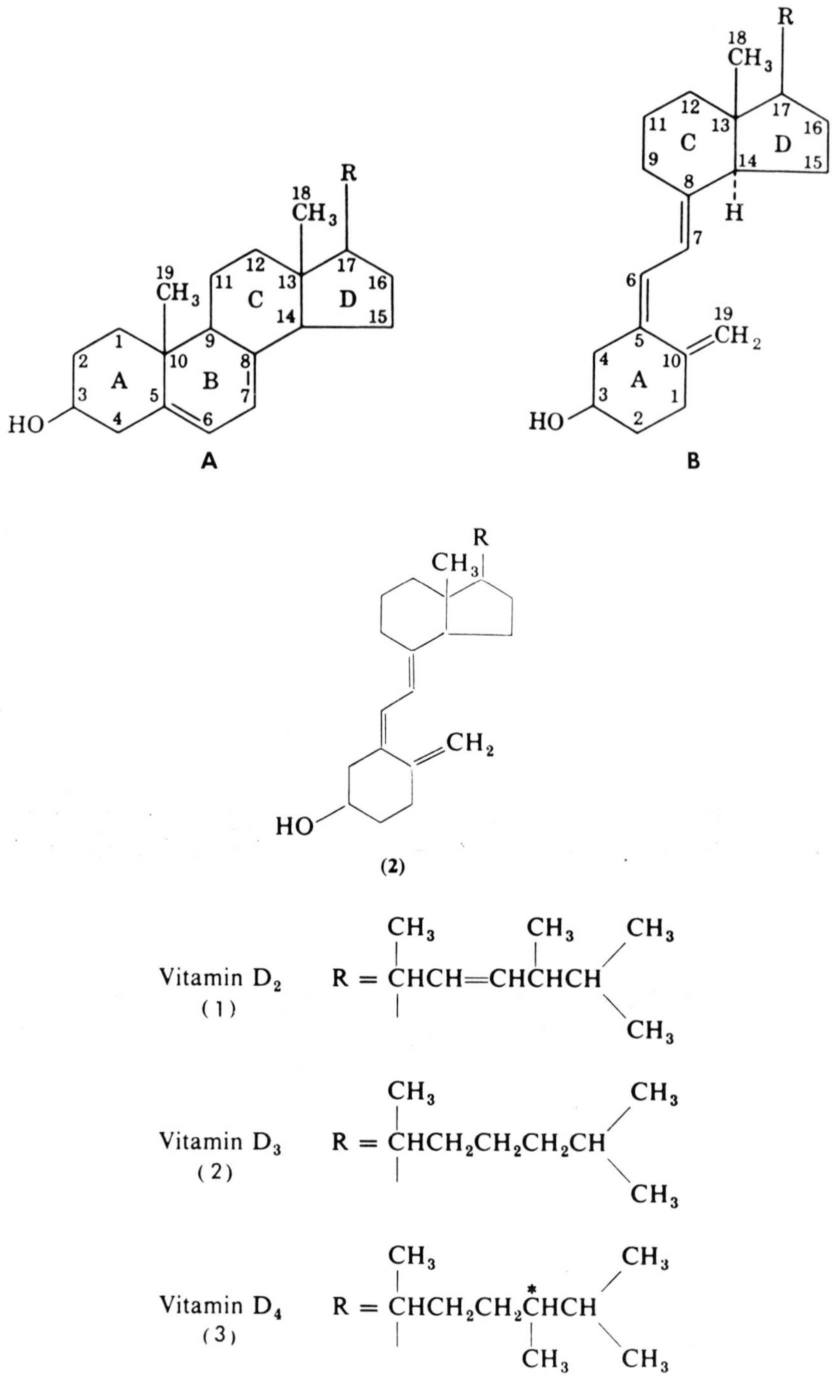

Figure 2–4. Structure of provitamins and vitamins D. *A*, The provitamins D are sterols with two double bonds in the B ring. *B*, The B ring is then opened by absorption of ultraviolet light energy forming vitamin D. *C*, The side chain (labeled R in parts *A* and *B*) on ring D differs (1, 2 and 3) in the various vitamins D.

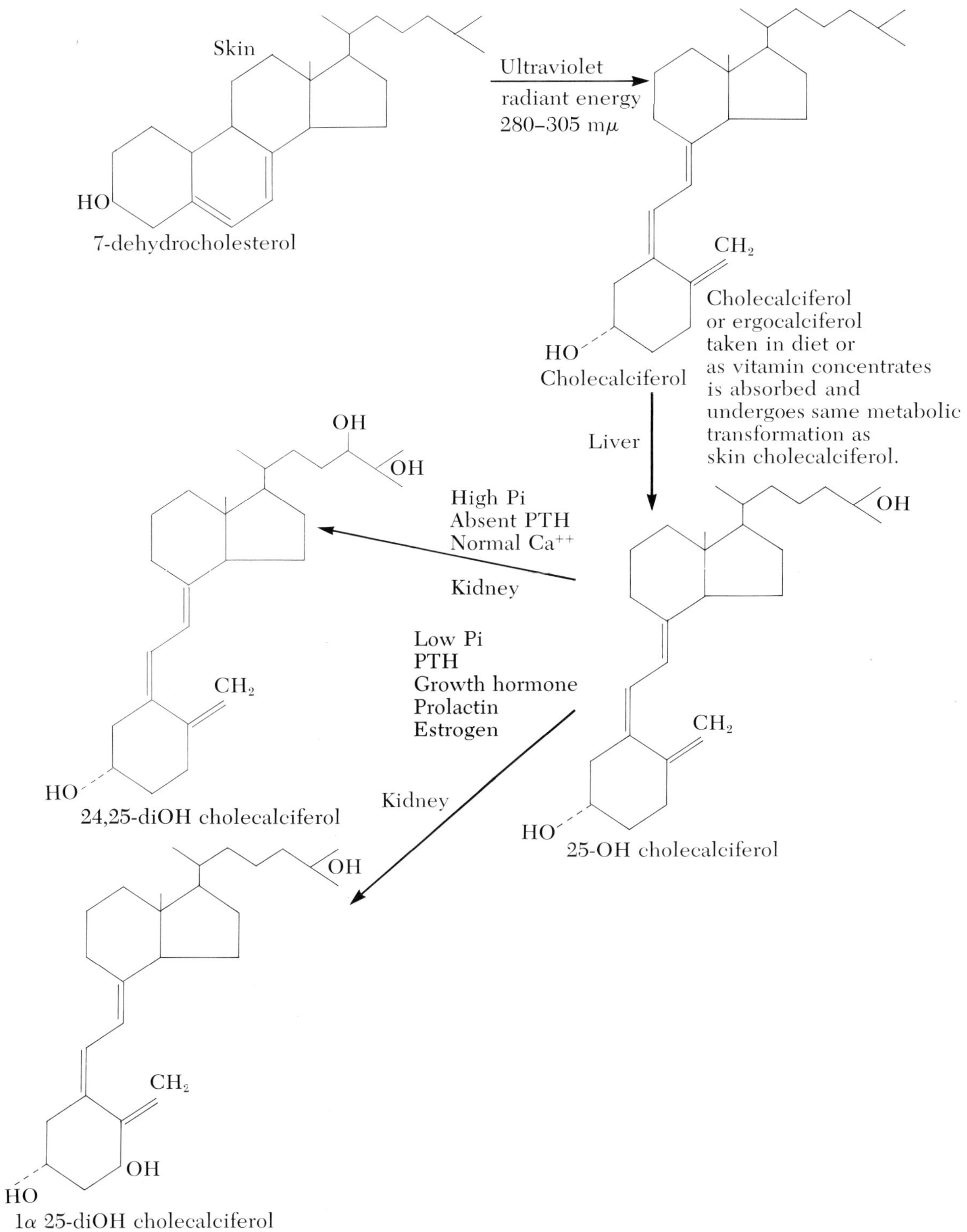

Figure 2–5. Metabolic cycle of vitamin D that leads to formation of active vitamin D hormone, 1,25-diOH vitamin D, which may have physiologic activity but is on the pathway to inactivation and elimination of vitamin D.

Table 2–1. Effect of Exposure to Sunshine on Serum Concentrations of 25-Hydroxyvitamin D

GROUP	NO.	AGE (YEARS)	CONSUMPTION OF D WEEKLY (UNITS)	WEEKLY EXPOSURE TO SUNLIGHT (HOURS)	PLASMA 25-HCC ng/ml
Normal Volunteers	40	30.2 ± 12.9	2230 ± 1041	8.8 ± 6.1	27.3 ± 11.8
Biliary Cirrhosis	4	1.5 – 55	2500 (est.)	–	6.4 ± 2.6*
Lifeguards	8	18.5 ± 2.0	2895 ± 677	53.0 ± 10.3	64 ± 8.7*

*p .001
Values represent mean ± SD
Modified from Haddad, J. C. and Chyn, K. J.: Competitive protein-binding radioassay for 25-hydroxycholecalciferol. J. Clin. Endocrinol. Metab. 33:992–995, 1971.

suggests that skin formation of cholecalciferol is limited. This could be due to exhaustion of the substrate, 7-dehydrocholesterol, but an important protective mechanism is probably the pigmentation of the skin resulting from formation of melanin aggregates. Highly pigmented skins are opaque to short wave ultraviolet radiation, and thus there is reduced conversion of provitamin D to vitamin D.

When vitamin D is given orally or parenterally in concentrated form, whether ergocalciferol or cholecalciferol, the concentration of serum 25-OH vitamin D can be increased to values much higher than those attained by exposure to ultraviolet radiation. This is the basis for the toxicity of vitamin D, since 25-OH vitamin D is an active compound and can produce hypercalcemia and the injurious effects of excessive concentrations of calcium in body fluids. Although serum 25-OH vitamin D concentrations are increased to tenfold the normal value by doses of vitamin D in the range of 2 to 3 mg per day, a considerable fraction of the vitamin D is not metabolized to 25-OH vitamin D but is stored in adipose tissue as unmetabolized vitamin D, which remains for months as an available substrate for continued 25-OH vitamin D formation. The quantities of vitamin D required to produce a given level of plasma 25-OH vitamin D are about ten times those of 25-OH vitamin D itself. This tissue storage of unmetabolized vitamin D is the basis for the prolonged toxicity of vitamin D. Once toxic concentrations of 25-OH vitamin D have been achieved, discontinuance of vitamin D does not lead to a rapid decrease of 25-OH vitamin D concentrations since there is a large store of substrate in the tissues.

Although 25-OH vitamin D is an active compound in high concentration it is not the potent vitamin D hormone, and at physiologic concentrations in plasma is not the compound of functional importance. Following its formation it is transported in plasma in bound form, is taken up by receptors in renal tubule cells and is there metabolized by a mitochondrial system to 1,25-diOH vitamin D, which is extraordinarily potent, being effective in amounts of 0.5 to 1μg/day. This kidney system is a regulated one and the production of 1,25-diOH vitamin D and its concentration in plasma is under close control. Such regulation of the 25-hydroxyvitamin D, 1-hydroxylase in the kidney is expected if the product 1,25-diOH vitamin D is the physiologically active principle that maintains calcium homeostasis.

A major factor controlling 1,25-diOH vitamin D synthesis in kidney is parathyroid hormone and, therefore, Ca^{++} ion concentration indirectly through parathyroid hormone. In addition, growth hormone, prolactin and estrogens act to increase 1,25-diOH vitamin D concentrations. Adaptation to low calcium intake with increased efficiency of intestinal calcium absorption is mediated through parathyroid hormone-stimulated increase of 1,25-di-OH vitamin D. The increased calcium absorption needed to maintain calcium balance during growth, pregnancy and lactation is also maintained by greater 1,25-diOH vitamin D formation. The half-life of 1,25-diOH vitamin D is quite short, so that not only is response to this hormone rapid but the effect is quickly dissipated when 1,25-diOH vitamin D production or its administration ceases. This makes it an efficient regulatory system responsive to physiologic needs for maintenance of calcium homeostasis.

Another regulating factor is extracellular phosphate concentration. When this is low, metabolism of 25-OH vitamin D is mainly to 1,25-diOH vitamin D, but when it is high 1-hydroxylation is inhibited and the major product is 24,25-diOH vitamin D; this may have some physiologic activity but is probably an initial step in the inactivation of 25-OH vitamin D by conjugation in the liver to form inactive glucuronides which are excreted in the bile. The reduction of 1,25-diOH vitamin D concentration secondary to hyperphosphatemia reduces intestinal absorption of phosphate, and probably also stimulates parathyroid hormone secretion, since there is some evidence that parathyroid gland cell activity is directly influenced by 1,25-diOH vitamin D as well as Ca^{++} ion concentration. The increased parathyroid hormone would reduce extracellular phosphate concentration by inhibition of tubular reabsorption of phosphate. This control of 1,25-diOH vitamin D formation by extracellular phosphate concentrations could, therefore, be of importance in the adaptation to high phosphate intakes. The increased formation of 1,25-diOH vitamin D when extracellular phosphate concentration is reduced also serves as a negative feedback control system, in this instance by both inhibiting parathyroid hormone secretion and increasing intestinal absorption of phosphate.

It is unlikely that the calciferols have physiologic activity unless metabolized to the 25-OH derivative. This is evidenced by the relative unresponsiveness to vitamin D of patients with hepatocellular injury, and the lack of effect of calciferol in in vitro systems in which the solubilization of bone mineral from calvarium or fetal bone is tested. 25-OH vitamin D does have physiologic activity since this compound is active in anephric subjects, and there is no evidence that 1-hydroxylation of 25-OH vitamin D can be performed except by a kidney system. However, a concentration of 25-OH vitamin D higher than the physiologic one is required for activity if the formation of 1, 25-diOH vitamin D is not possible. 1, 25-diOH vitamin D is the active compound, based on the evidence that it is the molecule taken up by receptors in the intestinal mucosal cell and bone cell and found to be bound to nuclear chromatin in these cells. The physiologic actions of 1,25-diOH vitamin D on intestinal absorption of calcium and phosphate, on bone cell function in the solubilization of bone mineral and on renal tubule reabsorption of phosphate will be discussed in the separate sections on these processes.

PARATHYROID HORMONE

Anatomy, Chemistry and Physiology

The parathyroid glands arise as paired structures from the entoderm of the third and fourth branchial pouches. The thymus also develops from the third branchial pouch and the parathyroids may descend with the thymus and be incorporated into it. The superior parathyroid glands coming from the fourth branchial cleft are attached to the posterior thyroid capsule or may be embedded in the thyroid. The inferior parathyroid glands from the third branchial pouch usually lie below the lower pole of the thyroid lateral to the trachea, but as indicated may migrate into the anterior mediastinum with the thymus. The parathyroid glands of children are made up primarily of cords of chief cells with few fat cells. At puberty fat cells begin to increase, and in the adult constitute a considerable portion of the volume of the glands. After puberty another cell type, the oxyphil cell, begins to increase in number in the parathyroid gland. The chief cells are the active secretory cells which synthesize and secrete parathyroid hormone.

Parathyroid hormone is a single chain peptide with a molecular weight of 9500. The hormone is biosynthesized in the cell as a larger molecule, proparathyroid hormone, which is converted intracellularly into parathyroid hormone by the splitting off of a hexapeptide at the aminoterminal end of the peptide chain. Following secretion of the hormone it is cleaved further into fragments, some of which are inactive, but one of the smaller fragments comprising the 34 amino acids starting at the aminoterminal end (peptide 1–34) possesses the full physiologic activity of parathyroid hormone. A larger carboxyl terminal fraction is biologically inactive but persists for a longer time in the circulation than the active fragments. This persistence of an inactive peptide fragment of parathyroid hormone has been useful in the diagnosis of hyperparathyroidism. Radioimmunoassay determination of parathyroid hormone can give different results depending on the specificity of the antibody used in the assay. An antibody that recognizes the inactive carboxyterminal fragment is commonly employed, since the concentration of this stable peptide in the serum can be increased to a greater extent in hyperparathyroidism than the concentrations of the intact hormone or the biologically active aminoterminal fragment, which have a short residence time in the plasma. The radioimmunoassay for this carboxyterminal fragment can be of value, therefore, in the diagnosis of chronic hypersecretion of parathyroid hormone. For physiologic studies in which the acute response of parathyroid gland cells to provocative stimuli is measured, radioimmunoassay with antibodies that recognize the short-lived aminoterminal peptide is more valuable.

The major controlling factor in determining the rate of hormone synthesis and release is the concentration of Ca^{++} ion in the fluid perfusing the parathyroid gland. Reduction of Ca^{++} ion stimulates gland cell synthesis of hormone and its secretion. Increase of Ca^{++} ion concentration inhibits both of these processes. Chronic hypocalcemia results in gland cell hyperplasia and hypertrophy with marked increase in hormone release following acute hypocalcemic challenge. Reduction of Mg^{++} ion concentration in the fluid perfusing the parathyroid gland stimulates release of parathyroid hormone but probably does not increase synthesis of the hormone. Profound Mg deficiency may indeed inhibit hormone synthesis and produce a picture of functional hypoparathyroidism. Increased concentration of phosphate by itself does not stimulate parathyroid hormone secretion, but may have an effect through reduction of Ca^{++} or Mg^{++} ion concentration. Parathyroid hormone probably is constantly secreted at a baseline level by parathyroid gland cells, so that an increased mass of parathyroid gland cells as in adenoma or parathyroid hyperplasia can secrete excessive hormone despite increased concentrations of Ca^{++} ion in plasma. Changes in secretion rate inversely proportional to the concentration of Ca^{++} ion can still be demonstrated. This phenomenon can be utilized to determine the state of the parathyroid glands by measuring concentrations of immunoreactive parathyroid hormone in serum following reduction of Ca^{++} ion concentration by Na_2EDTA infusion. If the glands are hyperactive a much greater increase of serum IPTH concentration will be obtained than with similar reduction of Ca^{++} ion concentration in subjects with normal parathyroid glands. A subnormal response indicates hypoplastic or hypofunctioning glands.

An additional controlling factor in parathyroid hormone secretion may be the concentration of 1,25-diOH vitamin D. The parathyroid gland cell membrane has receptors that bind this vitamin D metabolite, and there is evidence that administration of 1,25-diOH vitamin D to vitamin D-depleted subjects reduces parathyroid hormone concentrations before Ca^{++} ion concentration has risen to the normal range. The kidney and liver are the principal organs in the metabolism of parathyroid hormone, which normally is rapidly cleared from the circulation following its secretion. The marked increase in immunoreactive parathyroid hormone concentration in the serum of patients with renal insufficiency may reflect not only parathyroid hyperplasia with increased hormone synthesis, but also reduced rate of hormone degradation in the absence of normal kidney metabolism.

Following its secretion parathyroid hormone or its biologically active fragment is taken up by target cells, namely, bone cells and renal tubule cells. The initial effect is an activation of parathyroid hormone-responsive adenyl cyclase with increased formation of cyclic AMP from ATP. The cyclic AMP is the "second messenger" through which parathyroid hormone induces its effect in the specific responsive cells. The action of cyclic AMP is thought to be the activation of protein kinases which phosphorylate proteins with enzymatic activities. In the bone cells this action results in both increased metabolic activity of osteocytes and the formation of multinucleated osteoclasts from precursor cells. In consequence of their enzymatic activity and release of lysosomal enzymes, the osteocytes and osteoclasts solubilize bone mineral at the cell surface and dissolve the unmineralized collagen, causing loss of bone substance and release of calcium and phosphate ions into the extracellular fluid.

The mechanism by which the bone cells solubilize bone mineral is still obscure. It is known that the parathyroid hormone-stimulated activity of the bone cells increases cell formation of organic acids, primarily lactic acid and citric acid as well as CO_2, and release of these acids into the pericellular space. Bone mineral is presumably dissolved by the increased H^+ ion concentration in the cell environment as well as by the specific calcium complexing effect of citrate ion. The lysosomal enzymes are proteolytic, and in the acid environment act as collagenases to hydrolyze the collagen from which the mineral crystals have been dissolved. In the renal tubules the increased cyclic AMP concentration induced by parathyroid hormone action is followed by inhibition of tubular reabsorption of phosphate with phosphaturia, but increased reabsorption of calcium and magnesium with consequent conservation of these ions. Additional effects on renal excretion of ions are natriuresis and bicarbonaturia so that a more alkaline urine is excreted. This results in reduction of serum bicarbonate with a mild hyperchloremic acidosis.

An additional renal tubular action of parathyroid hormone is on mitochondrial function, which is reflected not in ion transport but in increased 1-hydroxylation of 25-OH vitamin D, resulting in higher concentrations of the vitamin D hormone, 1,25-diOH vitamin D. The action of parathyroid hormone in increasing intestinal absorption of calcium and phosphate may be a secondary phenomenon produced not by a direct action of the hormone on intestinal mucosa but by the effect of the increased synthesis of 1,25-diOH vitamin D, which is the most active agent controlling intestinal transport of calcium and phosphate.

The net result of parathyroid hormone action is to increase the entrance of calcium and phosphate into extracellular fluid from intestine and bone, with conservation of the calcium but elimination of the phosphate in urine, thus increasing calcium concentration but reducing phosphate concentration in extracellular fluid. The action of parathyroid hormone in increasing cyclic AMP concentration in renal tubule cells is associated with increased loss of cyclic AMP in urine, so that measurement of cyclic AMP excretion has been used as an index of parathyroid gland activity.

In the child another renal tubular effect of increased parathyroid hormone concentration is inhibition of tubular reabsorption of amino acids, causing a generalized increased excretion of all amino acids in urine, especially the essential amino acids which usually are almost completely reabsorbed so that their concentration in urine by normal subjects is extremely low. The generalized aminoaciduria of children with vitamin D deficiency is believed to be mainly due to secondary hyperparathyroidism. These children, as might be expected, also excrete increased amounts of cyclic AMP in urine.

CALCITONIN

This polypeptide hormone was discovered when Copp observed that after injections of crude parathyroid extract into animals there was an initial decrease of serum calcium concentration followed by the expected increase. He reasoned that there was a hypocalcemic factor in these extracts, which proved to be the case.

Calcitonin is secreted by the C cells or parafollicular cells of the thyroid. In the thyroid of man the C cells are concentrated in the posterior portion of the lateral lobes adjacent to the parathyroid glands. In birds, reptiles and amphibians the C cells are found in a separate gland, the ultimobranchial body. In mammals the C cells are carried during embryogenesis into the developing thyroid gland; however, their origin is not the entodermal cells of the pharyngeal pouch but neuroendocrine cells derived from the neural crest. Calcitonin is a 32 amino acid peptide, and although there are differences among the calcitonins of various animal species in their amino acid sequences, a considerable homology exists. The major stimulus for calcitonin secretion is an increased concentration of Ca^{++} ion in the perfusing plasma. However, oral administration of calcium salts causes an increased secretion of calcitonin without significant increase of Ca^{++} ion concentration in arterial plasma. This observation led to the finding that gastrointestinal hormones stimulate calcitonin secretion, and it is suggested that the effect of oral calcium feeding is to stimulate secretion of a hormone, possibly intestinal glucagon or gastrin, which is a secretagogue for calcitonin. Pancreatic glucagon has been demonstrated to stimulate calcitonin secretion, and a gastrin analogue, pentagastrin, is also effective.

The primary physiologic action of calcitonin is to inhibit bone resorption by an effect on osteocytes and osteoclasts. Calcitonin therefore has a hypocalcemic effect through reduction of the flow of Ca^{++} ion from bone to extracellular fluid. This hypocalcemic action occurs in the hypoparathyroid subject, so that it is not an inactivation or inhibition of parathyroid hormone action but a direct action on calcitonin-responsive bone cells. When populations of bone cells are studied in vitro it is possible to separate calcitonin-responsive and parathyroid hormone-responsive cells. Presumably they differ in their receptors for the specific hormone. Administration of calcitonin inhibits osteocytic osteolysis and causes reduction of the number of osteoclasts.

In young growing animals that have a high rate of bone turnover, calcitonin administration causes a rapid and marked decrease in concentration of serum calcium, but a comparable amount of calcitonin given to the adult animal adjusted for weight or surface area has much less of a serum calcium lowering effect. When the rate of bone mineral deposition is rapid, as in the growing animal, inhibition of bone mineral solubilization might be expected to cause more rapid decrease of serum calcium than during periods of slow bone mineral deposition. In addition, calcitonin has a renal tubular effect resulting in decreased phosphate reabsorption with phosphaturia and reduction of serum phosphate concentration. The action of calcitonin on calcium ion concentration is thus opposed to that of parathyroid hormone, but its action on phosphate concentration in extracellular fluid is in the same direction as that of parathyroid hormone.

Infusion of calcitonin slows gastric motility and increases gastric emptying time. This may explain one of the side-effects of calcitonin treatment of hypercalcemia, namely, nausea and vomiting. Another action of calcitonin on the intestinal tract is to increase secretion of water, sodium and chloride into the lumen, and this effect may be associated with decreased uptake of calcium and phosphate from the gut. Whether this is a physiologic action or a pharmacologic effect of excess calcitonin is uncertain. There is no evidence that calcitonin specifically alters metabolism of vitamin D, but it may do so indirectly through its hypophosphatemic action which should enhance 1-hydroxylation of 25-OH vitamin D.

The physiologic importance of calcitonin in calcium homeostasis is somewhat unclear. Theoretically it should protect against increases of Ca^{++} ion concentration, through either excessive absorption of calcium from the intestine or release of calcium from bone under conditions of increased resorption or reduced bone mineral deposition. However, thyroidectomized subjects show little abnormality of calcium homeostasis, indicating that the action of parathyroid hormone is the dominant factor and that the role of calcitonin in this function is subsidiary. Calcitonin levels are high in the newborn, and it is possible that this hormone may function in the fetus to inhibit bone resorption and thus to increase the net uptake of mineral by the growing skeleton. Since fetal plasma Ca^{++} ion concentrations are maintained by the placenta, skeletal control of this concentration is unnecessary during fetal life, and the hormonal regulation may be primarily directed toward maximal mineralization of growing bone.

Calcitonin in plasma can be determined by radioimmunoassay, and such determinations along with those of parathyroid hormone have been of importance in understanding the pathogenesis of hypocalcemia in the newborn. Human calcitonin has been synthesized, but the preparation used for treatment of hypercalcemic patients is synthetic salmon calcitonin, because this peptide is physiologically more active and has a more prolonged duration of action in man than has human calcitonin. For this reason it is clinically more useful than the latter.

CORTISOL

Although cortisol is not ordinarily regarded as a regulator of calcium and phosphate homeostasis, attention must be given to its role in bone mineralization. In the earliest description of Cushing's syndrome, the disease of overproduction of adrenal cortical hormones, osteoporosis with compression fractures of the vertebrae was noted. Cortisol in high dose does inhibit calcium absorption by the intestine, an action opposite to that of vitamin D. This is not simply due to inactivation of vitamin D, since experimental cortisol administration to vitamin D-depleted rats further reduces intestinal transport of calcium. However, clinically the action of cortisol on calcium absorption is reversed by increased doses of vitamin D metabolites. There is evidence that cortisol increases the rate of inactivation of vitamin D, possibly by induction of liver enzymes. It does not prevent the effect of vitamin D in the healing of experimental rickets, so that its antagonism to vitamin D is only partial. Cortisol administration to patients is followed by increased serum concentrations of immunoreactive parathyroid hormone, which is the expected response to diminished absorption of calcium and reduced Ca^{++} ion concentrations in plasma.

This increased secretion of parathyroid hormone maintains normocalcemia in the presence of cortisol excess. The hypercalcemia of hyperparathyroidism is not ordinarily reduced by suppressive doses of cortisol or other glucocorticoids. This would fit with the concept that the major effect of parathyroid hormone on calcium ion concentrations is an increase of the flow of calcium from bone mineral to extracellular fluid, which is not primarily inhibited by cortisol. On the other hand, the adrenocorticoids are effective in reducing the hypercalcemia of vitamin D intoxication or of sarcoidosis in which there is hyperresponsiveness to vitamin D. Since this hypercalcemia is in great part dependent on increased intestinal absorption of calcium, the response to cortisol is understandable. In the hypoparathyroid patient being treated with vitamin D, the onset of adrenal insufficiency can be associated with hypercalcemia, which is reduced by replacement therapy with physiologic amounts of cortisol or other glucocorticoids, indicating its modulating effect on the action of vitamin D in the absence of parathyroid hormone.

Glucocorticoids in excess increase renal excretion of calcium and of phosphate. The latter action does result in reduction of serum phosphate. This hypophosphatemic phosphaturia probably is also contributed to by the secondary increase in parathyroid hormone following glucocorticoid administration.

CALCIUM HOMEOSTASIS

CALCIUM ABSORPTION FROM THE INTESTINE

Calcium absorption is a regulated process. Unlike the monovalent cations, sodium and potassium, which under normal physiologic conditions are almost completely absorbed irrespective of body requirements, the absorption of calcium is incomplete and is controlled by the need for calcium. The physiologic importance of this control mechanism is clear when the renal excretion of the monovalent cations is contrasted to that of calcium. Even with great excesses of intake of sodium and potassium the renal excretion of the unnecessary ions is rapid and quantitative, although there may be temporary retention of sodium and water when sodium intake is abruptly increased. In the instance of calcium, unregulated intestinal absorption with excretion of excess calcium in the urine would produce serious problems owing to the relative insolubility of calcium phosphate and bicarbonate.

The calcium intake of man can vary enormously depending on the intake of milk or milk-derived products. The quantity of calcium phosphate that can be taken up by bone is limited even during periods of most rapid growth. If one examines the relationship of urine calcium output to calcium intake, as was done by Knapp (Fig. 2–6), it is evident that during childhood and adolescence urine excretion of calcium increases but slightly when intake is increased three- or fourfold. During adult life the increase of urine excretion of calcium with increase of calcium intake is greater than in the growing child, but is still much less than the increment of calcium intake. It may be that during growth calcium uptake by bone is somewhat greater with augmented calcium absorption, or that calcium absorption is more efficiently controlled in relation to the calcium need of the skeleton during the growing period.

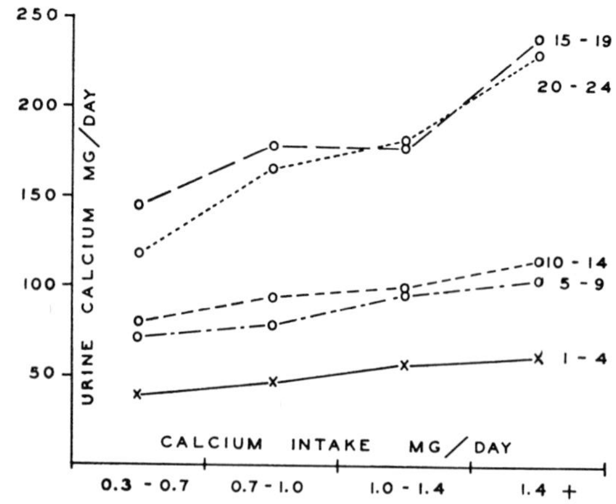

Figure 2–6. Urinary excretion of calcium at various ages as influenced by intake of calcium. (From Harrison, H. E. Reprinted from Federation Proceedings *18*:1085–1092, 1959.)

These differences between the growing child and adult, however, are minor compared with the relatively slight rise of urine output of calcium with large increments of diet calcium at all ages. Conversely, it has been found that the percentage absorption of calcium by individuals ingesting large amounts of calcium is low and increases when calcium intake is reduced. Comparison of percentage of food calcium absorbed by infants fed low and high calcium diets is shown in Figure 2–7. Similar data have been obtained in adults. The adaptation to low calcium intakes in some adults who have been accustomed to high calcium diets can be delayed, so that the regulatory processes may be inefficient in individual subjects, and there can also be true calcium malabsorption due to pathologic states of

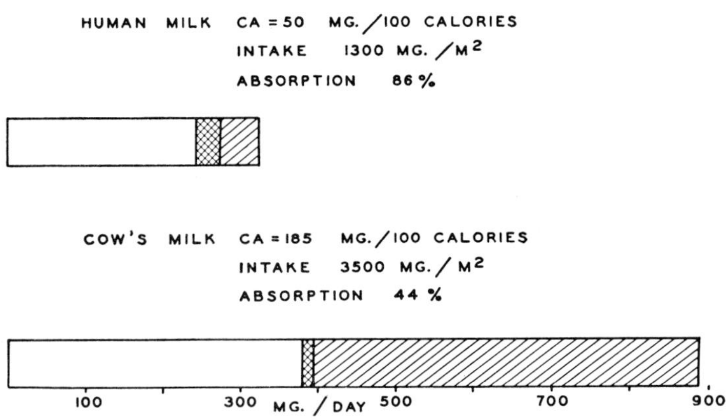

Figure 2–7. Adaptation of infants to low and high calcium intakes as indicated by percentage of food calcium absorbed from intestine. The length of the horizontal bar indicates calcium intake, the cross-hatched portion represents urine excretion, and the diagonally lined segment fecal excretion. Calcium retained is indicated by the clear portion. (From Harrison, H. E. Reprinted from Federation Proceedings *18*:1085–1092, 1959.)

intestinal function, which will be discussed separately.

The control of calcium absorption by the calcium need of the body depends chiefly on the concentration of 1,25-diOH vitamin. 25-OH vitamin D and related compounds such as 25-OH dihydrotachysterol can stimulate calcium absorption from the intestine, but only at concentrations much greater than the effective amount of 1,25-diOH vitamin D, so that they are pharmacologic agents rather than physiologic controlling factors. The adaptation to low calcium intakes is presumably through parathyroid hormone, since its secretion is stimulated by very slight reduction of extracellular calcium ion concentration. With increased parathyroid hormone output the formation of 1,25-diOH vitamin D in the kidney is accelerated and the efficiency of intestinal calcium absorption is increased. The increased calcium absorption during active growth, pregnancy and lactation to meet increased calcium needs can be mediated through the action not only of parathyroid hormone but of growth hormone, estrogens and prolactin through the control of 1,25-diOH vitamin D production.

The dependence of intestinal calcium absorption on vitamin D was an early physiologic observation based on studies in intact animals of varying vitamin D status. The role of 1,25-diOH vitamin D has been learned through investigations of calcium transport by intestine, using in vitro preparations such as the everted intestinal sac. These preparations transport calcium from the mucosal surface across the intestinal wall, but only if the donor of the intestine has received vitamin D or a vitamin D analogue or metabolite. It is clear that the most potent vitamin D metabolite in this system is 1,25-diOH vitamin D, and by the use of radioactive labeled vitamin D it has been found that the compound taken up by the intestinal mucosal cell is 1,25di-OH vitamin D. This compound is bound by cytosol proteins and is then transported into the nucleus, where it is bound to chromatin. Following entrance of the vitamin D metabolite into the nucleus there is increased production of mRNA, and it is believed that the action of vitamin D is to activate or derepress a segment of DNA, resulting in mRNA production and synthesis of new proteins that effect the transport of calcium across the mucosal cell into the intestinal interstitial space and the capillaries.

The exact nature and role of the proteins is still conjectural. Both a specific calcium binding protein and increased activity of an enzyme with alkaline phosphatase activity are found in the mucosa following vitamin D. Physiologic studies indicate that the action of the vitamin D metabolite increases the rate of transfer of calcium from luminal fluid into the mucosal cell, and also the rate of elimination of calcium from the mucosal cell into the interstitial fluid bathing the basal and lateral surfaces of the cell. Whether the primary action is to increase the rate of calcium permeation through the luminal surface, i.e., the brush border membrane of the intestinal mucosal cell, remains a question. The net effect, however, is that, even at high concentrations of calcium in the luminal fluid, calcium moves only slowly into the mucosal cell in the absence of the active vitamin D metabolite. This, then, is the controlling system that prevents excessive absorption of calcium with high intakes of calcium, and yet permits highly efficient absorption of calcium when intake of calcium is low or body needs for calcium are high.

The absorption of calcium is greater in hyperparathyroid states than in the normal, and is reduced following parathyroidectomy. It is also lower in hypoparathyroid subjects than in normals. This effect of parathyroid hormone on intestinal absorption of calcium has been observed in man and in experimental animals. The evidence, however, indicates that the parathyroid action is not a direct stimulation of calcium absorption by the hormone but is an indirect effect of the hormone through its stimulation of the 1-hydroxylation by kidney cells of 25-OH vitamin D, and that the active agent is 1,25-diOH vitamin D. Even in the absence of parathyroid hormone, intestinal absorption of calcium can be adequate to meet the calcium needs of the body, since children with congenital or acquired hypoparathyroidism can grow and have apparently normally mineralized skeleton despite chronic hypocalcemia. Their calcium absorption is somewhat reduced, but urine calcium excretion is very low because of reduced filtered calcium, so that the net balance of calcium can be normal.

Calcitonin lowers serum calcium concentration but does not have any known direct action on vitamin D metabolism. Indirectly the reduced calcium ion concentration following calcitonin injection stimulates parathyroid hormone output, which may then influence calcium absorption through its action on production of 1,25-diOH vitamin D. In addition, as mentioned, the hypophosphatemic action of calcitonin can stimulate 1,25-diOH vitamin D formation. There is some evidence that calcitonin has a slight effect in reducing intestinal calcium transport, but this probably is not a major action of this hormone.

Cortisol in high dosage, and other steroids with glucocorticoid properties when given in amounts that suppress inflammatory reactions, inhibit calcium absorption. There has been contradictory evidence concerning the mechanism of this action. In experimental studies it has been found that the adrenocorticoids inhibit calcium transport by a reaction that does not involve vitamin D. There is clinical evidence in patients, however, that the action of glucocorticoids in reducing calcium absorption can be overcome by increased intake of vitamin D. This may be through a compensatory action also involving parathyroid hormone, since increased parathyroid hormone output as measured by immunoreactive parathyroid hormone concentrations in serum is observed following glucocorticoid administration. Increased conversion of vitamin D to 1,25-diOH vitamin D under the influence of high parathyroid hormone concentrations should follow, with consequent compensation for the inhibitory effect of corticosteroids on calcium transport. It has also been suggested that the metabolic inactivation of vitamin D through hydroxylation and glycuronidation in the liver is accelerated by corticosteroids. This would reduce the concentration of 1,25-diOH vitamin D, but this reduction could be overcome by increased dosage of vitamin D. The osteoporosis resulting from prolonged treatment with suppressive doses of glucocorticoids is not explained simply by reduced intestinal absorption of calcium, but calcium malabsorption may be a contributing factor and should be corrected if present.

Calcium malabsorption can result from hypermotility of the intestine with reduced transit time, so that any chronic diarrhea such as in gluten-induced enteropathy, in Crohn's disease or following resection of a portion of the small intestine may be associated with poor absorption of calcium. This probably is also the mechanism of calcium malabsorption in hyperthyroidism, which can cause hypermotility and diarrhea. Impaired absorption of calcium is particularly caused by disorders of intestinal function in which there is fat malabsorption. There are two mechanisms for excessive calcium lossses in the feces in patients with steatorrhea. One is vitamin D deficiency due to losses of ingested vitamin D in the stool fat. Presumably this should be less important in patients who receive most of their vitamin D by skin exposure to ultraviolet radiation of sunshine or ultraviolet lamps. Whether endogenously formed vitamin D or its metabolites are secreted into the intestine and lost with the unabsorbed fat is unknown. Calcium malabsorption in steatorrhea can also occur through a physicochemical mechanism of precipitation of calcium in the intestinal lumen as insoluble calcium salts of fatty acids, so called calcium soaps. This should be found in those forms of steatorrhea in which lipolysis is not disturbed and the triglycerides are hydrolyzed to free fatty acids, but the latter are not adequately absorbed because of specific interference with mucosal transport as in gluten-induced enteropathy, tropical sprue or other intrinsic disease of intestinal epithelium.

Reduction of intestinal calcium absorption can also result from ingestion of very large amounts of inorganic phosphate, or especially of certain organic phosphates that precipitate or complex calcium and prevent its uptake by the intestinal mucosal cell. One of the organic phosphates is cellulose phosphate, which has been used as a pharmacologic agent to reduce absorption of calcium in patients with intestinal hyperabsorption of calcium and hypercalciuria. The other important organic phosphate is phytic acid, inositol hexaphosphate, which is found in cereal grains, particularly in the bran fraction. Phytic acid in whole wheat flour is broken down by yeast during the leavening process, so that ingestion of bread made with whole wheat flour does not affect calcium absorption unless it is unleavened. The unleavened whole wheat bread eaten by the Pakistani and other Asiatic peoples may contain sufficient phytic acid to reduce calcium absorption, particularly if vitamin D intake is minimal. The combination of increased intestinal motility due to high fiber content of food along with phosphate may be a factor in producing calcium malabsorption.

METHODS OF MEASURING CALCIUM ABSORPTION

The traditional method of determining calcium absorption by measurement of calcium in the food and in the feces and calculating net absorption is a difficult procedure. In addition to the technical problems this method does not estimate the contribution of endogenous calcium, i.e., the calcium entering the intestine from body fluids in the intestinal secretions and not reabsorbed. Endogenous calcium, however, is ordinarily small in comparison with diet calcium, and the over-all net retention of diet calcium is the desired value. Because of the methodologic problems of accurate balance studies, other techniques of determining calcium absorption have been developed using radiocalcium. The simple expedient of giving a standard oral dose of isotopic calcium and measuring plasma concentrations of radioactive calcium has been used as an index of calcium absorption.

There are two major criticisms of this method. The percentage of the radioactive isotope absorbed is inversely related to the quantity of stable calcium given along with the isotope, but if a constant amount of carrier calcium is given standard results can be obtained in control subjects which can be compared with findings in patients with calcium malabsorption. The concentration of radioactive calcium in serum, however, is a function not only of rate of calcium absorption but also of the rate of loss of radioactive calcium from plasma by exchange with stable calcium in bone, by deposition of bone mineral and by excretion in the urine. Despite this objection this method has been employed to study changes in calcium absorption with treatment.

A more accurate measure of calcium absorption requires a double isotope procedure. A given dose of isotopic calcium is injected intravenously and the concentrations of radioactive calcium in serum determined serially to obtain the disappearance rate of isotopic calcium by the various routes mentioned above. At a subsequent time a measured amount of radioactive calcium in a constant quantity of carrier calcium is given by mouth, and serial determinations of radioactive calcium in serum are made. Assuming that the disappearance rate of radioactive calcium is the same in both procedures, the quantity of calcium actually absorbed from the intestine can be calculated. The intravenous injection of radioactive calcium can also be used to determine endogenous calcium excretion by collection of the feces and measurement of radioactive calcium excreted. If serial determinations of radioactive and total calcium in serum are made or urine is collected during the period of fecal excretion, and radioactive calcium and total calcium determined, the average specific activity of radiocalcium can be estimated and the total calcium entering the intestine during the collection period can be calculated. Because of the restrictions on the administration of radioisotopes the isotopic methods of studying calcium absorption have not been useful in children. The expedient of giving an oral calcium load and measuring the increase of calcium concentration in serum at fixed times following the ingestion of the solution of calcium salt has been suggested as a crude measure of efficiency of calcium absorption, but this procedure is too inaccurate to be of much value as a clinical tool.

URINARY EXCRETION OF CALCIUM

The loss of calcium in the urine is determined predominantly by the quantity of filtered calcium, i.e., if renal function is normal, by the concentration of diffusible calcium in plasma. In patients with reduced glomerular filtration rate urine calcium excretion is low, because of reduced filtration rate and reduced concentration of diffusible calcium resulting from hyperphosphatemia. In the normal subject urine calcium excretion is increased by increased absorption of calcium from the intestine, whether by intake of soluble calcium salts, by high vitamin D levels or by low phosphate intake. In one form of idiopathic hypercalciuria abnormally high intestinal absorption of calcium is thought to be the basis of the excessive urinary output of calcium. This is discussed in greater detail in the chapter on renal stones (Chapter 6). Even at increased concentrations of diffusible calcium and excessive excretion of calcium in urine, tubular reabsorption of calcium is highly efficient. For example, in a normal 10 year old child weighing 30 kg with a diffusible calcium concentration averaging 5.0 mg/dl and a glomerular filtration rate of 75 ml/min or 108 liters per 24 hours, filtered calcium equals 5400 mg/24 hrs and urine calcium 50 mg/24 hrs, so that tubular reabsorption of calcium is somewhat over 99 percent of the filtered load. Even with a very high urine calcium output of 200 mg/24 hours, tubular reabsorption of calcium is still 96 percent of that filtered. This high tubular reabsorption of calcium is essential to prevent urine calcium concentrations from rising to values that would cause precipitation of calcium phosphate in the tubular contents or in the urine in the lower urinary tract.

There are, however, physiologic factors that influence tubular reabsorption of calcium, and therefore urine output of calcium independent of plasma concentration of diffusible calcium. An important controlling agent is parathyroid hormone, which acts to increase tubular reabsorption of calcium and thus reduce urine output of calcium. This is one of the controlling mechanisms through which parathyroid hormone operates to maintain normal concentrations of calcium ion in extracellular fluid. The urine output of calcium in hyperparathyroidism, therefore, is determined by the opposing forces of increased filtered load of calcium caused by elevated concentrations of calcium ion in extracellular fluid and increased tubular reabsorption of calcium under the influence of high concentrations of parathyroid hormone. This balancing of two effects may be apparent in some kindreds of familial hyperparathyroidism in which the affected individuals uniformly show low or normal excretions of calcium in urine despite considerable hypercalcemia. In other patients with hyperparathyroidism the influence of the increased filtered load of calcium dominates and high urine output of calcium results. These are the patients at risk for renal stone formation (see Chapter 4, *Hypercalcemic States*). As a general rule, at any given concentration of calcium in serum, patients with hyperparathyroidism show lower urine output of calcium than do those with comparable increase of serum calcium due to other causes such as hypervitaminosis D or sarcoidosis.

The tubular reabsorption of calcium is also influenced by sodium load. A high sodium load, and therefore increased quantities of sodium remaining in tubular contents and excreted in urine, increases urine output of calcium. This may be due in part to expansion of extracellular fluid volume and increased filtration rate, as well as reduced net removal of glomerular filtrate in the proximal tubule. It has also been proposed that there is some competition between cations for tubular transport systems, so that sodium in excess competes with calcium and reduces tubular reabsorption of calcium. The latter is also reduced by acidosis, which causes increased urine output of both sodium and calcium. Acidosis may increase urine output of calcium not only by influencing tubular reabsorption of calcium but also by increasing solubilization of calcium from bone mineral, and thereby increasing the concentration of diffusible calcium and possibly secondarily inhibiting parathyroid hormone secretion. Urine calcium excretion is augmented by high protein intake. High protein intake results in H^+ ion generation and increased urine excretion of anions, with consequent increased H^+ ion secretion by the tubule cells. The anion excess is predominantly sulfate, arising from the metabolism of the sulfur-containing amino acids of dietary protein. There is also an increased phosphate load contributed by the conversion of organic phosphates of animal and vegetable cell sources to inorganic phosphates. Whether the increased anion excretion and H^+ gradient in urine is the basis for the increased urine output of calcium has not been proved, but loading of patients with ammonium chloride or sulfate does lead to increased calciuria. Patients with renal tubular acidosis of the distal variety who have a defect in development of an H^+ ion gradient in tubular urine also show calciuria, which is reduced when the acidosis is corrected by sodium bicarbonate treatment or its equivalent, so that the state of acidosis rather than urine pH is apparently the factor in determining urine excretion of calcium.

As might be expected, certain diuretics that inhibit tubular reabsorption of sodium also inhibit tubular reabsorption of calcium, resulting in calciuresis. The carbonic anhydrase inhibitor, acetazolamide, which in essence prevents H^+ ion secretion and inhibits HCO_3^- resorption, produces a renal tubular acidosis and increases calcium excretion in urine as well as sodium and water. The diuretics that specifically inhibit sodium reabsorption, ethacrynic acid and furosemide, also inhibit calcium reabsorption and effectively increase calcium loss in urine. These diuretics are, therefore, useful in the emergency treatment of hypercalcemia (see Chapter 4). On the other hand, the benzothiazide diuretics such as chlorothiazide increase tubular reabsorption of calcium and reduce urine calcium excretion. This is sometimes associated with an actual rise of serum calcium concentration. Chlorothiazide or related diuretics are, therefore, useful in reducing urine calcium excretion in subjects with idiopathic hypercalciuria of renal tubular origin, i.e., patients who do not have primary hyperabsorption of calcium (see section on idiopathic hypercalciuria). Chlorothiazide is contraindicated, however, in hyperparathyroidism or other hypercalcemic states since its administration would augment serum calcium concentration through the reduction of urine output of calcium.

In summary, under normal physiologic conditions renal excretion of calcium plays only a minor role in calcium homeostasis. Although parathyroid hormone does influence renal tubular reabsorption of calcium, the net effect of this reduction of calcium excretion is less significant than the increased calcium absorption from the gut mediated by parathyroid hormone through its action in stimulating the formation of 1,25-diOH vitamin D from the substrate 25-OH vitamin D. The effects of sodium loading, high protein intake, metabolic acidosis, and the administration of acidifying salts and sodium-losing diuretics in increasing urine calcium output can usually be compensated for by increased calcium absorption from the intestine under the influence of the vitamin D hormone.

CONTROL OF Ca^{++} ION CONCENTRATION IN EXTRACELLULAR FLUID BY BONE CELL METABOLISM

The fine control of Ca^{++} ion concentration in extracellular fluid cannot be maintained by calcium absorption alone, since this is pulsatile, and during periods of no calcium intake there is no net calcium entrance from the intestine. Urine calcium excretion although low continues during starvation, as does endogenous fecal excretion, so that there is continued calcium loss from the body. The system that buffers the Ca^{++} ion concentration in extracellular fluid and prevents major fluctuations in this concentration, both following calcium loads and during periods of calcium deprivation, is the bone. The bone system consists of the insoluble bone mineral and the bone cells with the surrounding bone extracellular fluid. The bone cells – osteoblasts, osteocytes and

calcium homeostasis through their metabolic activities which are controlled by the major hormones of calcium homeostasis, parathyroid hormone, 1,25-diOH vitamin D and calcitonin. Following entrance of calcium and phosphate into extracellular fluid from the intestine, calcium phosphate is deposited in bone and cartilage matrix, removing calcium from extracellular fluid. A factor in this process is calcitonin, the hormone secreted by the parafollicular cells of the thyroid in response to increased concentration of calcium ion in the extracellular fluid. Simultaneously with the increased secretion of calcitonin there is also suppression of output of parathyroid hormone, and by these two responses the process of bone resorption is inhibited. Bone mineral deposition, therefore, predominates, thus reducing calcium ion concentration. An important additional controlling factor in this process is extracellular phosphate concentration. The inhibition of parathyroid hormone secretion is associated with increased phosphate reabsorption by renal tubules maintaining an increased concentration of phosphate in extracellular fluid. This is countered, however, by an effect of calcitonin in excess which inhibits tubular reabsorption of phosphate and lowers serum phosphate through increased excretion in urine. However, if the homeostatic mechanisms maintain serum phosphate above a minimal concentration, possibly 3 to 4 mg/dl, calcium entering the body is deposited in bone along with phosphate.

When calcium ceases to enter body fluids from the intestine the osteocytes and osteoclasts, through their metabolic activities, dissolve bone mineral, and calcium and phosphate return to the extracellular fluid. This latter process is stimulated by parathyroid hormone, the secretion of which increases in response to decrease of calcium ion concentration in extracellular fluid. 1,25-diOH vitamin D participates in this return of calcium from bone to extracellular fluid, both through its role in making possible the complete expression of parathyroid hormone action and by its direct action on bone cells. As pointed out in the section on vitamin D physiology, this action of 1,25-diOH vitamin D in facilitating the bone cell response to parathyroid hormone occurs beyond the activation of adenyl cyclase since the increase of cyclic AMP concentration following parathyroid hormone takes place in the absence of vitamin D, but the increased concentration of extracellular calcium does not occur.

The concentration of phosphate is also an important controlling factor in the flow of calcium from bone to extracellular fluid. For this reason the action of parathyroid hormone in inhibiting tubular reabsorption of phosphate, and thus reducing extracellular phosphate through loss of phosphate in urine, is an important one in increasing the efficiency of its role in maintaining extracellular calcium concentration by the flow of calcium from bone mineral to extracellular fluid. A high concentration of extracellular phosphate blocks this process, possibly by reprecipitation of calcium phosphate as it is dissolved.

In addition, hyperphosphatemia inhibits conversion of 25-OH vitamin D to 1,25-diOH vitamin D and thus impairs response to parathyroid hormone. The effect of hyperphosphatemia can be overcome by increased secretion of parathyroid hormone, which is one of the compensatory responses to phosphate loading. The direct stimulus to the parathyroid gland cell is not the increased phosphate concentration but reduction of calcium ion concentration. If parathyroid hormone output is limited as in the functional hypoparathyroidism of the newborn, phosphate loading causes persistent hypocalcemia. Similarly, in vitamin D deficiency the serum calcium concentration depends on serum phosphate concentration. A high phosphate load, either food phosphate or that of endogenous origin arising from the breakdown of tissue cells following fasting or increased metabolic rate as in fever, causes rapid decrease of serum calcium concentration and is the important mechanism of tetany in infants with vitamin D deficiency rickets.

In the absence of parathyroid hormone, serum calcium concentrations decrease to a low value that fluctuates depending on absorption of calcium from the gut and extracellular phosphate concentrations. Although the concentration of calcium in extracellular fluid is reduced, maintenance of extracellular calcium is still dependent on bone cell metabolism, which probably continues at a basal rate despite absence of stimulation of parathyroid hormone. The high concentration of phosphate in extracellular fluid in the hypoparathyroid state is an additional factor that reduces serum calcium concentration by increasing the flow of calcium into bone and reducing its return into extracellular fluid. The bone mineral density may be in the normal range in children with hypoparathyroidism, despite chronic hypocalcemia, since calcium absorption is only moderately reduced and this reduction is compensated for by reduced loss of calcium in urine. There have been occasional cases of osteomalacic and rickets-like lesions of bone associated with hypoparathyroidism, although these are the exception rather than the rule. Possibly the intestinal malabsorption of calcium and of phosphate in these particular patients is unusually severe, resulting in deficient mineral deposition in bone matrix.

The role of vitamin D in maintaining more normal serum calcium concentrations in hypoparathyroidism is primarily that of increasing intestinal absorption of calcium. The active agent is 1,25-diOH vitamin D, but since the formation of this hormone from 25-OH vitamin D is depressed in the absence of parathyroid hormone and the presence of hyperphosphatemia, very large amounts of vitamin D must be administered to the hypoparathyroid subject to accomplish this end, doses of the order of 2000 units vitamin D/kg per day (50 μg/kg). Smaller amounts of 25-OH vitamin D or dihydrotachysterol are required than of vitamin D, but the most active compound is 1,25-diOH vitamin D, which is effective in doses of 1/250 the amount of vitamin D. An additional effect of vitamin D, 25-OH vitamin D, dihydrotachysterol or 1,25-diOH vitamin D in the absence of parathyroid hormone is to reduce the concentration of serum phosphate by reducing tubular reabsorption of phosphate. This apparently contradictory action of the vitamin D hormone or related steroids is probably indirect and exerted through increase of extracellular Ca^{++} ion concentration, since the phosphate lowering effect lags behind the increase of serum calcium. Other evidence supporting this concept is the reduction of serum phosphate concentrations in patients with hypoparathyroidism by increase of serum calcium through continuous prolonged infusions of calcium gluconate without administration of vitamin D.

In addition to the action of vitamin D metabolites in increasing calcium absorption in the hypoparathyroid patient, there is in vitro evidence that 25-OH vitamin D and 1,25-diOH vitamin D have a direct action in stimulating solubilization of bone mineral. Their action in maintaining normal serum calcium concentrations in the absence of parathyroid hormone is partly due to stimulation of bone mineral resorption as well as intestinal calcium absorption. Much lower concentrations of 1,25-diOH vitamin D are needed to maintain efficient calcium absorption than to stimulate bone resorption, since the maintenance of serum calcium concentrations in the normal range in the hypoparathyroid patient requires only physiologic amounts of this steroid if the subject has an adequate calcium intake, whereas this amount of 1,25-diOH vitamin D fails to maintain normocalcemia in the fasting hypoparathyroid subject. The application of these physiologic actions of vitamin D in the absence of the parathyroid hormone is discussed in the section on treatment of hypoparathyroidism (Chapter 3).

CALCIUM NUTRITION

The possibility that inadequate calcium intake could be a cause of impaired bone mineralization and growth has often been suggested. This is based to a considerable extent on the fact that the intake of food rich in calcium, milk and milk products, or foods made with these products, differs markedly in different societies so that calcium intake varies strikingly. If the high calcium intakes of a milk-drinking population such as that of the United States and many European countries is accepted as the norm, then the much lower calcium intakes of many peoples living in Africa, Asia and Latin America represent calcium deficiency. There is no evidence, however, that nutritional calcium deficiency commonly occurs in man. As pointed out, the adaptation to varying amounts of dietary calcium is quite efficient, so that at low intakes of calcium the percentage of dietary calcium absorbed is high and is sufficient to provide for the growth of bone in the child and the loss of calcium in urine in both child and adult. This adaptation, of course, requires functioning parathyroid glands and sufficient vitamin D. Calcium deficiency can be produced in animals with normal vitamin D status by diets extremely low in calcium, and the bones of these animals are thin and undermineralized. Similar lesions have been seen in infants who have been given virtually calcium-free feeds by mistake. These instances have occurred under circumstances of the need for elimination of milk from the diet because of lactose intolerance or milk protein sensitivity, and the use of diets based on meat protein without the provision of a source of calcium. Aside from the instances of iatrogenic calcium malnutrition, it is doubtful whether with an ordinary mixed diet calcium intake is so low as to be deficient.

During the first year of life the breast-fed infant receives a relatively low calcium milk. Human milk contains 300 mg calcium per liter or 45 mg/100 calories, in comparison with cow's milk which contains 1200 mg calcium per liter or 180 mg/100 calories. The lower calcium intake of the breast-fed infant is sufficient for normal bone growth and mineralization during a period of rapid growth. During the first six months of life, when breast milk is the major food, the calcium intake is in the range of 50 to 60 mg/kg. A high percentage of this calcium is absorbed and little is excreted in the urine (Fig. 2–7). In contrast, infants given feedings based on cow's milk absorb a much lower percentage of the calcium intake, and although calcium balances appear to be greater with the higher calcium intake this may be an artifact of the balance method.

After the first six to twelve months of age when breast milk ceases to be the major source of calcium, the calcium requirements have been rather arbitrarily fixed at 800 mg/day during most of the growing period, with an increased amount during the adolescent growth spurt when, in the male, as much as 400 mg calcium may be retained as bone mineral per day. This high calcium requirement is characteristic of populations achieving a considerably greater adult height than was true in past generations, since skeletal mass is much greater. Calcium intakes of 800 to 1200 mg/day can still provide for these massive retentions of calcium if the adaptation processes are normal. In societies in which milk is less readily available, much lower calcium intakes are the rule without evidence of calcium malnutrition. In long-term balance studies in adults, some adapt poorly to low calcium intakes and might be calcium losers on low calcium diets. Some of these subjects have been found to have reduced calcium absorption during periods of emotional stress, which may influence intestinal function. There are also occasional individuals who have a high urine output of calcium, possibly owing to impaired renal tubular reabsorption of calcium. Those with this type of idiopathic hypercalciuria may also have increased calcium needs, but usually they show adaptation by increased intestinal absorption of calcium. Calcium deficiency is therefore an exceedingly rare problem, unless there is an intestinal disorder causing calcium malabsorption because of chronic diarrhea or fat malabsorption.

PHOSPHATE HOMEOSTASIS

INTRODUCTION

The control of inorganic phosphate concentrations in body fluids differs from that of calcium in several major respects. The intestinal absorption of phosphate is not as closely controlled and plays a relatively unimportant role in phosphate homeostasis; the renal mechanism for excretion of phosphate is the major controlling system, rather than intestine as in the case of calcium. In addition, the system for control of inorganic phosphate concentration in extracellular fluid is adjusted to protect against hyperphosphatemia preferentially, and the system for control of calcium concentration is set to protect most sensitively against hypocalcemia. In both instances the control of parathyroid hormone secretion and the interplay of parathyroid hormone and the vitamin D hormone are major regulating factors.

INTESTINAL ABSORPTION OF PHOSPHATE

The phosphate content of mixed diets is usually high and is considerably in excess of the requirements of either child or adult. Unlike calcium, which is present in high concentration in only a limited number of foods such as milk, milk derivatives, foods prepared with milk and some leafy vegetables, phosphate is supplied by a wide variety of foods in addition to milk. All foods derived from animal or vegetable cells are high in phosphate since organic phosphoric esters are essential constituents of all living cells. Although most food phosphate is not supplied as inorganic phosphate, the organic phosphoric compounds are rapidly converted in the intestine to inorganic phosphate and absorbed in that form. For example, a considerable portion of cow's milk phosphate is present as the phosphoprotein, casein. The phosphopeptides and phosphoamino acids formed by the enzymatic hydrolysis of the proteins are phosphorolyzed by the action of phosphatases in the intestinal mucosa, particularly the duodenum, and the inorganic phosphate released is rapidly absorbed. Absorption of phosphate is not complete, depending to some extent on the presence in the intestinal contents of divalent cations which form insoluble phosphates and thus precipitate the phosphate and make it unavailable. In food these cations are calcium, iron, strontium and possibly magnesium, with calcium the most important in terms of its role as a phosphate precipitator. The divalent cation, aluminum, can also precipitate phosphate, and aluminum hydroxide or carbonate given as medication can markedly reduce phosphate absorption. This property of aluminum is useful to prevent hyperphosphatemia in patients with renal insufficiency. A large excess of calcium can also be employed for the same purpose, and such a use of calcium is described in the chapter on hypocalcemic states (Chapter 3).

Phosphate absorption from the intestine is reduced in vitamin D deficiency and in hypoparathyroidism. In the latter instance the effect is mediated through a deficiency of the vitamin D hormone, 1,25-diOH vitamin D. However, the reduction of phosphate absorption is proportionately much less than the reduction of calcium absorption, and the increased fecal loss of phosphate is in part due to its binding with unabsorbed calcium. Intestinal absorption of phosphate even in vitamin D deficiency states is not ordinarily a limiting factor in phosphate homeostasis.

RENAL EXCRETION OF PHOSPHATE

The important control system in the regulation of inorganic phosphate concentrations in extracellular fluid is renal function. Since most of the phosphate in plasma is diffusible, the filtered phosphate is a function of serum phosphate concentration and filtration rate. At normal filtration rates and phosphate concentration in serum, the amount filtered is greatly in excess of the quantity of phosphate absorbed from the intestine, even at high loads of phosphate; thus, the extent of reabsorption of filtered phosphate by renal tubules determines the ability of the organism to maintain phosphate balance and a stable concentration of extracellular phosphate.

The control of renal excretion of phosphate and of extracellular inorganic phosphate concentrations can be explained by a reabsorptive tubular maximum function (T_mP). When the quantity of phosphate filtered decreases below the maximum reabsorptive capacity, tubular reabsorption of phosphate becomes more complete and essentially all of the filtered phosphate is recaptured by the tubule, so that the urine is virtually phosphate-free. Increase in serum phosphate concentration without change in glomerular filtration rate produces a proportionate increase in filtered phosphate, and as the latter increases above the maximum tubular reabsorptive capacity the excess is lost in the urine. There may be a third component of phosphate excretion in urine, namely, tubular secretion of phosphate, i.e , plasma to lumen flux of phosphate, but this is probably minor.

The two-component system, glomerular filtration and a maximum rate of tubular reabsorption, satisfactorily explains renal control of serum inorganic phosphate, since it operates to maintain this concentration close to the value at which filtered phosphate is approximately equal to the maximum capacity of the renal tubules to reabsorb phosphate. This is the so-called equilibrium value, $T_mP/_{GFR}$. When phosphate enters extracellular fluid, either exogenously from the intestinal tract or endogenously from tissue cells or by solubilization of bone mineral, as the concentration of extracellular phosphate rises above the equilibrium value excess phosphate is excreted in the urine, bringing the concentration down toward the equilibrium value. When extracellular phosphate is removed by uptake into cells or bone mineral, and filtered phosphate decreases below the tubular reabsorptive capacity, urine phosphate decreases to minimal values and there is efficient conservation of phosphate. There is some splay in the curve relating tubular reabsorption of phosphate to filtered phosphate so that the threshold concentration, i.e., the concentration at which the urine is essentially phosphate-free, is somewhat less than the equilibrium value. The tubular capacity to reabsorb phosphate does vary with glomerular filtration rate, so that contraction of extracellular fluid by sodium loss decreases filtered phosphate but also T_mP, so that the equilibrium value is unchanged. In man, fasting serum phosphate concentrations are usually somewhat lower than the equilibrium value but are highly correlated with the latter. It is therefore possible to equate changes in fasting concentration of serum phosphate with changes in the phosphate reabsorptive capacity of the tubule per volume of glomerular filtration rate, except under conditions of phosphate depletion with serum phosphate concentrations far below the equilbrium value. Such hypophosphatemia can be identified since simultaneously obtained urine samples are phosphate-free.

The equilibrium concentration ($T_mP/_{GFR}$) is influenced by the age and maturity of the subject. This explains the well-known variation of serum phosphate concentrations with age and sexual maturation (see Table 1–1). This age variation has been shown experimentally in the puppy, and the parallel variation of $T_mP/_{GFR}$ and serum phosphate in the postabsorptive state with age and growth is clearly shown in Figures 2–8 and 2–9. The high rate of tubular reabsorption of phosphate in relation to glomerular filtration rate of the infant and child with normal vitamin D status throughout the entire period of growth is the important factor in the maintenance of the high concentrations of serum phosphate during this age period. This high phosphate concentration is obviously an important factor in the need for rapid mineralization of growing cartilage and bone, since values of serum phosphate that would be normal for healthy adults would be too low to permit normal mineralization of growing bone and would result in rickets and osteomalacia.

Emphasis on the difference between serum phosphate concentration in the infant and child and the adult is justified, since occasional instances of hypophosphatemic rickets have been missed because serum phosphate concentrations between 3.5 and 4.0 mg/dl were considered normal. Conversely, hyperphosphatemia has been diagnosed mistakenly in young infants with serum phosphate concentrations in the range of 7 or 8 mg/dl. These can be normal values in infants on relatively high phosphate intakes such as cow's milk, particularly if the blood sample is obtained during the period following a feeding when phosphate is being absorbed from the intestine. Growth hormone may be a factor in the regulation of T_mP since concentrations of serum phosphate are higher in acromegalic adults than in normal adults, and lower in children with hypopituitarism than in normal children. Sexual maturation also influences serum phosphate concentration. Patients with testosterone deficiency have serum phosphate concentrations higher than normal adult values, and administration of testosterone to prepubertal boys re-

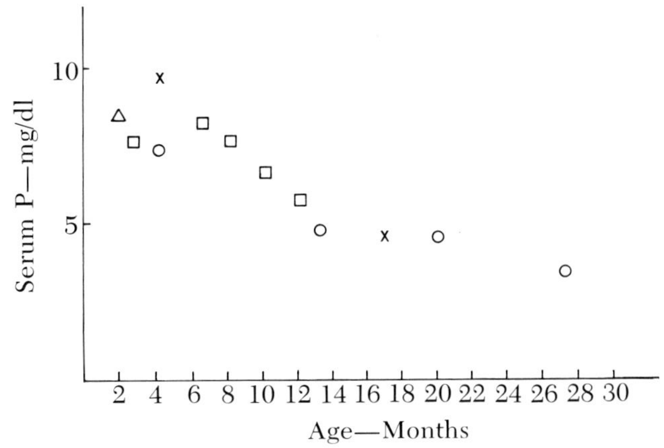

Figure 2–8. Variation with age of serum phosphate concentrations of young growing dogs.

Symbols △, x, □, and ○ represent individual dogs.

duces the concentration of serum phosphate. There are probably also intrinsic differences in the renal tubules during growth and maturation of the kidney, which account for the high $T_mP/_{GFR}$ in the young infant and growing child. An additional factor in the first few months of life is the low GFR in relation to surface area.

Tubular reabsorption of phosphate is chiefly a function of the proximal convoluted tubule, which is also the site of amino acid and glucose reabsorption as well as a major fraction of sodium and water reabsorption. Amino acid and glucose loading reduce tubular reabsorption of phosphate although there are separate transport systems for these three solutes. Expansion of extracellular fluid and plasma volume inhibits net proximal reabsorption of sodium and also reduces tubular reabsorption of phosphate, and there may be a linkage between sodium and phosphate transport by the renal tubule cell. There is also a component of tubular reabsorption of phosphate in the distal tubule, and possibly plasma to lumen backflow of phosphate. Distal tubular reabsorption of phosphate is quantitatively less important than proximal reabsorption, but may be of clinical significance in disorders of renal tubule transport of phosphate. (See *Primary Hypophosphatemia,* Chapter 5.)

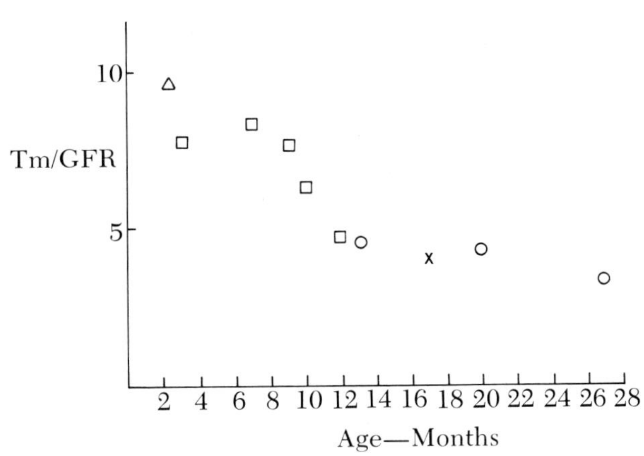

Figure 2–9. Variation with age of tubular reabsorption of phosphate in relation to glomerular filtration in the growing dog.

Symbols △, x, □, and ○ represent individual dogs.

The major hormones regulating phosphate transport by the renal tubules are parathyroid hormone and the vitamin D metabolites, 25-OH vitamin D and 1,25-diOH vitamin D. Parathyroid hormone inhibits tubular reabsorption of phosphate through its effect in activating adenyl cyclase and increasing cyclic AMP concentration. In the subject with intact parathyroids the vitamin D metabolites enhance tubular reabsorption of phosphate. This is in part through inhibition of parathyroid hormone secretion, but there is also a direct action of these metabolites on tubular reabsorption of phosphate, which is opposite to that of parathyroid hormone but does not involve inhibition of cyclic AMP production. A sharp increase in extracellular phosphate concentration due to intestinal absorption of phosphate or endogenous phosphate release by breakdown of tissue cells causes a decrease in extracellular Ca^{++} ion concentration sufficient to stimulate parathyroid hormone secretion. The increased parathyroid hormone concentration stimulates bone cell-mediated solubilization of bone mineral, adding both calcium and phosphate to extracellular fluid, but additionally the parathyroid hormone inhibits tubular reabsorption of phosphate, leading to an increased loss of phosphate in urine with return of extracellular phosphate to normal.

There is no comparable acute hormonal response to hypophosphatemia, and serum phosphate concentration can decrease markedly below normal as the result, for example, of aluminum hydroxide-induced malabsorption of phosphate without any compensating direct hormonal response. It is true that, if hypophosphatemia results in increased Ca^{++} ion concentration in extracellular fluid because of reduced rate of deposition of calcium phosphate in bone matrix, the secretion of parathyroid hormone is reduced and tubular reabsorption of phosphate is more complete. Moreover, a low concentration of inorganic phosphate enhances 1,25-diOH vitamin D formation in the kidney, thus causing more complete absorption of calcium and phosphate from the gut as well as increased tubular reabsorption of phosphate, so that there is some feedback effect of hypophosphatemia.

EQUILIBRIUM OF PHOSPHATE BETWEEN EXTRACELLULAR FLUID, INTRACELLULAR FLUID AND BONE

The body cells, chiefly the muscle mass, are a major depot of phosphate (as opposed to calcium) in the body. Skeletal phosphate comprises about 75 per cent of the total body phosphate, but internal phosphate adjustments are chiefly between extracellular fluid and tissue cells. Entrance of glucose into cells is accompanied by cell uptake of phosphate, so that extracellular phosphate can decrease to very low values following insulin or glucose loading. This phenomenon has been seen in infants and children who have not been receiving food for several days and have been maintained on intravenous glucose and saline infusions. Hypophosphatemia may also be seen in patients with diabetic ketoacidosis following treatment with insulin and redeposition of glycogen in liver and muscles. There is no evidence that skeletal phosphate is mobilized in response to this acute reduction of extracellular phosphate. Persistent hypophosphatemia theoretically should be followed by increased production of 1,25-diOH vitamin D, which increases phosphate absorption from the intestine and also increases tubular reabsorption of phosphate either directly or indirectly through suppression of parathyroid hormone output.

Rapid breakdown of cells releases large amounts of phosphate, and extreme hyperphosphatemia can occur, particularly if renal function is impaired. This is seen strikingly in patients with leukemia following initiation of cytotoxic drug treatment. It can also occur in acute rhabdomyolysis with myoglobinuria following heat stress. Because of associated uric acid nephropathy in the leukemic patient or myoglobinuric renal injury in heat stress, renal excretion of phosphate is limited, and marked hyperphosphatemia results with reduction of extracellular Ca^{++} ion concentration. The increased parathyroid hormone secretion which is the response to the reduced Ca^{++} ion concentration is ineffective in restoring Ca^{++} ion or phosphate homeostasis in the face of high extracellular phosphate and impaired renal function.

MEASUREMENT OF VITAMIN D AND ITS METABOLITES IN BLOOD AND TISSUES

The traditional method of determining vitamin D in blood and tissues depended on bioassay of the material, with the vitamin D-depleted weanling rat or chick as the test system. This was a laborious, time-consuming procedure that required large amounts of blood to determine the vitamin D content of blood serum in individuals of normal vitamin D status. This value was found to be one to two units of vitamin D activity per ml of blood serum, equivalent to 25 to 50 ng/ml. In patients with hypervitaminosis D a serum value of 100 times this amount can be found. In experimental hypervitaminosis D there is storage of vitamin D in excessive amounts in adipose tissue as well as circulating in plasma. It is now possible to measure the vitamin D metabolites, 25-OH vitamin D and 1,25-diOH vitamin D, in plasma by chromatographic separation and use of competitive protein binding assays. These determinations utilize specific binding proteins for these metabolites, and the displacement of 3H-labeled vitamin D metabolites from these proteins provides the estimate of the quantity of specific metabolite in blood serum. On the basis of these measurements it can be concluded that, in the normal state, 25-OH vitamin D makes up a major fraction of the material in serum with vitamin D potency by bioassay.

The values for 25-OH vitamin D differ somewhat in different populations and in the results obtained from various laboratories, but the usual range is between 10 and 50 ng/ml. This value is higher in subjects intensively exposed to sunshine and may be of the order of 100 ng/ml (Table 2–1). It has been suggested that the endogenous vitamin D produced by maximal exposure to sunshine or to artificially produced ultraviolet radiation is equal to doses of vitamin D of 8000 to 10,000 units per day. At this dosage considerable amounts of unmetabolized vitamin D are stored in adipose tissue and are only gradually metabolized to active metabolites. This could explain why exposure to sunshine during the summer months can supply a store of vitamin D sufficient to protect against the development of vitamin D deficiency in the winter season even if no vitamin D is ingested in food.

The concentration of 1,25-diOH vitamin D in serum is only about 40 picograms per ml or about 1/1000 that of 25-OH vitamin D. This ultramicro quantity of this extremely potent vitamin D metabolite is more closely controlled than that of 25-OH vitamin D, and does not increase to any great extent following large doses of vitamin D. It is, of course, low in vitamin D deficiency, vitamin D dependency and renal insufficiency. Its determination is much more difficult than that of 25-OH vitamin D, so that in most studies of vitamin D nutrition or metabolism the determination of 25-OH vitamin D has been used. In true vitamin D deficiency the concentration of 25-OH vitamin D in serum is markedly reduced and may be below 5 ng/ml. In hypervitaminosis D the concentration of serum 25-OH vitamin D is greatly increased to as much as ten times the normal value, so that this determination can be of practical value in the diagnosis of hypercalcemic states due to vitamin D excess.

VITAMIN D NUTRITION

Vitamin D requirements in the past have been determined with reasonable accuracy only in the rapidly growing infant under 1 year of age. In this age-group the biochemical and roentgenologic evidences of vitamin D deficiency and rickets could be measured in relation to vitamin D intake. Exposure to sunshine was also less important in this population, particularly in infants born in the late summer who received little outdoor exposure to sun until the following spring. Epidemiologic studies indicate that rickets was prevented in virtually 100 per cent of infants under 1 year of age by addition of vitamin D to the diet in a dose of 400 U/day (10 μ g). This was particularly effective if the vitamin D was dissolved in or mixed with infant's milk feedings, so that this dosage was distributed throughout the day. The dosage of 400 units also was found in calcium balance studies to be sufficient for maximal calcium retention and growth. This amount is also adequate in prematurely born infants despite the fact that they may be born with low serum concentrations of 25-OH vitamin D if their mothers have not been given vitamin D supplements. The recommended daily allowance of 400 units vitamin D per day provides a considerable margin of safety, and it is likely that 100 units (2.5 μ g) per day would prevent overt rickets in the large majority of infants.

Determination of vitamin D needs of older children and adults had been more difficult to ascertain since vitamin D deficiency rickets and osteomalacia is uncommon after the first year of life. This is in part due to the slow metabolic inactivation of vitamin D, so that once adequate body stores are built up by exposure to sunshine during the summer or by ingestion of vitamin D-containing foods, prolonged withdrawal of all vitamin D intake would be needed for the child to develop laboratory or clinical evidences of vitamin D deficiency. Such deficiency can occur in special circumstances, such as children with severe motor disabilities who do not get any outdoor exposure, women in certain societies who are not permitted to move freely outside the home after they are married, and elderly individuals who are on restricted diets and enjoy limited or no outdoor activity. On the basis of dietary histories in such persons whose exposure to sunshine is negligible, a minimum intake of 100 units vitamin D was considered necessary to prevent vitamin D deficiency. This value has now been confirmed by determinations of serum 25-OH vitamin D in these individuals as a means of evaluating vitamin D nutrition. It is clear that vitamin D is required throughout the life span, but that the body requirement of this compound is not related to weight, surface area or metabolic activity. It may be somewhat greater during the period of rapid growth but this has not been really proved. There probably are genetic differences among individuals with regard to vitamin D requirement. Whether this is related to differences in the formation of the active metabolite 1,25-diOH vitamin D is not proved, but this could be an important factor in determining vitamin D needs and susceptibility to rickets and osteomalacia when vitamin D intake is minimal.

REFERENCES

ESTIMATION OF EXTRACELLULAR CALCIUM

Bergman, L. and Isaksson, B.: Plasma calcium fractions in normal subjects from birth to adult ages. Acta Paediatr. Scand. *60*:630–636, 1971.

Brown, D. M., Boen, J. and Bernstein, A.: Serum ionized calcium in newborn infants. Pediatrics *49*:841–846, 1972.

Ladenson, J. H. and Bowers, G. N., Jr.: Free calcium in serum. I. Determination with the ion-specific electrode and factors affecting the results. Clin. Chem. *19*:565–574, 1973.

Ladenson, J. H. and Bowers, G. N., Jr.: Free calcium in serum. II. Rigor of homeostatic control, correlation with total serum calcium and review of data on patients with disturbed calcium metabolism. Clin. Chem. *19*:575–582, 1973.

Moore, E. W.: Ionized calcium in normal serum, ultrafiltrates, and whole blood determined by ion-exchange electrodes. J. Clin. Invest. *49*:318–334, 1970.

Pain, R. W., Rowland, K. M., Phillips, P. J. et al: Current "corrected" calcium concept challenged. Br. Med. J. *4*:617–619, 1975.

HORMONES OF CALCIUM AND PHOSPHATE HOMEOSTASIS

Vitamin D

Avioli, L. V. and Haddad, J. G.: Vitamin D: current concepts. Metabolism *22*:507–531, 1973.

Belsey, R., Clark, M. R., Bernat, M. et al: The physiologic significance of plasma transport of vitamin D and metabolites. Am. J. Med. *57*:50–56, 1974.

Daiger, S. P., Schanfield, M. S. and Cavalli-Sforza, L. L.: Group-specific component (Gc) proteins bind vitamin D and 25-hydroxyvitamin D. Proc. Natl. Acad. Sci. USA *72*:2076–2080, 1975.

DeLuca, H. F.: The kidney as an endocrine organ involved in the function of vitamin D. Am. J. Med. *58*:39–47, 1975.

DeLuca, H. F.: Metabolism of vitamin D: current status. Am. J. Clin. Nutr. *29*:1258–1270, 1976.

Fraser, D. R. and Kodicek, E.: Unique biosynthesis by kidney of a biologically active vitamin D metabolite. Nature *228*:764–766, 1970.

Fraser, D. R. and Kodicek, E.: Regulation of 25-hydroxycholecalciferol-1-hydroxylase activity in kidney by parathyroid hormone. Nature New Biol. *241*:163–166, 1973.

Gray, R. W., Weber, H. P., Dominguez, J. H. et al: The metabolism of vitamin D_3 and 25-hydroxyvitamin D in normal and anephric humans. J. Clin. Endocrinol. Metab. *39*:1045–1056, 1974.

Haddad, J. G. and Stamp, T. C. B.: Circulating 25-

hydroxyvitamin D in man. Am. J. Med. *57*:57–62, 1974.

Haussler, M. R. and McCain, T. A.: Basic and clinical concepts related to vitamin D metabolism and action. N. Engl. J. Med. *297*:974–983, 1041–1050, 1977.

Hill, L. F. and Stanbury, S. W.: Vitamin D and the kidney. Nephron *15*:369–386, 1975.

Holick, M. F., Schnoes, H. K. and DeLuca, H. F.: Identification of 1,25-dihydroxycholecalciferol, a form of vitamin D_3 metabolically active in the intestine. Proc. Natl. Acad. Sci. USA *68*:803–804, 1971.

Hughes, M. R., Brumbaugh, P. F., Haussler, M. R. et al: Regulation of serum 1α 25-dihydroxyvitamin D_3 by calcium and phosphate in the rat. Science *190*:578–580, 1975.

Huldschinsky, K.: Heilung von Rachitis durch künstliche Höhensonne. Dtsch. Med. Wochenschr. *45*:712–713, 1919.

Kodicek, E.: The story of vitamin D—from vitamin to hormone. Lancet *1*:325–329, 1974.

Lawson, D. E. M., Fraser, D. R., Kodicek, E. et al: Identification of 1,25-dihydroxycholecalciferol: a new kidney hormone controlling calcium metabolism. Nature *230*:228–230, 1971.

Mawer, E. B., Backhouse, J., Hill, L. F. et al: Vitamin D metabolism and parathyroid function in man. Clin. Sci. Mol. Med. *48*:349–365, 1975.

Mawer, E. B., Backhouse, J., Holman, C. A. et al: The distribution and storage of vitamin D and its metabolites in human tissues. Clin. Sci. *43*:413–431, 1972.

Mellanby, E.: An experimental investigation on rickets. Lancet *1*:407–412, 1919.

Omdahl, J. L. and DeLuca, H. F.: Regulation of vitamin D metabolism and function. Physiol. Rev. *53*:327–372, 1973.

Pechet, M. M. and Hesse, R. H.: Metabolic and clinical effects of pure crystalline 1α-hydroxyvitamin D_3 and 1α 25-dihydroxyvitamin D_3. Am. J. Med. *57*:13–20, 1974.

Rosenstreich, S. J., Rich, C. and Volwiler, W.: Deposition in and release of vitamin D_3 from body fat: evidence for a storage site in the rat. J. Clin. Invest. *50*:679–687, 1971.

Stamp, T. C. B. and Round, J. M.: Seasonal changes in human plasma levels of 25-hydroxyvitamin D. Nature *247*:563–565, 1974.

Suda, T., Hallick, R. B., DeLuca, H. F. et al: 25-hydroxydihydrotachysterol$_3$. Synthesis and biological activity. Biochemistry *9*:1651–1657, 1970.

Tanaka, Y., Castillo, L. and DeLuca, H. F.: Control of renal vitamin D hydroxylases in birds by sex hormones. Proc. Natl. Acad. Sci. USA *73*:2701–2705, 1976.

Parathyroid

Arnaud, C. D., Rasmussen, H. and Anast, C.: Further studies on the interrelationship between parathyroid hormone and vitamin D. J. Clin. Invest. *45*:1955–1964, 1966.

Arnaud, C. D., Sizemore, G. W., Oldham, S. B. et al: Human parathyroid hormone, glandular and secreted molecular species. Am. J. Med. *50*:630–638, 1971.

Berson, S. A. and Yalow, R. S.: Radioimmunoassays of peptide hormones in plasma. N. Engl. J. Med. *277*:640–647, 1967.

Bethune, J. E., Turpin, R. A. and Inoue, H.: Effect of parathyroid hormone extract on divalent ion excretion in man. J. Clin. Endocrinol. Metab. *28*:673–678, 1968.

Borle, A.: Calcium and phosphate metabolism. Annu. Rev. Physiol. *36*:361–390, 1974.

Chase, L. R. and Aurbach, G. D.: Parathyroid function and the renal excretion of 3′5′-adenylic acid. Proc. Natl. Acad. Sci. USA *58*:518–525, 1967.

Eisenberg, E.: Renal effects of parathyroid hormone. *In* Parathyroid Hormone and Thyrocalcitonin (Calcitonin) (Eds.: R. V. Talmage and L. F. Belanger), Excerpta Medica, Amsterdam, 1968, pp. 465–474.

Habener, J. F., Mayer, G. P., Dee, P. C. et al: Metabolism of amino- and carboxyl-sequence immunoreactive parathyroid hormone in the bovine. Evidence for peripheral cleavage of hormone. Metabolism *25*:385–395, 1976.

Harrison, H. E. and Harrison, H. C.: The renal excretion of inorganic phosphate in relation to the action of vitamin D and parathyroid hormone. J. Clin. Invest. *20*:47–55, 1941.

Harrison, H. E. and Harrison, H. C.: The interaction of vitamin D and parathyroid hormone on calcium, phosphorus and magnesium homeostasis. Metabolism *13*:952–958, 1964.

Harrison, H. E., Harrison, H. C. and Lifshitz, F.: Interrelation of vitamin D and parathyroid hormone: the responses of vitamin D depleted and of thyroparathyroidectomized rats to ergocalciferol and dihydrotachysterol. Parathyroid Hormone and Thyrocalcitonin (Calcitonin) (Eds.: R. V. Talmage and L. F. Belanger). Excerpta Medica, Amsterdam, 1968, pp. 455–460.

Harrison, H. C., Harrison, H. E. and Park, E. A.: Vitamin D and citrate metabolism. Effect of vitamin D in rats fed diets adequate in both calcium and phosphorus. Am. J. Physiol. *192*:432–436, 1958.

Lifshitz, F., Harrison, H. C. and Harrison, H. E.: Influence of parathyroid function upon the in vitro transport of calcium and phosphate by the rat intestine. Endocrinology *84*:912–917, 1969.

Popovtzer, M. M. and Robinette, J. B.: Effect of 25 (OH) vitamin D on urinary excretion of cyclic adenosine monophosphate. Am. J. Physiol. *229*:907–910, 1975.

Rasmussen, H.: Ionic and hormonal control of calcium homeostasis. Am. J. Med. *50*:567–588, 1971.

Rasmussen, H. and Feinblatt, J.: The relationship between the actions of vitamin D, parathyroid hormone and calcitonin. Calcif. Tissue Res. *6*:265–279, 1971.

Segre, G. V., Habener, J. F., Powell, D. et al: Parathyroid hormone in human plasma: immunochemical characterization and biological implication. J. Clin. Invest. *51*:3163–3172, 1972.

Sherwood, L. M., Herrman, I. and Bassett, C. A.: Parathyroid hormone secretion in vitro: regulation by calcium and magnesium ions. Nature *225*:1056–1058, 1970.

Wills, M. R., Wortsman, J., Pak, C. Y. C. et al: The role of parathyroid hormone in the gastro-intestinal absorption of calcium. Clin. Sci. *39*:89–94, 1970.

Calcitonin

Anast, C., Arnaud, C. D., Rasmussen, H. et al: Thyrocalcitonin and the response to parathyroid hormone. J. Clin. Invest. *46*:57–64, 1967.

Care, A. D., Duncan, T. and Webster, D.: Thyrocalcitonin and its role in calcium homeostasis. J. Endocrinol. *37*:155–167, 1967.

Copp, D. H., Cameron, E. C., Cheney, B. A. et al: Evidence for calcitonin—a new hormone from the parathyroid that lowers blood calcium. Endocrinology *70*:638–649, 1962.

Deftos, L. J.: Immunoassay for human calcitonin. I. The method. Metabolism *20*:1122–1128, 1971.

Haymovits, A. and Rosen, J. F.: Calcitonin: its nature and role in man. Pediatrics *45*:133–149, 1970.

Hirsch, P. F. and Munson, P. L.: Thyrocalcitonin. Physiol. Rev. *49*:548–622, 1969.

Luben, R. A., Wong, G. L. and Cohn, D. V.: Biochemical characterization with parathormone and calcitonin of isolated bone cells: provisional identification of osteoclasts and osteoblasts. Endocrinology *99*:526–534, 1976.

Milhaud, G.: Calcitonin 1975. Biomedicine *24*:159–162, 1976.

Sherwood, L. M.: Relative importance of parathyroid hormone and thyrocalcitonin in calcium homeostasis. N. Engl. J. Med. *278*:663–670, 1968.

Wallach, S., Chausmer, A., Mittleman, R. et al: In vivo inhibition of bone resorption by thyrocalcitonin. Endocrinology *80*:61–64, 1967.

Corticoids

Carré, M., Ayigbedé, O., Miravet, L. et al: The effect of prednisolone upon the metabolism and action of 25-hydroxy and 1,25-dihydroxyvitamin D_3. Proc. Natl. Acad. Sci. USA *71*:2996, 1974.

Fucik, R. F., Kukreja, S. C., Hargis, G. K. et al: Effect of glucocorticoids on function of the parathyroid glands in man. J. Clin. Endocrinol. Metab. *40*:152–155, 1975.

Harrison, H. E. and Harrison, H. C.: Transfer of Ca^{45} across intestinal wall *in vitro* in relation to action of vitamin D and cortisol. Am. J. Physiol. *199*: 265–271, 1960.

Lukert, B. P. and Adams, J. S.: Vitamin D metabolism in man. Effect of corticosteroids. Arch. Intern. Med. *136*:1241–1248, 1976.

Stern, P.: Inhibition by steroids of parathyroid hormone-induced Ca^{45} release from embryonic rat bone in vitro. J. Pharmacol. Exp. Ther. *168*:211–217, 1969.

Storey, E. The influence of adrenal cortical hormones on bone formation and resorption. Clin. Orthop. *30*:197–217, 1963.

Williams, G. A., Peterson, W. C., Bowser, E. N. et al: Interrelationship of parathyroid and adrenocortical function in calcium homeostasis in the rat. Endocrinology *95*:707–712, 1974.

CALCIUM HOMEOSTASIS

Calcium Absorption From Intestine

Avioli, L. V.: Intestinal absorption of calcium. Arch. Intern. Med. *129*:345–355, 1972.

Bullamore, J. R., Wilkinson, R., Gallagher, J. C. et al: Effect of age on calcium absorption. Lancet *2*: 535–537, 1970.

DeGrazia, J. A. and Rich, C.: Studies of intestinal absorption of $calcium^{45}$ in man. Metabolism *13*:650–660, 1964.

Harrison, H. E.: Factors influencing calcium absorption. Fed. Proc. *18*:1085–1092, 1959.

Heaney, R. P., Saville, P. D. and Recker, R. R.: Calcium absorption as a function of calcium intake. J. Lab. Clin. Med. *85*:881–890, 1975.

Heaney, R. P. and Skillman, T. G.: Secretion and excretion of calcium by the human gastrointestinal tract. J. Lab. Clin. Med. *64*:29–41, 1964.

Kemm, J. R.: Effect of previous dietary history of calcium intake on the skeleton and calcium absorption in the rat. J. Physiol. *230*:643–658, 1973.

Nicolaysen, R., Egg-Larsen, N. and Malm, O. S.: Physiology of calcium metabolism. Physiol. Rev. *33*:424–442, 1953.

Nordin, B. E. C., Wilkinson, R., Marshall, D. H. et al: Calcium absorption in the elderly. Calcif. Tissue Res. *21*:442–451, 1976.

Schwartz, M. K.: Recent advances in calcium and phosphorus metabolism. Ann. Clin. Lab. Sci. *55*:176–184, 1975.

Shah, B. G. and Draper, H. H.: Depression of calcium absorption in parathyroidectomized rats. Am. J. Physiol. *211*:963–966, 1966.

Methods of Measuring Calcium Absorption

Avioli, L. V., McDonald, J. E., Singer, R. A. et al: A new oral isotopic test of calcium absorption. J. Clin. Invest. *44*:128–139, 1965.

Curtis, E. K., Fellows, H. and Rich, C.: Estimation of human calcium absorption by external radioisotope counting. J. Lab. Clin. Med. *69*:1036–1041, 1967.

DeGrazia, J. A., Ivanovich, P., Fellows, H. et al.: A double isotope method for measurement of intestinal absorption of calcium in man. J. Lab. Clin. Med. *66*:822–829, 1965.

Nordin, B. E. C.: Measurement and meaning of calcium absorption. Gastroenterology *54*:294–301, 1968.

Parsons, V., Veall, N. and Butterfield, W. J. H.: The clinical use of orally administered ^{47}Ca for the investigation of intestinal calcium absorption. Calcif. Tissue Res. *2*:83–92, 1968.

Shimmins, J., Smith, D. A., Aitken, M. et al: The measurement of calcium absorption using an oral and intravenous tracer. 2. Clinical studies. Calcif. Tissue Res. *6*:301–315, 1971.

Urinary Excretion of Calcium

Axelrod, D. R.: Sodium chloride-induced calciuria in hyperparathyroidism. J. Clin. Endocrinol. Metab. *26*:207–212, 1966.

Davis, R. H., Morgan, D. B. and Rivlin, R. S.: The excretion of calcium in the urine and its relation to calcium intake, sex and age. Clin. Sci. *39*:1–12, 1970.

Duarte, C. G.: Effects of furosemide (F) and ethacrynic acid (ETA) on the renal clearance of phosphate (C_p) ultrafilterable Ca (CUFCa) and magnesium (CUF Mg). Clin. Res. *15*:357, 1967.

Grollman, A. P., Walker, W. G., Harrison, H. C. et al: Site of reabsorption of citrate and calcium in the renal tubule of the dog. Am. J. Physiol. *205*:697–701, 1963.

Jahan, I. and Pitts, R. F.: Effect of parathyroid on renal tubular reabsorption of phosphate and calcium. Am. J. Physiol. *155*:42–49, 1948.

Kleeman, C. R., Rockney, R. E. and Maxwell, M. H.: The effect of parathyroid extract (PTE) on the renal clearance of diffusible calcium. J. Clin. Invest. *37*:907, 1958.

Knapp, E. L.: Factors influencing the urinary excretion of calcium. J. Clin. Invest. *26*:182–202, 1947.

Litzow, J. R., Lemann, J. and Lennon, E. J.: The effect of treatment of acidosis on calcium balance in patients with chronic azotemic renal disease. J. Clin. Invest. *46*:280–286, 1967.

Margen, S. Chu, J.-Y., Kaufmann, N. A. et al: Studies in calcium metabolism. I. The calciuretic effect of dietary protein. Am. J. Clin. Nutr. *27*:584–589, 1974.

Nordin, B. E. C., Hodgkinson, A. and Peacock, M.: The measurement and the meaning of urinary calcium. Clin. Orthop. *52*:293–322, 1967.

Reidenberg, M. M.: Mechanism of hypercalciuria in acidosis. Clin. Res. *15*:328, 1967.

Walser, M.: Calcium clearance as a function of sodium clearance in the dog. Am. J. Physiol. *200*:1099–1104, 1961.

Widrow, S. H. and Levinsky, N. G.: The effect of parathyroid extract on renal tubular calcium reabsorption in the dog. J. Clin. Invest. *41*:2151–2159, 1962.

Williamson, B. J. and Freeman, S.: Effects of acute changes in acid-base balance on renal calcium excretion in dogs. Am. J. Physiol. *191*:384–392, 1957.

Control of Ca^{++} Concentration in Extracellular Fluid by Bone Cell Metabolism

Copp, D. H.: Parathormone, calcitonin and calcium homeostasis. *In* Mineral Metabolism. (Eds.: C. L. Comar and F. Bronner), Vol. III. Academic Press, New York, 1969, Chap. 7.

Lutwak, L.: Metabolic and biochemical considerations of bone. Ann. Clin. Lab. Sci. *5*:185–194, 1975.

Raisz, L. G.: Physiologic and pharmacologic regulation of bone resorption. N. Engl. J. Med. *282*:909–916, 1970.

Rasmussen, H. and Bordier, P.: The cellular basis of metabolic bone disease. N. Engl. J. Med. *289*:25–32, 1973.

Calcium Nutrition

Garn, S. M. and Wagner, B.: The adolescent growth of the skeletal mass and its implications to mineral requirements. *In* Adolescent Nutrition and Growth (Ed.: F. P. Heald). Appleton-Century-Crofts, New York, 1969.

Irwin, M. I. and Kienholz, E. W.: A conspectus of research on calcium requirements of man. J. Nutr. *103*:1019–1095, 1973.

Leitch, I. and Aitken, F. C.: The estimation of calcium requirement: A re-examination. Nutr. Abstr. Rev. *29*:393–411, 1959.

Pitkin, R. M.: Calcium metabolism in pregnancy. A review. Am. J. Obstet. Gynecol. *121*:724–737, 1975.

Symposium on Human Calcium Requirements. J.A.M.A. *185*:588–593, 1963.

Williams, M. L., Rose, C. S., Morrow, G., III et al: Calcium and fat absorption in neonatal period. Am. J. Clin. Nutr. *23*:1332–1330, 1970.

PHOSPHATE HOMEOSTASIS

Phosphate Absorption

Albright, F. and Sulkowitch, H. W.: Effect of vitamin D on calcium and phosphorus metabolism: studies on four patients. J. Clin. Invest. *17*:305–315, 1938.

Caniggia, A., Gennari, C., Bencini, M. et al: Intestinal absorption of radio phosphate in osteomalacia before and after vitamin D treatment. Calcif. Tissue Res. *2*:299–300, 1968.

Harrison, H. E.: Phosphorus. *In* Present Knowledge of Nutrition. 4th Ed. The Nutrition Foundation, New York, 1976, Chap. 24.

Harrison, H. E. and Harrison, H. C.: Intestinal transport of phosphate: action of vitamin D, calcium and potassium. Am. J. Physiol. *201*:1007–1012, 1961.

Renal Excretion

Agus, Z. S., Puschett, J. B., Senesky, D. et al: Mode of action of parathyroid hormone and cyclic adenosine 3′5′-monophosphate on renal tubular phosphate reabsorption in the dog. J. Clin. Invest. *50*:617–626, 1971.

Bijvoet, O. L. M.: Relation of plasma phosphate concentration to renal tubular reabsorption of phosphate. Clin. Sci. *37*:23–36, 1969.

Bijvoet, O. L. M., Morgan, D. B. and Fourman, P.: The assessment of phosphate reabsorption. Clin. Chim. Acta *26*:15–24, 1969.

Gekle, D., Stroder, J. and Rostock, D.: The effect of vitamin D on renal inorganic phosphate reabsorption of normal rats, parathyroidectomized rats, and rats with rickets. Pediatr. Res. *5*:40–52, 1971.

Harrison, H. E. and Harrison, H. C.: The renal excretion of inorganic phosphate in relation to the action of vitamin D and parathyroid hormone. J. Clin. Invest. *20*:47–55, 1941.

Klein, R. and Gow, R. C.: Interaction of parathyroid hormone and vitamin D on the renal excretion of phosphate. J. Clin. Endocrinol. Metab. *13*:271–282, 1953.

Knox, F. G. and Lechene, C.: Distal site of action of parathyroid hormone on phosphate reabsorption. Am. J. Physiol. *229*:1556–1560, 1975.

Lavender, A. R. and Pullman, T. N.: Changes in inorganic phosphate excretion induced by renal arterial infusion of calcium. Am. J. Physiol. *205*: 1025–1032, 1963.

Massry, S. G., Friedler, R. M. and Coburn, S. W.: Excretion of phosphate and calcium: physiology of their renal handling and relation to clinical medicine. Arch. Intern. Med. *131*:828–859, 1973.

Mercado, A., Slatopolsky, E. and Klahr, S.: On the mechanisms responsible for the phosphaturia of bicarbonate administration. J. Clin. Invest. *56*:1386–1395, 1975.

Puschett, J. B., Moranz, S. and Kurnick, W. S.: Evidence for a direct action of cholecalciferol and 25-hydroxycholecalciferol on the renal transport of phosphate, sodium and calcium. J. Clin. Invest. *51*:373–385, 1972.

Puschett, J. B., Beck, W. S. Jr. and Jelonek, A. et al: Parathyroid hormone and 25-hydroxyvitamin D:

synergistic and antagonistic effects on renal phosphate transport. Science *190*:473–475, 1975.

Suki, W. N., Martinez-Maldonado, M., Rouse, D. et al: Effect of expansion of extracellular fluid volume on renal phosphate handling. J. Clin. Invest. *48*:1888–1894, 1969.

Trohler, U., Bonjour, J-P and Fleisch, H.: Inorganic phosphate homeostasis. Renal adaptation to the dietary intake in intact and thyroparathyroidectomized rats. J. Clin. Invest. 57:264–273, 1976.

MEASUREMENT OF VITAMIN D AND ITS METABOLITES IN BLOOD AND TISSUES

Belsey, R. E., DeLuca, H. F. and Potts, J. T., Jr.: A rapid assay for 25-OH-vitamin D_3 without preparative chromotography. J. Clin. Endocrinol. Metab. *38*:1046–1051, 1974.

Braumbaugh, P. F., Haussler, D. H., Bressler, R. et al: Radioreceptor assay for 1,25-dihydroxyvitamin D_3. Science *183*:1089–1091, 1974.

Haddad, J. G. and Chyu, K. J.: Competitive protein-binding radioassay for 25-hydroxycholecalciferol. J. Clin. Endocrinol. Metab. *33*:992–995, 1971.

Haddad, J. G. and Stamp, T. C. B.: Circulating 25-hydroxyvitamin D in man. Am. J. Med. *57*:57–62, 1974.

Jones, G. and DeLuca, H. F.: High-pressure liquid chromatography: separation of the metabolites of vitamin D_2 and D_3 on small-particle silica columns. J. Lipid Res. *16*:448–453, 1975.

Kream, B. E. and DeLuca, H. F.: A specific binding protein for 1,25-dihydroxyvitamin D_3 in rat intestinal cytosol. Biochem. Biophys. Res. Commun. *76*:735–738, 1977.

Morris, J. F. and Peacock, M.: Assay of plasma 25-hydroxy vitamin D. Clin. Chim. Acta *72*:383–391, 1976.

Vitamin D Nutrition

Harrison, H. E.: Vitamin D Malnutrition. Progress in Food and Nutrition Science, Vol. 1, No. 5. Permagon Press, Great Britain, 1975, pp. 349–357.

Chapter Three

HYPOCALCEMIC STATES

Hypocalcemia ideally should be defined primarily in terms of the concentration of the physiologically active moiety of the serum calcium, i.e., the ionized calcium. Total serum calcium concentration may be low as the result of reduction of serum albumin concentration and therefore decrease of protein-bound undissociated calcium without reduction of the ionized calcium fraction. Since determination of serum ionized calcium concentration is not uniformly available, this value is usually estimated from total serum calcium and total serum protein concentrations. This estimation, first proposed by McLean and Hastings and for which they developed a nomogram (see Fig. 2–3), has been shown more recently to differ considerably in individual instances from actual determination of ionized calcium by the use of a calcium ion electrode. The data indicate that the calcium binding capacity of serum proteins is more variable than had been suggested by McLean and Hastings, and the factors responsible are not understood.

This variability has been of particular importance in the assessment of hypocalcemia in the newborn or young infant since symptoms in this age-period often do not correlate well with total serum calcium concentrations. When serum ionized calcium measurements are made it becomes apparent that symptoms are almost invariably present when the ionized calcium concentration is 2.5 mg/dl or less, and are often present at concentrations between 2.5 and 3.0 mg/dl. Since the normal range of concentration of serum ionized calcium is between 4.0 and 5.5 mg/dl with a mean value of 4.4 mg/dl, this means that serum ionized calcium must be reduced by at least one-third before overt symptoms occur. If ionized calcium concentration is estimated by the McLean-Hastings nomogram, at a total serum protein concentration of 6.0 gm/dl total serum calcium would be reduced to 6.0 mg/dl when serum ionized calcium was 3.0 mg/dl. Manifestations of hypocalcemia have been seen in infants and children at total serum calcium concentrations of 7.5 mg/dl, and even rarely at concentrations of 8.0 mg/dl. In a few instances determination of serum ionized calcium concentration has shown that this fraction was reduced below 3.0 mg/dl despite total serum calcium concentrations of 7.5 to 8.0 mg/dl and total serum protein concentrations in the normal range. For that reason total serum calcium concentrations of 7.5 mg/dl or less should be considered as possibly associated with symptoms and signs of hypocalcemia, even though in individual cases the concentrations of ionized calcium at this concentration are above 3.0 mg/dl and not accompanied by evidence of tetany.

On the basis of the physiology of calcium homeostasis, reduction of serum ionized calcium concentration theoretically can be classified as due either to deficiency of compensation by increased parathyroid hor-

Table 3–1. Hypocalcemic States

- I. FAILURE OF APPROPRIATE SECRETION OF PARATHYROID HORMONE
  - A. Neonatal hypocalcemia
    - 1. Early hypocalcemia – calcitonin – PTH imbalance
    - 2. Delayed post-feeding hypocalcemia – transient physiologic hypoparathyroidism
  - B. Congenital hypoparathyroidism
    - 1. Aplasia or hypoplasia of the parathyroids
    - 2. Pseudoidiopathic hypoparathyroidism
  - C. Acquired Hypoparathyroidism
    - 1. Idiopathic hypoparathyroidism
    - 2. Genetic autoimmune hypoparathyroidism
    - 3. Surgical hypoparathyroidism
  - D. Hypomagnesemia (see also inadequacy of target tissue response to PTH)
- II. INADEQUATE RESPONSIVENESS OF TARGET TISSUE TO PARATHYROID HORMONE
  - A. Vitamin D deficiency or block in vitamin D metabolic cycle resulting in deficiency of active vitamin D metabolite, 1, 25-dihydroxyvitamin D
  - B. Pseudohypoparathyroidism types I and II
  - C. Magnesium deficiency
  - D. Hypernatremia, hypokalemia and infections

mone output in response to appropriate stimuli (and this we shall call *hypoparathyroidism*), or to lack of responsiveness of target tissues to parathyroid hormone (in which is included the entity *pseudohypoparathyroidism*). The hypocalcemia of excess calcitonin secretion is normally compensated for by increased parathyroid hormone output, so that persistent hypocalcemia due to excess calcitonin should not occur if there is appropriate responsiveness of the PTH axis, which includes both parathyroid gland secretory activity and response of target tissues to the hormone. The classification of hypocalcemic states is summarized in Table 3–1.

SYMPTOMS AND SIGNS OF HYPOCALCEMIA

A major manifestation of hypocalcemic states is tetany, by which we mean hyperexcitability of the nervous system both central and peripheral. This hyperexcitability is thought to result from increased permeability of the plasma membrane surrounding cell or axon to sodium in the presence of reduced concentration of calcium ion or magnesium ion. Tetany can manifest itself by carpopedal spasm, i.e., spontaneous tonic contracture of the muscles of upper and lower extremities with the hands held with fingers extended, thumb adducted and wrist flexed (obstetric position) and the feet held adducted and extended (talipes equinovarus). The tonic contracture of hand muscles can also be provoked by ischemia (Trousseau's sign) produced by a blood pressure cuff inflated above systolic pressure for three minutes on the upper arm, reducing arterial blood flow.

Other evidences of spontaneous excitation of motor nerves or nerve cells are laryngospasm and convulsive seizures. The characteristic seizures of hypocalcemia are short generalized tonic and clonic seizures, repeated at brief intervals. During interconvulsion periods the patient is usually conscious and responsive. However, lateralized convulsions can also occur and after repeated convulsions postictal depression can result.

In addition to reduction of calcium ion concentration in extracellular fluid, other factors predisposing to tetany are reduction of magnesium ion concentration and of hydrogen ion concentration (alkalosis). Potassium ion concentration is also of importance in that a reduction of this ion inhibits the development of hypocalcemic tetany, so that a subject with concomitant hypocalcemia and hypokalemia may not have symptoms and signs of tetany until given potassium. Increased hydrogen ion concentration protects against tetany, and tetany can be precipitated by reduction of hydrogen ion concentration through increase of bicarbonate concentration or reduced pCO_2. Alkalosis does reduce ionized calcium concentration in serum, since the dissociation of the calcium protein complex is increased by increase of H^+ion concentration and reduced by decrease of this concentration. However, the reduction of serum ionized calcium concentration in hyperventilation tetany is not sufficient to cause symptoms on this basis alone. Reduction of magnesium ion concentration can cause tetany even at normal concentrations of calcium ion, but the coexistence of reduced calcium and magnesium ion concentration may be a factor in the initiation of severe tetany, particularly convulsive seizures. Hypokalemia does not protect against the tetany of hypomagnesemia. Because of the multiple factors influencing the excitability of the nervous system, tetany may fluctuate between the latent and manifest stages. Latent tetany may be diagnosed by evidences of a lowered threshold for mechanical stimulation of a motor nerve, such as the Chvostek sign or peroneal sign, in which the facial or peroneal nerves respectively are tapped and contraction of facial or peroneal muscles obtained. A lowered threshold for electrical stimulation of motor nerves can also be detected, Erb's sign. As mentioned above, carpal spasm can be induced by ischemia through cutting off of arterial circulation by a tourniquet applied for three minutes. Hyperventilation with lowering of pCO_2 and increase of pH may result in manifest tetany in occasional individuals with normal calcium ion concentration, and can precipitate manifest tetany in patients with latent tetany. Similarly, intravenous administration of sodium bicarbonate may cause tetany and convulsions in patients with low calcium ion concentrations who have been protected from tetany by metabolic acidosis, e.g., those with uremia.

The calcium ion concentration of extracellular fluid influences tissues and organs other than the central and peripheral nervous system. Hypocalcemia, therefore, can produce symptoms and signs in addition to those of hyperexcitability of the nervous system. One organ influenced by calcium ion concentration is the heart. Reduced concentration of calcium ion causes prolongation of electrical systole as measured by the Q — T interval. In some instances disturbances of conduction with heart block have been found in hypocalcemic infants. Since the Q — T interval varies with heart rate a corrected interval must be calculated, and this can be done by dividing the Q — T interval by the square root of the cycle length (R — R interval).

$$Q-T_c = \frac{Q-T}{\sqrt{R-R}}$$

The normal Q — T_c is usually given as 0.40 with a range from 0.36 to 0.44. It has been suggested that the Q — $_0T$, measured from the origin of the Q wave to that of the T wave, is more precise than the Q — T. The Q — $_0T$, corrected for heart rate as is the Q — T_c, is less than 0.20 seconds in premature infants and less than 0.19 in term infants. The prolonged Q — T or Q — $_0T$ interval in hypocalcemic states is more marked if hypomagnesemia is also present. Although prolongation of the Q — T interval may be a helpful diagnostic sign of hypocalcemia, it is not invariably present and can be within normal limits despite documented reduction of serum calcium concentration. It may also be prolonged in potassium deficiency states and as the result of myocarditis.

Hypocalcemia may also cause highly nonspecific symptoms such as lethargy, poor feeding, vomiting and cyanotic episodes, all of which resemble the symptoms and signs of early infection in the newborn. Therefore, we have determined serum calcium concentrations in infants with these symptoms of possible sepsis, and in several such infants hypocalcemia has probably been the primary cause of these findings, since blood and CSF cultures were negative and treatment of the hypocalcemia resulted in disappearance of the symptoms and complete recovery.

NEONATAL HYPOCALCEMIA

This is the most prevalent type of hypocalcemia encountered by the pediatrician. It can be divided into two main groups: (1) early hypocalcemia beginning in the first 48 hours of life before feedings have been given; and (2) late or phosphate-induced hypocalcemia beginning at the end of the first week of life in infants who have been taking feedings with high intake of phosphate. In both groups it has been suggested that lack of appropriate parathyroid hormone secretion is a major factor, and that hypocalcemia is precipitated by calcitonin excess in the early variety and by phosphate loading in the later form.

EARLY NEONATAL HYPOCALCEMIA

Pathophysiology

The predisposing factors in early hypocalcemia are prematurity, anoxia, acidosis and birth trauma. Infants born of diabetic mothers also have a high incidence of hypocalcemia whether or not they have respiratory distress. The studies of Tsang et al have suggested that increase of extracellular inorganic phosphate by breakdown of cell organic phosphate may be a factor in the development of hypocalcemia in the stressed infants with respiratory difficulty, anoxia and acidosis (Table 3–2). The phosphate effect suggests inadequate parathyroid gland response to the stimulus of reduced calcium ion concentration. This has been examined by radioimmune assays of circulating parathyroid hormone in plasma, and several investigators have shown that the concentrations of PTH are not increased as much as they should be with the stimulation of reduced calcium ion concentration. The alternative possibility has also been suggested that target cells are unresponsive to parathyroid hormone or to cyclic AMP, which is the "second messenger" through which PTH exerts its action in both bone cells and renal tubule cells. More recently, hypercalcitonemia has been demonstrated in the prematurely born infant, and it seems likely that an imbalance between calcitonin secretion and PTH output is the important factor.

A mechanism for poor parathyroid gland responsiveness in the newborn period can be proposed on theoretic grounds. The placenta transports calcium from maternal to fetal plasma, and the ionized calcium concentration in the fetal circulation is higher than that in the maternal circulation (Table 3–3). The calcium drain from mother to fetus, which occurs mainly during the last trimester of pregnancy, causes compensatory overactivity of maternal parathyroids as evidenced by increased levels of circulating parathyroid hormone (Fig. 3–1). The levels of ionized calcium in fetal plasma due to placental transport are probably high enough to depress fetal parathyroid activity and could account for the relative hypoparathyroidism of the neonatal period. This concept is supported by the incidence of transient hypoparathyroidism in infants born of mothers with true hyperparathyroidism, and conversely by evidence of fetal hyperparathyroidism in infants born of mothers with untreated hypoparathyroidism. The high calcitonin blood levels in the fetus could have physiologic value in reducing bone resorption and thus maintaining a high rate of net bone mineral deposition in the rapidly growing fetal skeleton. Since the fetal serum ionized calcium concentration is maintained by placental calcium transport, bone resorption would not be essential for calcium homeostasis during intrauterine life.

Table 3–2. Role of Serum Inorganic Phosphate Released Endogenously in Development of Early Neonatal Hypocalcemia

	TIME AFTER BIRTH	SUBJECTS *Hypocalcemic*	*Normocalcemic*
Serum Ca mg/dl	8 hrs	8.9	9.6
	29 hrs	6.3	9.0
Serum P mg/dl	8 hrs	7.6	6.3
	29 hrs	8.1	6.7

Taken From Tsang, R. C. and Oh, W.: Pediatrics *45*: 773, 1970.

Table 3–3. Comparison of Concentrations of Total Serum Calcium, Ionized Calcium, Protein and Phosphate of Maternal Blood at Time of Delivery and Cord Blood

PATIENT	TOTAL Ca	IONIZED Ca	PROTEIN	P
	mg/dl		gm/dl	mg/dl
Mother M-1	6.3	3.2	4.8	1.8
Cord Blood	9.7	5.2	4.7	4.9
Mother J	7.6	4.5	5.9	3.6
Cord Blood	10.2	5.6	5.3	5.6
Mother C	8.3	4.8	5.9	2.8
Cord Blood	9.5	5.2	5.6	4.1
Mother F	8.7	5.2	6.4	3.2
Cord Blood	10.5	6.7	5.3	5.2
Mother G	8.6	4.7	6.0	–
Cord Blood	11.1	6.0	5.5	–
Mother M-2	9.5	5.1	7.1	3.9
Cord Blood	10.6	4.2	6.5	4.8
Mother W	8.7	3.9	5.6	3.8
Cord Blood	9.0	3.8	5.3	7.7
Mother R	7.9	3.8	5.6	2.7
Cord Blood (A)	8.7	3.9	4.4	5.6
Cord Blood (B)	8.6	4.0	4.3	6.6
Mother M-3	9.0	5.0	4.9	3.4
Cord Blood (A)	10.6	6.8	6.2	6.7
Cord Blood (B)	11.0	7.0	6.6	5.9
Mother D	8.7	4.5	6.5	4.2
Cord Blood	11.4	6.8	3.7	7.8
Mother K	6.5	3.2	4.7	2.2
Cord Blood	9.2	5.2	6.3	5.2
Mother M-4	9.1	4.9	6.6	2.8
Cord Blood	9.4	6.0	4.1	5.6
Mother H	9.6	5.3	7.1	4.4
Cord Blood	9.9	6.5	4.8	7.9
Average Maternal	8.3	4.5	5.9	3.2
Average Cord Blood	10.0	5.5	5.2	6.0

Diagnosis

Early hypocalcemia, i.e., concentrations of total serum calcium below 7.5 mg/dl, occurs with great frequency in small, prematurely born infants weighing less than 1500 gm. These infants have a high frequency of respiratory distress, apneic episodes and hypoglycemia, among other problems, and it is difficult to determine to what extent the hypocalcemia contributes to their physiologic difficulties. Infants born of diabetic mothers may convulse owing to hypoglycemia as well as hypocalcemia, and similarly infants with birth trauma or anoxia may have convulsions due to subdural or intracerebral hemorrhage or other brain injury along with their hypocalcemia. When ionized calcium determinations become available on a more routine basis it may be possible to be more precise with regard to the role of calcium in the symptomatology of

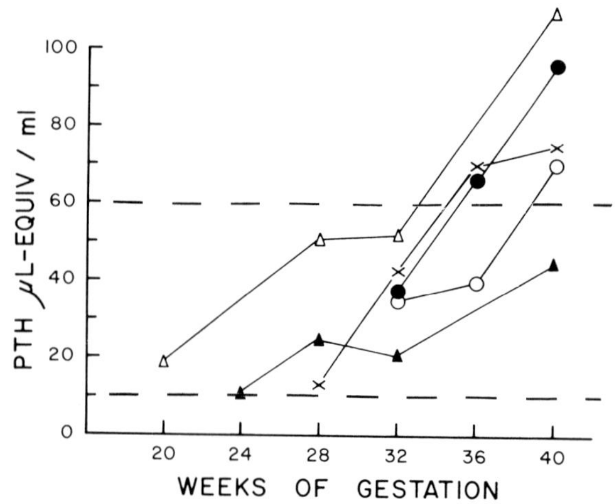

Figure 3–1. Concentrations of serum immunoreactive parathyroid hormone during pregnancy, illustrating rise of concentration during last trimester. (From Cushard, W. G., Jr., et al. J. Clin. Endocrinol. Metab. *34*:768, 1972. Reprinted with permission.)

these infants at risk. Rosen et al have suggested that convulsions occur only at calcium ion concentrations below 2.5 mg/dl.

Treatment

Because of the uncertainty of the contribution of hypocalcemia to the serious symptoms of the infants subject to early hypocalcemia, 10 per cent calcium gluconate is usually given to such infants by the intravenous route at the rate of 1 or 2 ml/kg body weight every six or eight hours, or a total of 4 to 6 ml/kg in the 24-hour infusion. One ml of 10 per cent calcium gluconate supplies 9 mg of calcium (calcium gluconate = 9 per cent calcium) so that 6 ml/kg/24 hrs would provide 54 mg calcium/kg. In a 1.5-kg infant this would be the equivalent of the calcium content of 250 ml of human milk. It is important that the calcium gluconate be given into a freely running intravenous line to avoid extravasation, since 10 per cent calcium gluconate is injurious to the soft tissues and causes severe necrosis with precipitation of calcium in the injured tissue. It goes without saying that it should never be given by intramuscular injection. If the calcium gluconate is given by "stat" injection, the heart rate should be monitored during the infusion, since rapid elevation of the calcium ion concentration in the right auricle will cause depression of the sinus node with bradycardia and even cardiac arrest. The rate of infusion should not be so rapid as to produce bradycardia; if this does occur the injection of calcium gluconate should be temporarily stopped until recirculation and resumption of normal heart rate has taken place. The duration of intravenous administration of calcium gluconate must be guided by repeated serum calcium determinations. When intravenous fluid administration is discontinued and these infants are given tube feedings, calcium lactate or gluconate can be added to the feeding mixture, as indicated in the section on delayed hypocalcemia.

Delayed or Late Neonatal Hypocalcemia (Transient Physiologic Hypoparathyroidism)

Pathophysiology

This definition is reserved for infants, usually full term but occasionally prematurely born, who have been started on feedings and have shown symptoms and/or signs of hypocalcemia only after several days to several weeks of feedings. Their early postdelivery course may have been entirely uneventful and there is no correlation with difficult delivery, birth trauma or anoxia. The responsible mechanism is relative hypoparathyroidism, and the hypocalcemia is precipitated by the high phosphate load of most feedings other than human milk. Human milk contains only 150 mg phosphorus per liter (700 calories) whereas cow's milk contains approximately 1000 mg per liter (700 calories). The various infant feeding preparations prepared from cow's milk whey or isolated soy bean protein provide about 500 mg phosphorus per liter of recommended dilution (20 cal/oz, 700 cal/liter). The low electrolyte feedings, Similac PM 60/40 and SMA-S-26, each contain approximately 170 mg P per liter.

Except for the occasional infant fed only at the breast or with the low electrolyte feeds, most infants receive relatively high phosphorus intakes. Many newborns are not well adapted to such phosphate loads, which is understandable since the physiologic feeding is a low phosphate food. In addition to the higher phosphate concentration of the various prepared infant feedings the phosphate intake is increased by the volume of feeding taken in the first week or ten days of life, which is often greater than the volume of breast milk available to the nursing infant. A high phosphate intake increases the extracellular phosphate concentration and reduces calcium ion concentration through the reaction:

$$3\ Ca^{++} + 2\ H\ PO_4^{=} \rightarrow Ca_3\ (PO_4)_2 + 2H^+$$

The appropriate physiologic response to this reduction of ionized calcium in extracellular fluid is increased parathyroid hormone release from the parathyroid gland cells. This results in solubilization of bone mineral with entrance of both calcium and phosphate into extracellular fluid. The phosphate is excreted in the urine through the phosphaturic action of PTH, and the calcium is retained by the action of PTH in increasing renal tubular reabsorption of calcium. If the response of the parathyroids is inadequate only a small fraction of the absorbed phosphate is excreted in the urine, serum phosphate concentration remains high, and serum calcium ion concentration progressively decreases because of inadequate solubilization of bone mineral in the face of high concentration of extracellular phosphate. The high phosphate concentration in the serum of infants in the first weeks of life is due not only to their low parathyroid hormone levels but also to the physiologically low glomerular filtration rate, so that tubular reabsorption of phosphate is high in relation to GFR.

Diagnosis

Infants who take high phosphate feedings usually seem perfectly well for the first four or five days of life or even longer. Early manifestations may be poor feeding, vomiting, lethargy or cyanotic episodes. In other instances there may be few premonitory symptoms and convulsive seizures may be the presenting complaint. These convulsive twitchings are most often generalized but can be lateralized; they usually are brief, lasting 30 seconds to a few minutes, but are repeated. Despite the recurrent seizures the infant is usually responsive during interconvulsive episodes. The convulsions may be precipitated by loud noises or vibration as in performing the Moro reflex. Occasionally laryngospasm with apnea may supervene. Aside from the brief duration of the seizures and the apparent conscious state of the infant, there is little to differentiate hypocalcemic seizures from other convulsions in the newborn. The diagnostic signs of tetany which are useful in older infants and children, namely, the Chvostek, peroneal or Trousseau sign, are rarely elicited in the very young infant. The electrocardiographic sign of hypocalcemia, the prolonged Q — T or $Q — {}_0T$ interval, is not invariably present.

The diagnosis is, therefore, dependent on the determination of serum calcium concentration. Since this frequently is not available for some time the therapeutic test of intravenous administration of 10 per cent calcium gluconate can be employed in a convulsing infant. This solution is given at a dose of 2 ml/kg (18 mg Ca/kg) and as a stat. infusion, but at a rate controlled by the heart rate. If given too rapidly the concentration of calcium in the right auricle rises to excessive concentrations with depression of the rhythmicity of the sinus node. As little as 1 ml of 10 per cent calcium gluconate injected rapidly into an arm vein of a small infant can cause cardiac arrest. Doses that temporarily slow the heart rate are not sufficient, however, to raise the calcium ion concentration in the extracellular fluid of the central nervous system high enough to suppress the hyperexcitability resulting from calcium deficiency. By monitoring the heart rate and waiting for recirculation to reduce the concentration of calcium in the right auricle, it is possible to give the required dose of calcium without circulatory complications. If the convulsion or other evidences of hyperexcitability of the nervous system are due to hypocalcemia there should be immediate improvement. Lack of response to a proper dose of intravenous calcium gluconate solution indicates some other cause for the abnormalities of the nervous system. Usually these infants have already been given hypertonic glucose solution because of the possibility of hypoglycemia. Inasmuch as primary hypomagnesemia can produce CNS hyperexcitability resembling that of hypocalcemia, infants unresponsive to intravenous calcium gluconate can then be given magnesium sulfate as a therapeutic trial and diagnostic test. The solution can be given intramuscularly as a 25 per cent solution of the anhydrous salt (50 per cent solution of $MgSO_4 \cdot 12\ H_2O$) in a dosage of 0.2 ml/kg body weight.

Once hypocalcemia has been established the differential diagnosis rests between transient physiologic hypoparathyroidism, the rarer congenital hypoparathyroidism, and, in the United States, the even less common congenital vitamin D deficiency. In both transient and congenital hypoparathyroidism the serum phosphate concentrations are elevated, usually above 8 mg/dl and occasionally as high as 12 mg/dl. In congenital vitamin D deficiency the serum phosphate concentrations are more often normal or reduced. More importantly the serum concentrations in the mother provide diagnostic help. In congenital vitamin D deficiency the concentrations of calcium and phosphate in the maternal serum should be reduced, whereas the calcium concentration of the serum of the mother of an infant with transient physiologic or congenital hypoparathyroidism should be normal or even somewhat increased, since some of these mothers have asymptomatic hyperparathyroidism. Roentgenograms of the long bones of the infant with congenital vitamin D deficiency could show undermineralization and possibly irregularity of the metaphyseal ends of the bone. The differential diagnosis between transient physiologic hypoparathyroidism and congenital hypoparathyroidism may not be possible until there has been time to evaluate the response to treatment. The usual age of onset of physiologic hypoparathyroidism is between 5 and 10 days of age, but may be as late as 6 weeks of age (*see case reports*). The duration of physiologic hypoparathyroidism is commonly less than three to four weeks, but in occasional cases the physiologic deficiency of parathyroid secretion may persist for several months, evoking the possibility of persistent hypoparathyroidism.

The parathyroid glands arise from the third and fourth entodermal pouches. Since the third entodermal pouches are the source of the thymus, congenital aplasia of the parathyroids may coexist with thymic aplasia (DiGeorge syndrome), so that a chest X-ray may be of value in determining the presence of the thymus. Similarly, abnormalities of development of the heart (especially the great vessels) may occur along with parathyroid aplasia, so that cardiac murmurs, enlargement or cyanosis coexisting with hypocalcemia would strongly suggest congenital hypoparathyroidism.

Treatment

The response to treatment is also helpful in the differentiation of transient physiologic hypoparathyroidism from congenital aplasia of the parathyroids. In the former parathyroid hormone secretion is not completely absent but is inadequate to compensate for the increased hormone required to maintain calcium ion concentrations in the face of high phosphate intake and hyperphosphatemia. Reduction of serum phosphate concentrations, therefore, should be adequate to bring serum calcium concentrations to normal. This is in fact the case. The serum phosphate concentration can be reduced by preventing phosphate absorption from the intestine, which is accomplished by addition of a soluble calcium salt to the feeding so that the ratio of calcium to phosphorus by weight is 4. This provides an excess of calcium, which precipitates the diet phosphate in the intestinal lumen as insoluble $Ca_3(PO_4)_2$ (Ca:P ratio of approximately 2) and prevents phosphate absorption. The soluble salts of calcium usually employed for this purpose are calcium lactate and calcium gluconate. Calcium chloride should not be used since solutions of calcium chloride are irritating to the gastric mucosa and the large chloride intake almost always results in hyperchloremic acidosis if continued for several days.

Calcium lactate powder is 13 per cent calcium by weight whereas calcium gluconate is 9 per cent calcium. Calcium gluconate is available as a prepared solution, syrup of NeoCalglucon, which provides 90 mg calcium per teaspoon, and because of its convenience is frequently used to provide the calcium supplement to the feeding needed to increase the Ca:P ratio to 4. The large amount of sugar in this syrup may cause diarrhea in the young infant. Calcium lactate powder added to milk goes into solution readily, and we have not observed any gastrointestinal disturbances in infants fed as much as 12 gm of calcium lactate daily (1.5 gm calcium). The only drawback to its use is that the requisite amount of powder must be weighed out by the pharmacist. If calcium lactate tablets are substituted the results may be unsatisfactory, since the compressed tablets are poorly soluble in milk and even if broken up may not go into solution, so that the calcium dissolved is inadequate for treatment. When the powder has been employed and the proper amount given as calculated, we have seen no failures in the treatment of transient physiologic hypoparathyroidism. The patients are followed by determinations of serum calcium and phosphate concentrations at three- or four-day intervals. When the serum calcium concentration has reached a normal value the calcium supplement is gradually decreased. It should not be abruptly discontinued, since a normal or even high serum calcium in the face of a low serum phosphorus does not mean that parathyroid hormone output is adequate to maintain calcium homeostasis when the serum phosphate concentration is permitted to rise again (*see case reports*). The calcium supplement is gradually decreased and if serum calcium concentrations remain in the normal range may be discontinued after a total of two to three weeks.

Other modes of treatment of transient physiologic hypoparathyroidism have also been suggested. In the instances of prolonged hypocalcemia lasting for several weeks, treatment with vitamin D, dihydrotachysterol or experimentally with 1,25-dihydroxyvitamin D has been tried. The treatment is therefore similar to that of congenital aplasia of the parathyroids (*see section on treatment of this entity*). Although treatment with vitamin D is successful we believe it unnecessary unless properly administered calcium salt supplementation has failed. We have not had such failures in transient physiologic hypoparathyroidism, and an inadequate response indicates true congenital hypoparathyroidism. We have not seen the development of any intolerance to calcium lactate. However, if refusal of feeds, vomiting or diarrhea should occur, treatment with activated sterols would be indicated. The repeated injection of magnesium sulfate has also been suggested in the treatment of infantile hypocalcemia; this controls the symptoms but has to be continued at intervals for a number of days. The addition of calcium lactate or gluconate to the feedings seems much simpler and specifically controls the deficiency of calcium ion.

CASE REPORTS

Transient Physiologic Hypoparathyroidism in One of Fraternal Twins

Patient McC.-B. was one of dizygotic twins delivered by cesarean section somewhat prematurely. Birth weight was 2000 gm. The infant had no respiratory difficulties and took a prepared feeding (Similac) well. On the tenth day of life hyperirritability and muscular twitchings were noted. The serum calcium concentration on that day was 6.9 mg/dl and further studies are

Table 3–4. Patient McC.–B.

AGE DAYS	Ca	SERUM P mg/dl	Mg	TREATMENT
10	6.9	–	–	
12	7.1	9.2	0.86	Calcium lactate, Ca:P = 4
14	9.2	8.3	0.81	
19	14.0	3.0	1.64	Calcium lactate discontinued, Ca:P = 1.5
22	6.9	8.8	0.77	Calcium lactate, Ca:P = 3
32	10.7	6.2	1.80	Calcium lactate, Ca:P = 2

listed in Table 3–4. The nonidentical twin had normal serum calcium and phosphate concentrations throughout, and the mother's serum calcium concentration at the 12th postpartum day was 11.1 mg/dl, and serum phosphate 3.5 mg/dl. Further studies on the mother were not made.

COMMENTS: If the intrauterine environment, specifically the concentration of calcium in fetal blood as influenced by maternal parathyroid function and placental calcium transport, is the responsible factor in suppressing fetal parathyroid activity, the occurrence of transient hypoparathyroidism in only one of nonidentical twins would suggest that there was a difference between the two placentas in their calcium transport activities. More likely, genetic differences between the two twins altered their response to intrauterine environment. Otherwise, this is a classic case of neonatal hypocalcemia with onset on the tenth day of life after feeding of modified cow's milk. The hypocalcemia was rapidly controlled by oral calcium lactate supplement. An important associated finding was the hypomagnesemia coincident with hyperphosphatemia, with return to normal when the serum phosphorus concentration was lowered by the oral calcium lactate. Another phenomenon in the patient was the hypercalcemia that developed when the serum phosphorus was reduced to 3.0 mg/dl. This did not indicate recovery of parathyroid responsiveness, however, since discontinuance of the calcium lactate supplement resulted in a sharp rise of serum phosphorus to 8.8 mg/dl and a coincident decrease of serum calcium and magnesium concentration. This emphasizes the importance of gradual reduction of the calcium supplement even though the serum calcium has risen above the normal range. It also indicates the major role of extracellular phosphate concentration in the control of extracellular concentrations of both calcium and magnesium.

Transient Physiologic Hypoparathyroidism with Late Onset in Consecutive Siblings

L.E. was the second child of an apparently healthy woman. The previous child, born the year before, had had hypocalcemic convulsions starting at 6 weeks of age with complete recovery. Pregnancy was normal. Serum calcium determinations had not been made in this mother. Delivery was spontaneous at term, and birth weight 3100 gm. On the third day the infant was sent home apparently well on breast feedings plus supplementary evaporated milk feedings. His weight at 4 weeks was 4230 gm. At 30 days of age he was found by the mother to be convulsing and apneic. She had noticed no premonitory signs. She resuscitated him and brought him to the emergency room. During the trip to the hospital he had three more brief convulsions. Examination revealed a well-nourished infant with no unusual neurologic signs except for hyperactive deep reflexes. Blood was drawn for serum calcium determination. During the next few hours he had another brief convulsion and the Chvostek and peroneal signs were demonstrated. Another blood sample was drawn and the patient was given 6 ml of 10 per cent calcium gluconate. The serum calcium prior to this treatment was 5.0 mg/dl. The following day the infant was given a diluted evaporated milk feeding with 12 gm calcium lactate powder added to 540 ml of the feeding mixture, providing a Ca:P ratio of 4. The serum calcium and phosphate concentrations are shown in Table 3–5. The further clinical course was uneventful except that asymptomatic hypocalcemia recurred when the calcium lactate was abruptly discontinued on day 11.

COMMENTS: There are several points to be emphasized. A previous sibling had had hypocalcemia, also starting after the first month of life. The mother's serum calcium concentration was not measured during pregnancy. The possibility of hyperparathyroidism in this woman should have been considered and serum calcium concentrations should have been determined both in the mother and in the infant, L.E., prior to discharge from the newborn unit. The mother's serum calcium was measured following admission of L.E. to the hospital and was minimally elevated. Because of her reluctance to have further studies done, a definitive diagnosis of hyperparathyroidism in the mother could not be made. The infant was maintained in the hospital on a high phosphate cow's milk diet, but addition of a large excess of calcium lactate was well tolerated and was successful in reduction of serum phosphate concentration and elevation of serum calcium concentration. Discontinuance of the calcium supplement at 42 days of age resulted in relapse of hypocalcemia, but at 61 days of age the infant no longer required calcium supplementation to maintain normal serum calcium and phosphate concentrations despite a high phosphorus intake.

Transient Physiologic Hypoparathyroidism in Siblings with Documented Hyperparathyroidism in the Mother

E. McC. was a full term infant born after an uneventful pregnancy and normal delivery. He was given a standard infant feeding preparation, Enfamil, which contains 500 mg P per liter of standard dilution (20 cal/oz). He fed well and there was no vomiting and no other premonitory symptoms. On the seventh day he had several brief convulsive twitchings for which he was taken to a community hospital. The family history was important in that a sibling now 7 years of age had convulsions due to hypocalcemia in the second week of life controlled by calcium administration. Subsequently another child was born who did not have symptomatic hypocalcemia. The serum calcium concentration was found to be 7.0 mg/dl and attempts were made to give the infant intravenous calcium gluconate. Only 1 or 2 ml were given, however, without control of the convulsions. He was then transferred to the

Table 3–5. Patient L. E.

AGE DAYS	SERUM *Ca* (*mg/dl*)	SERUM *P* (*mg/dl*)	SERUM *BUN* (*mg/dl*)	TREATMENT
31	5.4	8.2	7.6	
31	5.0	10.5	8.6	6 ml 10% calcium gluconate IV
32	–	–	–	Calcium lactate 12 gm per day, Ca:P = 4
34	9.9	–	–	
42	–	–	–	Calcium lactate discontinued
48	7.3	10.1	–	Calcium lactate resumed, Ca:P = 4
50	10.0	7.6	–	
61	–	–	–	Calcium lactate discontinued
65	10.2	6.6	–	

Johns Hopkins Hospital. The hypocalcemia was confirmed and the seizures stopped by the intravenous injection of 4 ml of 10 per cent calcium gluconate. He was then placed on a low phosphorus feed, Similac PM 60/40, and calcium added to the feed to provide a Ca:P ratio of 4. The calcium was first given as syrup of Neo-Calglucon, but when the infant began having loose stools calcium lactate powder was substituted. The feedings were taken without difficulty and the course of the serum calcium and phosphorus concentrations is summarized in Table 3–6.

Determinations of serum calcium, ionized calcium and phosphorus were made in the mother, and she was found to have hypercalcemia with an ionized calcium concentration of 6.00 mg/dl and a total calcium concentration of 11.7 mg/dl. Urine calcium excretion was 440 mg per 24 hours. The serum IPTH concentration was increased. The diagnosis of hyperparathyroidism was made in the mother and confirmed at operation by the finding of an adenoma of one parathyroid. Since operation the mother's serum calcium concentration has been normal. If we assume that the previous sibling with hypocalcemia who is now 7 years of age had transient hypoparathyroidism because of maternal hyperparathyroidism, this woman has had this disorder for at least seven to eight years with no renal stones, gastric distress or skeletal demineralization.

COMMENTS: The value of determination of ionized calcium concentration in this case is illustrated, since the first measurement of 2.88 mg/dl after some treatment had been given showed that the infant was still in danger of symptomatic tetany. This is also a documented instance of transient physiologic hypoparathyroidism in an infant born of a woman with asymptomatic hyperparathyroidism. An important clue was a previous infant with hypocalcemia. However, maternal hyperparathyroidism has been diagnosed without multiple instances of infantile hypocalcemia, so that determination of serum calcium concentration should be done regularly in mothers of infants with transient physiologic hypoparathyroidism.

Phosphate Load Hypocalcemia in Infant Who Swallowed Maternal Blood

During the course of delivery of this infant profuse vaginal bleeding occurred as a result of marginal sinus rupture. The infant was delivered vaginally and had no respiratory distress or evidence of aspiration, but had swallowed a considerable quantity of blood. He vomited bloody material, and further blood was removed by gastric aspiration. Nevertheless, he passed stools with large quantities of blood which was of maternal origin since the hemoglobin was of the adult type. On the second day of life the infant had convulsive twitchings, and hypocalcemia and hyperphosphatemia were found, as shown in Table 3–7. The blood urea nitrogen con-

Table 3–6. Patient E. McC.

AGE DAYS	SERUM *Total Ca*	*Ionized Ca*	*P*	*Mg*	TREATMENT
		mg/dl			
8	6.5	–	–	–	
9	6.2	2.88	10.6	–	Calcium gluconate, Ca:P = 4
13	6.8	3.30	9.2	1.5	Calcium lactate, Ca:P = 4
17	8.4	4.00	7.9	–	
18	10.2	5.00	8.2	–	
		Mother			
	11.7	6.06	3.0	–	
		Sibling with History of Hypocalcemia			
	9.7	4.38	4.7	–	

Table 3–7. Patient B. H.

AGE DAYS	Ca	SERUM P	Urea N	REMARKS
		mg/dl		
1	–	–	–	Swallowed large quantity of blood
2	–	–	–	Feedings started
3	6.4	9.0	34	Tremors
3	7.1	7.8	28	Calcium lactate added, Pa:P = 4
4	8.4	6.6	7	
7	10.6	5.6	5	
8	11.0	4.6	–	Calcium lactate reduced, Ca:P = 3
9	11.0	3.7	–	Calcium lactate reduced, Ca:P = 2
11	11.1	8.3	–	
14	9.7	6.8	–	Calcium lactate discontinued Subsequent course uneventful

centration was elevated; however, this was not due to renal injury but to the very high protein intake in the large blood meal. By 4 days of age the urea nitrogen concentration was in the normal range. This infant was also treated with oral calcium lactate and recovered over a period of two weeks.

COMMENTS: This is a striking example of the effect of a high phosphate load, in this instance coming from swallowed blood, precipitating hypocalcemia within two days in an infant whose parathyroid glands presumably were unable to respond to the stress of the induced hyperphosphatemia and hypocalcemia by secretion of the greatly increased amounts of parathyroid hormone that would have been necessary to eliminate the phosphate in the urine and maintain calcium homeostasis. The hypocalcemia in this instance occurred earlier than in the ordinary case of phosphate load hypocalcemia, because of the unusual source of the phosphate load, but the course was otherwise similar to that seen in infants with transient physiologic hypoparathyroidism who are fed cow's milk preparations.

Transient Physiologic Hypoparathyroidism Without Convulsions or Other Evidences of Hyperexcitability of Nervous System

Although most infants with hypocalcemia are diagnosed because of hyperirritability with muscular tremors or convulsions, some may show only depressed responsiveness, poor feeding and vomiting which resembles the picture of sepsis. Baby O. was a full term infant who had no apparent difficulties at birth. He was fed in the usual fashion with a standard feeding preparation and was taking this well until the seventh day of life, when he seemed lethargic and fed poorly. A "sepsis work-up" was done including a lumbar puncture. Blood and CSF cultures showed no growth. Because of a short systolic murmur a congenital heart defect was considered and an electrocardiogram was taken. This showed a prolonged $Q - {}_0T$ interval, and the serum calcium concentration was then determined and found to be 5.2 mg/dl. When this was confirmed he was started on a calcium lactate supplement calculated to give a Ca:P ratio of 4 in his feeding. He gradually improved without other treatment, and the course of serum calcium and phosphate concentration is given in Table 3–8.

COMMENTS: In this patient the only manifestations of hypocalcemia were lethargy and poor feeding. The determination of serum calcium concentration was made because of abnormality of the $Q - {}_0T$ interval on electrocardiography. It is worth noting that such nonspecific manifestations of hypocalcemia as poor feeding, vomiting and lethargy should be considered as possible evidences of hypocalcemia warranting serum calcium measurements. Although the ECG can provide suggestive evidences of hypocalcemia, prolongation of the $Q - {}_0T$ interval is not invariably

Table 3–8. Patient Q.

AGE DAYS	SERUM Ca	SERUM P	REMARKS
	mg/dl		
8	5.2	–	Lethargic, poor feeding
9	5.0	10.5	Calcium lactate supplement, Ca:P = 4
11	6.1	–	
14	9.4	5.7	
16	12.6	3.6	Calcium lactate reduced, Ca:P = 2.5
18	11.1	4.1	Calcium lactate discontinued
26	10.0	7.6	
38	10.5	7.4	

present in patients with low serum calcium concentrations.

CONGENITAL HYPOPARATHYROIDISM

The parathyroids arise from pharyngeal pouches III and IV, and congenital aplasia or hypoplasia of the parathyroids can occur either associated with defective development of other structures developing from these pharyngeal pouches (namely, thymus and the great vessels of the heart) or as a solitary developmental abnormality. This defect of embryogenesis is usually sporadic without known cause, but multiple cases have been seen in one family, indicating a genetic factor. Congenital hypoparathyroidism may be complete, with total absence of parathyroid function. Such infants have hypocalcemia and convulsions in the neonatal period, and usually are regarded initially as examples of transient physiologic hypoparathyroidism. They do not respond to the treatment described above, i.e., addition of a calcium salt to the feed to provide a large excess of calcium over phosphate in the diet (Ca:P = 4). Further investigation of such a nonresponsive infant is then required. An x-ray examination of the chest should be made to determine presence or absence of the thymus. Studies of T cell function are also now available as a further method of ascertaining inadequate thymic development. IPTH concentrations in serum are measured, and in these infants are undetectable or extremely low in the face of hypocalcemia and hyperphosphatemia. Evaluation of cardiac status should also be made. Congenital vitamin D deficiency as a cause of hypocalcemia in the first weeks of life might also be considered since such patients are unresponsive to calcium salt supplementation, but this is extremely rare except in situations in which mothers are completely prevented from normal skin exposure to sunshine. This diagnosis would be suggested by the finding of hypocalcemia and hypophosphatemia in the mother with skeletal evidences of osteomalacia. Serum 25-OH vitamin D concentration in mother and infant would also be extremely low, <5 ng/ml. Skeletal radiographs in the infant might also show defective mineralization, whereas this would be normal in congenital hypoparathyroidism.

Congenital hypoparathyroidism, however, may not be total, i.e., showing aplasia of the parathyroids. Hypoplasia with partial hypoparathyroidism also may occur. These infants, therefore, may not show signs or symptoms of hypocalcemia until later in infancy or childhood at a time when they are stressed by a febrile illness or operative procedure with associated starvation. Such later onset is usually diagnosed as idiopathic hypoparathyroidism, which in many instances may well be a manifestation of congenital hypoplasia of the parathyroids. The stress of a febrile illness can provoke hypocalcemia by breakdown of intracellular phosphoric acid esters, with increase of extracellular inorganic phosphate causing hypocalcemia if the parathyroids are incapable of increasing hormone output. The evidence for such incomplete parathyroid development is inferential, based on the findings in patients with partial hypoparathyroidism associated with congenital heart defects (*see case reports*). The presenting manifestation is almost always a convulsive seizure that leads to determination of serum calcium and phosphorus concentrations.

Treatment

If the parathyroid deficiency is complete, i.e., showing aplasia of the parathyroids, the hypocalcemia and hyperphosphatemia cannot be corrected by the sequestration of phosphate in the intestine by an excess of calcium. In the total absence of parathyroid hormone there is also failure of formation of 1,25-diOH vitamin D from 25-OH vitamin D, so that both solubilization of bone mineral in response to hypocalcemia and absorption of calcium from the intestine are depressed. Treatment depends on providing a substitute for the deficient 1,25-diOH vitamin D. This can be accomplished by large doses of vitamin D which is converted to 25-OH vitamin D; high concentrations of this compound are effective as replacement for 1,25-diOH vitamin D, and at high concentration of substrate 1,25-diOH vitamin D is possibly formed in the absence of PTH. This last premise remains to be determined.

An alternative treatment is the use of a vitamin D-like compound, dihydrotachysterol, which has advantages in the treatment of hypoparathyroidism since it is not dependent on 1-hydroxylation by the kidney for its activity. It is hydroxylated at the 25 carbon by hepatic enzymes and the active compound is 25-OH dihydrotachysterol. Dihydrotachysterol is more rapidly metabolized than is vitamin D, so that it both is more potent than vitamin D on a weight basis and has a shorter duration of action. The dosage therefore can be more readily adjusted, with less possibility of cumulative effect and prolonged hypercalcemia. 25-OH vitamin D can also be used in treatment since it is an effective agent in lower doses than the precursor vitamin D and its residence time in the body is much shorter than that of vitamin D, avoiding the cumulative toxicity of the latter. The most potent preparation, of course, is 1,25-diOH vitamin D. Both 25-OH vitamin D and 1,25-diOH vitamin D are still investigational drugs not yet available for routine treatment; 1,25-diOH vitamin D has been approved and released for treatment only of patients with renal insufficiency. Dihydrotachysterol or vitamin D, therefore, must be the basis of treatment at this time. The details of dosage are discussed in the section on treatment of idiopathic hypoparathyroidism.

CASE REPORTS

Congenital Hypoparathyroidism Associated with Congenital Heart Defect

L. H. was born at term following an uneventful pregnancy. At 6 weeks of age he had "pneumonia" and congestive heart failure which responded to antibiotics and digoxin. At 10 months he had a convulsion lasting ten minutes following an injection of penicillin for a febrile illness. Serum calcium determination was not made. At 18 months cyanosis and clubbing were evident, and at age 5 years cardiac catheterization showed a truncus arteriosus, type IV, with the right pulmonary arteries and one left pulmonary artery coming off the descending aorta. During all this time no signs or symptoms of tetany were present and there were no repeat convulsions. At age 13 during hospitalization for repeat cardiac studies the serum calcium and phosphorus concentrations were found to be 8.0 and 9.1 mg/dl respectively. These findings were not evaluated and further studies were not done. It was noted that he was mentally slow and school performance was poor. At age 16 he suddenly had a tonic clonic convulsion lasting several minutes with postictal somnolence. He was again hospitalized and at this time the serum calcium concentration was 6.4 mg/dl and serum phosphorus 9.4 mg/dl. On examination Chvostek and Trousseau signs were not obtained. There were no lenticular opacities. The EEG was abnormal, slow and disorganized with frequent seizure discharges. There was no consistent focus of origin. An Ellsworth-Howard test was done and Table 3–9 shows the findings (this is a test primarily of the phosphaturic action of PTH following the intravenous infusion of 200 units of PTE).

Because of the transient phosphaturia following intravenous parathyroid extract, a more rigorous PTH response test was given, namely, 200 units PTE intramuscularly q six hours for three days, with the results shown in Table 3–10.

This test showed sensitivity to parathyroid hormone, with induced hypercalcemia as well as phosphaturia and decrease of serum phosphate concentration. Rapid decrease of serum calcium concentration followed cessation of PTE injection. Pseudohypoparathyroidism was eliminated by this response to PTH. Treatment with dihydrotachysterol was instituted and normocalcemia has been maintained with a dose of 1 mg daily.

COMMENTS: This case is probably one of congenital hypoplasia of the parathyroids associated with maldevelopment of the great vessels of the heart. The earliest documented hyperphosphatemia and hypocalcemia were at age 13 years. However, the convulsion at age 10 months might have been due to hypocalcemia at this time. It is presumed that the parathyroids were hypoplastic and either failed to grow or became atrophic as the result of overstimulation, with inability to respond in the normal fashion by hyperplasia. There were no evidences of associated thymic aplasia, T cell deficiency or autoimmune disorder as a cause of the hypoparathyroidism. Of further interest is the response to PTE in the studies done to rule out pseudohypoparathyroidism. The Ellsworth-Howard test resulted in a transient phosphaturic response. When PTE was given intramuscularly, 200 units q six hours, a more dramatic response was obtained with rise of serum calcium and decrease of serum phosphorus concentration. The serum calcium concentration was not monitored daily and the PTE treatment was continued for three days, by which time hypercalcemia had developed. This regressed rapidly when PTE was discontinued, and no major toxic effects occurred. It is desirable with this test for responsiveness to PTH to monitor the serum calcium concentration daily, and usually only a 48-hour period of treatment is required to demonstrate a rise of serum calcium concentration.

J. C. This was an adopted female child who had a birth weight of 6 lb, 6 oz. Pregnancy history was unavailable. A diagnosis of cyanotic heart disease was made at age 2½ months. A Waterston shunt was done at age 3 months with a preoperative diagnosis of pulmonic atresia and ventricular septal defect. Immediately following surgery a convulsion occurred and the serum calcium concentration was found to be 4.5 mg/dl. She was treated with oral calcium supplements and dihydrotachysterol, but the steroid treatment was discontinued and serum calcium concentration remained in the normal range on syrup of

Table 3–9. Patient L. H.: Phosphaturic Response to PTE

Time	Serum		Urine		
	Ca	*P*	*Volume*	*P*	
	mg/dl		*ml*	*mg.*	*mg/hour*
10:00–12:00	5.8	–	210	63	31.5
12:00–13:00	PTE 200 U I V				
12:00–13:00			400	116	116
13:00–15:00	6.4	8.4	70	22	11

Table 3–10. Patient L. H.: Calcemic Response to Course of PTE

DAY	SERUM Ca	SERUM P	URINE Volume	URINE Ca	URINE P
	mg/dl		ml/24 hr	mg/24 hr	
0	6.7	8.5	800	3.2	736
*1			1600	6.4	1440
*2			1200	13.2	1490
*3	13.1	6.4	1000	18.0	1560
5	9.6	4.7			
6	7.5	–			
7	6.4	–			

*200 U PTE I.M. q 6 hrs.

Neo-Calglucon, 8 ml four times daily (approximately 700 mg calcium per day). When the calcium supplement was discontinued serum calcium decreased, but returned to normal values with resumption of treatment. It was felt that this child had partial hypoparathyroidism with presumed hypoplasia of the parathyroids, but enough residual function to maintain normocalcemia when a moderate excess of calcium was added to her daily diet. T cell function was found to be normal, ruling out thymic aplasia. She responded to PTH treatment with a sharp rise in serum calcium concentration. At age 6 years a definitive correction of the congenital heart defect was scheduled; it was thought preferable to maintain the serum calcium with dihydrotachysterol rather than rely on oral calcium supplement, since on the latter regime serum calcium fluctuated considerably and dropped rapidly when a few doses of calcium gluconate were omitted. Dihydrotachysterol, 0.1 mg/day, without calcium gluconate supplementation maintained the serum calcium concentration at 9.0 mg/dl and serum phosphorus at 5.9 mg/dl. Unfortunately she did not tolerate the surgical procedure for total correction of the complex cardiac abnormalities, and died postoperatively.

COMMENTS: This is another example of partial hypoparathyroidism in a child with congenital heart defect. It seems most likely that an abnormality of development of the parathyroid glands is the basis for this dysfunction, with hypoplasia and inadequate secretory capacity. In the absence of stress the serum calcium could be maintained by oral calcium supplement alone, although marked fluctuations occurred between 7.5 and 9.0 mg/dl. Under the stress of operation at 3 months, however, the serum calcium concentration dropped and a convulsion resulted. For this reason, before the final operative procedure, regulation of serum calcium concentrations with dihydrotachysterol was decided on to prevent this complication. Dihydrotachysterol acts not only by increasing calcium absorption from the gut but also by stimulating solubilization of bone mineral through an effect on bone cell metabolism. This provides a more stable concentration of calcium in serum during fasting periods but still does not substitute for endogenous parathyroid hormone, which can increase markedly in concentration under the stimulus of hypocalcemia.

Congenital Hypoparathyroidism as an Isolated Anomaly

J. G. was born at term in a community hospital; in the newborn nursery there had been symptoms of hypocalcemia for which he was treated with supplementary calcium lactate. Hypocalcemia persisted, however, and a diagnosis of congenital hypoparathyroidism was made for which he was given 50,000 units of vitamin D per day in the form of Drisdol. After one year of treatment the serum calcium concentrations were found to be in the range of 13 to 15 mg/dl and vitamin D dosage was reduced to 5000 units per day. Despite this reduced dose hypercalcemia persisted and vitamin D was discontinued. Serum urea nitrogen concentrations were elevated at this time. An IVP and cystourethrogram were unremarkable. A diagnosis of recurrent urinary tract infection was considered because of periodic low grade fever and proteinuria with increased numbers of WBC in urine; he was treated with Gantrisin and Mandelamine. However, urine cultures were not positive and the urinary abnormalities were probably the result of renal calcinosis and scarring secondary to hypercalcemia. Over the next several years renal injury progressed. Hypocalcemia again necessitated treatment with dihydrotachysterol. The patient was finally transferred to the Baltimore City Hospital at age 5 years because of end-stage renal disease with severe hypertension which had its onset at age 4. The renal insufficiency required repeated hemodialysis, but despite dialysis he succumbed to an episode of pulmonary edema. No post mortem was obtained.

COMMENTS: This infant with prolonged hypocalcemia in the neonatal period was unresponsive to usual treatment with adequate amounts of calcium lactate dissolved in milk feedings. The diagnosis of transient physiologic hypoparathyroidism was changed to congenital aplasia or hypoplasia of the parathyroids as an isolated developmental abnormality without thymic aplasia or defect of development of the great vessels. Unfortunately, treatment with vitamin D resulted in hypervitaminosis D with rapidly progressive renal injury. Doses of vitamin D of greater than 2000 U/kg are rarely necessary in the treatment of hypoparathyroidism in childhood, and this child received a considerably higher dose. Because of the slow rate of metabolic breakdown of vitamin D it accumulates in the adipose tissue and plasma, and once toxic levels are developed many months may elapse before the vitamin D is eliminated. For this reason careful repeated monitoring of serum calcium concentrations is needed for proper treatment of infants and children with hypoparathyroidism, and serum calcium concentrations preferably should be maintained in the range of 9 to 10 mg/dl. The rapidly progressive renal disease in this child represents a severe form of renal calcinosis and fibrosis. This renal lesion may be mistaken for chronic pyelonephritis as in this case.

P.B. was referred at the age of 17 months because of obesity and mental retardation. Prolonged jaundice had been present during the first month of life. He had had constant diarrhea but despite this had become obese, weighing 16 kg at 17 months. He seemed unresponsive to his environment and could not support himself in a sitting position. Respirations were noisy, suggesting laryngospasm. Slight twitching movements of eyelids were noted. Tetany was not initially suspected, but serum calcium concentration was 7.4 mg/dl and serum phosphate 12.4 mg/dl. A positive Chvostek was then found. There were no lenticular opacities. Dental radiography showed large pulp cavities and defective dentine and enamel mineralization. Long bone x-rays were reported as showing minimal osteopenia. The EEG showed a pattern compatible with generalized seizure disorder. Following admission the patient had severe diarrhea with dehydration and acidosis. When he recovered from this episode the serum calcium concentration had decreased to 4.1 mg/dl and there was severe tetany with laryngospasm and carpopedal spasm, which was controlled by intravenous calcium gluconate. He was given parathyroid extract, 100 units per day for five days, with increase of serum calcium concentration from 5.3 to 7.1 mg/dl and decrease of serum phosphate from 9.1 to 7.6 mg/dl. Treatment with vitamin D was begun. He was given 200,000 units per day (12,500 U/kg) for nine days, at the end of which time serum calcium concentration had increased to 9.0 mg/dl. The vitamin D dosage was reduced to a maintenance dose of 50,000 units per day (3000 U/kg). This dose maintained relatively normal concentrations of serum calcium and phosphate with complete relief of tetany. The diarrhea also ceased following correction of hypocalcemia. The EEG following five months of treatment showed a normal pattern. Improvement of mental function also occurred and at 3 years of age the patient walked and talked, although I.Q. was only 78.

COMMENTS: This is an example of congenital hypoparathyroidism in which overt convulsions were not present and severe cortical dysfunction was the major neurologic manifestation. The dental changes are evidence of hypocalcemia during early infancy. The chronic diarrhea exhibited by this patient is of interest. There was no evidence of autoimmune disease and the complete recovery with vitamin D treatment suggests that the intestinal hypermotility was a manifestation of hypocalcemia. The degree of recovery of cortical function following restoration of normal extracellular calcium concentrations was impressive. The infant had evidently been in constant electrical status epilepticus, as evidenced by the EEG, and this had been the basis of the cortical dysfunction. Because of mental retardation, obesity and short stature, pseudohypoparathyroidism was considered. He did not have the characteristic brachydactyly and the response to PTE injection was thought to rule out this diagnosis.

D.M. was an apparently normal full term newborn, weight 7 lb. 8 oz. He had been discharged from the newborn unit at 4 days of age although noted to be feeding poorly and to have jerking movements of the extremities. At 5 days of age cyanotic episodes occurred during feedings and clonic movements of the extremities were seen. At 6 days he was admitted to a hospital where he was observed to be cyanotic, with a hoarse voice and a positive Chvostek sign. Serum calcium concentration was 5 mg/dl. He was then transferred to the Johns Hopkins Hospital. Cyanosis, hoarse weak cry and intermittent tremors of the extremities were present, and Chvostek and peroneal signs. Serum calcium concentration was 5.7 mg/dl, serum phosphate 8.8 mg/dl and serum magnesium 0.96 mg/dl. Shortly after admission there was a brief apneic episode. Five ml of 10 per cent calcium gluconate was given intravenously, with improvement. He was then placed on a low phosphate feeding with added calcium salt to provide a Ca:P ratio of 4. He seemed clinically improved, although at discharge at 13 days of age serum calcium concentration was only 7.3 mg/dl with a serum phosphate of 7.3 mg/dl.

Following discharge the calcium salt supplement was tapered and discontinued and a low phosphate feeding was continued. At 3 months of age the patient was considered to be doing well and serum calcium concentration was 8.8 mg/dl. At 6 months he was admitted elsewhere because of fever and bulging fontanelle. At 7 months his feedings were changed to whole cow's milk. When 8 months old he had three convulsive seizures with fever. The seizures were treated with phenobarbital and he was referred for readmission when 8½ months old. He was at the third percentile for height and had nonpitting edema of hands and feet. He was lethargic, fed poorly and development was delayed. He could not roll over nor sit unsupported. Trousseau and peroneal signs were positive but a Chvostek was not elicited. Slit lamp examination showed fine zonular opacities in both lenses. Dental radiography revealed hypomineralization of all deciduous teeth. Following admission he had a right-sided seizure. At that time serum calcium concentration was 4 mg/dl. Parathyroid extract, 1 to 2 ml day, was injected intramuscularly for eight days. Serum calcium concentration rose to 12 mg/dl and phosphate dropped from 7 to 4 mg/dl. The infant was placed on vitamin D, 25,000 U/day, which was then decreased because of hypercalcemia. At 11 months of age he was crawling, rolling over and socially responsive.

COMMENTS: This was an infant with congenital hypoplasia of the parathyroids who presented in the neonatal period with manifestations compatible with transient physiologic hypoparathyroidism of the newborn. His response to calcium supplement treatment was partial, but serum phosphate concentration was not reduced below 7.3 mg/dl and serum calcium concentration increased only slightly to 7.3 mg/dl. This poor response was suggestive of more complete parathyroid insufficiency than usually found in transient physiologic hypoparathyroidism. This was borne out by the subsequent course with poor motor development, zonular opacities of the lens, impaired dental mineralization and hypocalcemic convulsions. The response to PTH treatment ruled out pseudohypoparathyroidism, which had been considered because of the patient's short pudgy appearance and apparent mental retardation. The moderate nonpitting edema of hands and feet has been seen in infants with hypocalcemia. The basis for the accumulation of fluid in the subcutaneous tissue is unknown. This child also showed rapid improvement in his developmental milestones once the hypocalcemia was corrected.

CONGENITAL PSEUDOIDIOPATHIC HYPOPARATHYROIDISM

This term has been used to describe a young adult with hypoparathyroid manifestations (hypocalcemia, hyperphosphatemia, tetany and cataracts), normal to high concentrations of IPTH in serum, and responsiveness to exogenous parathyroid hormone as measured by increase in urine excretion of cyclic AMP and phosphate. It was postulated that an abnormal PTH was secreted which was biologically inactive, although immunologically sufficiently similar to normal PTH to be recognized by the antibodies used in the radioimmunoassay for PTH. This patient showed no clinical features that might differentiate his condition from idiopathic hypoparathyroidism. The diagnosis depends on the determination of serum IPTH, which has not been done in most of the cases diagnosed as idiopathic hypoparathyroidism in the past, so that we cannot say how many such cases have not been properly diagnosed. The treatment and response to treatment do not differ, so that there are no practical consequences of this possible difference in diagnosis. In the particular patient reported, a 20 year old male, 100,000 units of vitamin D daily brought the serum calcium and phosphorus concentrations into the normal range.

ACQUIRED HYPOPARATHYROIDISM

IDIOPATHIC HYPOPARATHYROIDISM

We reserve this diagnosis for hypoparathyroidism as a solitary endocrine lesion without associated endocrinopathies, as contrasted to autoimmune hypoparathyroidism in which multiple abnormalities may be present. It is possible that idiopathic hypoparathyroidism as so defined may really be a form of congenital hypoplasia of the parathyroids, with onset of recognizable symptoms of hypocalcemia in later childhood. The impairment of enamel development sometimes seen in these children indicates hypocalcemia during the period of mineralization of the permanent dentition. Another clue to pre-existing hypocalcemia prior to evident tetany or convulsions is cataract formation, which is a complication of prolonged hypocalcemia. Additional evidence of unrecognized hypocalcemia is the appearance of ectodermal changes involving skin and nails, particularly the latter, which become thickened with longitudinal ridges. Somatic growth and physical development can be normal. Although most cases of idiopathic hypoparathyroidism are sporadic, familial incidence does occur with both dominant and recessive modes of inheritance.

A reported example of congenital parathyroid insufficiency remaining undiagnosed until late childhood concerned a 10 year old girl who was seen primarily because of impaired vision starting at 9 years of age resulting from bilateral cataracts. This child had had several convulsions at 6 months, 8½ years and 10 years. Abnormalities of nails and teeth were noted at 2 years. She was physically normally developed with severe malformations of finger- and toenails, enamel hypoplasia with caries and malocclusion, and bilateral cortical cataracts. On neurologic examination a positive Chvostek was found but Trousseau and peroneal signs were negative. Serum calcium concentration was 4.7 mg/dl and phosphate 12.3 mg/dl. Skeletal radiographs were normal but dental radiographs showed defects of enamel and zones of hypocalcification in the incisors and six-year molars. The calcification of these permanent teeth begins in the first months after birth, so that it is possible to date the onset of hypocalcemia at shortly after birth. The convulsion at 6 months, therefore, was probably the result of hypocalcemia despite the fact that a further convulsion did not occur until she was 8½ years of age. The abnormalities of finger- and toenails were noted at 2 years of age, additional evidence of prolonged pre-existing hypocalcemia. This is, therefore, a presumed case of congenital hypoplasia of the parathyroids which was not diagnosed until the patient was 10 years of age and might have been considered an instance of acquired idiopathic hypoparathyroidism if it were not for the enamel and nail abnormalities. An unfortunate consequence of the delayed diagnosis was the development of the bilateral cataracts.

The usual presenting manifestation of hypocalcemia in childhood is a convulsive seizure. This may be precipitated by infection and fever and can be misdiagnosed as a febrile seizure unless serum calcium determinations are made. The peripheral signs of tetany are often difficult to obtain in infants and young children. The EEG in hypocalcemia shows gross abnormalities with seizure discharges, and cannot be differentiated from that of epilepsy with grand mal seizures. As one of the following case reports indicates, hypocalcemic seizures have been misdiagnosed as epilepsy. This is a serious error since such seizures are not well controlled by anticonvulsant drugs such as phenobarbital and phenytoin, and prolonged hypocalcemia with recurrent seizures leads to mental deterioration. Although this may be reversible, we have seen children with undiagnosed hypoparathyroidism with mental retardation persisting after specific treatment was given. It has already been pointed out that additional consequences of prolonged hypocalcemia are ectodermal defects of nails and cataracts. If hypocalcemia is recognized and treated early in the course of cataract formation, some regression of the cataracts may occur and visual acuity remains unimpaired. When the cataracts are advanced following prolonged hypocalcemia, lens removal is the only treatment.

Another CNS manifestation of hypoparathyroidism is increased intracranial pressure with papilledema resembling pseudotumor cerebri. The mechanism of this is obscure, but when associated with the general abnormalities of cortical function secondary to hypocalcemia it may lead to this mistaken diagnosis of brain tumor. For example, a 6 year old boy was reported who at 5 years of age showed changes of mood, irritability, anorexia and sleeplessness. Shortly thereafter he showed loss of sense of balance and began to drop things. When examined following two convulsive seizures he was exceedingly irritable, uncooperative and screaming. His gait was unsteady. Examination of the eyes showed hemorrhages in and around the papilla and papilledema of three diopters. Separation of the sutures was found on skull x-ray examination. Carotid arteriography was unrevealing. While blood pressure was being determined a positive Trousseau sign was noted followed by a tonic seizure. The Chvostek sign was positive and the ECG indicated a prolonged Q–T interval. Electroencephalopathy showed dysrhythmia, with many slow waves. The serum calcium concentration was 4.1 mg/dl and serum phosphate 10.4 mg/dl. A cisternal puncture produced clear fluid, initial pressure 335 mm H_2O. Cerebral spinal fluid protein was 19 mg/dl, there were no cells, and calcium was 4.2 mg/dl. Following six weeks of vitamin D treatment CSF pressure had decreased to 180 mm H_2O. In pseudotumor cerebri of other types there is evidence that absorption of CSF is impaired. Whether Ca^{++} ion plays any role in this process has not been determined.

An unusual finding on skull x-ray examination in patients with hypoparathyroidism is calcification of the basal ganglion region (Fig. 3–2). Histologically the calcium deposits are in the perivascular region. This lesion is more commonly seen in pseudohypoparathyroidism (*see below*) than in true hypoparathyroidism, but it can also occur in the latter.

Since the changes of affect and cortical function resulting from hypocalcemia are nonspecific, and the convulsions can be associated with fever and mistaken for febrile seizures, it is important to determine serum calcium in all children with seizures. The signs of peripheral motor nerve hyperexcitability, such as the Chvostek and Trousseau or spontaneous carpopedal spasm, often are not obtainable or are indeterminate. The EEG is not instructive and resembles that of patients with grand mal seizures. For this reason, patients with hypoparathyroidism on

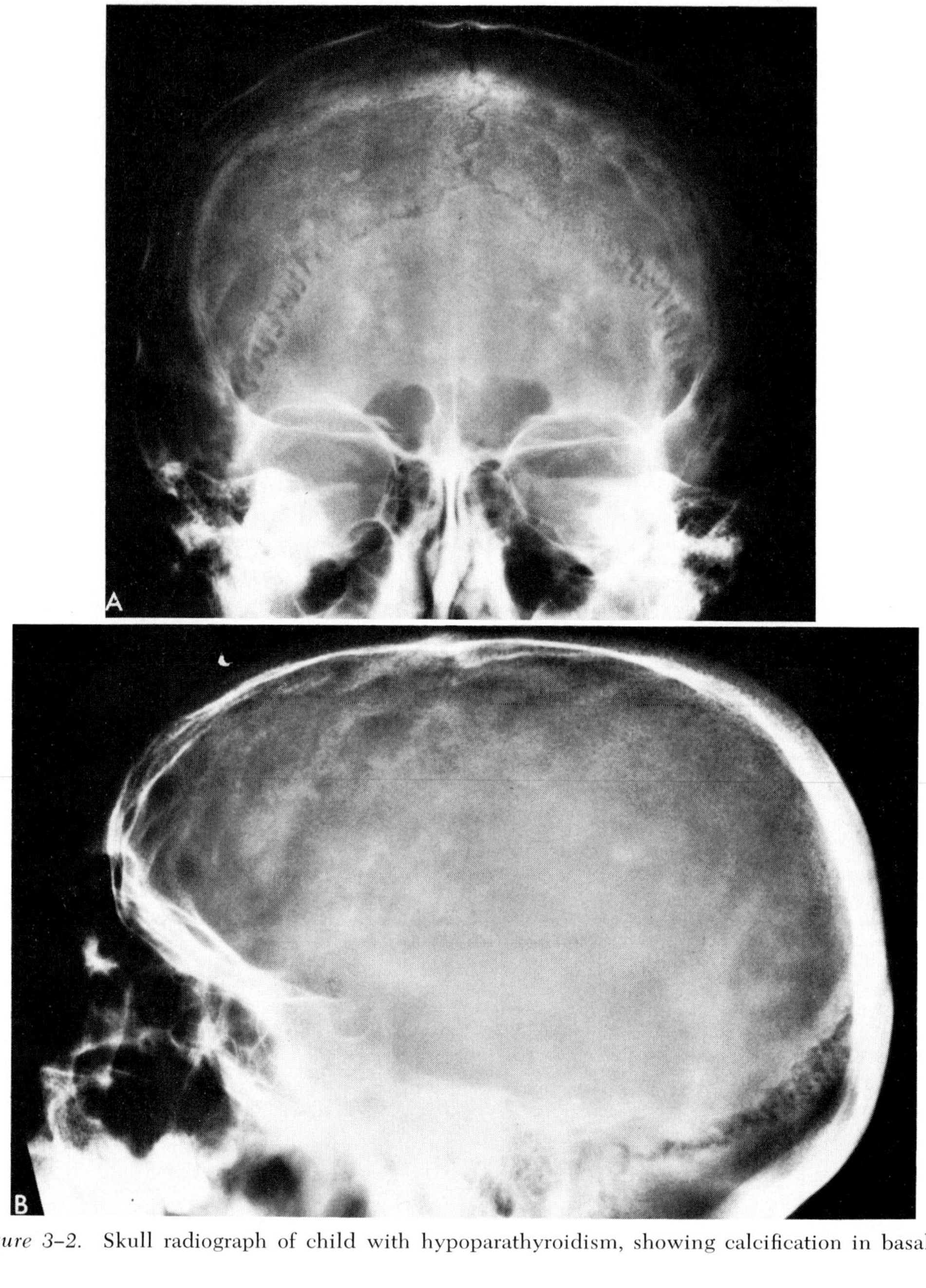

Figure 3–2. Skull radiograph of child with hypoparathyroidism, showing calcification in basal ganglia region.

occasion have been given anticonvulsant drugs with a diagnosis of epilepsy. The results of such treatment can be tragic since, as mentioned before, hypocalcemic seizures do not respond well to usual anticonvulsant drug treatment, and the persistent hypocalcemia and uncontrolled electrical seizure activity may result in mental deterioration which if prolonged may be irreversible. In addition, cataracts may occur. Admittedly, hypoparathyroidism is uncommon compared with other causes of seizures and many years may pass before this diagnosis is made even in a large seizure clinic. Nevertheless, the value of making this diagnosis is so great that it is worth doing thousands of serum calcium determinations to detect a single patient with hypoparathyroid hypocalcemia and seizures (see case reports).

GENETIC AUTOIMMUNE HYPOPARATHYROIDISM

This entity can usually be differentiated from congenital hypoplasia of the parathyroids or idiopathic hypoparathyroidism, which may be a variant of congenital hypoplasia because of the multiplicity of sites of autoimmune tissue injury in addition to loss of parathyroid gland function. Autoimmune hypoparathyroidism is often familial and the inheritance is that of an autosomal recessive gene. In addition to the parathyroid injury, the other endocrine glands most commonly affected are the adrenals and less often the thyroid. Antibodies to gastric parietal cells result in intrinsic factor deficiency and pernicious anemia. Severe intestinal malabsorption with steatorrhea is frequently seen in this disorder. Whether this represents mucosal cell injury is uncertain since hypocalcemia itself may cause impaired fat absorption. The steatorrhea in autoimmune hypoparathyroidism is more severe and more frequently observed than in idiopathic hypoparathyroidism, suggesting that intestinal mucosal cells may also be affected by the autoimmune process. A striking immunologic defect in these patients is lack of immune reaction to Candida, so that severe intractable moniliasis of nails, skin and mucous membranes of the gastrointestinal tract is an important and unfortunate manifestation of this type of hypoparathyroidism. There is no generalized T cell dysfunction, however, and the thymus is present. An additional ectodermal lesion also thought to result from autoimmune injury is lymphocytic infiltration and atrophy of the hair follicles, with resulting alopecia which may be total.

The hypoparathyroidism is one of the early manifestations of this process and almost invariably precedes the hypoadrenocorticism. In the absence of a family history of a previous sibling with this disorder the initial impression may be of idiopathic hypoparathyroidism. However, alopecia, steatorrhea and moniliasis, when present, indicate an abnormality of the immune mechanism. Lymphocytic thyroiditis and especially adrenocortical insufficiency may follow many years after the initial diagnosis of hypoparathyroidism. Because of the frequency of associated adrenocortical insufficiency repeated tests of adrenocortical function should be made. Serum cortisol concentrations determined in the early morning, when cortisol concentrations should be highest, can be used as a screening procedure. If concentrations of less than 5 μg/ml are found, the response to ACTH can be determined by measurement of serum cortisol concentrations after intramuscular injection of ACTH. If there is inadequate rise of serum cortisol concentration, early adrenal insufficiency is present.

The parathyroids evidently are particularly susceptible to autoimmune destruction in this familial disorder since hypoparathyroidism is often the earliest manifestation. Intestinal malabsorption, alopecia and moniliasis may also occur relatively early. Adrenocortical insufficiency, on the other hand, is ordinarily a much later phenomenon and may appear only many years after onset of parathyroid dysfunction. In some instances adrenal insufficieny does not supervene, and the autoimmune pathogenesis of the hypoparathyroidism is inferred on the basis of associated alopecia and moniliasis, and possibly also of pernicious anemia and intestinal malabsorption. The severity of such malabsorption and the resulting steatorrhea may be such as to suggest a primary intestinal injury such as gluten-induced enteropathy, with fat malabsorption and vitamin D deficiency as the basis for the hypocalcemia (see case report A. O'M.).

The diagnosis of hypoparathyroidism can be made more readily now in such a patient by measurement of serum IPTH and urine excretion of cyclic AMP. The hypocalcemia of vitamin D deficiency secondary to steatorrhea would be associated with secondary hyperparathyroidism, and in such patients serum IPTH concentration and urine cyclic AMP excretion are elevated. Hypoparathyroid hypocalcemia, on the other hand, would be accompanied by low concentration of serum IPTH in relation to serum calcium concentration and low-to-normal urinary excretion of cyclic AMP. Clinically, autoimmune hypoparathyroidism would be suggested if alopecia or moniliasis were present, since these lesions are not ordinarily seen in vitamin D deficiency hypocalcemia. Cataracts have rarely been seen associated with the hypocalcemia of vitamin D deficiency. Response to parathyroid extract might also separate these entities. In the hypocalcemia of primary fat malabsorption and vitamin D deficiency, relative unresponsiveness of the serum calcium concentration to PTE would be expected, whereas in hypoparathyroidism a marked rise of serum calcium concentration should result after injection of 200 units of PTE every eight hours for 48 hours. Finally, primary intestinal malabsorption with steatorrhea and secondary vitamin D deficiency should be associated with definite roentgenographic evidences of impaired bone mineralization with rickets or osteomalacia. In primary hypoparathyroidism with associated malabsorption, however, either bone structure is normal or there may be some degree of undermineralization, but not the marked changes seen in vitamin D-deficient states with intact hyperplastic parathyroids.

The autoimmune injury of gastric parietal cells with intrinsic factor deficiency may also appear years after the onset of hypoparathyroidism. The onset of macrocytic anemia in such a patient has been misinterpreted as being caused by folate deficiency secondary to chronic intestinal malabsorption and diarrhea *(see case report A. O'M.)*. Such errors should no longer be likely with our current capability of measuring the concentration of serum folate directly as well as serum cobalamin.

Autoimmune thyroiditis (lymphocytic thyroiditis or Hashimoto's thyroiditis) has also been observed in patients with this genetic disorder. In one patient chronic active hepatitis has occurred. Whether this is an associated autoimmune lesion or a coincidental finding remains undecided.

The advent of adrenal insufficiency in these patients must be anticipated and carefully looked for, since its onset is gradual and insidious, and unless anticipated may appear as circulatory collapse when such a patient is exposed to the stress of a severe infection or trauma. Another method of presentation may be the development of hypercalcemia in a patient who has been normocalcemic for a long time on a stable dose of vitamin D or dihydrotachysterol. The adrenal glucosteroids in physiologic concentrations counteract the effect of vitamin D or its metabolites, and withdrawal of adrenocorticoids results in an enhanced action of the vitamin D hormone with resulting hypercalcemia. Since vitamin D is cumulative in its action because of slow metabolic inactivation, hypercalcemia can also result from gradual accumulation of vitamin D in adipose tissue with excessively high concentration of the liver metabolite, 25-hydroxyvitamin D in plasma. Hypercalcemia, therefore, can occur without change in adrenocortical function. Nevertheless, in autoimmune hypoparathyroidism, the onset of hypercalcemia should be an indication for investigation of adrenocortical function by measurement of serum cortisol concentrations in the early morning when such concentrations are at their highest, as well as by determination of urinary excretion of 17-hydroxysteroids. Response of serum cortisol to ACTH stimulation would also be of importance in assessing adrenocortical function.

SURGICAL HYPOPARATHYROIDISM

Hypoparathyroidism following subtotal thyroidectomy for hyperthyroidism has been a rare occurrence in our own experience, although it has been a significant complication in some series of adult hyperthyroidism treated surgically. Even total thyroidectomy for thyroid carcinoma does not result in hypoparathyroidism if the parathyroid glands in the thyroid capsule are carefully preserved. However, in some instances the vascular supply to the parathyroid glands is compromised, causing their eventual atrophy. Such hypoparathyroidism is often partial, with hypocalcemia and overt tetany occurring only during periods of stress such as infection and fever with associated poor calcium intake.

The transient hypocalcemia commonly seen during the first day post thyroidectomy probably is not due to parathyroid dysfunction but is more likely caused by release of thyrocalcitonin during the manipulation of the thyroid. The serum calcium concentration may decrease to values between 7 and 8 mg/dl during the first six to 12 hours postoperatively, but hypocalcemia of a degree sufficient to produce tetany or convulsions is rarely seen. This transient hypocalcemia is corrected spontaneously within 24 to 48 hours if the parathyroids are intact.

Surgical hypoparathyroidism has also been produced by the intentional total parathyroidectomy for the treatment of hyperparathyroidism caused by hyperplasia of all of the parathyroids (see Chapter 4, *section on familial hyperparathyroidism*). In such instances hypocalcemia develops by 18 to 24 hours postoperatively and persists, requiring treatment.

ACQUIRED HYPOPARATHYROIDISM CAUSED BY IRON ACCUMULATION

A rare cause of hypoparathyroidism is parathyroid injury by iron pigment. Iron deposition in large excess occurs in patients with transfusion-dependent diseases such as thalassemia major and hypoplastic anemia. The endocrine glands including the parathyroids are sites of hemosiderin deposition, but parathyroid insufficiency is relatively uncommon. Injury to the islets of Langerhans is the most common endocrine disorder, and myocardial insufficiency resulting from iron accumulation in the heart muscle is probably the major complication.

CASE REPORT

An 18 year old white female was admitted because of congestive heart failure secondary to hemosiderosis and coincidental hypocalcemia and hyperphosphatemia. Following injection of furosemide for treatment of heart failure, tetany developed. As a young infant she was found to have congenital hypoplastic anemia diagnosed as Blackfan-Diamond syndrome. This failed to respond to steroid treatment, and from infancy on she required blood transfusions every six to eight weeks to maintain an adequate red blood cell volume. At the time of admission she was very pale and in respiratory distress. She was markedly undersized and sexually immature. Studies of growth hormone response to provocative stimuli showed no abnormality and serum thyroxine concentration was normal, so that pituitary and thyroid injury were not present. Adrenal corticoids were also normal, as was the glucose tolerance test, so that parathyroid insufficiency was the only demonstrable endocrinopathy. Response of serum calcium concentration to PTE injection was normal, ruling out target cell refractoriness, and the treatment with dihydrotachysterol in low dosage plus calcium carbonate was successful in preventing further hypocalcemia. The onset of tetany in this patient following furosemide injection may have been the result of the increased calcium excretion in the urine produced by this diuretic, which inhibits both renal tubular sodium and calcium reabsorption, causing further reduction of the already low extracellular calcium concentration.

Treatment of Hypoparathyroidism

The goal of treatment in hypoparathyroidism is the maintenance of serum calcium concentrations in a range that will prevent both the nervous system manifestations of hypocalcemia and the hypocalcemia-induced changes in ectodermal tissues (particularly cataract and alopecia or trophic changes of the skin and nails), but will not cause hypercalcemia and hypercalciuria with resulting renal injury. In addition, treatment must take into account the associated endocrinopathies such as adrenocortical insufficiency and hypothyroidism, and the complications of pernicious anemia and steatorrhea. The degree of hypoparathyroidism is an important consideration in this type of treatment.

Partial Hypoparathyroidism. TRANSIENT PHYSIOLOGIC HYPOPARATHYROIDISM OF NEWBORN. The term *partial hypoparathyroidism* connotes the continuance of some degree of parathyroid hormone output but one insufficient to meet the needs for calcium homeostasis, especially under conditions of high phosphate load and elevated serum phosphate levels. One form of this, seen in the newborn period, we have referred to as transient physiologic hypoparathyroidism of infancy. As described earlier, these infants develop hypocalcemic symptoms, particularly convulsive seizures, when given a high phosphate intake as in cow's milk feedings. The principle of treatment is the reduction of the serum phosphate concentrations by sequestration of phosphate in the intestinal lumen and prevention of its absorption, combined with increased absorption of calcium from the intestine. This can be accomplished simply by adding a calcium salt to the diet so that calcium is in great excess, resulting in precipitation of calcium phosphate in the gut. A Ca:P ratio of 4:1 by weight provides a sufficient excess of calcium so that phosphate absorption is reduced and serum inorganic phosphate concentrations are reduced. In the infant, calcium lactate and calcium gluconate are the most suitable calcium salts because they are soluble in milk and do not produce gastric or intestinal irritation nor any metabolic disturbance. Calcium chloride is undesirable since it is irritating to the gastric mucosa if present in too high concentration (above 3 per cent) and will produce a metabolic acidosis owing to the practically complete absorption of chloride with only partial absorption of calcium; the net effect, therefore, is that of absorption of HCl. Calcium makes up 13 per cent of the weight of the calcium lactate powder and 9 per cent of the weight of calcium gluconate, so that 7.7 gm of the former and 11.1 gm of the latter supply 1 gm of calcium. The results of such treatment have been documented in the case reports.

As was seen in infant C. Mc.C the serum calcium concentration can rise to values above the normal as the serum phosphate is decreased. This does not indicate recovery of parathyroid function, since the serum calcium concentration plummets again when the calcium supplement is withdrawn and the serum phosphorus concentration rises. The elevation of the serum calcium concentration above normal, secondary to excessive decrease of the serum phosphate, is probably explained by the fact that the calcium absorbed from the intestine is not removed by the bones at low concentrations of extracellular phosphate and remains in the extracellular fluid. This can easily be corrected by reduction of the extra calcium supplement so that the serum phosphate concentration rises sufficiently to prevent hypercalcemia. It is also of interest that the serum magnesium concentrations are inversely related to the serum phosphate. In the transient physiologic hypoparathyroidism of the infant, the treatment outlined may be needed for only a week or ten days, or as long as four to six weeks. The treatment regime must be based on repeated determinations of serum calcium and phosphorus concentrations.

Before institution of oral calcium salt supplement, emergency treatment of the hypocalcemic convulsion or other symptoms such as laryngospasm may be needed. The convulsions of hypocalcemia are relatively refractory to the usual anticonvulsant agents, phenobarbital or Dilantin. The elevation of the extracellular calcium ion concentration by intravenous injection of 10 per cent calcium gluconate solution is the specific treatment. This solution provides 9 mg of calcium per ml (100 mg calcium gluconate), and 5 to 10 ml are needed to raise the extracellular calcium in the fluids bathing the nervous system. The precautions required for the safe administration of calcium gluconate intravenously have been described in the section on neonatal hypocalcemia.

PARTIAL HYPOPARATHYROIDISM BEYOND INFANCY. Partial hypoparathyroidism may occur in older patients as the result of parathyroid injury or partial ablation during thyroidectomy, or may result from congenital hypoplasia of the parathyroids. The intravenous infusion of 20 to 30 ml of 10 per cent calcium gluconate when symptoms of hypocalcemia develop, or when serum calcium concentrations are found to be persistently below 7.5 mg/dl in the postthyroidectomy patient, can be followed by feeding of calcium salts as the first therapeutic approach. If sufficient calcium is to be given to the older child to reduce serum phosphate by sequestration of phosphate in the intestinal lumen, at least 4 to 5 gm of calcium must be given daily; this means over 30 gm of calcium lactate powder or 10 to 12 gm of calcium carbonate. The smaller bulk of calcium carbonate needed to provide an equivalent weight of calcium is useful in the patient who can take an insoluble powder as a suspension or in capsules.

According to Thomas et al, the feeding of 10 to 18 gm of calcium carbonate daily in four divided doses (at mealtimes and before bedtime) to untreated adult hypoparathyroid patients resulted in increases of serum calcium concentrations of 2.5 to 5 mg/dl. An interval of four to seven days elapsed before maximal concentrations of calcium were reached, and during the first few days of treatment symptoms of hypocalcemia were increased, possibly owing to the mild metabolic alkalosis resulting from the treatment. As the serum calcium rose, however, symptoms disappeared although the metabolic alkalosis persisted. It has been suggested that there may be some special therapeutic value of calcium carbonate by analogy with the milk alkali syndrome, in which hypercalcemia results from the combination of a high intake of calcium as milk along with sodium bicarbonate sufficient to produce an alkalosis.

Calcium chloride has also been successfully used in the treatment of posthyroidectomy hypoparathyroidism in adults and some adult patients with idiopathic hypoparathyroidism. One regime is to prescribe a 40 per cent solution of calcium chloride which is taken in amounts of 5 to 10 ml four times daily, well diluted in at least 100 ml of diluent per 5 ml of solution. The diluent can be fruit juice or other flavored beverage, since the calcium chloride solution has an unpleasant taste that is difficult to disguise. As mentioned previously, the solution of calcium chloride is irritating to the gastric mucosa unless well-diluted. Thirty ml of the 40 per cent calcium chloride solution would provide slightly over 4 gm of calcium per day. Calcium chloride in these amounts will produce a mild acidosis in most patients, and in those with a defect of urine acidification it will cause a pronounced metabolic acidosis. Calcium lactate does not alter the acid base equilibrium, and aside from the large quantity of powder needed to supply the 4 gm of calcium has no disadvantages or untoward side effects. Calcium lactate compressed into tablets may be poorly dissolved and absorbed and is less effective than calcium lactate powders, which are dissolved and taken in solution. If patients can be maintained free of tetany on treatment with calcium salts alone, no other treatment may be necessary even though the serum calcium remains below the normal value. Careful examination of the lens is necessary, however, to ensure that cataract does not develop. The presence of lenticular opacities would be an indication for more intensive treatment to maintain serum calcium concentrations between 8.5 and 10 mg/dl.

Treatment of Severe or Complete Hypoparathyroidism with Activated Sterols, Ergocalciferol or Dihydrotachysterol

Treatment with calcium salts alone does not suffice for patients with more complete parathyroid insufficiency, such as the infant with congenital absence of the parathyroid, or most patients with idiopathic or autoimmune hypoparathyroidism. Some patients with postsurgical hypoparathyroidism also fall into this category. Substitution therapy with parathyroid hormone fortunately is not necessary. The only parathyroid extracts available are relatively impure and quite dilute with respect to PTH. Although aparathyroid patients can be maintained normocalcemic for short periods with repeated intramuscular injections of PTE, the injections are painful and it is likely that refractoriness to the extract will eventually develop. Little is known about the long-term treatment with PTE since this has never been used as a standard treatment procedure.

In the 1930's, preparations of irradiated ergosterol became available which made it possible to give large doses of ergocalciferol (vitamin D_2) in this form, and it was quickly determined that the hypocalcemia of hypoparathyroidism was correctable by pharmacologic doses of vitamin D. The amount of vitamin D needed by the euparathyroid individual to prevent a deficiency state is of the order of a few micrograms per day, and the standard prophylactic dose of vitamin D is 10 μg (400 units) daily in infants and older children alike. The quantity of vitamin D required to raise the serum calcium of the hypoparathyroid subject is of the order of 0.025 to 0.05 mg (1000 to 2000 u)/kg/day.

One explanation for this huge requirement of vitamin D is the role of PTH in the formation of the active vitamin D hormone, 1,25-dihydroxyvitamin D. This theory is supported by the observation that physiologic doses of 1,25-dihydroxyvitamin D, 0.5 to 1 μg/day, will maintain normocalcemia in a hypoparathyroid subject. A deficiency of 1,25-dihydroxyvitamin D is, therefore, part of the pathophysiology of hypoparathyroidism. The effectiveness of high dose therapy with vitamin D in hypoparathyroidism is presumably through the development of high concentrations of 25-hydroxyvitamin D in the plasma, since 25-hydroxylation in the liver is not dependent on PTH. High concentrations of 25-hydroxyvitamin D apparently can substitute for 1,25-dihydroxyvitamin D with respect to enhancement of intestinal calcium absorption, as well as stimulation of bone cell metabolism and solubilization of bone mineral. High dose vitamin D treatment will maintain calcium homeostasis by increased absorption of calcium from the intestine and increased mobilization of calcium from bone, and by raising serum calcium concentration will also indirectly reduce tubular reabsorption of phosphate. This latter effect lags behind the effect on calcium absorption and mobilization, so that the serum calcium will be raised before the serum phosphate concentration begins to decrease to the normal range. In fact, the serum phosphate concentration may continue to remain somewhat elevated in these patients although the serum calcium is brought within normal limits by treatment.

Treatment with vitamin D must be individualized. In infants and children with congenital hypoparathyroidism or with acquired idiopathic hypoparathyroidism, an initial dose of 0.125 mg (5000 units)/kg/day until the serum calcium concentration begins to rise, followed by a maintenance dose of approximately 0.05 mg (2000 units)/kg/day, has been appropriate treatment. Krane managed most of his adult patients on maintenance doses between 0.625 and 5.0 mg per day, whereas Ireland et al found that the mean dose of ergocalciferol required in the treatment of surgical hypoparathyroidism was slightly above 2.0 mg/day with no patient controlled on less than 1.25 mg/day. In Ireland's series every patient receiving more than 2.5 mg/day became hypercalcemic at some time or other during treatment.

This observation emphasizes the problem: the dose of ergocalciferol must be sufficient to maintain the serum calcium in the asymptomatic range without causing hypercalcemia and hypercalciuria, with resultant renal injury. This requires close study of the patient with serial determinations of serum calcium and, if possible, urine calcium excretion. In children we prefer to maintain the serum calcium concentrations between 8.5 and 10 mg/dl, and urine calcium below 3 mg/kg/24 hours. A rise of serum calcium above 10.5 mg/dl calls for discontinuance of vitamin D treatment and resumption at a lower dose after the serum calcium has begun to drop. Because of the slow decay of vitamin D activity it has a cumulative effect, and a dose that has maintained a patient normocalcemic for several months may finally result in hypercalcemia. When the treatment is then discontinued the serum calcium concentrations may not fall for many months, giving the false impression that parathyroid activity has returned to normal. Eventually the serum calcium will drop, however, necessitating resumption of therapy at a lower dose than that which caused the excessive rise of serum calcium concentration.

The use of oral calcium salts in addition to vitamin D may not be necessary or desirable. We have preferred to treat children with vitmain D or dihydrotachysterol (*see below*) without supplemental calcium salts. This simplifies treatment, since it eliminates one variable and reduces the possibility of large variations in calcium intake caused by inconsistencies and neglect in the intake of calcium salts. In most patients who are responsive to activated sterol therapy, adequate regulation of serum calcium concentrations can be achieved without calcium salt supplementation. Some patients may do better with calcium lactate or gluconate supplement, 5 to 10 gm daily, and some who are relatively refractory to vitamin D treatment may have to be given large doses of calcium salts, as in the treatment of incomplete hypoparathyroidism, in addition to unusually large amounts of vitamin D.

The phenomenon of hypocalcemic hypercalciuria must also be taken into account. Hypoparathyroid patients on activated sterol therapy may excrete excessively large amounts of calcium, above 250 mg per 24 hours, even though the serum calcium concentration is below 9.0 mg/dl. This is explained by the action of parathyroid hormone in increasing tubular reabsorption of calcium. In the absence of PTH, hypercalciuria may result when serum calcium concentration is brought to normal by activated sterol treatment. This is of importance because: (1) it invalidates the urine Sulkowitch test for calcium as a guide to activated sterol dosages and makes serum calcium determinations necessary for proper supervision of treatment; and (2) these patients may develop renal calculi because of the excessive concentration of calcium in the urine. The administration of calcium salts along with vitamin D therapy may enhance the hypercalciuria, and in such patients the urinary calcium excretion should be measured and calcium supplementation discontinued to determine whether serum calcium concentrations can be maintained in the asymptomatic range with vitamin D treatment alone without massive hypercalciuria. Thiazide diuretics, which increase tubular reabsorption of calcium and reduce calciuria, have been suggested as a possible addition to treatment, but there is not sufficient experience to warrant a recommendation.

Dihydrotachysterol in the Treatment of Hypoparathyroidism

Soon after development of the process of irradiation of ergosterol for the production of vitamin D_2 in the laboratory, it was found that certain preparations of irradiated ergosterol had hypercalcemic effects greatly disproportionate to their antirachitic properties as assayed in the rachitic rat. This suggested that among the products of the irradiation of ergosterol might be compounds capable of elevating the serum calcium concentration in non-vitamin D-deficient subjects even though they were not effective as replacements for vitamin D. Holtz undertook the separation and identification of these compounds with the purpose of finding a hypercalcemic compound, "calcinosis" factor, that would be effective in the treatment of the hypocalcemia of hypoparathyroidism. Two compounds, toxisterol and tachysterol, were identified as having hypercalcemic properties. The latter was purified but was unstable. However, it can be reduced by hydrogenation to dihydrotachysterol, which is stable and has marked hypercalcemic effects in animals. Holtz named this preparation A.T. — 10 (antitetanische — 10) and studied its action in hypoparathyroid patients. His studies were encouraging and convinced him that dihydrotachysterol was superior to calciferol in the treatment of hypoparathyroidism on the basis of more rapid onset of action and possibly greater potency. He predicted in a review article published in 1941 that dihydrotachysterol would be the standard preparation for treatment of parathyroid tetany. This prediction has not come true as of this date and it is worthwhile exploring the reasons. The commercial preparation A.T. — 10 purporting to be Holtz' compound, dihydrotachysterol, was quickly tried by a number of investigators with conflicting results. In this country, A.T. — 10 was renamed Hytakerol without any change in its properties. Most investigators found unexplained inconsistencies in its action. Different batches showed different potencies in patient studies although the preparation supposedly was assayed in terms of its dihydrotachysterol content. Since Hytakerol was also much more costly than amounts of ergocalciferol which were equivalent in their ability to maintain the serum calcium concentrations of hypoparathyroid patients, it was generally stated that dihydrotachysterol (Hytakerol) was not indicated in the treatment of these patients except on a trial basis in the occasional one who appeared refractory to usual amounts of ergocalciferol.

The puzzle of the erratic response to Hytakerol and its relative lack of potency compared with Holtz' original preparation was solved by Terepka, Chen and Jorgensen. They found that Hytakerol was not a solution of a pure compound but contained a mixture of steroids of which only a small fraction was dihydrotachysterol. The compound present in greatest concentration was an isomer, dihydrovitamin D II, which is much less potent than dihydrotachysterol in its hypercalcemic properties. Pure dihydrotachysterol was available in crystalline form from several sources and this compound was then tested in hypoparathyroid patients. It was found to be much more effective than Hytakerol when compared in terms of the supposed content of dihydrotachysterol in the latter preparation. Since that time the manufacturers of Hytakerol have conceded that the original assay of their preparation was incorrect and that the Hytakerol solution contained only about one-fifth the effective concentration of dihydrotachysterol it was stated to possess.

The studies of crystalline dihydrotachysterol have indicated that it has certain superiorities over ergocalciferol in the treatment of hypoparathyroidism. Its calcium-raising effect is achieved more rapidly than with equivalent doses of vitamin D, but more importantly its action is much less cumulative than that of vitamin D because of a more rapid disappearance, either through inactivation or excretion. This provides a greater margin of safety. Ergocalciferol is very slowly inactivated, and it is not an uncommon clinical experience to have a patient become hypercalcemic on a constant dose of vitamin D which for several months was maintaining the serum calcium concentration in a suitable range. When the vitamin D is then discontinued months may elapse before the serum calcium decreases again, suggesting the persistence of its activity. Hypercalcemia due to overdose of dihydrotachysterol can be corrected within a week or two by cessation of treatment because of the rapid elimination or inactivation of the steroid, and treatment can quickly be resumed at a lower dose. Since both the onset of action and disappearance of effect are quicker, adjustment of dosage and avoidance of hypercalcermia are more readily attained.

In our experience dihydrotachysterol is also several times more potent than ergocalciferol on a weight basis (Fig. 3–3). This is in agreement with experimental studies in parathyroidectomized rats in which dihydrotachysterol is about three to five times as potent as ergocalciferol. The basis for this increased potency may be the stereochemical difference between this compound and

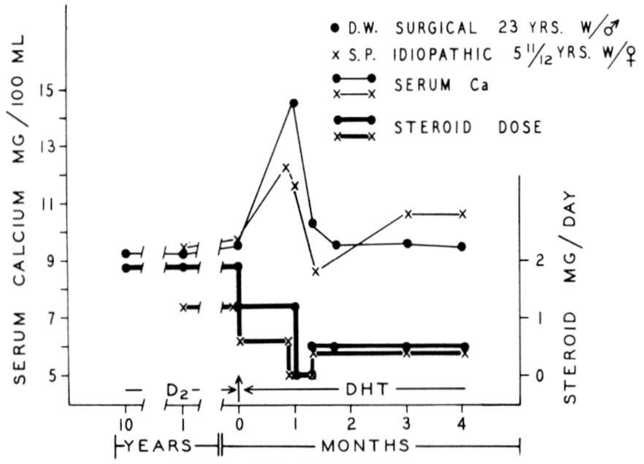

Figure 3–3. Comparison of calciferol and dihydrotachysterol with respect to calcemic effect in patients with hypoparathyroidism. (From Harrison, H. E., Lifshitz, F. and Blizzard, R. M. Reprinted by permission from the New England Journal of Medicine 276:895, 1967.)

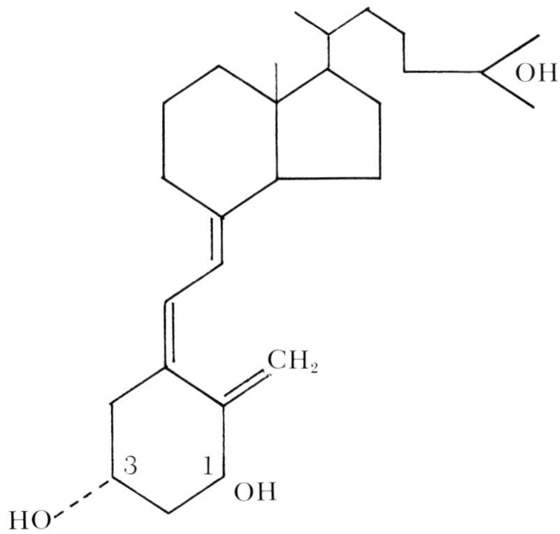

1,25-diOH cholecalciferol

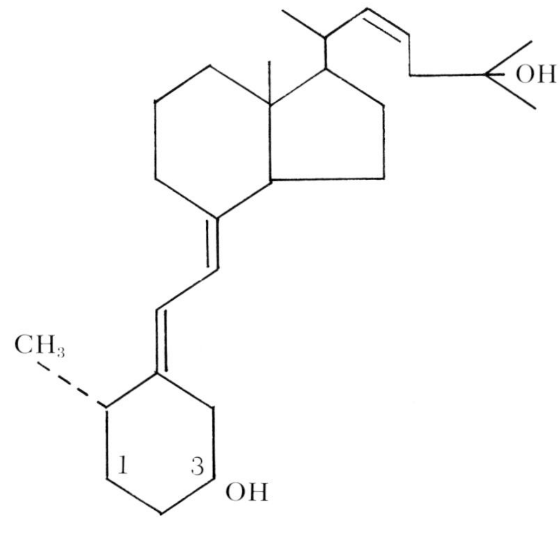

25-OH dihydrotachysterol$_2$

Figure 3–4. Comparison of ring structure of dihydrotachysterol with calciferol metabolite indicating the 180° rotation of the A ring of dihydrotachysterol, so that the OH group at the 3-carbon is in the pseudo 1-carbon position when viewed in relation to the configuration of the whole molecule.

vitamin D. The A ring of dihydrotachysterol is rotated 180° in comparison with that of vitamin D (Fig. 3–4). This puts the OH group on the 3 carbon in the geometric position of the 1-OH of 1,25-diOH vitamin D, so that dihydrotachysterol does not undergo 1-hydroxylation in the kidney. It is acted upon by the liver and the active compound is presumably 25-OH dihydrotachysterol. The greater potency is of no practical importance, however, in comparison with the ease of adjustment of the dosage and prevention of overdosage with prolonged hypercalcemia.

The pharmaceutical preparation Hytakerol (1 ml = 0.25 mg dihydrotachysterol) is still more expensive in terms of necessary dosage than preparations of ergocalciferol. Crystalline dihydrotachysterol, however, can be purchased relatively cheaply,* and can be prepared for the patient by making a solution in vegetable oil, either sesame or peanut oil. The solution that we use contains 1 mg crystalline dihydrotachysterol per ml of solution. The usual daily dose in infants has been 0.05 to 0.2 mg, and in older children and adults 0.2 to 1 mg. On a weight basis these doses are of the order of 0.01 to 0.02 mg/kg/day or about 20 per cent to 40 per cent of the dose of vitamin D. In patients who are refractory to activated sterol treatment we have employed as much as 5 mg per day and even larger amounts for short periods. Refractory patients show greater fluctuations in requirement and much more lability of serum calcium concentrations.

*Philips Roxane and Co., 100 E. 42nd Street, New York, New York.

Hypoparathyroidism Refractory to Ergocalciferol or Dihydrotachysterol

Although most patients with hypoparathyroidism can be managed as described with either ergocalciferol or dihydrotachysterol, occasional individuals show a refractoriness to ergocalciferol or dihydrotachysterol or both. This may take several forms. Some may respond initially to conventional doses of activated sterol and then become progressively decompensated despite increase of dosage to extraordinarily high levels. One subgroup of such patients described by Dent et al was refractory to ergocalciferol and responded more satisfactorily to another activated sterol, either cholecalciferol (vitamin D_3) or dihydrotachysterol. We have seen this type of specific refractoriness, but more commonly a more general resistance to sterols, so that patients unresponsive to ergocalciferol have also been refractory to cholecalciferol and to dihydrotachysterol.

The mechanism for failure of response to activated sterol is obscure and possibly there are multiple causes. In the case of the patients described by Dent the refractoriness to one of a group of molecules with vitamin D activity, and not to all, suggested some specific interference with physiologic activity, perhaps by development of an antibody to the sterol that resulted in inactivation. In the instance of the more general failure of response to all of the activated sterols, one possible cause is malabsorption. Steatorrhea is one of the recognized complications of autoimmune hypoparathyroidism and is aggravated by hypocalcemia. This leads to a vicious cycle of effects: steatorrhea with poor absorption of the activated sterol and persistent hypocalcemia with more marked steatorrhea. The cycle must be attacked by attempts at increase of serum calcium and reduction of the fecal fat. A diet low in fat, particularly the long chain fatty acids, combined with a high intake of calcium salts plus massive doses of ergocalciferol or preferably dihydrotachysterol (*see below*), must be employed until an effect is obtained. Fat can be given as medium chain triglycerides, which are more completely absorbed than ordinary fats. Severe hypocalcemia with recurrent episodes of tetany must be treated by repeated intravenous injections of calcium gluconate solution or sometimes with injections of parathyroid extract.

Magnesium deficiency has also been proposed as a factor in resistance to the serum calcium-raising action of vitamin D. Hypomagnesemia is a common concomitant of hypocalcemia in the patient with hypoparathyroidism, and the serum magnesium concentration rises along with the calcium concentration following injection of parathyroid extract. The serum magnesium also increases when the hypoparathyroidism is treated with vitamin D, but less rapidly, and seems inversely related to the serum phosphorus concentration. It may remain low in some patients even though the serum calcium has risen into the normal range. It has been suggested that some of the neuromuscular manifestations in hypoparathyroid patients that persist despite elevation of the serum calcium concentration to a level that should prevent tetany might well be the result of reduction of magnesium ion concentration. The possibility that poor response to activated sterol might be secondary to magnesium deficit has also been proposed, and an improved response has been reported following magnesium administration. We have given magnesium as magnesium chloride (in doses up to 4 mEq/kg) to hypoparathyroid patients who have had poor response to massive doses of activated sterols. In several patients increase in serum calcium concentrations with reduced calciferol requirement has resulted. In others, however, there has been no dramatic improvement. It would seem that a number of as yet undetermined factors influence the response to vitamin D and related sterols, and the treatment of vitamin D refractoriness must take into account the multiplicity of factors: (1) possible specific inactivation of one form of vitamin D; (2) steatorrhea and malabsorption of the fat soluble sterols; (3) hypomagnesemia; and (4) as yet undetermined causes.

Treatment with 1,25-diOH Vitamin D

When the role of parathyroid hormone in the metabolism of vitamin D was understood, the possibility was investigated that 1,25-dihydroxyvitamin D, the vitamin D hormone that requires PTH for its synthesis, might be the definitive treatment for hypoparathyroidism. Several workers have found that daily doses of 0.5 to 1 μg of 1,25-dihydroxyvitamin D are effective in raising the serum calcium of hypoparathyroid subjects. This drug is now available but not for the routine treatment of hypoparathyroidism. It was hoped that 1,25-dihydroxyvitamin D might be effective in those subjects resistant to other forms of vitamin D or dihydrotachysterol. However, there are now several known instances of patients refractory to ergocalciferol and dihydrotachysterol who also are unresponsive to amounts of 1,25 dihydroxycholecalciferol five times or more in excess of the usual dose. At least one of these patients did respond to doses of dihydrotachysterol of between 5 and 10 mg/day, whereas another responded to 1,25-diOH vitamin D in a dose of 8 μg per day. The problem of refractoriness to steroid treatment of hypoparathyroidism remains unsolved. It is seen more commonly in the autoimmune form of hypoparathyroidism, and may represent the development of antibodies to some portion of the vitamin D ring structure as a hapten, since this ring structure is common to all the steroids with vitamin D activity including 1,25-dihydroxyvitamin D.

Treatment of Associated Autoimmune Disorders

In young patients with hypoparathyroidism the possibility of adrenocortical insufficiency developing at a later date must be kept in mind. Treatment of such insufficiency with corticoids can complicate the management of hypoparathyroidism, since the serum calcium rises in adrenocortical insufficiency and is decreased by corticosteroid therapy. Patients with the combined disease are, therefore, more likely to show considerable fluctuations of serum calcium concentrations, requiring careful attention to regulation of the dose of activated sterols, whether ergocalciferol, dihydrotachysterol or 1,25-dihydroxyvitamin D.

Another of the disorders associated with idiopathic hypoparathyroidism is deficiency of intrinsic factor production by gastric mucosa, with consequent pernicious anemia. Since steatorrhea, as mentioned above, is also a known complication of autoimmune hypoparathyroidism, patients with hypocalcemia, steatorrhea and megaloblastic anemia on occasion have been mistakenly thought to have folate deficiency on the basis of intestinal malabsorption. The treatment of the pernicious anemia complicating hypoparathyroidism is of course vitamin B_{12}, as in any other form of intrinsic factor deficiency.

Intractable moniliasis with involvement of mucous membranes, skin and nails is one of the perplexing and trying complications of autoimmune hypoparathyroidism, and is more apt to occur in patients with associated adrenocortical insufficiency or pernicious anemia. Once the moniliasis is established, it is extremely difficult to eradicate even though serum calcium concentrations are maintained in the normal range. The usual local treatment with Mycostatin along with oral Mycostatin to reduce intestinal colonization with Monilia may reduce the lesions, but rarely eradicates the disease. Local hypochlorite treatment has been suggested on the basis of the observation that the moniliasis of skin and nails was improved in patients who spent a considerable amount of time in chlorinated swimming pools. Attempts have been made to treat this disorder with transfer factor prepared from lymphocytes of normal individuals immunized to Candida. It is not yet possible to evaluate this treatment, but it is worth considering in view of the extremely distressing nature of the cutaneous and mucous membrane lesions in severe cases.

CASE REPORTS

Idiopathic Hypoparathyroidism

L. B. This girl was admitted to the Baltimore City Hospital at 6½ years of age because of convulsions of one year's duration with progressive serious mental deterioration. At the age of 5 she had a febrile illness and was seen in another hospital. The previous medical history had been uneventful and motor development, speech and hearing abilities were normal. Following this initial seizure she had repeated seizures. Skull x-rays were negative and the EEG was reported as showing characteristic grand mal seizure discharges. The child was therefore treated with anticonvulsants, including phenobarbital and phenytoin. She continued to have convulsions, however, and began to show progressive mental deterioration. She had to be taken out of school, ceased to speak, became incontinent of feces and urine, and lay unresponsive with constant drooling. She was referred to us for possible degenerative brain disease.

On admission she was indifferent to her environment and drooling at the mouth. There were no localized neurologic signs. As part of the initial laboratory evaluation, serum calcium and phosphate determinations were made, and were 4.5 and 8.9 mg/dl respectively. This unexpected finding was confirmed on a repeat examination. Attempts to elicit specific manifestations of hypocalcemic tetany resulted in equivocal peroneal and Trousseau signs. The child was started on vitamin D_2, 100,000 units (2.5 mg) per day, and the serum calcium concentration rose progressively with gradual improvement in the mental status. The vitamin D_2 dose of about 5000 units per kg was not reduced when the serum calcium concentration began to rise, and after a year of treatment the serum calcium was found to be 14.7 mg/dl with a serum phosphate of 3.6 mg/dl. Vitamin D was discontinued and a short course of prednisone, 2 mg/kg/day, was given, following which the serum calcium was reduced below 10 mg/dl. She was then treated with vitamin D_2, 25,000 units (0.625 mg) per day. On this regime serum calcium remained about 9 mg dl.

Two years after initial diagnosis dihydrotachysterol was substituted for vitamin D and she has been well maintained on daily doses of 0.4 increasing to 1.2 mg dihydrotachysterol. There have been no recurrences of hypercalcemia. The child has had no further convulsions and her mental development progressively improved so that she returned to school and became a good student, graduating from high school with high standing. She is at present a junior at college and is successful in her studies. Physical growth and maturation have been normal. There are no other endocrine abnormalities, unless mild hypertension noted at 16 years of age is evidence of such. Renal function is normal. There are no cataracts or other ectodermal abnormalities.

COMMENTS: This is an example of the potential tragic effects of failure to consider hypocalcemia as a cause of repeated convulsive seizures precipitated initially by a febrile illness. The severe mental deterioration is characteristic of chronic hypocalcemia with persistent electrical cortical seizure activity. Fortunately in this case the mental deterioration was reversible. The diagnosis was idiopathic hypoparathyroidism without known associated abnormalities of immune mechanism or cardiac development. The possibility remains that this was due to congenital hypoplasia of parathyroids. During treatment with large doses of vitamin D, about 5000 u/kg, hypercalcemia developed which was reversed by discontinuance of vitamin D and the use of prednisone suppressive treatment. The patient was then well maintained on a smaller dose of vitamin D and finally on dihydrotachysterol. The current mild hypertension is controlled by hydrochlorothiazide. The possibility of renal vascular injury secondary to vitamin D hypercalcemia has been considered, but renal function tests show no abnormality and there is no calcification visible by x-ray. This is one of four children we have seen who showed deterioration of mental function while being treated for convulsions without diagnosis of hypocalcemia. In two instances they remained severely retarded despite treatment for hypoparathyroidism. The importance of doing serum calcium determinations on all children presenting with convulsions is emphasized, since the specific signs of hypocalcemic tetany may be lacking or difficult to elicit.

J. G., a white male, was age 15 years at the time of referral. He had been treated with phenobarbital and phenytoin for convulsive seizures for four years before the diagnosis of hypoparathyroidism was considered. During this period he had become dull and emotionally disturbed, and failed badly in school. A psychometric examination gave an I.Q. of 75. When admitted the serum calcium concentration was 7.3 mg/dl and phosphate concentration 12.8 mg/dl. Lenticular opacities were present and skull x-rays revealed calcification in the basal ganglion region. Treatment was begun with 100,000 units of vitamin D daily and the anticonvulsant drugs were discontinued. As the serum calcium concentration rose toward normal he became more alert and his school work improved. He graduated from high school and successfully completed a junior college course. A repeat psychometric study after five years of treatment gave an I.Q. of 99. Following four years of stable vitamin D dosage at 2.5 mg (100,000 units) per day the serum calcium concentration began to rise above the normal range. Vitamin D dosage was reduced and then discontinued completely. The serum calcium and phosphate concentrations remained in the normal range for approximately six months before hypocalcemia again developed, requiring resumption of vitamin D treatment. The lenticular opacities remained stationary and visual acuity has not been impaired.

COMMENTS: This is an example of idiopathic hypoparathyroidism without evidence of autoimmune disease. This patient showed mental deterioration during a four-year period of convulsions treated with a standard anticonvulsant drug regime. Lenticular opacities and intracranial calcification were present at the time the diagnosis of hypoparathyroidism was made. With vitamin D treatment he showed improved cortical function and the lens changes did not progress. Of unusual interest is the development of hypervitaminosis D after more than four years of continuous treatment at a stable dose of vitamin D, necessitating discontinuance of such treatment. The stores of vitamin D in the tissues were evidently sufficient to maintain normocalcemia for over six months without treatment. Such prolonged periods of storage of vitamin D following its over-use have been seen in other patients, indicating the prolonged biologic half-life of vitamin D itself when stored in adipose tissue.

Autoimmune Hypoparathyroidism

D. H. is a 13 year old white female. Two siblings have autoimmune hypoparathyroidism with associated adrenal insufficiency and moniliasis. One of these children died at age 5 of viral encephalitis.

The patient had pneumonia with pleural effusion at age 2 years. Six months later she was found to have mucocutaneous candidiasis; serum calcium concentration was 6.8 mg/dl and serum phosphate 10.2 mg/dl. Following six intramuscular injections of parathyroid extract of 100 U each the serum calcium was raised to 12.5 mg/dl and the serum phosphate decreased to 5.7 mg/dl. She was then placed on 50,000 units vitamin D_2 daily. The serum calcium concentration has fluctuated between 8 and 9 mg/dl. Permanent dentition shows enamel hypoplasia and severe caries. Mucocutaneous candidiasis has been a serious problem ever since its development at age 2½ years. Treatment with Mycostatin was unavailing. Skin hypersensitivity to Candida antigen is positive despite failure of the immune reaction to eliminate the disease.

At age 4 years, increased thirst, craving for salt and polyuria were noted. Urinary keto and 17-OH steroids were found to be normal at this time. At age 4 years, 10 months, however, the patient lapsed into shock following tonsillitis and a diagnosis of adrenal insufficiency was made. Since that time, she has been treated with cortisol, 15 mg daily, and Florinef, .075 mg per day. At age 5 years, 9 months an elevated bilirubin was first found. A year later she had obvious jaundice, vomiting and anorexia with evidences of elevated transaminases. Australia antigen was not present. Liver biopsy was reported as showing chronic active hepatitis with postnecrotic cirrhosis. Liver function tests have remained abnormal despite a course of prednisone, 60 mg/day tapering to 20 mg/day over a three-month period. Subsequently, azathioprine was tried without any major response. Since age 9 years, 5 months, the patient has lost her hair, and at age 13 has almost complete lack of scalp hair.

Immunologic studies have shown normal concentrations of immunoglobulins and a positive skin hypersensitivity to Candida antigen. Other tests such as NBT, opsonization, phagocytosis and leucocyte inhibition were normal. At age 13, the patient was somewhat cushingoid in appearance with alopecia and candidiasis of mouth and nails. She also had slight jaundice; the thyroid was not enlarged.

COMMENTS: This patient showed a number of the usual features of autoimmune hypoparathyroidism, namely, familial incidence compatible with autosomal recessive genetic transmission, early onset of hypoparathyroidism and mucocutaneous candidiasis as major presenting manifestations, with adrenal insufficiency developing some years later. The chronic active hepatitis is an unusual feature, however, and has not been reported in other cases. The alopecia is a well-recognized manifestation of this syndrome. This girl did not show steatorrhea, intrinsic factor deficiency with pernicious anemia or autoimmune thyroiditis (Hashimoto's), which have been reported in other patients. The mechanism of the failure of healing of the Monilia infection remains unclear, since the skin test for delayed sensitivity to Monilia antigen is positive, as has been noted in other patients. (We are indebted to Dr. Leo Stern, Rhode Island Hospital, for permission to include this case.)

A. O'M., a white female, was noted to have photophobia and blepharospasm in the first year of life. At 3 years of age chronic diarrhea with foamy foul stools developed, and the diagnosis of celiac disease was made. She failed to respond to dietary treatment. At 4 years of age she had a convulsive seizure followed by carpopedal spasm. At that time serum calcium and phosphate concentrations were 4.0 and 10.3 mg/dl respectively. The alkaline phosphatase was normal, 12 Bodansky Units. She was given dihydrotachysterol (A.T. – 10) and the serum calcium concentration increased to 11.1 mg/dl, with decrease of serum phosphate to 6.1 mg/dl. The photophobia improved and diarrhea subsided. There was poor compliance, however, and treatment lapsed with return of symptoms.

At the age of 5, hypocalcemia was severe. The diagnosis of hypoparathyroidism had not been made previously and the hypocalcemia had been attributed to malabsorption. The significance of the hyperphosphatemia had not been appreciated. When the child returned at age 5, hypoparathyroidism was diagnosed. At this time alopecia and abnormalities of fingernails were present. Skull films also showed symmetric calcification in the region of the caudate nuclei. Treatment with vitamin D, 175,000 units per day, was begun which corrected the hypocalcemia, but follow-up supervision was not adequate and hypercalcemia, serum calcium concentration 15.1 mg/dl, developed at 6 years of age. Treatment was discontinued and the child was not brought back for almost five years. During this period she had severe photophobia and blepharospasm, with development of keratitis and corneal opacities. Vision was markedly impaired. She also had recurrent generalized convulsions.

At 12 years of age she returned to the Endocrine Clinic at the Johns Hopkins Hospital. In addition to corneal opacities there were lenticular opacities in the anterior subcapsular region. Scalp hair was sparse. The teeth were poorly formed with enamel hypoplasia. Dental x-rays showed enlarged pulp chambers, poorly formed roots and poorly mineralized enamel. Extensive moniliasis of finger- and toenails and the mucous membranes of mouth and vagina were present. Serum calcium concentration was 5.4 mg/dl and that of phosphate 8.6 mg/dl. Parathyroid extract, 5 ml q six hours, was given intramuscularly for four doses, with increase of serum calcium concentration to 7.3 mg/dl and decrease of serum phosphate to 4.4 mg/dl. Urine phosphate excretion increased appropriately following PTE. Serum calcium concentration then decreased to 4.6 mg/dl, with rise of serum phosphate to 9.5 mg/dl. Treatment with vitamin D was begun. 400,000 units per day were given for 10 days with only slight increase of serum calcium to 6.6 mg/dl. Calcium lactate, 15 gm per day (2 gm calcium), was then added. Serum phosphate concentration decreased from 8.8 to 6.3 mg/dl during the next ten days and serum calcium increased to 9.9 mg/dl. The child was subsequently maintained on 100,000 units per day, with considerable fluctuations in serum calcium concentration requiring changes of dose. There were several episodes of hypercalcemia, at which time vitamin D was discontinued, sometimes for two to three months.

At age 12 years the child was also discovered to be anemic, with hemoglobin 8.2 gm/dl and R.B.C. 2.86 million, MCV 90μ^3. The anemia did not respond to iron supplementation. Folic acid, 10 mg per day, corrected the anemia and was continued. However, 2½ years later she began to stagger and was unable to walk, and neurologic examination revealed combined system disease of posterior and lateral columns of the spinal cord. A Schilling test revealed inability to absorb B_{12} unless intrinsic factor was given. Following intensive treatment with intramuscular vitamin B_{12}, there was marked improvement in the neurologic signs.

COMMENTS: This child with autoimmune hypoparathyroidism has five manifestations of autoimmune disease: hypoparathyroidism; alopecia; intractable moniliasis; steatorrhea; and intrinsic factor deficiency, probably due to autoimmune injury of gastric mucosa. The hypoparathyroidism was so early in onset that it suggested congenital hypoparathyroidism, but the other manifestations are more in accord with an autoimmune disorder. The severe steatorrhea was striking and initially confused the issue, since primary intestinal malabsorption of vitamin D and calcium was thought to be the diagnosis. The hyperphosphatemia should have been the diagnostic clue, since vitamin D and calcium malabsorption would result in secondary hyperparathyroidism with hypophosphatemia. The severe blepharospasm seen in this patient has been a rare manifestation of hypocalcemia. The importance of recognizing the various complications of autoimmune hypoparathyroidism is underscored by the development of intrinsic factor deficiency with vitamin B_{12} depletion. The patient was given folic acid in the mistaken notion that the macrocytic anemia was due to folate deficiency secondary to the intestinal malabsorption. This resulted in masking the anemia of vitamin B_{12} deficiency and the consequent development of the neurologic complications of pernicious anemia.

The difficulties of management of hypoparathyroidism complicated by steatorrhea are evident in this patient. Response to vitamin D has been irregular, with periods of hypocalcemia and hypercalcemia requiring frequent modifications of treatment. The concurrent administration of calcium lactate was apparently useful in the initial treatment since serum phosphate decreased and serum calcium rose rapidly when calcium was added to the vitamin D treatment.

M. H., a white male, had symptoms starting at 2½ years of age. At this time he had "shaking spells" and probably laryngospasm. Similar manifestations with pains in arms and legs recurred at age 5. When he was 11 years old a diagnosis of hypoparathyroidism, adrenal insufficiency and moniliasis was made. The family history was important in that a sibling who had cutaneous moniliasis had died suddenly at 5 years of age.

When admitted at age 11 the patient had total alopecia (Fig. 3–5), zonular cataracts by slit lamp examination and calcification of basal ganglion region by skull x-ray. Serum calcium concentration was reported as 8.9 mg/dl, and serum phosphate 8.2 mg/dl. An ACTH stimulation test showed no rise of 17 hydroxysteroid concentrations in plasma. Treatment with cortisone acetate and vitamin D was begun. Hypercalcemia developed, which required discontinuance of the vitamin D and resumption

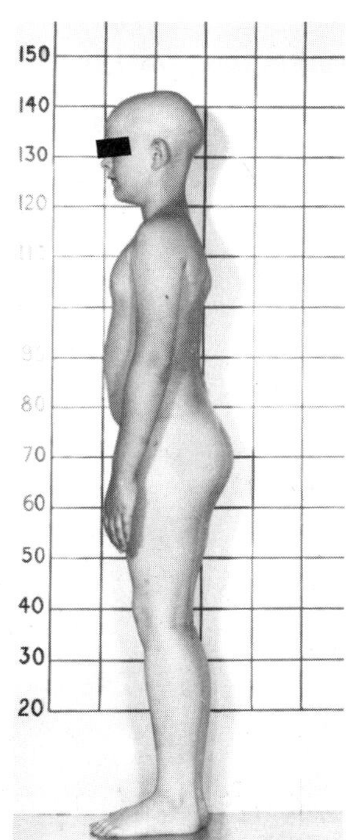

Figure 3–5. Total alopecia in a child with autoimmune hypoparathyroidism.

at a lower dose. At 14 years of age signs of tetany were present, and the boy admitted that he had not been taking the prescribed vitamin D. Serum calcium concentration was 5.7 mg/dl and serum phosphate 8.9 mg/dl. Serum magnesium concentration was also low, 0.85 mg/dl. He was again placed on vitamin D with rise of serum calcium concentration to 11.0 mg/dl and decrease of serum phosphate to 4.0 mg/dl. Serum magnesium remained low, 0.6 mg/dl.

COMMENTS: This child presented with the triad of hypoparathyroidism, adrenal insufficiency and moniliasis, and a family history of a sibling with moniliasis who died suddenly, probably with adrenal insufficiency. Although the serum calcium concentration was said to be 8.9 mg/dl this was probably a laboratory artifact, since serum phosphate was 8.2 mg/dl and there were evidences of hypocalcemia dating back to early childhood, i.e., symptoms of seizures and laryngospasm, dental maldevelopment and cataract formation. Subsequent serum calcium concentrations were low, as would be expected in hypoparathyroidism. In this autoimmune disease hypoparathyroidism almost invariably antedates the adrenal insufficiency, often by many years, and it would be highly unlikely that adrenal insufficiency was the first endocrinopathy with hypoparathyroidism following. The marked hypomagnesemia seen in this patient has been found in others with hypoparathyroidism. We had thought that reduction of serum magnesium concentration was secondary to hyperphosphatemia, and in this patient the serum magnesium increased following PTE injection, with decrease in serum phosphate. However, on vitamin D treatment serum magnesium remained low although serum calcium and phosphate concentrations were brought into the normal range. The child had no evidences of intestinal malabsorption, nor was he refractory to vitamin D treatment. It has been proposed that magnesium deficiency is one of the mechanisms for resistance of hypoparathyroidism to vitamin D, but this correlation was not present in this patient.

PSEUDOHYPOPARATHYROIDISM

Pseudohypoparathyroidism in the context of this discussion refers to all the entities in which parathyroid gland cells apparently secrete parathyroid hormone in normal or increased amounts but in which the target cells, bone and renal tubule cells, show inadequate response to the hormone, so that hypocalcemia and hyperphosphatemia result, with the symptoms of peripheral and central hyperexcitability of the nervous system characteristic of reduced calcium ion concentration. Such failure of response of the target cells can be due to inborn metabolic errors, as in pseudohypoparathyroidism types I and II, or to acquired metabolic disturbances that interfere with response to PTH, such as vitamin D deficiency, phosphate poisoning and possibly magnesium deficiency.

PSEUDOHYPOPARATHYROIDISM TYPE I

The physiologic abnormality in this disorder is recognized by a failure of the specific adenyl cyclase of renal tubule cells to respond to parathyroid hormone. The normal response is measured by a manifold increase in the excretion of cyclic AMP in the urine following the intravenous infusion of a standard dose of PTH (PTE, 200 units).

In addition to this physiologic marker these patients also have anatomic markers to distinguish them from those with true hypoparathyroidism. The characteristic subject is short in stature, has a round face and is brachycephalic (Fig. 3–6). Most striking is the brachydactyly (Fig. 3–7*a*). All the fingers are short and pudgy, but in addition there is a striking growth failure of the fourth and fifth metacarpals and metatarsals. As a result the fourth and fifth fingers and toes appear unusually short in comparison to the other digits. The shortness of the fourth and fifth metacarpals results in the absence of the knuckles of these digits when the hands are clenched (Fig. 3–7*b*). In addition to the physical abnormalities these patients are usually mentally retarded,

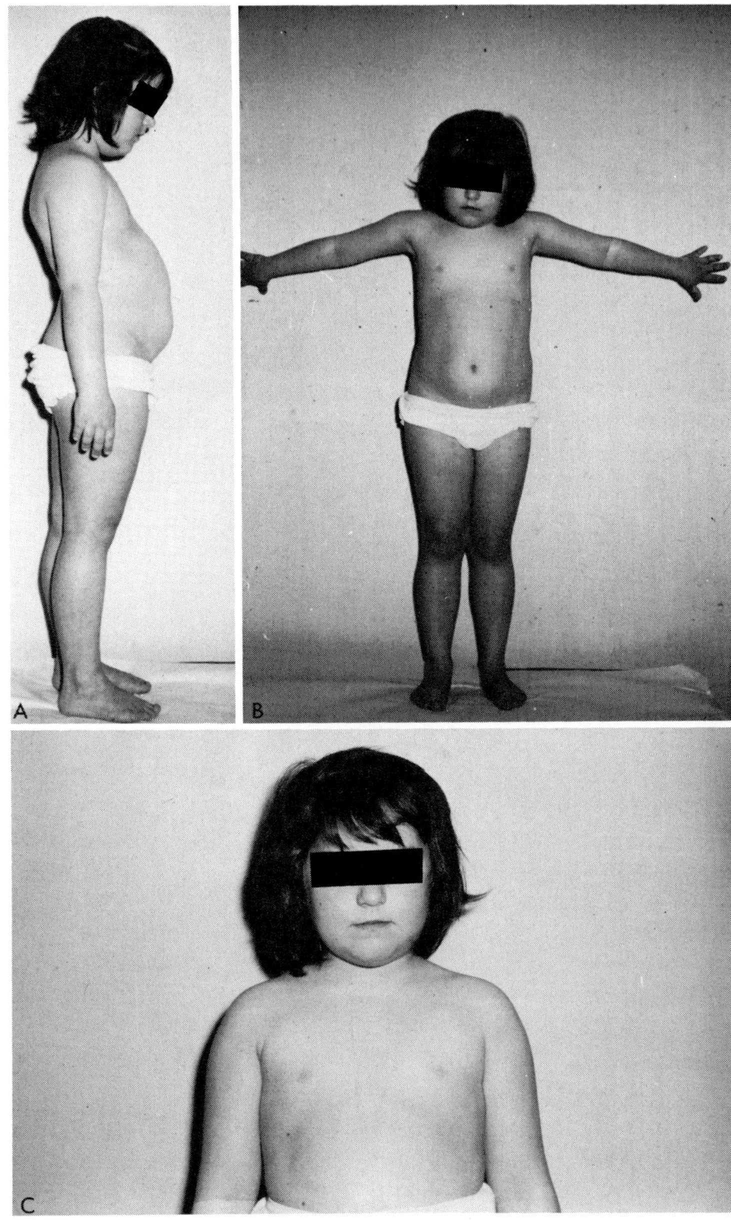

Figure 3–6. Pseudohypoparathyroidism, type I (Albright's hereditary osteodystrophy) illustrating short stocky stature, rounded face and brachycephaly.

some having been found in mental institutions. Since many of these subjects have a history of convulsive seizures starting in early infancy, the retardation may be due both to maldevelopment of brain cells and to acquired injury secondary to untreated hypocalcemia.

The serum findings are with one exception the same as those in hypoparathyroidism, a reduced concentration of calcium and an elevated concentration of phosphate in serum. The serum IPTH, which is very low or undetectable in patients with hypoparathyroidism, is, however, in the upper normal range or elevated in subjects with pseudohypoparathyroidism. The clinical manifestations in addition to the anatomic findings are those attributable to the reduced concentration of calcium ion in extracellular fluid: tetany and convulsions. In addition, however, these patients are more likely to deposit calcium phosphate in soft tissues, particularly in the periarticular tissues and in the perivascular spaces in the

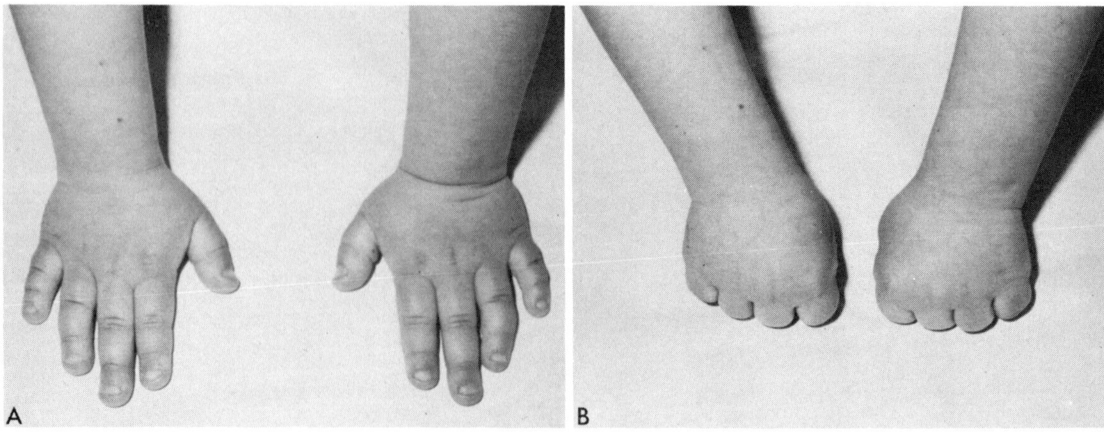

Figure 3–7. Hands of patient shown in Figure 3–6 showing brachydactyly with particular shortening of fourth and fifth metacarpals.

basal ganglion region of the brain. Although such ectopic calcific deposits are sometimes seen in true hypoparathyroidism, they are much more frequently found in pseudohypoparathyroidism type I.

Pseudohypoparathyroidism type I is genetically determined by an autosomal dominant gene with variable penetrance. Familial incidence, therefore, is often seen, but children with this disorder may also be found whose parents are apparently unaffected. In the same kindreds, however, there may be subjects with the anatomic abnormalities of short stature and brachydactyly but without hypocalcemia and hyperphosphatemia. In some instances hypocalcemia has been documented which later disappeared. It has been suggested, therefore, that the genetically determined abnormalities and the resistance to PTH action are linked, but that the latter may not be completely expressed. The entity of short stature, moon facies, brachydactyly and sometimes mental retardation, with normal serum calcium concentrations, has been termed *pseudo-pseudohypoparathyroidism*, a clumsy name but one established in the literature.

Diagnosis

The diagnosis of pseudohypoparathyroidism type I, therefore, depends on recognizing the combination of the specific anatomic features plus the finding of hypocalcemia and hyperphosphatemia. The response to PTH should be tested; the most specific test is the measurement of urine cyclic AMP excretion during a control period and then following the intravenous infusion of 2 ml (200 units) PTE (Lilly). The responses of the hypoparathyroid patient and of the subjects with pseudohypoparathyroidism type I are shown in Table 3–11. In addition to the rise in urine cyclic AMP the phosphaturic effect of PTH is also measured. If urine cyclic AMP determinations are not available the latter alone may be tested, since this phosphaturic action is absent in pseudohypoparathyroidism. Because of the variability of the phosphaturic action of various parathyroid extracts it may be necessary to study simultaneously a control subject to ensure that the particular extract gives a positive response in the normal. Another technique is to inject PTE intramuscularly in repeated doses of 2 ml (200 units) every eight hours over a period of 48 hours to measure the calcemic action of this hormone. This test has been positive in some patients with hypoparathyroidism who gave only an equivocal phosphaturic response to a single intravenous injection of 2 ml of PTE. Patients with pseudohypoparathyroidism are refractory to the calcemic action of PTE as well as to the renal effects. Following vitamin D treatment there may be greater calcemic response, but these patients still do not show the phosphaturic effect of PTE injection.

Treatment

The treatment of pseudohypoparathyroidism does not differ from that of hypoparathyroidism. Although patients with the former are refractory to ordinary amounts of PTH, their response to vitamin D, vitamin D metabolites or dihydrotachysterol does not differ from that of the hypoparathyroid subject. Vitamin D, 2000 units (0.05 mg) per kg, or dihydrotachysterol, 0.02 mg/kg, are given as daily maintenance doses. An initial loading treatment of vitamin D, 5000 units (0.125 mg)/kg/day or dihydrotachysterol, 0.05 mg/kg/day, can be given at the start of treatment, but this dose should be decreased to the maintenance dose as soon as serum calcium concentrations begin to increase ordinarily within seven to ten days following commencement of treatment. As in hypoparathyroidism, an attempt is made to bring the serum calcium concentration into the range of 8.5 to 9.5 mg/dl, and if it rises to 10 mg/dl or higher the dose of vitamin D or dihydrotachysterol is decreased. The serum phosphate concentrations also decrease with treatment, but this decrease lags behind the rise of serum calcium and some degree of hyperphosphatemia may persist. Although calcium salts have been given along with the activated sterol it is possible to treat these patients without supplementary calcium, and this is preferable in order to simplify treatment and improve compliance.

A few patients with pseudohypoparathyroidism have been treated with 1,25-dihydroxyvitamin D or 1-αhydroxyvitamin D, and have shown rise of serum calcium concentration with doses of only 1 to 2 μg per day. It is presumed that 1α-hydroxyvitamin D is metabolized by the liver to 1,25-dihydroxyvitamin D and that it is thus equivalent to the vitamin D hormone. Once these sterols with such marked potency are available as approved drugs they will presumably be the treatment of choice, if they can be manufactured and sold at a price competitive with dihydrotachysterol. At this time we believe that the latter is the treatment of choice, although some clinics still feel that vitamin D_2 gives equally good results and have continued its use.

PSEUDOHYPOPARATHYROIDISM TYPE II

There are no anatomic markers for this form of pseudohypoparathyroidism, so that it cannot be suspected until a response test to parathyroid extract is done in a patient with hypocalcemia and hyperphosphatemia, or measurements of serum IPTH are made. Unlike patients with hypoparathyroidism, these subjects have an elevated serum IPTH. Physiologically the subjects are differentiated from pseudohypoparathyroidism type I by the response to PTH. Type II patients respond with a prompt increase

Table 3–11. Parathyroid Hormone and Urinary Excretion of Cyclic AMP in Normal Subjects. Patients with Hypoparathyroidism and Patients with Pseudohypoparathyroidism Type I

		Cyclic 3′5′-AMP n mols/min		
	Control		Post PTH	
Subjects	*Period*	*0–30 min*	*30–60 min*	*60–90 min*
Normal	3.3	67.9	40.4	7.3
Hypoparathyroidism				
Idiopathic	2.8	79.4	42.2	9.6
Surgical	3.5	85.8	25.0	4.5
Pseudohypoparathyroidism Type I	2.7	6.4	4.5	3.0

From Chase, L. R., Melson, G. L. and Aurbach, G. D.: J. Clin. Invest. *48*:1832–1844, 1969.

in urine cyclic AMP excretion but without a phosphaturic effect. The physiologic defect is, therefore, in the response of the renal tubular cells to increased concentrations of cyclic AMP, rather than in failure of the cyclic AMP mechanism to be stimulated by PTH. The number of reported patients with pseudohypoparathyroidism type II is quite low and the genetic transmission unknown. It is possible, however, that some patients who have been diagnosed as having idiopathic hypoparathyroidism, but in whom response to PTE was not measured or serum IPTH concentrations were not determined, might have this disorder.

The treatment of this form of pseudohypoparathyroidism also is identical with that of idiopathic hypoparathyroidism, since the response to vitamin D, vitamin D metabolites or dihydrotachysterol is not different from that of patients with hypoparathyroidism.

VITAMIN D DEFICIENCY HYPOCALCEMIA

This entity is included under pseudohypoparathyroidism since vitamin D-deficient subjects are hypocalcemic not only because of impaired absorption of calcium from the intestine, but because of lack of responsiveness to parathyroid hormone. Convulsions were a common manifestation of rickets in early infancy, and the fact that these resulted from hypocalcemia became evident as soon as methods for determination of calcium in small amounts of serum became available. It was also noted that infants with rickets who died, usually as a result of infections, had enlarged parathyroid glands. This association of hyperplasia of the parathyroids and the hypocalcemia of vitamin D deficiency was not accorded much attention, since it was generally believed that the action of vitamin D was entirely on the intestinal absorption of calcium and possibly phosphate.

In experimental studies of vitamin D deficiency in animals, the vitamin D-deficient rachitic chick was also found to be hypocalcemic. The findings in rats, however, were found to depend on the calcium and phosphate content of the diet. The vitamin D-depleted rat fed a high calcium, low phosphate diet had normal serum calcium concentrations with marked hypophosphatemia and skeletal lesions of rickets. The animal fed a diet containing amounts of calcium and phosphate in suitable proportions and quantities for normal nutrition showed, when vitamin D-depleted, marked hypocalcemia with normal serum phosphate concentrations and absence of classical rickets in the bones. These animals, like the rachitic infants, also had enlarged parathyroid glands. It was postulated, therefore, that vitamin D depletion made the animals unresponsive to PTH. When these hypocalcemic rats were injected with PTE they showed no elevation of serum calcium concentration and minimal phosphaturic effect. Following a small dose of vitamin D complete responsiveness to PTE returned.

Similar unresponsiveness to PTE has been observed in the hypocalcemic vitamin D deficient infant. The compound necessary for the response to PTH is presumably the vitamin D hormone, 1,25-dihydroxyvitamin D, since a physiologic dose of this compound given to vitamin D-depleted subjects restores serum calcium concentrations to normal. The failure of the vitamin D-deficient child to respond to the calcemic action of PTH is not due to a block at the adenyl cyclase stage, since urine cyclic AMP excretion in vitamin D-deficient children is elevated above normal compatible with their state of secondary hyperparathyroidism. It has been suggested that PTH action requires entrance of calcium into cells across the plasma membrane, and that vitamin D is required for this transmembrane calcium transport. There have been some studies indicating that elevation of extracellular calcium concentrations in the vitamin D-deficient animal without provision of vitamin D restores some degree of responsiveness to PTH. This would be compatible with the above hypothesis, since at high extracellular concentrations of calcium movement of calcium into cells might occur even in the vitamin-depleted subject.

Hypocalcemia secondary to vitamin D deficiency is an early manifestation, and can be seen in infants as early as 2 to 3 months of age if vitamin D is lacking in the diet or there is vitamin D malabsorption, or if a block exists in the conversion of vitamin D to the active metabolite. In the instances of deficient vitamin D intake or impaired absorption, there must also be failure of skin exposure to sunshine if vitamin D deficiency is to result.

Hypocalcemia in infants during the first weeks can also be caused by vitamin D deficiency under the unusual circumstances of vitamin D deficiency in the mother. Such deficiency in the pregnant woman is seen in societies in which women are secluded and therefore receive little exposure to sunshine. This entity has been seen in recent years in England and Scotland in infants born of Pakistani women. The combination of diets that lack vitamin D, climatic conditions at the northern latitudes and social customs precluding normal skin exposure to sunshine may produce true deficiency of vitamin D in the mothers, and consequently their infants are born without tissue stores of vitamin D. In the United States such vitamin D deficiency in adults is rare, but there have been studies in New York City indicating low concentrations of serum 25-hydroxyvitamin D in women whose infants were born prematurely. These were women of poor socioeconomic status who probably received little prenatal nutritional advice and had little sunshine exposure in crowded slum areas of New York.

Most of the hypocalcemia due to vitamin D deficiency is seen in infants over 2½ to 3 months of age who are either breast-fed or fed milk not enriched in vitamin D. In one instance reported to us the infant had been fed goat's milk (obtained from a local farmer) because of suspicion that he could not tolerate cow's milk. Malabsorption of vitamin D secondary to excessive loss of fat in stools can also be a cause of vitamin D deficiency.

An important cause of lack of the active vitamin D hormone is liver disease of infancy. In congenital atresia of the bile ducts, malabsorption of vitamin D occurs because of lack of bile salts in the intestinal lumen. In addition, as liver cirrhosis secondary to cholestasis develops, the liver dysfunction causes failure of 25-hydroxylation of vitamin D, with resulting deficiency of the active vitamin D hormone. Neonatal hepatitis, even without persistent cholestasis, can also result in failure of 25-hydroxyvitamin D formation and hypocalcemia owing to the block in vitamin D metabolism (Table 3–12). This has been seen in the liver disease of cytomegalic inclusion body virus infection, as well as in neonatal hepatitis of unknown etiology. The earliest evidence of such block in vitamin D metabolism may be hypocalcemic convulsions.

Vitamin D-dependent rickets, in which the block in metabolism is believed to be at the kidney level with a defect of the renal 25-hydroxyvitamin D-1 hydroxylase system, is almost invariably associated with hypocalcemia. This is usually later in onset than the hypocalcemia of vitamin D deficiency or of neonatal liver disease, and is discussed in Chapter 5.

The diagnosis of hypocalcemia due to vitamin D deficiency, vitamin D malabsorption, liver disease or congenital defects of the vitamin D metabolic system depends on

Table 3–12. Hypocalcemia in Infant with Neonatal Hepatitis—Etiology Undetermined

AGE DAYS	SERUM				
	Bilirubin Direct *mg/dl*	*Bilirubin* Total *mg/dl*	*Ca* *mg/dl*	*P* *mg/dl*	*Alk. P'tase Bodansky Units*
3	2.0	11.4	–	–	–
69*	3.3	5.0	7.2	6.7	29.8
76	2.6	3.9	7.2	5.9	37.3
81**	1.8	3.1	7.0	5.6	41.2
92	1.0	1.3	10.0	5.9	45.3
124	–	1.6	11.5	7.3	17.7

*Infant was being fed a prepared infant feeding supplying 400 units vitamin D per day.
**Vitamin D supplement started, 5000 units per day.

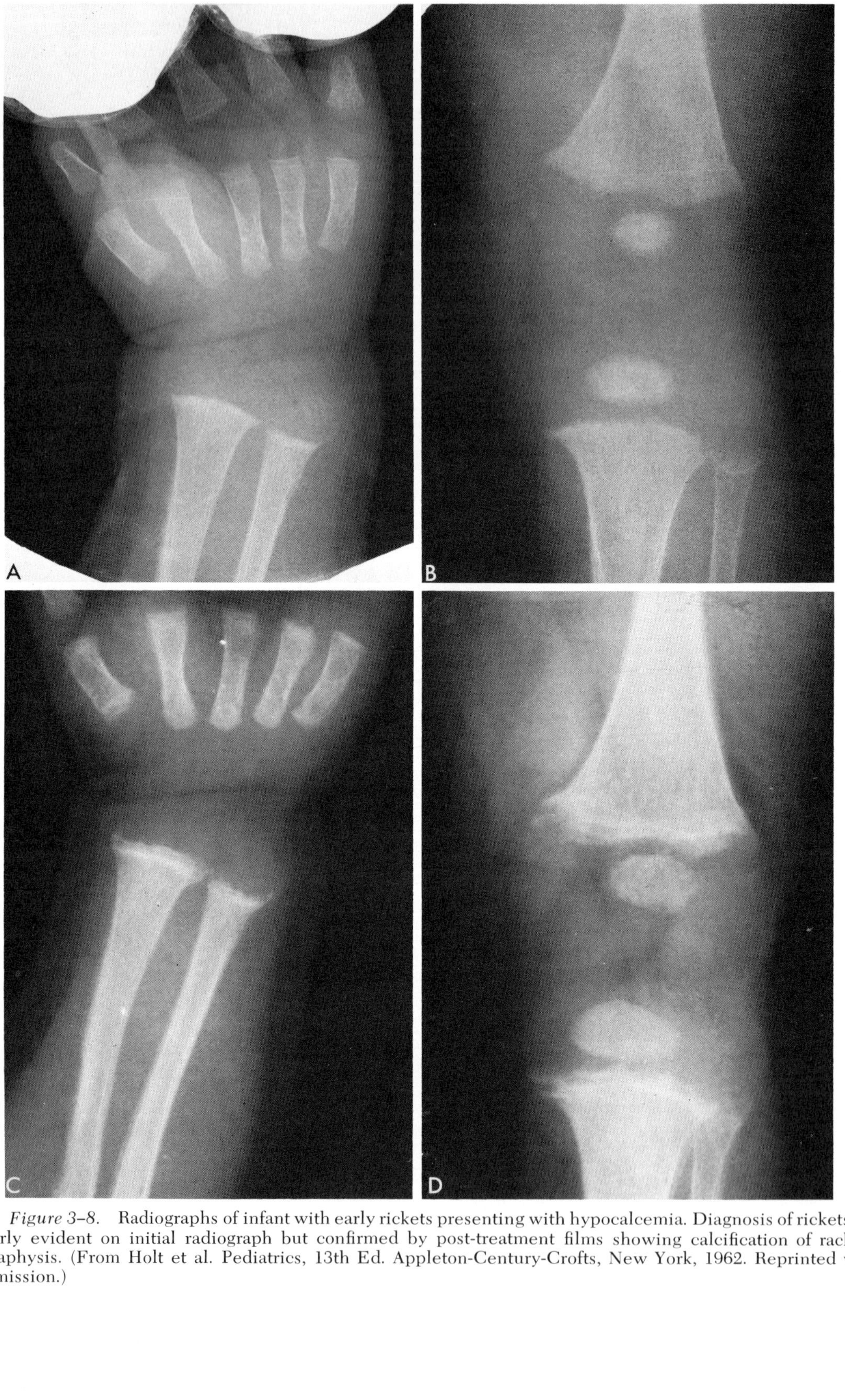

Figure 3–8. Radiographs of infant with early rickets presenting with hypocalcemia. Diagnosis of rickets not clearly evident on initial radiograph but confirmed by post-treatment films showing calcification of rachitic metaphysis. (From Holt et al. Pediatrics, 13th Ed. Appleton-Century-Crofts, New York, 1962. Reprinted with permission.)

both history and laboratory investigation. The serum phosphorus concentrations are in the low range, whereas they are increased in the hypocalcemia of hypoparathyroidism or pseudohypoparathyroidism. The serum alkaline phosphatase activity is usually increased in states of vitamin D deficiency and is usually normal in hypoparathyroidism, although it may be slightly increased in some infants with transient physiologic hypoparathyroidism. Serum IPTH concentrations are increased in vitamin D deficiency. Urine amino acid concentrations are also increased in vitamin D deficiency states associated with secondary hyperparathyroidism, whereas they are normal in hypoparathyroidism. Serum 25-hydroxyvitamin D concentrations are reduced in vitamin D deficiency if this is due to lack of vitamin D intake, malabsorption of vitamin D or hepatic injury. In vitamin D-dependent rickets serum 25-hydroxyvitamin D concentrations should be normal or even high. Roentgenographic examination of the bones in early vitamin D deficiency may show only minimal changes at a time when hypocalcemia is already a problem. The initial radiographic diagnosis may not be clear-cut, therefore, but following vitamin D treatment healing changes may make possible a retrospective diagnosis of early rickets (Fig. 3–8).

Treatment

The emergency treatment of hypocalcemic convulsions due to vitamin D deficiency is the same as that in hypoparathyroidism, namely, intravenous 10 per cent calcium gluconate. Oral administration of calcium salts is ineffective, however. The definitive treatment is oral vitamin D in relatively high dosage. Parenteral vitamin D preparations are not readily available in the United States and are not necessary. Patients with fat malabsorption can absorb high doses of vitamin D in the form of vitamin D_2 dissolved in propylene glycol, Drisdol (8000 units per ml), sufficiently to raise serum vitamin D concentrations with correction of the hypocalcemia. Doses of 8000 to 16,000 units (0.2 to 0.4 mg) per day for ten days can be given, following which the dosage may be reduced depending on the problem. In patients with fat malabsorption or atresia of the bile ducts, continued high doses are required of the order of 4000 to 8000 units per day. In infants with liver disease 25-hydroxyvitamin D would be the suitable mode of treatment; however, this vitamin D metabolite has not yet been approved by the Food and Drug Administration and therefore is not readily available except for investigational use. 1,25-dihydroxyvitamin D could be given in physiologic doses of 0.5 μg per day but vitamin D in high dosage is effective. Children with vitamin D dependent rickets may require quite large amounts of vitamin D because of the block in hydroxylation, and in these subjects the dosage must be determined on an individual basis.

HYPOCALCEMIA SECONDARY TO PHOSPHATE OVERLOAD

Rapid elevation of serum inorganic phosphate concentrations by absorption of phosphate from the intestinal tract, following either oral ingestion of concentrated solutions of inorganic phosphate or injection of such solutions into the colon, has resulted in severe hypocalcemia with tetany, convulsions and heart block. The commonly used Fleet enema is a solution of sodium phosphate. When given to patients with pre-existing renal disease, to children with Hirschsprung's disease or to infants, enough phosphate can be absorbed by the colon to raise the serum phosphate concentration to 20 mg/dl or greater. This exceeds the solubility product of $CaHPO_4$, and calcium ion is thus removed and is deposited in bone or soft tissues. At these high concentrations of phosphate and with the high rate of entrance of phosphate into extracellular fluid from the phosphate depot in the gut, parathyroid hormone cannot act effectively to mobilize calcium from bone, since the calcium is reprecipitated as calcium phosphate and the renal phosphaturic action of PTH is not sufficient to remove phosphate from extracellular fluid as rapidly as it enters. The patients at most risk are those with low glomerular filtration rates, whether due to renal disease or to immaturity of kidney development as in the young infant. However, children with Hirschprung's disease with normal renal function have been thrown into hypocalcemic tetany by sodium phosphate enemas. Phosphate cathartics have also been reported to cause hypocalcemic convulsions in infants as well as in older subjects. In one such instance moderate hypomagnesemia was also found, which might be expected in view of the negative correlation between serum magnesium and phosphate concentrations referred to previously. Administration of magnesium as intramuscular $MgSO_4$ solution may be advisable in such patients if hyperexcitability of the nervous system is not controlled by intravenous injection of 10 per cent calcium gluconate, 1 to 2 ml/kg (9 to 18 mg Ca/kg).

HYPOCALCEMIA IN HEAT STROKE WITH RHABDOMYOLYSIS

Hyperpyrexia can cause muscle membrane injury, with leakage of myoglobulin into extracellular fluid and breakdown of intramuscular organic phosphates with hyperphosphatemia and hypocalcemia. In one instance of hyperpyrexia with myoglobinemia and myoglobinuria, severe hypocalcemic tetany occurred despite low serum phosphate concentrations. The mechanism of hypocalcemia in this patient is obscure. Whether parathyroid stimulation of bone cell metabolism was impaired by the toxic effects of myoglobinemia or high body temperature cannot be determined from available observations.

HYPOMAGNESEMIA WITH FUNCTIONAL HYPOPARATHYROIDISM

The association of hypocalcemia and hyperphosphatemia with states of magnesium deficiency has been recognized for some time, but only since methods for measuring parathyroid hormone in blood have become available has it been possible to determine the mechanism. It is known that reduction of Mg^{++} ion stimulates PTH secretion, at least under in vitro conditions, and in the albino rat magnesium deficiency is associated with manifestations of hyperparathyroidism, namely, hypercalcemia and renal calcinosis. In man and certain other animals, however, this association is not seen. One hypothesis is that magnesium deficiency in man results in target cell resistance to PTH, producing a form of pseudohypoparathyroidism. There is some evidence of poor response of hypomagnesemic subjects to PTE. However, determinations of immunoreactive PTH concentrations in the serum of severely magnesium-deficient patients have shown low concentrations despite hypocalcemia, so that it is postulated that magnesium deficiency in man alters the capacity of the parathyroid gland to synthesize or secrete PTH. Hypomagnesemia, therefore, can be associated with hypocalcemia and hyperphosphatemia as the result of functional hypoparathyroidism.

NEONATAL HYPOMAGNESEMIA

Specific Magnesium Malabsorption

The infants in this group have all been males. There is no history of maternal abnormalities, and pregnancies and deliveries have not been unusual. Presenting symptoms in the first weeks of life are hyperirritability, "jitteriness" and convulsions. Serum magnesium levels have been very low, 0.2 to 0.6 mg/dl, and serum calcium concentrations have also been reduced to between 5.5 and 7.6 mg/dl. Specific magnesium malabsorption has been documented by balance studies or by isotopic magnesium uptake determinations. This apparently is an isolated specific transport dysfunction, since other indices of intestinal function have been normal. Treatment with vitamin D or calcium salts has not influenced the hypomagnesemia or the symptoms, nor has injection of parathyroid hormone been effective, even though serum calcium concentration is increased. Initial therapy can be the intramuscular injection of $MgSO_4$, 0.2

ml/kg of 25 per cent $MgSO_4$ (50 per cent $MgSO_4 \cdot 12\ H_2O$), repeated at six-hour intervals. Chronic treatment requires oral magnesium supplementation. Even though the specific intestinal magnesium transport system is defective, high concentrations of magnesium in the intestinal lumen evidently result in increased absorption. This may indicate that the functional lesion of the specific transport system is such that the K_m of the system is markedly increased, or that passive diffusion of magnesium across the brush border membrane occurs when it is present in high concentration in the solvent phase in contact with the membrane.

HYPOMAGNESEMIA IN TRANSIENT PHYSIOLOGIC HYPOPARATHYROIDISM

As indicated in the appropriate sections, reduction of serum magnesium concentration is seen in hypoparathyroidism, particularly in the transient physiologic hypoparathyroidism of the newborn. A suggested mechanism is that the high concentration of extracellular phosphate reduces extracellular magnesium concentrations, either by altering the equilibrium between intracellular and extracellular magnesium or by uptake of magnesium phosphate by bone mineral. In Table 3–4, describing the course of an infant with transient physiologic hypoparathyroidism treated with calcium lactate supplementation of the feed, it can be noted that before treatment serum P_i concentration was 9.2 mg/dl and serum Mg concentration 0.86 mg/dl. When the serum P_i was reduced to 3.0 mg/dl by calcium sequestration of phosphate in the gut, the Mg concentration in serum rose to 1.64 mg/dl. When the calcium supplement was abruptly discontinued, however, because of hypercalcemia the serum P concentration rose sharply to 8.8 mg/dl and the concentration of serum Mg again decreased to 0.77 mg/dl. This reciprocal relationship between serum Mg and P_i concentrations has also been noted in animal studies. This type of hypomagnesemia does not require specific magnesium therapy, since reduction of serum P_i concentration by feeding an excess of calcium in our experience has sufficed to bring serum Mg concentrations into the normal range (early infancy 1.5 to 1.9 mg/dl).

In persistent hypoparathyroidism, whether congenital, idiopathic or autoimmune, hypomagnesemia can also occur. In some instances this is not corrected by treatment with vitamin D or dihydrotachysterol, and it has been suggested that resistance to activated sterol treatment can be secondary to extracellular magnesium deficiency. This has been discussed in the section on treatment of hypoparathyroidism.

ACQUIRED HYPOMAGNESEMIA WITH HYPOCALCEMIA

Intestinal Malabsorption

Magnesium deficiency can arise from excess loss of magnesium in stools resulting from chronic diarrhea and steatorrhea. This has been seen particularly in extensive disease of the distal small intestine (Crohn's disease) or following resection of a considerable portion of the small bowel. Associated with the hypomagnesemia is hypocalcemia, which is quite refractory to treatment unless the magnesium deficiency is corrected. These patients usually present with marked tetany, and in younger patients convulsions may be the initial manifestation.

Treatment consists of increased magnesium intake in the form of oral magnesium chloride, citrate or lactate, 2 to 4 mEq/Mg/kg/day = 24 to 48 mg/kg/day in divided doses up to a maximum of 1 gm Mg per day. If these doses of Mg salts aggravate the diarrhea in patients with resection or disease of small intestine, it may be necessary to resort to intramuscular injection of $MgSO_4$ in amounts that raise serum Mg sufficiently to maintain the secretory function of the parathryoid glands. The initial dose is 0.2 ml/kg of a 25 per cent solution of $MgSO_4$ which provides 10 mg or 0.8 mEq Mg/kg. This dose may have to be given twice daily until the magnesium deficiency is repaired.

Renal Wastage of Magnesium

Magnesium deficiency with hypomagnesemia and hypocalcemia can also result from abnormalities of renal tubule function with failure of magnesium reabsorption and excessive loss of magnesium in urine. This may be on the basis of a rare inborn error of renal tubule function, or more commonly in recent years as the result of renal tubule injury following treatment with aminoglycoside antibiotics such as gentamicin. In most instances the nephrotoxicity has resulted from combinations of gentamicin and cephalosporin, but gentamicin alone may produce this tubular dysfunction, particularly in patients receiving antitumor drugs. Many of these patients, therefore, are suffering from leukemia and have received gentamicin along with other antibiotics because of high fever with suspicion of superimposed gram-negative bacillary infection. However, severe infection and gentamicin treatment may result in this syndrome, as exemplified by the case report. The treatment is, of course, discontinuance of the nephrotoxic antibiotic and repair of the magnesium deficiency by intramuscular injection of $MgSO_4$. In the case of the chronic disorders of renal tubular function with magnesium wastage, increased oral intake of magnesium is required as in the instances of magnesium malabsorption.

CASE REPORT

C. McC., a male, aged 11 years, had severe suppurative bronchiectasis necessitating a pneumonectomy. Following operation, he had empyema for which he was treated with gentamicin IV, 120 mg/day and ampicillin IV, 6 gm/day. After a month of treatment he showed symptoms and signs of tetany, and the serum electrolyte determinations revealed reduced serum potassium concentration, hypochloremic alkalosis and hypocalcemia. Gentamicin was discontinued and intravenous injections of calcium gluconate were given. However, hypocalcemia and tetany persisted and a serum magnesium determination revealed a concentration of 0.7 mg/dl. He was given intramuscular $MgSO_4$ and the course is summarized in Table 3–13.

COMMENTS: This boy with severe intrathoracic infection became markedly hypocalcemic with hypokalemia and hypochloremic alkalosis, despite potassium supplementation, while on antibiotic treatment with gentamicin and ampicillin. The gentamicin was discontinued on the day that the boy showed obvious manifestations of tetany with carpopedal spasm and painful muscle contractions. Hypocalcemia was discovered, which was not corrected by intravenous injection of calcium gluconate. This was explained when the serum magnesium concentration was discovered to be markedly reduced, 0.7 mg/dl. Following intramuscular injection of $MgSO_4$ the serum calcium concentration increased even though serum magnesium was only temporarily increased. It is possible that intracellular magnesium deficit was replaced to a degree sufficient to

Table 3–13. Hypomagnesemia and Hypocalcemia in a Child with Severe Infection Treated with Gentamicin and Ampicillin

DATE	SERUM *Na*	*Cl*	*HCO₃* *mEq/l*	*K*	*Ca*	*P* *mg/dl*	*Mg*
2–10	130	89	32	4.3	–	–	–
	Patient on gentamicin IV 120 mg/day and ampicillin IV 6 gm/day						
3– 7	134	65	46	1.5	–	–	–
3–10	133	81	37	3.0	–	–	–
	Gentamicin discontinued						
3–11	–	–	–	2.6	6.8	5.0	–
	Calcium gluconate IV						
3–12	131	–	–	2.6	6.8	–	–
3–14	–	–	–	–	–	–	0.7
	$MgSO_4$ IM						
3–16	134	87	36	3.0	8.3	–	0.7
3–17	–	–	–	–	9.2	–	1.2
3–23	–	–	32	3.5	8.6	–	0.8

restore more adequate parathyroid hormone secretion. The hypomagnesemia in this patient was probably the result of both low magnesium intake and the nephrotoxic action of gentamicin. This nephrotoxicity presumably was also responsible in part for the marked hypokalemia that was associated with the hypomagnesemia. As mentioned above, renal tubular dysfunction with hypomagnesemia has been seen most frequently in patients treated with gentamicin and cephalosporin.

Unexplained Hypocalcemia in Association with Severe Infections and Hypernatremic Dehydration

Reduction of serum calcium concentrations has been seen on occasion in patients with severe infections who are not receiving nephrotoxic antibiotics. The mechanism for this is unexplained. Since serum magnesium determinations usually were not made because of poor methodology prior to atomic absorption spectrophotometry, we do not know whether some of this hypocalcemia was associated with magnesium deficiency. As the following case report indicates, hypokalemia is also present and it is possible that deficiency of magnesium, another predominantly intracellular cation, occurs along with that of intracellular potassium.

CASE REPORT

This 11 month old infant was admitted because of fever and convulsions with post-convulsive stupor. On examination the anterior fontanelle was bulging, but cerebrospinal fluid obtained by lumbar puncture did not show evidence of purulent meningitis. A subdural tap was productive of thick fluid loaded with polymorphonuclear leukocytes. The serum electrolytes, including calcium and phosphorus concentrations, are summarized in Table 3–14.

COMMENTS: The child, with a severe infection that was undoubtedly present for many days, had water deficit with hypernatremia, severe potassium deficiency and marked hypocalcemia. Although the hypernatremia and potassium deficiency were corrected by appropriate water and electrolyte administration, the hypocalcemia was unchanged. Unfortunately the serum magnesium concentration was not determined. Whether this is an instance of hypomagnesemia associated with severe potassium deficiency, as in the case of the boy treated with gentamicin, cannot be known. The hypocalcemia associated with severe infection may also represent a form of parathyroid insufficiency, but appropriate studies have not been done to solve this problem.

The incidence of hypocalcemia in infants with diarrhea and hypernatremic dehydration, or infants who have received excessive sodium-containing fluids, is well recognized. In the infant with hypernatremic dehydration the hypocalcemia usually develops after he has been given large volumes of fluid with the goal of reducing the serum sodium concentration into the normal range. This form of treatment is known to predispose to convulsive seizures due to brain edema, but hypocalcemia must be considered as a possible predisposing cause. The mechanism of the hypocalcemia was studied experimentally by Finberg et

Table 3–14. Hypocalcemia in Infant with Severe Infection

HOSPITAL DAY	SERUM CO_2	*Cl* (*mEq/l*)	*Na*	*K*	*Ca*	*P* (*mg/dl*)	*Urea N*
1	36.4	97.6	155	2.2	6.0	–	14
2	35.4	94.4	153	2.3	5.9	–	–
3	32.5	97.6	147	3.1	6.2	3.0	–
8	21.1	99.6	146	3.2	5.8	4.8	–
12	21.0	100.8	144	5.5	6.8	6.0	–

Table 3–15. Hypocalcemia in Infant with Sodium Overload

HOSPITAL DAY	WEIGHT gm	CO_2	Cl mEq/l	SERUM Na	UREA N	Ca mg/dl	P	TOTAL PROTEIN gm/dl
1		19.8	—	—	39	—	—	—
2	4380	25.6	114.8	150	—	—	—	—
7	6200	—	—	—	—	4.1	2.5	3.9

al, who found that serum calcium concentrations were reduced in experimental hypernatremia of rats if the rats were also hypokalemic. If potassium was given during the development of hypernatremia the serum calcium concentrations were not reduced. This effect of potassium deficiency has not been adequately examined. Whether potassium deficiency leads to loss of intracellular magnesium and thus to parathyroid hormone deficiency, or whether potassium deficiency per se causes failure of appropriate PTH output or lack of bone cell responsiveness to PTH, is as yet unknown.

The following case is an example of hypocalcemia resulting from administration of a great excess of sodium and water to a dehydrated infant.

CASE REPORT

This 2 month old infant was admitted with severe dehydration resulting from profuse watery diarrhea for several days. Salmonella enteritidis was cultured from the stools. The course is summarized in Table 3–15.

On day 7 the child was obviously edematous and had gained 1820 gm or 40 per cent of his admission weight. On this day he began to have convulsive seizures and the serum calcium was found to be 4.1 mg/dl. Treatment with intravenous calcium gluconate controlled the convulsions, and recovery occurred following restriction of sodium and water intake, with delivery of the excess extracellular fluid.

COMMENTS: This type of hypocalcemia following excessively large loads of sodium and water without potassium has been seen in the past, but is no longer a problem. Dodd and Rapaport referred to this phenomenon as the postacidotic state, but the role of acidosis is probably only that of a stimulus for administration of excessive amounts of sodium as sodium bicarbonate. The serum potassium was not available in the patient under discussion but was almost certainly reduced, since it is well-known that hypokalemia does result when patients with severe diarrhea are treated only with large amounts of sodium salts and water without replacement of potassium losses. Again, this type of hypocalcemia has not been studied sufficiently to propose a mechanism for the causation, but is included here since it resembles in some respects the hypocalcemia seen in hypomagnesemic states.

REFERENCES

SYMPTOMS AND SIGNS OF HYPOCALCEMIA

Colletti, R. B., Pan, M. W., Smith, E. W. et al: Detection of hypocalcemia in susceptible neonates: The Q – oT$_c$ interval. N. Engl. J. Med. *290*: 931–935, 1974.

Engel, F. L., Martin, S. P. and Taylor, H.: On the relation of potassium to the neurological manifestations of hypocalcemic tetany. Bull. Johns Hopkins Hosp. *84*:285–301, 1949.

Harrison, H. E.: Hypoparathyroidism. Mod. Treat. *7*:636–648, 1970.

Harrison, H. E.: Tetany. *In* Textbook of Pediatrics (Eds.: V. C. Vaughan, III and R. J. McKay). W. B. Saunders Co., Philadelphia, 1975, p. 1391.

Harrison, H. E., Harrison, H. C., Tompsett, R. R. et al: Potassium deficiency in a case of lymphosarcoma with the sprue syndrome. Am. J. Med. 2:131–143, 1947.

Lewis, T.: Trousseau's phenomenon in tetany. Clin. Sci. *4*:361–364, 1942.

Sorell, M. and Rosen, J. F.: Ionized calcium: serum levels during symptomatic hypocalcemia. J. Pediatr. *87*:67–70, 1975.

NEONATAL HYPOCALCEMIA

Bergman, L.: Early neonatal hypocalcemia. Acta Paediatr. Scand. (Suppl.) *248*:5–25, 1974.

Capen, C. C. and Young, D. M.: Thyrocalcitonin: evidence for release in a spontaneous hypocalcemic disorder. Science *157*:205–206, 1967.

Craig, W. S., and Buchanan, M. F. G.: Hypocalcemic tetany developing within 36 hours of birth. Arch. Dis. Child. *33*:505–511, 1958.

Cushard, W. G., Jr., Creditor, M. A., Canterbury, J. M. et al: Physiologic hyperparathyroidism in pregnancy. J. Clin. Endocrinol. Metab. *34*:767–771, 1972.

David, L. and Anast, C. S.: Calcium metabolism in newborn infants. The interrelationship of parathyroid function and calcium, magnesium, and phosphorus metabolism in normal, "sick" and hypocalcemic newborns. J. Clin. Invest. *54*:287–296, 1974.

Gardner, L. I.: Tetany and parathyroid hyperplasia in the newborn infant: influence of dietary phosphate load. Pediatrics *9*:534–543, 1952.

Garel, J.-M., Barlet, J.-P. and Kervran, A.: Metabolic effects of calcitonin in the newborn. Am. J. Physiol. *229*:669–675, 1975.

Hillman, L. S. and Haddad, J. G.: Human perinatal vitamin D metabolism. I. 25-hydroxyvitamin D in maternal and cord blood. J. Pediatr. *84*:742–749, 1974.

Hillman, L. S. and Haddad, J. G.: Perinatal vitamin D metabolism. II. Serial 25-hydroxyvitamin D concentrations in sera of term and premature infants. J. Pediatr. *86*:928–935, 1975.

Keen, J. H.: Significance of hypocalcemia in neonatal convulsions. Arch. Dis. Child. *44*:356–361, 1969.

Kooh, S. W., Fraser, D. and Toon, R.: Response of protracted neonatal hypocalcemia to 1-α 25 dihydroxyvitamin D. Lancet *2*:1105–1107, 1976.

Lealman, G. T., Logan, R. W., Hutchison, J. H., et al: Calcium, phosphorus and magnesium concentrations in plasma during first week of life and their relation to type of milk fed. Arch. Dis. Child. *51*:377–384, 1976.

Linarelli, L. G. with technical assistance of Bobik, J. and Bobik, C.: Newborn urinary cyclic AMP and developmental renal responsiveness to parathyroid hormone. Pediatrics *50*:14–23, 1972.

Peshin, J., Ornoy, A. and Menczel, J.: Transplacental effects of thyrocalcitonin on intestinal calcium-binding protein, alkaline phosphatase activity and ossification of long bones in rat fetuses. Isr. J. Med. Sci. *12*:248–256, 1976.

Purvis, R. J., MacKay, G. S., Cockburn, W. J., et al: Enamel hypoplasia of the teeth associated with neonatal tetany: a manifestation of maternal vitamin-D deficiency. Lancet *2*:811–814, 1973.

Roberts, S. A., Cohen, M. D. and Forfar, J. O.: Antenatal factors associated with neonatal hypocalcemic convulsions. Lancet *2*:809–811, 1973.

Rosen, J. F., Roginsky, M., Nathenson, G. et al: 25-hydroxyvitamin D—plasma levels in mothers and their premature infants with neonatal hypocalcemia. Am. J. Dis. Child. *127*:220–223, 1974.

Rösli, A. and Fanconi, A.: Neonatal hypocalcemia. "Early type" in low birth weight newborns. Helv. Paediatr. Acta *28*:443–457, 1973.

Snodgrass, G. J. A. I., Stimmler, L., Went, J. et al: Interrelations of plasma calcium, inorganic phosphate, magnesium and protein over the first week of life. Arch. Dis. Child. *48*:279–285, 1973.

Tsang, R. C., Donovan, E. F. and Steichen, J. J.: Calcium physiology and pathology in the neonate. Pediatr. Clin. North Am. *23*(4):611–626, 1976.

Tsang, R. C., Gigger, M., Oh, W. et al: Studies in calcium metabolism in infants with intrauterine growth retardation. J. Pediatr. *86*:936–941, 1975.

Tsang, R. C., Wen-Chen, I. and Friedman, M. A.: Parathyroid function in infants of diabetic mothers. J. Pediatr. *86*:399–404, 1975.

Tsang, R. C., Wen-Chen, I., Friedman, M. A. et al: Neonatal parathyroid function: role of gestational age and postnatal age. J. Pediatr. *83*:728–738, 1973.

Tsang, R. C., Chen, I. W. McEnery, P. et al: Parathyroid function tests with EDTA infusions in infancy and childhood. J. Pediatr. *88*:250–256, 1976.

Turner, T. L., Cockburn, F. and Forfar, J. O.: Magnesium therapy in neonatal tetany. Lancet *1*:283–284, 1977.

Watney, P. J. M., Chance, G. W., Scott, P. et al: Maternal factors in neonatal hypocalcemia: a study in three ethnic groups. Br. Med. J. *2*:432–436, 1971.

Zetterstrom, R. and Arnhold, R. G.: Impaired calcium phosphate homeostasis in newborn infants of diabetic mothers. Acta Paediatr. *47*:107–112, 1958.

CONGENITAL HYPOPARATHYROIDISM

Benson, P. F. and Parsons, V.: Hereditary hypoparathyroidism presenting with oedema in the neonatal period. Q. J. Med. *33*:197–208, 1964.

DiGeorge, A. M.: Congenital absence of the thymus and its immunologic consequences: concurrence with congenital hypoparathyroidism. Birth Defects *4*(1):116–121, 1968.

Moshkowitz, A., Abrahamov, A. and Pisanti, S.: Congenital hypoparathyroidism, simulating epilepsy with other symptoms and dental signs of intra-uterine hypocalcemia. Pediatrics *44*:401–409, 1969.

Olambiwonnu, N. O., Ebbin, A. J. and Frasier, S. D.: Primary hypoparathyroidism associated with ring chromosome 18. J. Pediatr. *80*:833–835, 1972.

IDIOPATHIC HYPOPARATHYROIDISM

Forbes, G. B.: Clinical features of idiopathic hypoparathyroidism in children. Ann. N. Y. Acad. Sci. *64*:432–455, 1956.

Harrison, H. E.: Idiopathic hypoparathyroidism. Pediatrics *17*:442–448, 1956.

Nusynowitz, M. L., Frame, B. and Kolb, F. O.: The spectrum of the hypoparathyroid states: a classification based on physiologic principles. Medicine *55*:105–119, 1976.

Robinson, P. K., Carmichael E. A. and Cumings, J. N.: Idiopathic hypoparathyroidism: a study of three cases. Q. J. Med. *23*:383–402, 1954.

Strom, L. and Winberg, J.: Idiopathic hypoparathyroidism. Acta Paediatr. *43*:574–581, 1954.

Watkins, E., Jr., Bell, G. O., Snow, J. C. et al: Incidence and current management of post-thyroidectomy hypoparathyroidism. J. A. M. A. *182*:140–146, 1962.

AUTOIMMUNE HYPOPARATHYROIDISM

Craig, J. M., Schiff, L. H. and Boone, J. E.: Chronic moniliasis associated with Addison's disease. Am. J. Dis. Child. *89*:669–684, 1955.

Kahn, A., Snapper, I. and Drucker, A.: Corticosteroid-induced tetany in latent hypoparathyroidism. Arch. Intern. Med. *114*:434–438, 1964.

Russell, R. I.: Hypoparathyroidism and malabsorption. Br. Med. J. *3*:781–782, 1967.

Spinner, M. W., Blizzard, R. M. and Childs, B.: Clinical and genetic heterogeneity in idiopathic Addison's disease and hypoparathyroidism. J. Clin. Endocrinol. Metab. *28*:795–804, 1968.

Treatment of Hypoparathyroidism

Chesney, R. W., Horowitz, S. D., Kream, B. E. et al: Failure of conventional doses of 1α,25-dihydroxycholecalciferol to correct hypocalcemia in a girl with idiopathic hypoparathyroidism. N. Engl. J. Med. *297*:1272–1275, 1977.

Dent, C. E. and Friedman, M.: A comparison between A. T. 10 and pure dihydrotachysterol in controlling hypoparathyroidism. Lancet *2*:164–169, 1964.

Dent, C. E., Morgans, M. E., Harper, C. M., et al: Insensitivity to vitamin D developing during the treatment of postoperative tetany. Its specificity as regards the form of vitamin D taken. Lancet *2*: 687–690, 1955.

Harrison, H. E., Lifshitz, F. and Blizzard, R. M.: Comparison between crystalline dihydrotachysterol and calciferol in patients requiring pharmacologic vitamin D therapy. N. Engl. J. Med. *276*:894–900, 1967.

Homer, L.: Hypoparathyroidism requiring massive amounts of medication with apparent response to magnesium sulfate. J. Clin. Endocrinol. Metab. *21*:219–223, 1961.

Ireland, A. W., Clubb, J. S., Neale, F. C. et al: The calciferol requirements of patients with surgical hypoparathyroidism. Ann. Intern. Med. *69*:81–89, 1968.

Jorgensen, H. and Vogt, J. H.: 1α-hydroxycholecalciferol in the treatment of hypoparathyroidism. Acta Med. Scand. *201*:3–7, 1977.

Kind, H. P., Prader, A., DeLuca, H. F. et al: 1,25-dihydroxycholecalciferol in hypoparathyroidism and pseudohypoparathyroidism. Lancet *1*: 1145–1146, 1975.

Kooh, S. W., Fraser, D., DeLuca, H. F. et al: Treatment of hypoparathyroidism and pseudohypoparathyroidism with metabolites of vitamin D: evidence for impaired conversion of 25-hydroxyvitamin D to 1α,25-dihydroxyvitamin D. N. Engl. J. Med. *293*:840–844, 1975.

Rosen J. F., Fleischman, A. R., Finberg, L. et al: 1,25-dihydroxycholecalciferol: its use in the long term management of idiopathic hypoparathyroidism in children. J. Clin. Endocrinol. Metab. *45*: 457–468, 1977.

Russell, R. G. G., Walton, R. J., Smith, R. et al: 1,25-dihydroxycholecalciferol and 1α-hydroxycholecalciferol in hypoparathyroidism. Lancet *2*:14–17, 1974.

Schneider, A. B., and Sherwood, L. M.: Progress in endocrinology and metabolism. Pathogenesis and management of hypoparathyroidism and other hypocalcemic disorders. Metabolism *24*:871–898, 1975.

HYPOMAGNESEMIA

Anast, C. S., Mohs, J. M., Kaplan, S. L. et al: Evidence for parathyroid failure in magnesium deficiency. Science *177*:606–608, 1972.

Estep, H., Shaw, W. A., Watlington, C. et al: Hypocalcemia due to hypomagnesemia and reversible parathyroid hormone unresponsiveness. J. Clin. Endocrinol. Metab. *29*:842–848, 1969.

Medalle, R., Waterhouse, C. and Hahn, T. H.: Vitamin D resistance in magnesium deficiency. Am. J. Clin. Nutr. *29*:854–858, 1976.

Muldowney, F. P., McKenna, T. J., Kyle, L. H. et al: Parathormone-like effect of magnesium replenishment in steatorrhea. N. Engl. J. Med. *282*: 61–68, 1970.

HYPOCALCEMIA SECONDARY TO DEFICIENCY OF 1,25-diOH VITAMIN D

Arnaud, C., Rasmussen, H. and Anast, C.: Further studies on the interrelationship between parathyroid hormone and vitamin D. J. Clin. Invest. *45*:1955–1964, 1966.

Drezner, M. K., Neelon, F. A., Haussler, M. et al: 1,25-dihydroxycholecalciferol deficiency: the probable cause of hypocalcemia and metabolic bone disease in pseudohypoparathyroidism. J. Clin. Endocrinol. Metab. *42*:621–628, 1976.

Harrison, H. E.: Parathyroid hormone and vitamin D. Yale J. Biol. Med. *38*:393–409, 1966.

Harrison, H. E. and Harrison, H. C.: The interaction of vitamin D and parathyroid hormone on calcium, phosphorus and magnesium homeostasis in the rat. Metabolism *13*:952–958, 1964.

Harrison, H. E., Harrison, H. C. and Lifshitz, F.: Interrelation of vitamin D and parathyroid hormone: responses of vitamin D depleted and of thyroparathyroidectomized rats to ergocalciferol and dihydrotachysterol. *In* Parathyroid Hormone and Thyrocalcitonin (Calcitonin) (Eds.: R. V. Talmage and L. Belanger). Excerpta Medica, Amsterdam, 1968, p. 455.

PSEUDOHYPOPARATHYROIDISM

Albright, F., Burnett, C. H., Smith, P. H. et al: Pseudohypoparathyroidism—an example of "Seabright-bantam syndrome." Endocrinology *30*:922–932, 1942.

Albright, F., Forbes, A. P., Henneman, P. H. et al: Pseudo-pseudohypoparathyroidism. Trans. Assoc. Am. Physicians *65*:337–350, 1952.

Chase, L. R., Melson, G. L. and Aurbach, G. D.: Pseudohypoparathyroidism: defective excretion of 3′5′-AMP in response to parathyroid hormone. J. Clin. Invest. *48*:1832–1844, 1969.

Drezner, M., Neelon, F. A. and Lebovitz, H. E.: Pseudohypoparathyroidism type II: a possible defect in the reception of the cyclic AMP signal. N. Engl. J. Med. *289*:1056–1060, 1973.

MacGregor, M. E. and Whitehead, T. P.: Pseudohypoparathyroidism. A description of three cases and a critical appraisal of earlier accounts of the disease. Arch. Dis. Child. *29*:398–418, 1954.

Miles, J. and Elrich, H.: Pseudo-pseudo-hypoparathyroidism. J. Clin. Endocrinol. Metab. *15*:576–584, 1955.

Monn, E., Osnes, J. B., Oye, I. et al: Pseudohypoparathyroidism—a difficult diagnosis in childhood. Acta Paediatr. Scand. *65*:487–493, 1976.

HYPOCALCEMIA SECONDARY TO PHOSPHATE OVERLOAD

Chesney, R. W. and Haughton, P. B.: Tetany following phosphate enemas in chronic renal disease. Am. J. Dis. Child. *127*:584–586, 1974.

Davis, R. F., Eichner, J. M., Bleyer, A. et al: Hypocalcemia, hyperphosphatemia, and dehydration following a single hypertonic phosphate enema. J. Pediatr. *90*:484–485, 1977.

Hebert, L. A., Lemann, J., Jr., Petersen, J. R. et al: Studies of the mechanism by which phosphate infusion lowers serum calcium concentration. J. Clin. Invest. *45*:1886–1894, 1966.

Levitt, M., Gessert, C. and Finberg, L.: Inorganic phosphate (laxative) poisoning resulting in tetany of an infant. J. Pediatr. *82*:479–481, 1973.

Osmond, S. C., O'Bell, J. and Grupe, W. E.: Severe tetany in an azotemic child related to a sodium phosphate enema. Pediatrics *53*::105–106, 1974.

Smith, M. S., Feldman, K. W. and Furukawa, C. T.: Coma in an infant due to hypertonic sodium phosphate medication. J. Pediatr. *82*:481–482, 1973.

HYPOCALCEMIA ASSOCIATED WITH HYPERNATREMIA AND HYPOKALEMIA

Finberg, L.: Experimental studies of the mechanisms producing hypocalcemia in hypernatremic states. J. Clin. Invest. *36*:434–439, 1957.

Finberg, L. and Harrison, H. E.: Hypernatremia in infants. An evaluation of the clinical and biochemical findings accompanying this state. Pediatrics *16*:1–14, 1955.

Harrison, H. E. and Finberg, L.: Hypernatremic dehydration. Pediatr. Clin. North Am. *6*:193–199, 1959.

Chapter Four

HYPERCALCEMIC STATES

Hypercalcemia in infants and children may be defined as a total serum calcium concentration consistently greater than 11.0 mg/dl and an ionized calcium concentration greater than 5.6 mg/dl. For accurate measurement of total serum calcium the blood should be drawn at least six hours after the last calcium intake and taken with a minimum of venous stasis. Since close to 50 per cent of the total serum calcium is nondiffusible protein-bound calcium, the loss of protein-free fluid from the plasma because of increased venous pressure will increase the total serum calcium concentration. Theoretically, this can be corrected by measurement of total serum protein. However, it has been shown that there are variations among individuals in the binding of calcium by protein, so that although an average correction for increased protein concentration has some value it is preferable to collect venous blood with a minimum of venous stasis to avoid error. The determination of ionized calcium rather than total calcium has been recommended, since the ionized calcium fraction is the physiologically active component the concentration of which is closely regulated by homeostatic mechanisms.

This determination can now be made by the use of a calcium ion electrode. For such measurement the blood and separated serum or plasma must be protected from loss of CO_2, as binding of calcium by protein is pH-dependent and increase of pH resulting from loss of CO_2 reduces calcium ion concentration. The various colorimetric methods for measurement of calcium ion concentration, depending on the formation of calcium complexes with various dyes, are more cumbersome and less accurate than the calcium electrode technique. The diffusible calcium fraction obtained by membrane ultrafiltration consists not only of ionized calcium but also of a small fraction of diffusible calcium complexes with anions such as citrate. For practical purposes, however, the calcium electrode method for determination of calcium ion concentration is not yet available in most clinical laboratories, so that reliance must be placed on the concentration of total serum calcium determined as accurately as possible. The reference method considered most reliable is the determination by atomic absorption spectrophotometry. Emission flame photometric determination of serum calcium can also be accurate with proper instrumentation but has been largely superseded by the atomic absorption system. A variety of titrimetric or colorimetric techniques, depending on the formation of fluorescent or colored calcium complexes, are also used, particularly for routine measurements with automated systems. The variability of these determinations is greater than with atomic absorption spectrophotometry, and interfering factors such as lipemia can give erroneously high values. It is thus important to be sure of the methodology used in a given laboratory before making a definitive diagnosis of hypercalcemia.

The important causes of hypercalcemia in childhood to be considered are shown in Table 4-1, and are classified in terms of the major physiologic alteration, either increased intestinal absorption of calcium or excessive solubilization of bone mineral that may coexist with increased intestinal uptake of calcium. Whether calcium is added to extracellular fluid by absorption from the intestine or by solubilization of bone mineral, or both, the major homeostatic control is at the site of entrance into extracellular fluid rather than at the excretory site in the kidney. The renal excretion of calcium does increase to some extent with increasing concentration of ionized calcium in plasma and consequently increased filtered load. The percentage of filtered calcium reabsorbed by the renal tubules is high, however, and although the percentage reabsorbed is reduced when the concentration of diffusible calcium in plasma is increased it usually is still greater than 95 per cent. The increment of urine calcium with increase of serum diffusible calcium concentration is less than the increment of filtered load. This protects the kidney against a concentration of calcium in tubular urine which would be so high that precipitation of calcium phosphate in the tubule would result. It prevents, however, the removal of excess calcium in the urine at a rate equaling the entrance of this ion into extracellular fluid when the latter rate is excessively high.

Table 4–1. Hypercalcemic States

I. INTESTINAL HYPERABSORPTION OF CALCIUM
- A. Hypervitaminosis D
- B. Idiopathic hypercalcemia
- C. Sarcoidosis
- D. Hyperparathyroidism
- E. Phosphate depletion

II. INCREASED MOBILIZATION OF BONE MINERAL WITH OR WITHOUT HYPERABSORPTION
- A. Hyperparathyroidism
- B. Tumoral hypercalcemia
  1. Tumors secreting parathyroid-like peptides (pseudohyperparathyroidism)
  2. Tumors secreting prostaglandins
  3. Metastatic tumors with osteolysis (osteoclast-stimulating factor)
  4. Myeloma
- C. Hyperthyroidism
- D. Immobilization

HYPERVITAMINOSIS D

PHYSIOLOGY AND CLINICAL MANIFESTATIONS

The intestinal absorption of calcium is a saturable process (*see Chapter 2*) that, if under normal homeostatic control, is limited so that even at intakes of calcium greatly above the normal requirement the quantity absorbed, over an extended period, does not exceed the rate at which calcium is removed from extracellular fluid either by uptake into skeleton or excretion in urine. This homeostatic control depends chiefly on the vitamin D cycle since the major factor determining the efficiency of calcium absorption is the vitamin D hormone, 1,25-diOH vitamin D.

The negative feedback control of the vitamin D cycle (see Chapter 2), particularly at the step of 1-hydroxylation of 25-hydroxyvitamin D by kidney mitochondria, is effective in preventing excessive formation of the active vitamin D hormone, 1,25-dihydroxyvitamin D, when the major source of vitamin D is the endogenous production of cholecalciferol in the skin as the result of exposure to short wave ultraviolet radiation, supplemented by the small amounts of vitamin D in foods such as fish and egg yolk. Hypervitaminosis D, therefore, does not result from maximum exposure to sunshine, with the possible exception of the patient with sarcoidosis, who will be discussed separately. When concentrated forms of vitamin D in the form of irradiated ergosterol or irradiated cholecalciferol are ingested the control system of the vitamin D cycle is no longer effective in preventing its toxic effects. It may be that very high concentrations of the unmetabolized vitamin D itself can produce hypercalcemia. It is more likely that 25-hydroxyvitamin D, the formation of which is not as closely controlled as that of 1,25-dihydroxyvitamin D, can accumulate in concentrations sufficient to produce the manifestations of excessive vitamin D hormone action even though 1,25-dihydroxyvitamin D is not present in excessive concentration.

It is established that in the nephrectomized rat or the anephric patient large doses of vitamin D or of 25-hydroxyvitamin D can increase intestinal transport of calcium and raise serum calcium concentrations, presumably in the absence of formation of 1,25-dihydroxyvitamin D, since the kidney is the only organ in which 1-hydroxylation of 25-hydroxyvitamin D takes place. Plasma 25-hydroxyvitamin D concentrations in such subjects are many times the normal concentrations, suggesting that this compound in supernormal concentration can produce the effects of an excess of the physiologic vitamin D hormone.

Vitamin D activity in serum can be measured by a bioassay in the vitamin D-depleted rat which determines the unmetabolized ergo- or cholecalciferol as well as the 25-hydroxy and 1,25-dihydroxy derivatives. The data suggest that, in patients receiving moderately large amounts of vitamin D, 25-hydroxyvitamin D represents the major compound determined in the serum by bioassay. 25-hydroxyvitamin D can be ascertained separately by a competitive protein-binding method utilizing either a rat serum or kidney cytosol preparation, which has a much greater affinity for the 25-hydroxyderivative than for the unmetabolized vitamin. The concentration of 1,25-dihydroxyvitamin D in serum is so much less than that of 25-hydroxyvitamin D that it provides a quantitatively insignificant fraction of the total 25-hydroxyvitamin D measured.

As a result of bioassays of vitamin D in patients and animals receiving large amounts of this compound, it has been learned that in the mammal most of the vitamin D is accumulated in adipose tissue and plasma. Liver storage is unimportant, in contrast to the findings in certain species of fish in which large amounts of vitamin D are accumulated in the liver. The significance of this difference with regard to the physiologic action of vitamin D is unknown. Vitamin D is only slowly metabolized and degraded in the body, so that when amounts greatly in excess of the physiologic requirements are ingested a cumulative effect is obtained as the concentration in tissues and plasma progressively increases. When toxic effects are recognized the amount stored in adipose tissue is sufficiently great for serum concentrations to remain high for months after vitamin D administration is discontinued. This cumulative effect due to storage of vitamin D is the basis for the serious consequences of continued intake of vitamin D in excessive amounts. The toxic dose of vitamin D in man varies, partly as a result of apparent marked individual differences in response to vitamin D intake, but to a great extent because of the cumulative nature of vitamin D over dosage.

At the moment there have been too few determinations of 25-hydroxyvitamin D in serum of patients receiving large doses of vitamin D to permit definition of a toxic concentration of this vitamin D metabolite. Many of the studies of serum 25-hydroxyvitamin D concentration have been in patients receiving pharmacologic doses of vitamin D for such disorders as hypoparathyroidism or hypophosphatemic rickets, rather than in subjects given high doses of vitamin D as a prophylaxis against vitamin D deficiency, and concentrations as high as five to ten times the normal value have been reported. Toxic effects of high vitamin D dose have been seen both in patients given pharmacologic amounts of vitamin D for diseases for which it is indicated and in otherwise normal subjects whose vitamin D intake was high owing to a mistaken enthusiasm for vitamins. With the reservations in mind about the difficulty of establishing a toxic dose of vitamin D, it seems likely that for older children and adults 2000 units/kg body weight per day given over a period of many months will produce hypercalcemia and hypercalciuria in most subjects. In infants, 20,000 to 40,000

Table 4–2. Response of Vitamin D-Deficient Infant to "Stosstherapy" with Vitamin D

	R. T. AGE 6 MONTHS SERUM		
DAY OF STUDY	P_i *mg/dl*	*Ca* *mg/dl*	*Alk. P'tase Bodansky Units*
1	3.2	7.7	27
3	3.2	7.2	27
4	3.2	–	24
5	600,000 units Vitamin D_2 in oil given in 24 hours		
8	4.1	9.3	16
11	4.7	9.0	16
18	5.5	10.8	21
26	6.3	10.4	11

The rapid increase of serum calcium concentration following this large dose of vitamin D is noteworthy. The serum phosphate concentration also shows an early rise but increases progressively after the serum calcium concentration has plateaued. No hypercalcemia results from this massive dose treatment.

units per day has caused fatal hypervitaminosis D, and some may be injured by much less than this amount.

The importance of duration of vitamin D therapy in causing toxicity is emphasized by the fact that a single large dose of vitamin D can be tolerated and has been used in the prevention and treatment of rickets in infancy, so-called "stosstherapy." A dose of 600,000 units (15 mg) of vitamin D given at one time, or in several divided doses during a 24-hour period, can be used for the rapid treatment of vitamin D deficiency rickets (Table 4-2). In some European countries this dose was used for the prevention of rickets in infants whose social circumstances made it likely that the normal daily requirement of vitamin D would not be given regularly. This large dose was tolerated, possibly because when given as a bolus it was not completely absorbed, and much of it was inactivated by liver metabolism and eliminated in the bile as inert derivatives. Enough vitamin D was retained to provide protection against rickets for many months. The danger of such massive dose vitamin D prophylaxis lay in the possibility that repeat doses at too frequent intervals might be given through inadvertence or because of a mistaken clinical impression of early rachitic manifestations in the infant.

In some countries in Europe, North Africa and the mid-East highly concentrated preparations of vitamin D have been available without a prescription. As a result of ignorance concerning the dangers of vitamin D overdosage, and because of the feeling that very high vitamin intake might be beneficial in promoting growth and resistance to infection, many infants and children received excessive amounts of vitamin D owing to this availability. In the United States concentrated preparations of vitamin D are prescription drugs, and the content of vitamin D in multivitamin preparations sold without prescription is limited to 400 units per recommended daily intake of the vitamin preparation. This type of hypervitaminosis D, therefore, has been much less common in the United States than in some other parts of the world.

The major toxicity of vitamin D is due to its action on intestinal absorption of calcium, with hyperabsorption resulting from excessive dose. Additionally, at high concentrations of vitamin D a PTH-like effect on bone cells with an enhanced rate of solubilization of bone mineral also results, so that the large amounts of calcium taken up from the gut are not retained in bone and the concentration in extracellular fluid rises despite increase of urine calcium excretion. The action on bone cells is probably mediated by the vitamin D metabolites rather than the unmodified vitamin D. At least experimentally, 25-hydroxyvitamin D (but not vitamin D) added to in vitro systems stimulates release of calcium from bones in tissue culture preparations analogous to the action of parathyroid hormone. As hypercalcemia develops calcium phosphate is precipitated in soft tissues, and this metastatic calcification is the major toxic effect of chronic hypervitaminosis D.

The earliest clinical evidence of hyperabsorption of calcium due to excessive intake of vitamin D is increased urine excretion of calcium, which is a valuable index of calcium absorption in the absence of direct determination of intestinal uptake of this ion. The simplest method for measuring intestinal absorption of calcium, i.e., the determination of uptake of an orally administered radioactive isotope of calcium (see Chapter 2), cannot be employed in children owing to restrictions on the administration of radioactive isotopes to children. Calcium balance studies are extremely difficult to perform. The determination of urine calcium excretion is, therefore, of importance in ascertaining response to vitamin D in patients receiving pharmacologic doses of vitamin D, particularly those with primary hypophosphatemic rickets who are at great risk for hypervitaminosis D if not closely supervised.

The excretion of calcium in the urine of normal children and young adults has been determined by Knapp and the data are plotted in Figure 2-6. In the growing child the urine excretion of calcium varies only slightly with intake of calcium. On a moderately high calcium intake of 1000 to 1400 mg/day the average daily calcium output of a child under 4 years of age is approximately 25 to 50 mg/day, and of children between 5 and 14, 75 to 100 mg/day. In young adults on a similar total intake the excretion is between 150 to 200 mg/day. On a weight basis the urine excretion of calcium in children is normally in the range of 2 to 4 mg/kg per day, or if expressed in terms of creatinine excretion the upper limit of normal is 0.25 mg calcium per mg creatinine. Urine excretion of calcium above these normal values in subjects receiving large doses of vitamin D would be a danger signal indicative of hyperabsorption of calcium and requiring reductions of vitamin D dosage. Total serum calcium concentrations in the fasting state might still be in the normal range at a time when urine calcium excretion is increased sufficiently to warrant attention. If accurate collection of 24-hour samples of urine is not possible an approximate estimate of urine calcium output can be made by determining calcium and creatinine concentrations in as much urine as can be collected, and calculating calcium as mg per mg creatinine. In children, a value greater than 0.25 mg calcium per mg creatinine is evidence of sufficient hypercalciuria to require reduction or elimination of vitamin D treatment.

The early symptoms of vitamin D-induced hypercalcemia are those of any form of hypercalcemia: anorexia; constipation; nausea and vomiting; polydipsia; polyuria; weight loss; and growth failure. The polyuria is caused in part by unresponsiveness to antidiuretic hormone induced by the hypercalcemia. Dehydration due to excessive loss of water in the urine, plus poor water intake resulting from the anorexia and nausea of hypercalcemia, may cause a precipitous rise of serum calcium concentration with severe disorder of CNS function, giving rise to abnormalities of mentation, drowsiness, stupor and coma, or in some instances convulsions. In addition to direct action of calcium on nerve cells, CNS manifestations of hypercalcemia may be the result of calcium-induced contraction of vascular smooth muscle with hypertension, hypertensive encephalopathy and cerebral ischemia. Marked elevation of serum calcium concentration is almost invariably associated with CNS manifestations that sometimes can be misdiagnosed as psychosis. The hypertension is probably due to direct action of calcium in increasing the tone of smooth muscle of arteries and arterioles, but additionally may be intensified by increased activity of the renin-angiotensin system resulting from renal arteriolar constriction.

Renal insufficiency may arise from the renovascular effects of hypercalcemia, but more importantly from precipitation of calcium phosphate in the interstitial tissue of the kidney with resultant tissue reaction, fibrosis and obliteration of nephrons. The marked hypercalciuria may also cause intratubular precipitation of calcium salts or formation of stones in the more distal urinary tract. Metastatic calcification (Fig. 4-1) can also occur in soft tissues: skin; subcutaneous tissue; falx cerebri; media or large

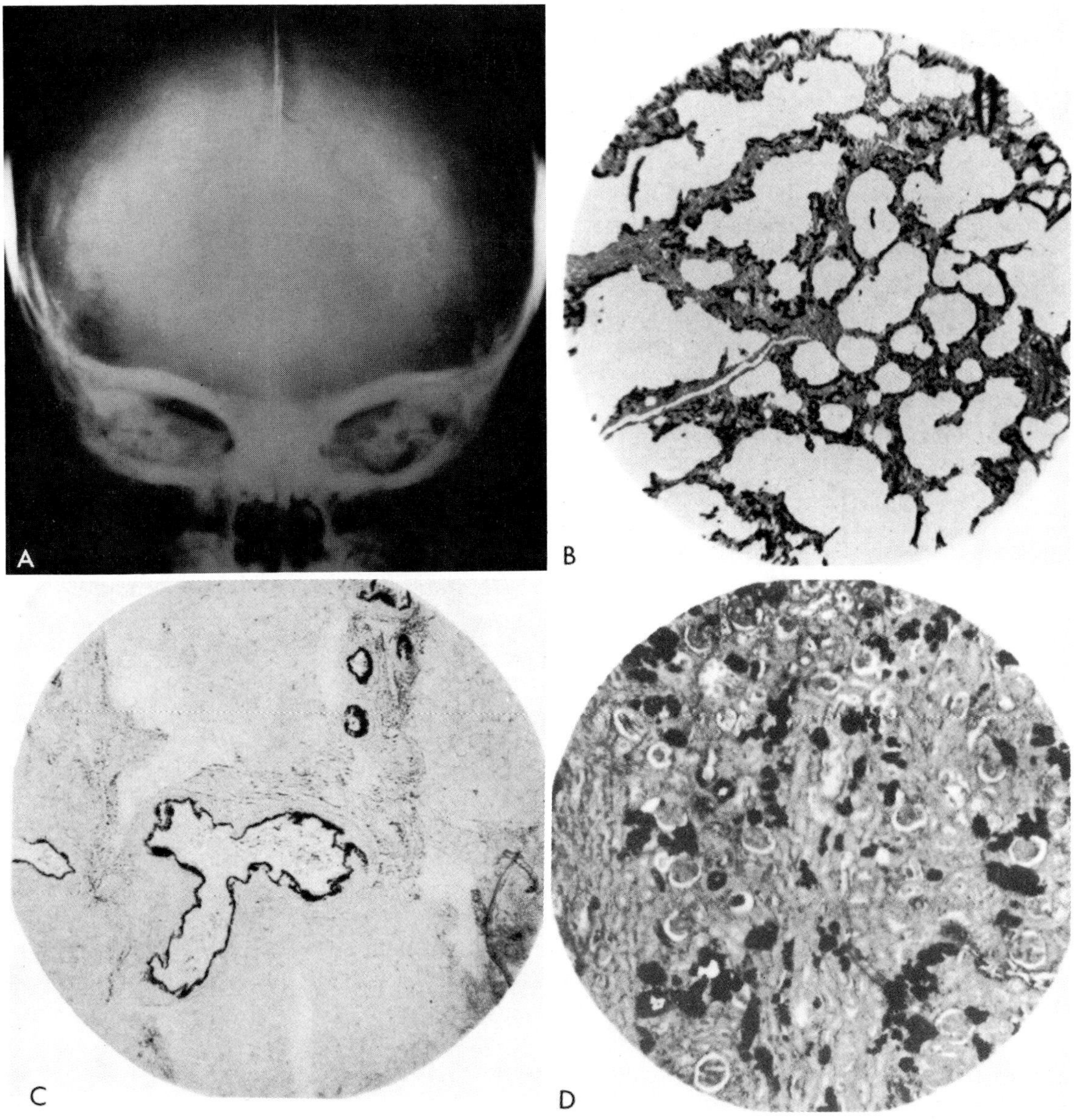

Figure 4–1. Metastatic calcification in hypervitaminosis D: (*a*), falx cerebri; (*b*), lungs; (*c*), intima of artery; (*d*), kidney. (*b*, *c* and *d* from Ross, S. G.: Vitamin D intoxication in infancy. J. Pediatr. *41*:815–822, 1952. Reprinted with permission.)

arteries; and in the most severe instances the lung, myocardium and gastric mucosa. Since calcium stimulates release of gastrin, which stimulates gastric hydrochloric acid secretion, hyperchlorhydria and multiple peptic ulceration may contribute to the gastrointestinal symptoms of vitamin D intoxication. The major danger, however, is the renal injury, which can be irreversible by the time the hypercalcemia is detected and treatment instituted. This kidney damage can be insidious in onset and may progress to severe renal insufficiency in the absence of the more dramatic symptoms of hypercalcemia. For this reason, patients treated with pharmacologic amounts of vitamin D for resistant rickets or hypoparathyroidism must be monitored closely to prevent irreversible renal injury.

Another manifestation of chronic hypercalcemia is band keratopathy resulting from deposition of calcium phosphate crystals in the cornea beneath Bowman's membrane. Subconjunctival calcium deposits may also be seen. These foreign bodies set up an inflammatory reaction with hyperemia and photophobia. The ocular signs of hypercalcemia have been seen more frequently in adults, perhaps because of the greater chronicity of the disorder in older patients.

The roentgenographic findings in hypervitaminosis D are of importance in the differential diagnosis of hypercalcemia in the infant and young child. There is characteristically an intense line of mineralization of the provisional zone of calcification at the metaphyseal end of the long bones (Fig. 4-2). This is not only denser but usually wider than the line of calcified cartilage seen in normal growing bone. At a later stage the submetaphyseal bone may be excessively reabsorbed so that a zone of rarefaction is seen proximal to the metaphyseal plate. The skull shows increased density of mineralization of the base and the orbits.

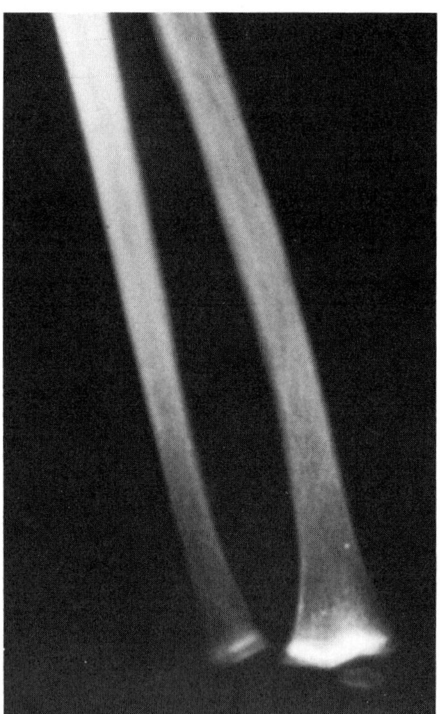

Figure 4–2. Roentgenographic changes in long bones of infant with hypervitaminosis D, showing broad zone of increased density at growing end of bone.

Metastatic calcification of the falx cerebri can also be visualized. As mentioned above, soft tissue shadows of metastatic calcification are also found, including stippled calcification in the kidney if the nephrocalcinosis has become marked.

Analysis of the serum reveals increase of serum calcium concentration and variable changes in the serum phosphate concentration that may be normal or high, depending on the state of renal function, but not reduced as in primary hyperparathyroidism. The alkaline phosphatase concentration is usually low, but if the vitamin D intoxication has been severe and prolonged with marked demineralization of bone, secondary repair and hyperphosphatasia may be found. Serum IPTH concentrations should be low. There usually is marked hypercalciuria unless renal insufficiency has progressed to the stage of severe reduction of glomerular filtration rate. Corticosteroids in suppressive amounts should reduce the serum calcium concentration, differentiating hypervitaminosis D from hyperparathyroidism, and this response combined with low concentrations of serum IPTH should clearly differentiate vitamin D toxicity from hyperparathyroidism. Since the hypercalcemia of sarcoidosis and that associated with certain tumors may also show response to the corticosteroid suppression test along with reduced concentrations of IPTH, the history of vitamin D intake and the roentgenographic findings are of importance in the differential diagnosis. Finally, it is now feasible to determine 25-hydroxyvitamin D in the serum, and such measurements in patients receiving pharmacologic doses of vitamin D have shown concentrations many times greater than the values found in individuals on usual doses of vitamin D. This, therefore, can be the definitive determination in the establishment of the diagnosis of excessive vitamin D intake.

TREATMENT

The treatment of hypervitaminosis D depends on the stage of toxicity and starts obviously with the elimination of vitamin D intake, specifically the concentrated vitamin D preparations that have resulted in overdosage and also vitamin D-fortified milk. Since most other commonly ingested foods are not major sources of vitamin D, other restrictions are not necessary. Intensive exposure to sunshine should be avoided although it is doubtful whether this will be a major factor in response to therapy. Calcium intake also is reduced since vitamin D excess is associated with hyperabsorption of calcium. Elimination of milk reduces both calcium and vitamin D intake. Since vitamin D is accumulated in the adipose tissue, manifestations of excessive vitamin D function may persist for months after its intake is discontinued. If the symptoms of hypercalcemia are severe rapid reduction of serum calcium is essential.

One method for rapid removal of calcium from extracellular fluid is the use of a calciuretic diuretic such as furosemide or ethacrynic acid. Diuretics of this type inhibit tubular reabsorption of calcium as well as sodium and water, and in combination with a high sodium and water intake they markedly increase urine calcium output. Benzthiazide diuretics such as chlorothiazide must not be used since they increase tubular reabsorption of calcium and are hypercalcemic in their effect. By combining intravenous furosemide with a high intake of sodium and potassium, maximal calcium diuresis can be accomplished without depletion of body sodium and potassium. If, along with furosemide, 1 mg/kg given intravenously at six- to eight-hour intervals, a solution of one-third 5 per cent glucose and two-thirds Ringer's lactate or saline, with potassium chloride added to make a final concentration of potassium of 30 mEq/liter, is infused at a rate equal to the urine output plus estimated evaporative water loss, a high sodium and water excretion can be maintained without body water deficit, and maximum calciuresis is achieved. Such treatment can reduce serum calcium concentrations appreciably. At the same time cortisone, 10 mg/kg up to 300 mg/day, or prednisone, 2 mg/kg up to 60 mg/day, is begun, and the furosemide diuresis is discontinued as soon as the serum calcium concentration is reduced below dangerous concentrations, i.e., 12 mg/dl or less. A high sodium and water intake should be maintained to promote continued large urine volume and calcium excretion. The low calcium, low vitamin D diet is also continued and the steroid dose gradually reduced after ten days to two weeks, the rate of reduction being dependent on the stability of the serum calcium concentrations.

In some severe hypercalcemic states calciuretic treatment may not suffice, and additional modes of therapy may be required. Another therapeutic approach to reducing serum calcium concentrations at this stage is a high phosphate intake, given orally in the form of a buffered solution of sodium and potassium phosphates such as Neutra-Phos. Sufficient amounts of these phosphate solutions are given to supply phosphate equivalent to 1 to 2 gm of phosphorus per day. In adults a slow intravenous infusion of sodium phosphate solution followed by oral sodium phosphate has been effective in reducing serum calcium concentrations. Experience with this treatment has been entirely in adults, and the recommended dose is 50 m mols of phosphate as sodium phosphate in an isotonic solution given slowly intravenously over six to eight hours, followed if possible by oral sodium or potassium phosphate, providing 1 to 3 gm of phosphorus per day in divided doses. Serious immediate reactions have been reported with the intravenous phosphate treatment, possibly owing to formation of aggregates of colloidal calcium phosphate intravascularly as the result of increased phosphate concentration. At the moment, this treatment cannot be recommended for children.

Either as a supplement or an alternative to calciuretic diuretics, calcitonin preparations can be given to reduce serum calcium concentrations through inhibition of the solubilization of bone mineral, thus reducing the delivery of calcium and phosphate from bone into extracellular fluid and resulting in a net increase of uptake of calcium and phosphate by bone. Synthetic salmon calcitonin has more prolonged action than mammalian calcitonins, and therefore is more effective in the treatment of hypercalcemia than human or porcine calcitonin. Salmon calcitonin is now available as the preparation Calcimar, which can be injected either intravenously or intramuscularly. Intramuscular injection of 5 to 8 MRC units of calcitonin per kg body weight at six- to twelve-hour intervals has been successful in the reduction of serum calcium concentrations. Corticosteroid therapy can be started with the calcitonin injections; after several days of treatment the latter can be discontinued, and corticosteroids plus a low calcium diet should be adequate to prevent recurrence of hypercalcemia.

Mithramycin, an antibiotic that inhibits DNA-dependent mRNA synthesis, also has a hypocalcemic action, but because of its toxicity has been reserved for special needs. It has been used mainly in the treatment of the hypercalcemia associated with metastatic tumors or parathyroid carcinoma.

In hypercalcemic patients with poor kidney function the intravenous treatment with furosemide and sodium chloride solution cannot be used, and intravenous or oral sodium phosphate would be particularly dangerous in such subjects because of nephrotoxicity of precipitated calcium phosphate. Salmon calcitonin injections and, if necessary, peritoneal dialysis with a calcium-free dialysis solution would probably be the most suitable treatment, corticosteroids being started simultaneously with the emergency therapy and continued as outlined above.

The acute nervous system manifestations of hypercalcemia are dramatic and there is real danger of death during the acute crisis. Following recovery, chronic brain damage with mental retardation can ensue, particularly in infants.

The renal injury of hypercalcemia can be exceedingly insidious in its appearance, and presents a major danger since the seriousness of the problem may be missed until renal injury has reached the stage of progressive irreversible change. This progression may be slow, and two or three decades may elapse before end/stage renal insufficiency results. It is therefore important to determine renal function over a prolonged period following an episode of hypervitaminosis D. Even if the hypercalcemia is recognized and treated promptly the renal injury can progress because of scarring, so that eventual renal failure takes place. This has been an unfortunate sequel in patients receiving pharmacologic vitamin D treatment for primary hypophosphatemic rickets, hypoparathyroidism or the hypocalcemia of renal insufficiency, and necessitates careful monitoring of serum calcium concentrations in such patients.

Because of the slow metabolic inactivation or excretion of vitamin D and its storage in adipose tissue, hypervitaminosis D may only appear after months of treatment at a constant dose, so that monitoring of serum and urine calcium concentrations cannot be discontinued because of months of apparent stability. As mentioned above, the quantitative measurement of urine calcium output in a patient with normal renal function is a sensitive method for gauging excess absorption of calcium from the gut and therefore impending vitamin D toxicity. If 24-hour specimens can be collected, an output of greater than 2 mg/kg per day should be closely followed and vitamin D or dihydrotachysterol dose reduced. If output reaches 4 mg/kg per day, the vitamin D treatment should be discontinued and restarted at a lower dose when urine calcium excretions have decreased below 2 mg/kg. If accurate 24-hour collections are not feasible, the urine calcium-to-creatinine ratio can be used as a rough index, and a calcium concentration of greater than 0.25 mg per mg creatinine should be a signal for more thorough study of calcium concentrations in serum and excretion in urine.

Because of the cumulative nature of the toxicity of large doses of vitamin D, and the prolonged time required to eliminate the large stores of vitamin D in the body, it is desirable to substitute steroids with a shorter biologic half-life in the treatment of patients who require pharmacologic doses of vitamin D. In primary hypophosphatemic rickets, hypoparathyroidism and renal insufficiency, dihydrotachysterol$_2$, a derivative of irradiated ergosterol, is available and can be used effectively in place of vitamin D. The value and use of this compound is discussed in detail in the sections on these disease entities. The vitamin D hormone, 1,25-dihydroxyvitamin D, and a synthetic analog, 1-αhydroxyvitamin D, have been used experimentally in these three disorders. They are effective in much smaller amounts than vitamin D or dihydrotachysterol and also have a short biologic half life. 1,25-dihydroxyvitamin D has recently become available.

IDIOPATHIC HYPERCALCEMIA (WILLIAMS SYNDROME)

CLINICAL MANIFESTATIONS

Idiopathic hypercalcemia is still a poorly understood entity, although it generally is explained as a subgroup of hypervitaminosis D in infants unusually reactive to amounts of vitamin D that would not produce hypercalcemia in the ordinary individual. This type of hypercalcemia was first noted in Great Britain during and after World War II, when more than 200 cases were reported. The early cases recognized were infants who ate poorly, failed to grow and gain weight, and when studied further proved to have impaired kidney function and hypercalcemia that did not seem to fit the pattern of hyperparathyroidism. The serum phosphate concentrations were variable, but not usually reduced as in hyperparathyroidism, and the roentgenographic changes in the skeleton were markedly different from those of hyperparathyroidism. The bones did not show demineralization but rather osteosclerosis, particularly the bones of the base of the skull which arise from cartilage. The long bones also showed dense bands of mineralization at the metaphyseal ends resembling the lesions of hypervitaminosis D. However, these children had been receiving only the vitamin D routinely prescribed for children in the British Isles at that time.

As investigations continued it was recognized that during and after World War II the vitamin D intake of infants and children in Britain had been increased in the concerted effort to prevent nutritional deficiencies in children subject to the dislocations of wartime. The national dried milk preparation available without cost to families was fortified with vitamin D in a concentration of 1800 units per equivalent quart of reconstituted milk. Because the manufacturers were anxious that the vitamin D potency on bioassay should not decrease below the 1800 units per quart during storage, the actual fortification was appreciably in excess of this concentration. In addition, the routine administration of cod liver oil to infants and young children was continued. Cereals were also fortified with vitamin D. In consequence of all of these factors many infants and children received 3000 to 4000 units of vitamin D per day.

Of the large number of children taking this amount only a small percentage developed the overt manifestations of hypercalcemia. The basis of this hyperreaction remains unsolved. It is now known that vitamin D is metabolized by 25-hydroxylation in the liver to 25-hydroxyvitamin D, and that the latter is the substrate for the 25-hydroxyvitamin D 1-hydroxylase system in the kidney, the product of which is the vitamin D hormone, 1,25-dihydroxyvitamin D. Since in man as little as 1 μg of 1,25-dihydroxyvitamin D per day is adequate to provide the vitamin D requirement, it is not difficult to imagine that the daily intake of 50 to 100 μg, 2000 to 4000 units of the vitamin D precursor, might produce hypervitaminosis in those who, because of individual metabolic idiosyncrasy, produced excessive amounts of vitamin D hormone, the production of which is usually controlled by regulatory systems dependent on calcium concentrations in extracellular fluid. This is hypothetic, however, since determinations of 25-hydroxyvitamin D or 1,25-dihydroxyvitamin D have not yet been reported in patients with idiopathic hypercalcemia during the acute phase of the disease.

We have talked thus far primarily in terms of the postnatal intake of vitamin D in infants and children. As a result of the epidemiologic studies of idiopathic hypercalcemia by investigators in Britain, the amount of vitamin D added to fortified milk was reduced and other sources of vitamin D discontinued, so that the daily intake of vitamin D of most infants and chidren was reduced to approximate the quantity routinely given in the United States, i.e., 400 units per day. There is no doubt that the incidence of idiopathic hypercalcemia has decreased and that it is now again a rare entity, which suggests that the increased quantity of vitamin D given to large numbers of infants and children was the responsible factor.

In apparent contradiction to the above statement is the fact that the physical findings of one form of severe idiopathic hypercalcemia (Williams syndrome) may be present at, or soon after, birth, suggesting a problem of intrauterine development. These physical features consist of maldevelopment of the facial bones, producing the so-called elfin facies, i.e., a depression of the nasal bridge, a prominent maxilla and a receding mandible. The nose is short and upturned and the upper lip is prominent. Hypertelorism, strabismus, a wide epicanthic fold and large, low-set ears may also contribute to the distinctive facies. An additional skull abnormality is craniostenosis, but the time of onset of premature suture fusion is not known. Failure of weight gain and growth starting in early infancy also suggest an early metabolic error. Finally, a considerable percentage of these infants have distinctive congenital defects of the vascular system, supravalvular aortic stenosis or peripheral pulmonic artery stenosis.

Although it has not always been possible to date the onset of these various manifestations of idiopathic hypercalcemia of the Williams type, murmurs were heard at birth in several patients. Most of these infants have also shown delayed functional development, and psychometric tests done in later infancy have usually indicated appreciable reduction of the developmental quotient. All these features suggest that the abnormality has its onset in intrauterine life, but there is no evidence that the vitamin D intake of the mothers has been excessive: when the mothers are studied at the time the diagnosis is entertained, which is usually many months after delivery, their serum calcium concentrations have been found to be normal. There have been a few instances of affected siblings but the familial incidence is actually uncommon. It is therefore difficult to hypothesize an error of metabolism of vitamin D in the mother, although in one family studied by us in which two siblings were affected the mother showed a suggestive hyperresponsiveness to small doses of vitamin D.

Experimentally, vascular lesions similar to those seen in Williams syndrome have been produced in rabbit pups born of dams given large amounts of vitamin D in pregnancy. Nevertheless, the relationship of vitamin D in pregnancy to the syndrome of idiopathic hypercalcemia remains a mystery. At the time the diagnosis is made, which may be when the infant is many months of age, the infant's serum calcium concentration may be normal, and it can only be assumed that hypercalcemia was once present either prenatally or in the early postnatal months. Further, osteosclerosis of the skull, and in some patients nephrocalcinosis or extraskeletal calcification in blood vessels or other tissues, strongly suggest that hypercalcemia was once present. It is known that the calcium concentration of fetal plasma is higher than that of the maternal plasma, and it is believed that the placenta has a calcium transport system that actively pumps calcium into the fetal circulation. The role of vitamin D and parathyroid hormone in this placental system is not known. It is theoretically possible that fetal hypercalcemia might result from hyperfunction of the placental calcium transport system in response to unknown factors, but this also can only be a speculation.

The major conclusion to be reached is that excessive amounts of vitamin D should be avoided both by the mother during pregnancy and for the infant postnatally. The recommended daily allowance of 400 units of vitamin D (10 μg) is adequate for the pregnant woman, particularly since exposure of the skin to sunshine in ordinary environments also provides a source of vitamin D. The same is true of the infant, whether full term or premature. The rare instances of idiopathic hypercalcemia seen today in infants whose mothers have not taken excessive vitamin D, and who have not been given more than the recommended quantity of vitamin D, remain as an unexplained metabolic problem.

TREATMENT

The treatment of idiopathic hypercalcemia is similar to that of hypervitaminosis D. A diet low in vitamin D with minimum calcium intake is prescribed. It is possible to prepare a low calcium diet for infants by substituting for milk a preparation consisting of strained meat, vegetable oil and glucose. A preparation we have used had the following composition:

1½ cups strained lamb (prepared infant food);
2⅔ tbs sugar;
½ tsp salt;
2 tsp olive oil;
32 oz water;
2⅓ tbs rice flour.

These ingredients are mixed in a Waring blender and fed in the usual volume for prepared infant feeding.

Corticosteroid treatment is also instituted in the severely hypercalcemic infant, and continued until serum and urine calcium concentrations have returned to normal. Following gradual discontinuance of the steroid the calcium intake is progressively increased as tolerated by the patient. Eventually these children can take an ordinary diet without recurrence of the hypercalcemia and hypercalciuria. It is recommended that vitamin D intake continue to be minimal but there is no reason to proscribe ordinary exposure to sunshine. In fact, one of the mysteries of the condition is that following recovery these children apparently can tolerate usual vitamin D intakes without relapse of the hypercalcemia.

CASE REPORTS

S.H., a white female, was admitted at 5 years of age with complaints of facial weakness; unilateral, intermittent abdominal pain; poor weight gain; and a history of documented hypercalcemia.

She was born at term, birth weight 2610 gm following a normal pregnancy. The mother had taken supplemental calcium. There was one male sibling in good health. The infant was fed a vitamin D-fortified milk and given additional vitamin D, 1000 units per day. She vomited frequently from birth on, gained weight slowly, was constipated and had frequent respiratory illnesses. At 10 months of age she was hospitalized because of failure to thrive, and serum calcium concentrations were increased, ranging from 12.2 to 16.2 mg/dl, with serum phosphate concentrations from 4.0 to 5.0 mg/dl. Renal function tests gave normal results and radiographs of the long bones showed some increased density of the metaphyseal plates. No specific therapy was given. She continued to vomit, and weight gain and growth were subnormal. At 3 years of age anorexia (except for milk) was a problem, and polyuria and polydipsia were now noted. During the fourth year she began to complain of abdominal pain, and two months before admission to the Johns Hopkins Hospital she had a laparotomy at another hospital because of abdominal pain, tenderness, muscle spasm, fever and leukocytosis. At laparotomy calcified plaques were found on the peritoneal surfaces.

At admission she was a thin, apathetic child, 11.5 kg in weight and 96 cm in height. Blood pressure was 120/90. The conjunctivae were reddened and slit lamp microscopy revealed crystalline deposits beneath the basement membrane. The teeth were discolored. Abdominal examination did not reveal any masses. Laboratory data showed serum calcium concentration 17.1 mg/dl, serum phosphate 4.8 mg/dl, alkaline phosphatase 15.4 Bodansky units, nonprotein nitrogen 60 mg/dl and serum bicarbonate 23 mEq/l; hemoglobin concentration was 11.0 gm/dl; and urine specific gravity 1.014, with a trace of albumin. Roentgenograms of the abdomen showed intra-abdominal calcification (Fig. 4-3a). Skull examination revealed calcification of the falx cerebri (Fig. 4-3b) and the internal carotid arteries. The long bone x-rays showed poor mineralization with submetaphyseal radiolucency and relative increased density of metaphyseal plates.

A determination of vitamin D in serum was made by bioassay in vitamin D-depleted rats and the value of 23 units per ml was obtained, far in excess of the normal value of 1 to 2 units per ml. Several days

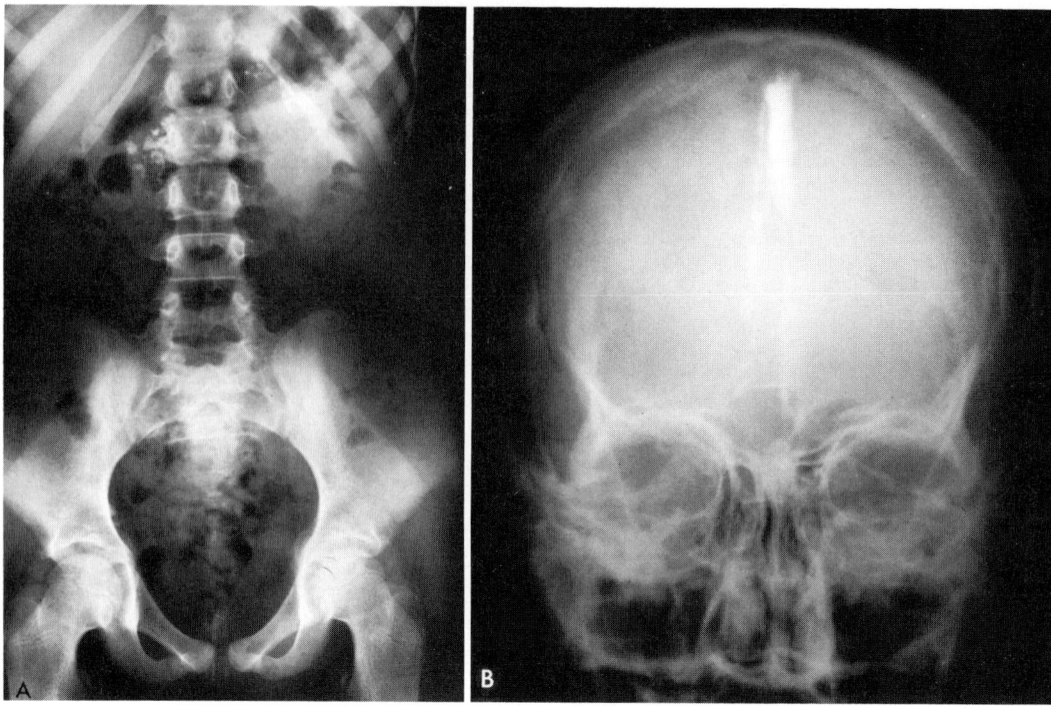

Figure 4–3. *a*, Intra-abdominal metastatic calcification found in a patient with idiopathic hypercalcemia (S.H.) *b*, Calcification of falx cerebri.

after admission following a period of poor fluid intake the patient had acute abdominal pain, vomited bloody material and had a convulsive seizure followed by coma. Blood pressure was 210/160. The serum calcium concentration was 16.0 mg/dl and that of phosphate 6.8 mg/dl. She was treated with intravenous Na_2EDTA (a calcium chelating agent), 1000 mg, over several hours. She regained consciousness and began taking fluids orally. The hypertension persisted and did not respond to intramuscular injection of magnesium sulfate. The possibility of a parathyroid adenoma with parathyroid crisis was considered, and an exploration of neck and anterior mediastinum failed to reveal an enlarged parathyroid gland.

Treatment with cortisone, 100 mg/day, was started with a decrease in serum calcium concentration, but when cortisone administration was reduced serum calcium rose again. During this time the calcium intake had not been reduced because the child refused most foods, except milk. Because of persistent hypercalcemia she was finally placed on a low calcium diet, less than 100 mg/day, based on a strained lamb formula. Vitamin D administration was also eliminated. Despite no dietary vitamin D a serum vitamin D assay 21 months later still showed an elevated concentration, 5 units/ml. Following two years of low calcium vitamin D-free diet, serum calcium concentrations decreased to the normal range and the diet was gradually liberalized. At this time blood pressure had fallen to 106/70. Renal function tests showed fixed specific gravity and moderate elevation of serum nonprotein nitrogen concentration.

This girl has been followed for over 20 years and is now 25 years of age. She is short in stature and moderately retarded in mental development. She has had several convulsive seizures despite normocalcemia and has been treated with anticonvulsants. Serum calcium and phosphate concentrations are persistently normal, as is serum 25-OH vitamin D concentration. The most recent values are: calcium, 10.8 mg/dl; phosphate, 4.0 mg/dl; and 25-OH vitamin D, 40 ng/ml. She has continued on a moderately low calcium diet without added vitamin

D, but with no limitation on outdoor activities in the summer, so that most of the vitamin D presumably is of endogenous origin as a result of exposure to sunshine. She is not hypertensive and renal function shows no progressive deterioration.

COMMENTS: This infant born in 1949 had hypercalcemia of unknown pathogenesis. This entity was just beginning to be documented and discussed, and the possible relationship to an abnormality of vitamin D metabolism was not recognized. By the age of 5 when she suffered severe manifestations of hypercalcemia including ectopic calcification in the abdomen, nephrocalcinosis and hypertension with neurologic sequelae, assays of vitamin D in the serum indicated unusually high concentrations despite the fact that the vitamin D intake, although generous, had never been in the range accepted as toxic. Treatment with corticosteroids was only partially successful until she was placed on an extremely low calcium intake and a diet virtually free of vitamin D.

The subsequent course has been of unusual interest in that the patient became normocalcemic and has remained so on a diet that is not greatly restricted but is moderately low in calcium, since only minimal amounts of milk are taken. She suffered unfortunate consequences of prolonged hypercalcemia, namely, stunting of growth, some degree of brain damage, and moderate but not progressive renal injury. At the present time serum 25-OH vitamin D concentrations are in the normal range, but no attempt has been made to determine the response to increased vitamin D intake because of understandable opposition to the risk of hypercalcemia and further injury. As the subsequent case report indicates, the abnormality in this patient was familial and presumably genetic, but the specific disturbance of vitamin D metabolism remains to be unraveled. The somewhat low birth weight and symptoms, particularly marked irritability dating back to the neonatal period, suggest that hypercalcemia was present at birth or in intrauterine life, but the earliest documented increase of serum calcium concentration was at 10 months of age.

K.H., a sibling of S.H., was born 7 years after her sister. The pregnancy history was not unusual. The mother took 400 units vitamin D per day and 250 mg calcium in a vitamin mineral preparation. The infant was born at 38 weeks' gestation weighing 2200 gm. Aside from her small size, she seemed normal at birth, was not hyperirritable and fed without difficulty. Her feeding was a diluted evaporated milk preparation providing 400 units per quart of vitamin D. In addition she received a vitamin preparation with vitamin D, 1000 units per day. At the age of 3 months she was studied because of the history of hypercalcemia in her sister. The serum calcium concentration was reported as 14.5 mg/dl. At 3 months of age she weighed 4500 gm and was 60 cm in length. She did not have an "elfin" facies and no cardiac murmur was heard. Blood pressure was 75/50 mm Hg. X-ray examination of the skull and long bones showed suggestive increase of density at the base of the skull and metaphyseal ends of the long bones. Serum calcium concentrations remained elevated until she was fed a low vitamin D milk diet, and during this diet serum vitamin D concentrations measured by bioassay in the rachitic rat were in the range of 2 units per ml, which is at the upper limits of normal. However, when a feeding of vitamin D-fortified milk was resumed, 400 units per day, serum vitamin D concentration rose to 8 units per ml and serum calcium increased to 12.5 mg/dl. Urine calcium concentrations were also high, up to 118 mg/day or more than 20 mg/kg. A low vitamin D diet was then resumed with decrease of serum calcium concentration, of urine calcium excretion and of serum vitamin D concentration.

At 1 year of age the child weighed 7.1 kg and was 71 cm in length. At 18 months she weighed 9.1 kg and was 76 cm in length. Serum calcium concentration was 10.7 mg/dl at this time and mental development apparently normal. Since that time she has been on an unrestricted diet except that vitamin D-fortified milk was limited and no extra vitamin D was given. There has been no restriction of outdoor exposure to sunshine. She is now 17 years of age and an apparently normal young adult. Renal function and blood pressure have been normal and hypercalcemia has never recurred. At this time serum 25-OH vitamin D was 65 ng/ml, which is within the normal range although on the high side.

COMMENTS: The familial incidence of idiopathic hypercalcemia does suggest an inborn error of metabolism. The moderately high concentration of vitamin D by bioassay in the serum is compatible with accumulation of a physiologically active metabolite of vitamin D. The puzzling aspect of the case is the apparent spontaneous recovery and lack of hypersensitivity to vitamin D after infancy.

This child with mild manifestations of idiopathic hypercalcemia could well have been missed if serum calcium determinations had not been done on the basis of the finding in the previous sibling. The only clinical finding was some retardation of growth and weight gain in the first year of life. The relative infrequency of reports of multiple cases of idiopathic hypercalcemia may be due to a high ratio of mild, clinically unapparent cases to the more severe instances of severe disease.

Infantile hypercalcemia has also been seen in association with subcutaneous fat necrosis. In these infants indurated masses develop in the subcutaneous tissue within the first weeks of life, with bluish red discoloration of the skin overlying them. The masses are often at sites of pressure, and histologically contain fatty acid crystals with lymphocytes, histiocytes and giant cells in the subcutaneous adipose tissue. Ectopic calcification may be present. The symptoms of hypercalcemia are irritability, anorexia, constipation and failure of weight gain. The basis of the hypercalcemia is still speculative, but the possibility of prostaglandin excess has been considered as part of the picture of precipitation of saturated fatty acids and foreign body reaction in the subcutaneous tissue. There is evidence that prostaglandins do increase bone mineral solubilization by bone cell activity. If excess prostaglandin production is a factor, treatment with an inhibitor of prostaglandin synthetase such as aspirin or ibuprofen should be of value. The few cases reported have been treated primarily by corticosteroids and low calcium diet.

CASE REPORT

J.A.M. was born at term weighing 3540 gm following a normal pregnancy and delivery. The amniotic fluid was meconium-stained and the Apgar score at birth was 3. Blood glucose was too low to determine and IV glucose was given. At 36 hours of life the infant had a convulsive seizure and serum calcium concentration was 4 mg/dl. She was given both I.M. $MgSO_4$ and IV calcium gluconate. At 48 hours she still had hypoglycemia but serum calcium concentration was now 13 mg/dl. She was treated with ACTH. On the third day renal failure was evident and peritoneal dialysis was started and continued for two days, following which renal function improved. Repeat serum calcium concentrations were always high. The child remained irritable, fed poorly and was admitted to the Children's Hospital of Pittsburgh at 11 weeks of age. At that time she weighed 3900 gm, a thin infant who did not appear to be in acute distress. Several large subcutaneous nodules were present on the upper arms and thighs, with scattered small nodules over chest and abdomen and on calves. The pertinent laboratory data were serum: Ca 15.4 mg/dl; P 4.4 mg/dl; creatinine 1.6 mg/dl; BUN 47 mg/dl. Urine showed a trace protein.

Radiographs showed abnormal density at the base of skull and increased density of the long bones, with a zone of rarefaction proximal to the metaphyseal plate (Fig. 4–4). Skin biopsy of a nodule was reported to show subcutaneous fat necrosis. The infant was treated with prednisone, 1 mg/kg/day, which was increased after seven days to 2 mg/kg/day. She was fed a low calcium diet based on strained lamb. The mother's serum calcium and phosphate were 9.8 and 3.0 mg/dl respectively.

COMMENTS: This infant with subcutaneous fat necrosis and hypercalcemia represents a variant of idiopathic hypercalcemia. No tests were made of hyperresponsiveness to vitamin D because of the severity of the child's symptoms. The radiographic appearance was compatible with hypervitaminosis D although there was no history of excessive vitamin D intake during pregnancy.

The possibility has been considered that

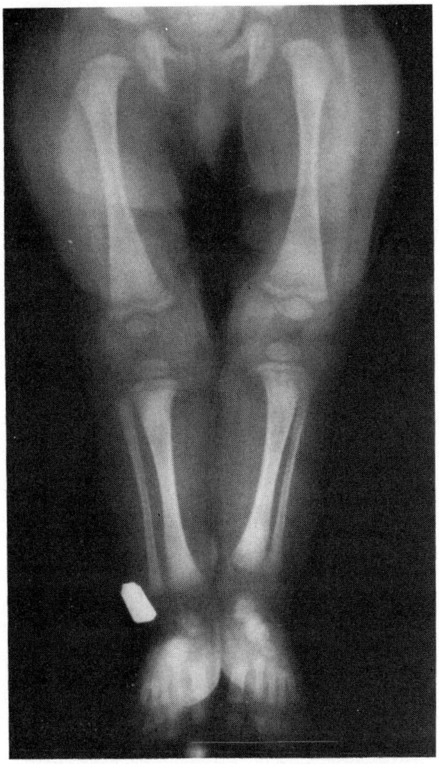

Figure 4–4. Idiopathic hypercalcemia associated with subcutaneous fat necrosis. Long bones show submetaphyseal zone of rarefaction.

prostaglandin excess could be a cause of hypercalcemia associated with subcutaneous fat necrosis, but there is no proof of this hypothesis. Treatment of such patients with inhibitors of prostaglandin synthetase such as aspirin or ibuprofen might be successful if prostaglandin excess were the basic problem. This child responded satisfactorily to suppressive corticosteroid treatment plus a low calcium feeding.

We are indebted to Dr. Thomas P. Foley, Jr., Children's Hospital of Pittsburgh, for permission to include this case.

HYPERCALCEMIA ASSOCIATED WITH SARCOIDOSIS

Sarcoidosis uncommonly has its onset in childhood, and hypercalcemia of sarcoid origin is very rare in the younger age-groups. This, however, is an extremely interesting form of hypercalcemia because there is evidence that patients with sarcoidosis are unusually sensitive to vitamin D, and the hypercalcemia can be alleviated by strict elimination of vitamin D in diet and protection of the skin from the short wave ultraviolet radiation that activates 7-dehydrocholesterol. Serial studies of sarcoid patients have shown significant increase of serum calcium concentration during the summer months when exposure to sunshine was not avoided. This sensitivity to vitamin D suggests a similarity to the problem of children with idiopathic hypercalcemia. Neither in sarcoidosis nor idiopathic hypercalcemia has the mechanism of the hyperresponsiveness to vitamin D been explained. The possibility has been suggested of increased rates of formation of the active vitamin D hormone, 1,25-diOH vitamin D, but there are no definitive studies that demonstrate such a metabolic change. There are other possibilities such as increased affinity of target cell receptors for 1,25-diOH vitamin D, or increased activation of specific DNA by the vitamin D hormone in the intestinal mucosal cell nucleus with greater formation of the specific proteins active in calcium and phosphate transport. It is hoped that further investigation in patients with sarcoidosis and hypercalcemia will provide some insight into the question of hyperreactivity to vitamin D.

The diagnosis of sarcoid hypercalcemia obviously depends on searching for such evidence as involvement of lymph nodes, lungs, liver and spleen. The characteristic hyperglobulinemia should be a helpful clue. The hypercalcemia of sarcoidosis responds rapidly to treatment with corticosteroids, as does that of hypervitaminosis D, so that prednisone, cortisol or dexamethasone in appropriate suppressive doses can be used both for diagnosis and for treatment. Treatment with corticosteroids may be required for an extended period until the granulomatous lesions of sarcoidosis become inactive. The duration of treatment of the hypercalcemia can be shortened by scrupulous reduction of vitamin D intake in diet and protection of the skin against exposure to sunshine. Since the foods providing vitamin D are limited in number, i.e., vitamin D-fortified milk, egg yolk, and such fish as sardines, herring, salmon and tuna fish, it is not difficult to devise a diet low in vitamin D for these patients.

HYPERCALCEMIA ASSOCIATED WITH DISORDERS OF THYROID FUNCTION

Strangely enough, hypercalcemia has been seen both in infants with severe hypothyroidism—athyreotic cretinism—and in hyperthyroidism. In the former instance the hypercalcemia is associated with hypercalciuria, and renal calculus formation has been described in cretinism. This suggests hyperabsorption of calcium as a basis for the hypercalcemia since bone density by radiograph does not indicate demineralization. The growth failure of the hypothyroid infant may contribute to the hypercalcemia since there is little skeletal uptake of bone mineral. The mechanism is unclear but it has been suggested that vitamin D inactivation is impaired in hypothyroidism. Whether the hypercholesterolemia in these patients in any way is connected with the postulated abnormality of vitamin D metabolism is unknown. A deficiency of calcitonin could also be a factor in the development of hypercalcemia. Treatment is, of course, the correction of the thyroxine deficiency. Corticosteroids should be effective in reducing the hypercalcemia but probably are unnecessary.

The hypercalcemia of hyperthyroidism has been reported mainly in adults and is apparently due to accelerated resorption of bone. This is thought to be the result of the increased metabolic activity of bone cells. Immunoreactive parathyroid hormone concentrations in serum are low in thyrotoxic patients with increased serum calcium concentration, so that the effect is not mediated through excessive secretion of PTH. The hypercalcemia ordinarily is not marked, and in some instances only determinations of ionized calcium concentration in serum show an increase and total serum calcium concentrations may be in the high normal range. Treatment is the specific treatment of the hyperthyroidism.

PHOSPHATE DEPLETION HYPERCALCEMIA

In experimental animals depletion of phosphate by diets extremely low in phosphate is associated with hypercalcemia and hypercalciuria. To what extent hyperabsorption of calcium from the intestine is responsible is unknown, but it seems likely that the absorbed calcium accumulates in extracellular fluid since the phosphate concentrations are too low to permit deposition of calcium phosphate in cartilage and bone matrix. Dietary phosphate depletion of this type is unknown in man, as phosphate is found in practically all foods and reduction of phosphate intake to deficiency levels does not occur. However, phosphate depletion can be produced by prevention of phosphate absorption by a great excess of either calcium or aluminum. Aluminum-induced hypophosphatemia has been seen in adults who have taken a great excess of aluminum antacids because of gastric symptoms. This problem has not been reported in children but, as mentioned in the section on transient physiologic hypoparathyroidism, hypercalcemia has resulted when hypophosphatemia was induced by a large excess of calcium lactate added to the feed. This is easily corrected by reducing the calcium lactate supplement and allowing the serum phosphate to rise into the normal range.

HYPERCALCEMIA CAUSED BY HYPERVITAMINOSIS A

The syndrome of poisoning by vitamin A includes rarefaction of the bones, which in some instances may be accompanied by hypercalcemia. The mechanism of the imbalance between deposition and solubilization of bone mineral resulting from vitamin A intoxication is not known. If the hypervitaminosis A is the result of ingestion of fish liver oils with a high concentration of vitamin A there may also be an accompanying increased intake of vitamin D. Hypervitaminosis A has become rare since the dangers of excessive doses of this fat-soluble vitamin have been publicized. Treatment is essentially discontinuance of intake of vitamin A, which should be followed by spontaneous recovery.

BLUE DIAPER SYNDROME

Hypercalcemia has been found in infants with an unusual inborn error of tryptophane metabolism, the blue diaper syndrome. Because of a block in tryptophane metabolism excessive amounts of indole derivatives are excreted in the urine, including indican which gives the blue stain of the urine-soaked diaper. The mechanism of the hypercalcemia remains obscure, and so few instances of this metabolic error have been found that it is not possible to generalize about the course and treatment.

HYPERPARATHYROIDISM

CLINICAL MANIFESTATIONS

Symptomatic hyperparathyroidism is uncommon in infants and children, but since its familial incidence has been recognized more cases have been discovered, with onset either at birth, early infancy or later childhood, but presumably of congenital origin. Familial hyperparathyroidism with both autosomal dominant transmission and autosomal recessive gene transmission has been reported. In both types of inheritance the pathologic lesion found has been chief cell hyperplasia of all the parathyroids. Adenoma of a single parathyroid is a rare cause of hyperparathyroidism in children. Familial hyperparathyroidism of the dominant variety also includes the entity of multiple endocrine adenomas type I in which pituitary, thyroid, parathyroid, adrenal and pancreatic tumors occur in various combinations. Although this is presumably a genetic disorder, manifestations of multiple endocrine adenoma rarely have their onset in childhood. Most children with diagnosed familial hyperparathyroidism have not shown evidences of adenomas of other endocrine glands, and as stated above the usual pathologic finding has been chief cell hyperplasia of the parathyroids without adenomatous change.

The physiologic abnormalities of primary hyperparathyroidism stem in part from the action of parathyroid hormone in activating cyclic AMP production in bone cells, which in turn stimulates the resorptive function of osteocytes and osteoclasts and results in increased solubilization of bone mineral and dissolution of bone matrix. This leads to increased entrance of calcium and phosphate into extracellular fluid. Another action of PTH is the activation of parathyroid-responsive adenyl cyclase in the membranes of proximal renal tubules. The increased formation of cyclic AMP in this cell inhibits renal tubular reabsorption of phosphate and stimulates renal tubular reabsorption of calcium and magnesium. Consequently there is enhanced loss of phosphate in the urine and retention of calcium and magnesium. The net result is increase of serum calcium and decrease of serum phosphorus concentrations. Serum magnesium concentrations are variable and consistent increase is not seen in hyperparathyroidism, although in hypoparathyroidism serum magnesium concentrations are usually reduced.

An additional action of PTH is the stimulation of the formation of the active vitamin D metabolite, 1,25-dihydroxyvitamin D, in the kidney. This metabolite acts to increase calcium and phosphate absorption from the intestine. Because of this effect and the skeletal reaction to increased bone resorption by stimulation of osteoblastic proliferation, there may be no net change in bone mineral despite a marked increase in rate of bone mineral turnover. Progressive demineralization of the skeleton with osteitis fibrosa cystica, therefore, is not a necessary manifestation of primary hyperparathyroidism either in children or adults. The symptoms of hyperparathyroidism may be primarily due to hypercalcemia, to hypercalciuria with stone formation or renal calcinosis, and less commonly to bone demineralization.

Hypercalcemia causes a variety of symptoms, many of which are nonspecific, so that the diagnosis cannot be made without determination of serum calcium concentration. These symptoms are anorexia, weakness, fatigability, constipation and polyuria. The last is due to vasopressin resistance. Acute hypercalcemia, above 15 or 16 mg/dl, causes hypertension and can result in hypertensive encephalopathy with obtunded state of consciousness, amblyopia, convulsions and coma. Even without marked hypertension, apathy progressing to stupor and coma, with or without convulsions, may result from marked elevation of extracellular calcium concentration.

These manifestations of severe hypercalcemia may be precipitated by the dehydration that results from the obligatory polyuria, plus reduced water intake because of anorexia, nausea and lethargy. Renal function deteriorates during hypercalcemia, not only because of water deficit but because of reduction of glomerular filtration rate as a probable result of constriction of the renal afferent arterioles. The consequent increased renin production also contributes to the hypertension of hypercalcemia. The spasm of the smooth muscle of arterioles is believed to be a direct effect of increased calcium ion concentration in the muscle. Further impairment of kidney function may also result from renal calcinosis with interstitial fibrosis and disappearance of nephrons. Renal calculus formation causes renal colic, hematuria and hydronephrosis.

The manifestations of bone demineralization are bone pains, particularly in the back and lower extremities, probably owing to microfractures. Collapse of the demineralized vertebrae may occur, and also pathologic fractures of the long bones due to osteopenia or to the "brown tumors" of osteitis fibrosa cystica in which the destroyed bone is replaced by fibroblasts and connective tissue. Eventually, severe deformities result from the multiple fractures and the inability of the demineralized bone to resist the stress of weight-bearing.

The roentgenographic characteristics of the skeletal lesions of hyperparathyroidism are of importance in the diagnosis. The diffuse demineralization cannot be distinguished from that of osteoporosis, i.e., osteopenia of unknown pathogenesis. However, in addition to diffuse demineralization, localized subperiosteal resorption of bone may be present. This is often most pronounced in the short tubular bones of the hand, metacarpals and phalanges, which show irregularity and thinning of the bone cortex and sometimes erosion of the terminal phalanges. The lateral ends of the clavicle are also sites of predilection for the demineralizing action of parathyroid hormone. The lamina dura of the tooth sockets is essentially a line of denser bone deposited under the periosteum at the boundary between bone and tooth. Disappearance of the lamina dura is another sign of subperiosteal resorption of bone and is frequently found in hyperparathyroidism. In the growing bone the metaphyseal ends of the bone become ragged as

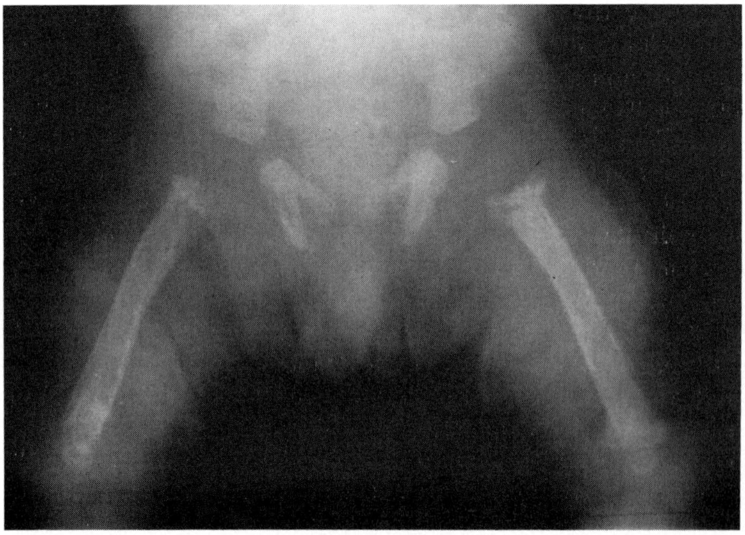

Figure 4–5. Destructive lesions of metaphyseal ends of long bones due to primary hyperparathyroidism.

the result of the resorptive action of excess PTH, and the irregularity of the metaphyseal ends may suggest severe rickets or the lesions of hypophosphatasia (Fig. 4–5). It must be emphasized that the skeletal abnormalities of primary hyperparathyroidism cannot be differentiated by x-ray examination from those of secondary hyperparathyroidism, although in the latter there may also be a considerable osteomalacic component histologically.

The renal lesions of hyperparathyroidism may be divided into two categories: (1) calculus formation resulting from hypercalciuria with obstruction and secondary infection; and (2) destruction of renal parenchyma with calcification of kidney tissue. As mentioned above, the vasospasm and ischemia of the kidney owing to the elevated calcium ion concentration may be an important factor in the chronic progressive renal disease. Not all patients with hyperparathyroidism have increased urinary calcium excretion. One of the actions of PTH is to increase tubular reabsorption of calcium, thus reducing urine calcium excretion at a constant filtered load of calcium. The increased concentration of diffusible calcium in the plasma, on the other hand, increases the filtered load if glomerular filtration rate remains constant. There are, therefore, three variables: GFR, concentration of diffusible calcium, and tubular reabsorption of calcium, all of which may be altered in hyperparathyroidism. The end-result, the loss of calcium in the urine, therefore, varies depending on the interrelationship among these variables.

The diagnosis of hyperparathyroidism rests mainly on the repeated finding of increased serum calcium concentrations, and particularly increased serum calcium ion concentrations. Several reports have indicated that in some patients with hyperparathyroidism total serum calcium concentrations may be variable and only slightly increased, with some determinations within the normal range but with consistent increase of calcium ion concentrations above 5.6 mg/dl. In addition to the elevation of serum calcium concentration, serum phosphate concentrations are reduced, although this may be inconsistent, particularly in patients with impairment of renal function and reduced GFR.

Determinations of tubular reabsorption of phosphate theoretically should provide more definitive evidence of PTH excess, but this is technically difficult, particularly in children, and has not been clinically useful. Since the action of PTH is mediated through cyclic AMP, and its effect on the renal tubule is associated with increased excretion of cyclic AMP of renal tubular origin in the urine, measurements of this compound in the urine have been made in patients with hyperparathyroidism, and increased amounts found. This determination has not yet become a readily available method for clinical purposes.

It is now possible to ascertain circulating PTH by a radioimmune assay, but there are still problems in the determinations of IPTH and the interpretation of the results. The PTH peptide is synthesized as a larger molecule, proparathyroid hormone, which is converted to the initial circulating PTH by splitting off of several amino acids. Following its secretion further degradation of the PTH molecule occurs, and a number of smaller peptides are found in the circulation. Only some of the smaller fragments are biologically active.

The various antisera produced against PTH and which are used in the radioimmune assay recognize these various peptides differently, so that the results obtained in various laboratories with a variety of antisera are not uniform. Although the aminoterminal peptide fractions are the biologically active peptides, they are less stable than the carboxyterminal peptide fractions which are inactive. As a rule, antisera that recognize the more stable carboxyterminal fragments are more useful in the diagnosis of the increased PTH output of hyperparathyroidism.

It is also important that the determination of IPTH be made and properly interpreted with simultaneous determination of serum calcium concentration. An IPTH value in the high normal range, with concomitant increased serum calcium concentration, can be indicative of excess PTH secretion since this secretion should be inhibited by calcium

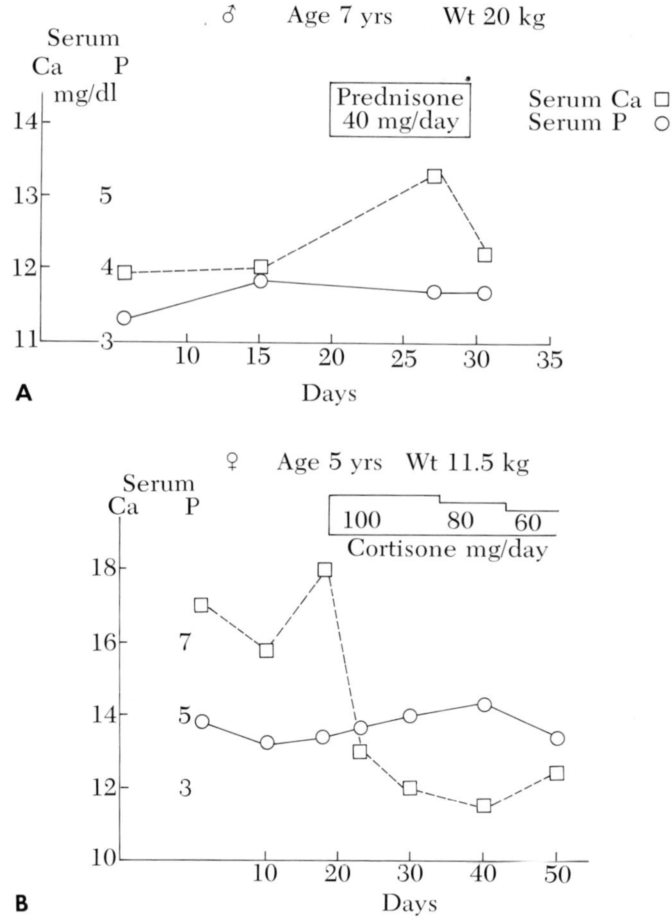

Figure 4–6. *a,* Refractoriness of hypercalcemia of primary hyperparathyroidism to suppressive effect of corticosteroid treatment, in contrast to *b,* reduction of serum calcium concentration by corticosteroids in a patient with idiopathic hypercalcemia.

concentrations above the normal range. Arnaud has also described a diurnal variation in serum IPTH concentrations so that, in theory at least, blood samples for diagnostic purposes should be taken at a uniform time. The determination of serum IPTH can be made more diagnostic by combining it with the response to reduction of ionized calcium concentration. In this test a calcium chelating agent, sodium EDTA, is infused to reduce calcium ion concentration, and the parathyroid gland response is monitored by IPTH measurements.

A helpful procedure in the differentiation of the hypercalcemia of hyperparathyroidism from other types of hypercalcemia is the adrenocorticoid suppression test. Suppressive doses of cortisone, or related steroids such as prednisone or dexamethasone, can reduce the hypercalcemia caused by vitamin D intoxication, sarcoidosis or tumors, but not that caused by hyperparathyroidism (Fig. 4–6). One of the actions of the corticoids is to reduce intestinal absorption of calcium by an action which opposes that of vitamin D. There may also be inhibition by corticoids of the bone cell metabolic processes which determine the rate of solubilization of bone mineral.

Familial Hyperparathyroidism with Autosomal Recessive Inheritance

Several kindreds with this variety of hyperparathyroidism have been described in reports by Hillman et al and Goldbloom et al. Most of the reported instances of congenital hyperparathyroidism have been in this category. The presenting symptoms in the affected infants have been mainly the nonspecific manifestations of hypercalcemia resulting in failure to thrive because of anorexia, lethargy, constipation, vomiting and dehydration. Some of the affected infants died without diagnosis because serum calcium determinations were not done and the process was only suspected after postmortem findings were available. Because of the familial incidence, subsequent infants in the affected families with similar symptoms were followed with serum calcium levels, found to have marked hypercalcemia and treated by parathyroidectomy. In addition to the symptoms of hypercalcemia, roentgenographic changes in the skeleton were found in several patients.

Since the lesion is diffuse hyperplasia of all the parathyroids the surgical treatment poses a dilemma. Removal of three or even three and one-half of the four identified glands has not necessarily resulted in a permanent cure because the remaining hyperplastic parathyroid tissue was capable of producing enough hormone to maintain the hypercalcemic state. In several patients a second operation was necessary to remove the remaining gland. Total parathyroidectomy is curative if all the parathyroid glands can be found, but it results in permanent hypoparathyroidism which then requires treatment for the remainder of the patient's life. In desperately ill children, however, this may be a necessary expedient to avoid the risk of an unsuccessful partial parathyroidectomy.

Another technique has recently been attempted by Wells with some success, namely total parathyroidectomy with autotransplantation of some of the removed parathyroid gland into the patient's forearm. The technique is not complex, requiring sectioning of one of the removed parathyroids into small segments and implantation into muscles of the forearm. This location is chosen since the function of the parathyroid graft can then be monitored by sampling venous blood from the antecubital vein with determination of IPTH. If this method should prove successful in further experience it would offer the best solution to the treatment of familial hyperparathyroidism due to parathyroid gland hyperplasia. By this procedure the parathyroid gland graft can easily be removed in part if eventually it should also become hyperplastic and cause hyperparathyroidism. The basis of the hyperplasia is unknown, but since parathyroid gland cell growth and function is normally stimulated by reduction of calcium ion concentration and inhibited by increase of this concentration, there may be an abnormality of the sensing mechanism for calcium ion concentration which may be located in membrane receptor groups.

The nonspecificity of many of the symptoms of hypercalcemia in infancy makes the early diagnosis of congenital hyperparathyroidism difficult unless serum calcium concentrations are measured. Goldbloom et al have summarized the findings and results in 11 patients ranging from 7 days to 30 months of age at the time of diagnosis. Of these only four survived. Of the seven fatalities one died following parathyroidectomy, and the remainder without operative treatment. Ten of the 11 patients had had roentgenographic examinations of the bones, and eight of these showed severe demineralization, with subperiosteal resorption or areas of bone destruction in two patients and pathologic fractures in an additional two. Serum phosphate concentrations were reduced in seven patients, in two they were within normal limits for the age, and in one the determination was not made. Serum IPTH was determined in only one patient during a relapse following subtotal parathyroidectomy. This was found to be elevated when related to the serum calcium concentration of 15.5 mg/dl. Of the five patients subjected to subtotal parathyroidectomy one died and two relapsed, requiring reoperation to remove the remaining hyperplastic parathyroid tissue. These latter two patients were then hypoparathyroid but responded well to treatment with vitamin D, 5000 and 15,000 units per day respectively, plus oral calcium supplements as calcium gluconate.

FAMILIAL HYPERPARATHYROIDISM TRANSMITTED AS AN AUTOSOMAL DOMINANT CHARACTERISTIC

Unlike the recessive variety in which most of the reported cases have been symptomatic in early infancy, this type of genetic hyperparathyroidism usually is not manifest until late childhood or adult life. However, in one kindred an instance of congenital hyperparathyroidism has been seen (*see Case Report C.D.*). There is a wide spectrum of severity of the disorder, and many subjects have been known to have hypercalcemia for years without obvious symptoms. Such cases have usually been detected because of family studies following discovery of an index case. Again, as in the recessive variety, multiple involvement of the parathyroid glands is the rule rather than a single adenoma. Because hyperparathyroidism may be the predominant feature of multiple endocrine adenomatosis, which is also transmitted by the dominant mode of inheritance, the question has been raised whether the dominant variety of familial hyperparathyroidism is not simply a subset of the former entity. However, there have been several carefully studied kindreds in which diffuse parathyroid hyperplasia without important involvement of other endocrine glands has been found, and this warrants separation of this entity from multiple endocrine adenomatosis, at least for clinical purposes.

We have divided the kindreds of dominantly inherited familial hyperparathyroidism into two major groups, those with hypercalcemia but with normal or reduced urinary excretion of calcium, and those with hypercalcemia and hypercalciuria. In any particular kindred the urinary calcium excretion appears to follow a uniform pattern, i.e., all members of the kindred have either low to normal urine calcium output or increased urine calcium excretion. The factor that determines whether hyperparathyroid hypercalcemia is associated with hypercalciuria is unknown, since the degree of elevation of the concentrations of serum total cacium does not differ between hypocalciuric and hypercalciuric patients. This distinction may be important in decisions concerning treatment. In the normo- or hypocalciuric patient kidney stone formation is not a risk, and watchful observation in the relatively asymptomatic patient seems justified. In the hypercalciuric patient, particularly one who has had evidence of stone formation, parathyroidectomy is indicated to avoid recurrent stone formation, urinary tract infection and renal insufficiency. Our own experience is summarized in the case reports that follow.

CASE REPORTS

Familial Hyperparathyroidism—Dominant Inheritance

Neonatal Familial Hyperparathyroidism with Severe Skeletal Manifestations

C.D. was a female infant born at term of a healthy primigravida. She weighed 3.3 kg at birth. Although the delivery was uncomplicated the infant was hypotonic, had a high-pitched cry and was tachypneic. The chest was unusually narrow, suggesting the abnormality of asphyxiating thoracic dystrophy of Jeune. A chest x-ray, however, showed extreme demineralization of the ribs (Fig. 4–7). Further roentgenographic study of the skeleton revealed evi-

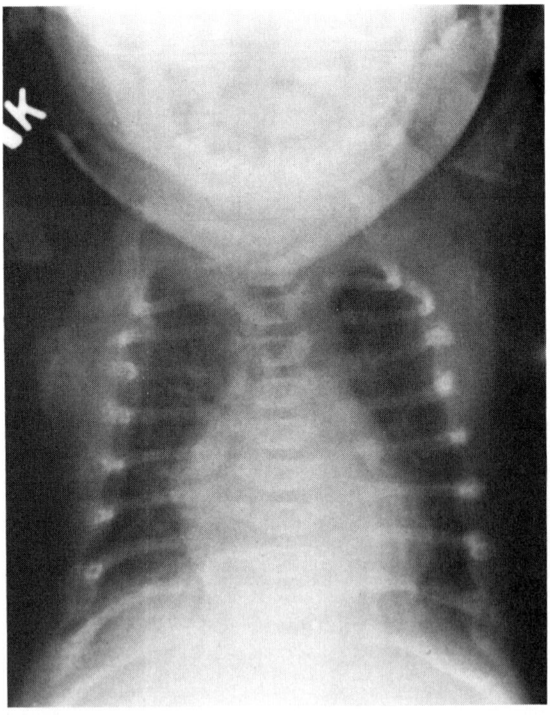

Figure 4–7. Chest radiograph showing demineralization and compression of the ribs (C.D.).

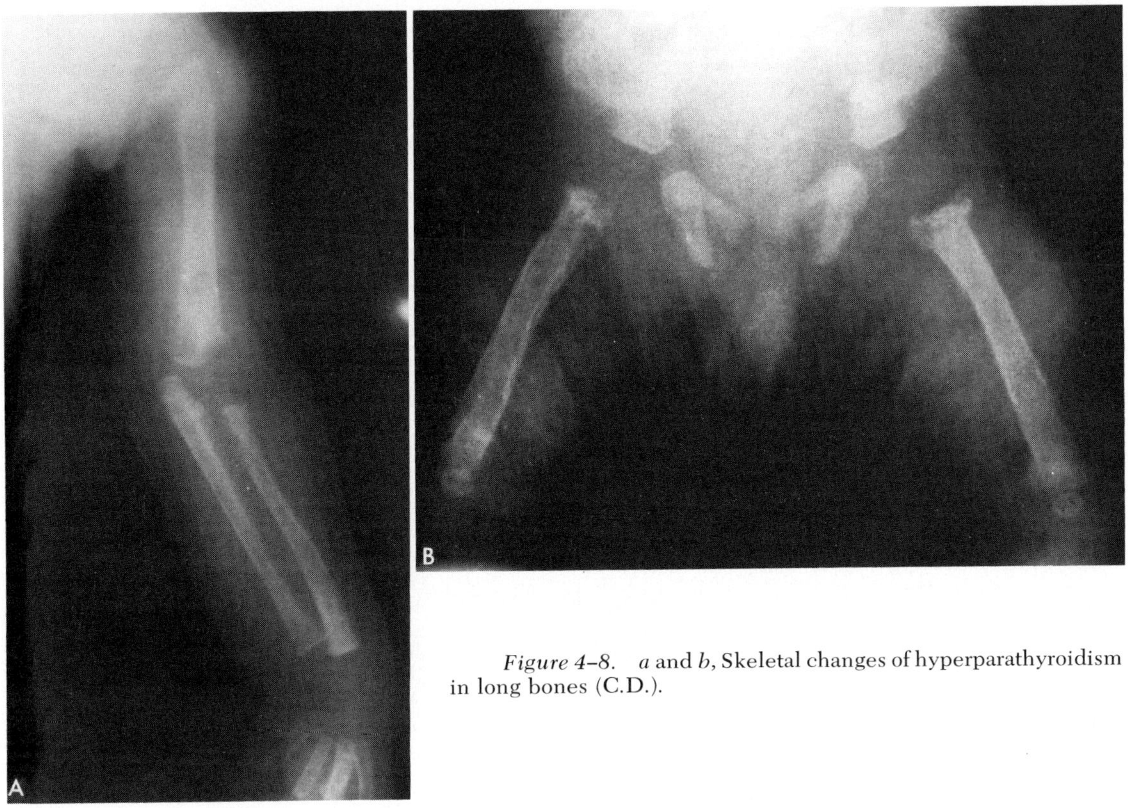

Figure 4–8. *a* and *b*, Skeletal changes of hyperparathyroidism in long bones (C.D.).

dences of bone resorption characteristic of severe hyperparathyroidism involving the growing skeleton (Fig. 4–8). Serum studies showed hypercalcemia and hypophosphatemia, and the infant was transferred to the Baltimore City Hospital. At age 2 days the concentration of serum calcium was 13.9 mg/dl (ionized calcium 7.0 mg/dl), serum phosphate was 3.4 mg/dl and alkaline phosphatase 30.2 Bodansky units. The serum IPTH concentration was markedly increased, 2.3 ng/ml. (We are indebted for this determination to Dr. Giraud Foster, Department of Physiology, Johns Hopkins University School of Medicine.) The serum concentrations of the mother were: calcium, 9.2 mg/dl and phosphate, 3.7 mg/dl. A diagnosis of congenital hyperparathyroidism was made.

Further family studies were not available at that time but it was thought that the underlying lesion probably was diffuse hyperplasia of all the parathyroids. For this reason it was decided to do a total parathyroidectomy. As mentioned above, in cases of congenital hyperparathyroidism with autosomal recessive inheritance Goldbloom et al had reported that removal of three of the hyperplastic parathyroids failed to alleviate the hypercalcemia, and repeat operation was necessary to remove the remaining gland. After two weeks of symptomatic treatment, including addition of phosphate to the feedings, total parathyroidectomy was performed by Dr. J. Alex Haller. Four parathyroids were identified and removed. The total weight was 90 mg. Histologically the lesion was thought to be diffuse chief cell hyperplasia (Fig. 4–9). The postoperative course was smooth and the serum calcium concentration was maintained by intravenous calcium gluconate as needed, or supplements of calcium lactate to the feed. Two weeks postoperatively the serum calcium and phosphate concentrations were 7.3 and 4.0 mg/dl respectively. It was thought that the child had had a total parathyroidectomy, and therapy with dihydrotachysterol was begun.

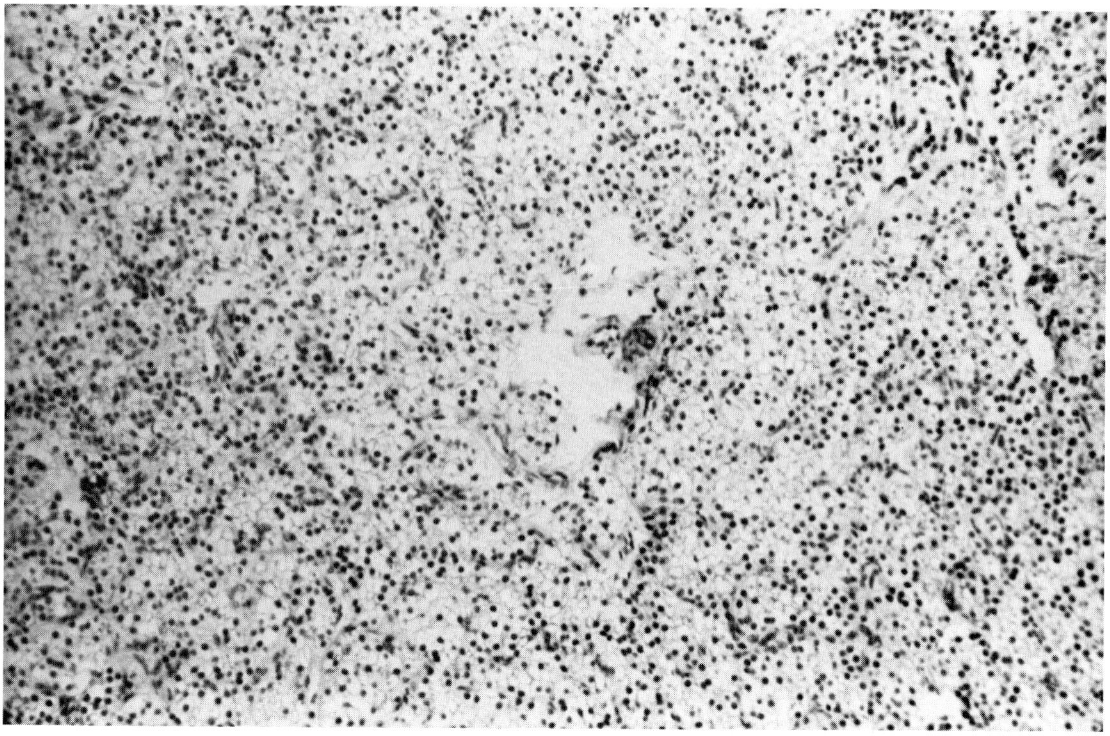

Figure 4–9. Section of parathyroid gland showing chief cell hyperplasia (C.D.).

She is now 6 years of age and is growing along the 10th percentile for height and 25th percentile for weight. Serum calcium concentrations have been maintained between 9 and 10 mg/dl, with one episode of hypercalcemia necessitating revision of the dihydrotachysterol dosage. The current dose is 0.25 mg per day. Aside from severe dental caries secondary to poor mineralization of enamel, the child shows no abnormalities. Remineralization of the skeleton has been complete (Fig. 4–10) and there are no residual abnormalities.

Subsequent study of the family in collaboration with Dr. Allen M. Spiegel of the National Institute of Arthritis, Metabolism and Digestive Diseases provided an extremely interesting story. As mentioned above, the mother's serum calcium concentration was normal. The father, however, was found to have a serum calcium concentration of 12.5 mg/dl, with serum phosphate 1.8 mg/dl; he was discovered to be related to a large kindred of subjects with hyperparathyroidism who had been under investigation at the N.I.H. This pedigree is diagrammed in Figure 4–11. Of the 15 subjects with definite hypercalcemia (Table 4–3) six had been explored, including C.D. In all instances the lesion was chief cell hyperplasia of all the parathyroids. C.D. alone had an initial total parathyroidectomy and was the only one whose hypercalcemia was corrected by the first operation. Three of the five subjects who had persistent hypercalcemia following the first operation were reoperated. In addition to C.D. there were four other subjects with hypercalcemia under 16 years of age. None of these four had symptoms characteristic of hypercalcemia. D.D., however, a 3 year old child, was growing well below the third percentile and because of marked hypercalcemia, 13.6 mg/dl, and hypophosphatemia, 3.1 mg/dl, was subjected to parathyroidectomy. Unfortunately this was not total since only three parathyroids could be identified. Following operation the serum calcium concentrations continued to be markedly elevated, indicating that hyperplastic parathyroid gland tissue remained. Reoperation was not possible because of parental objection. It is also of interest that

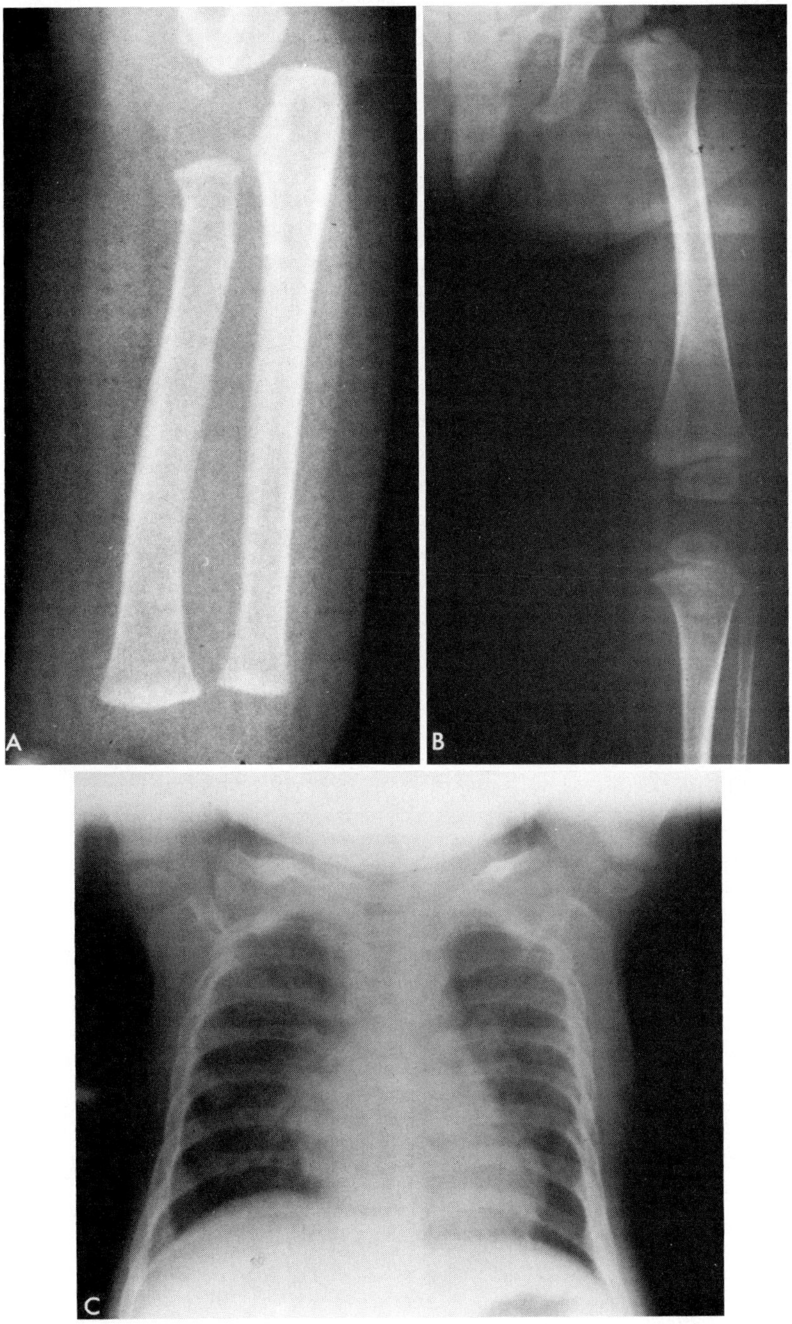

Figure 4–10. *a*, *b*, and *c*, Remineralization of ribs and long bones following total parathyroidectomy (C.D.).

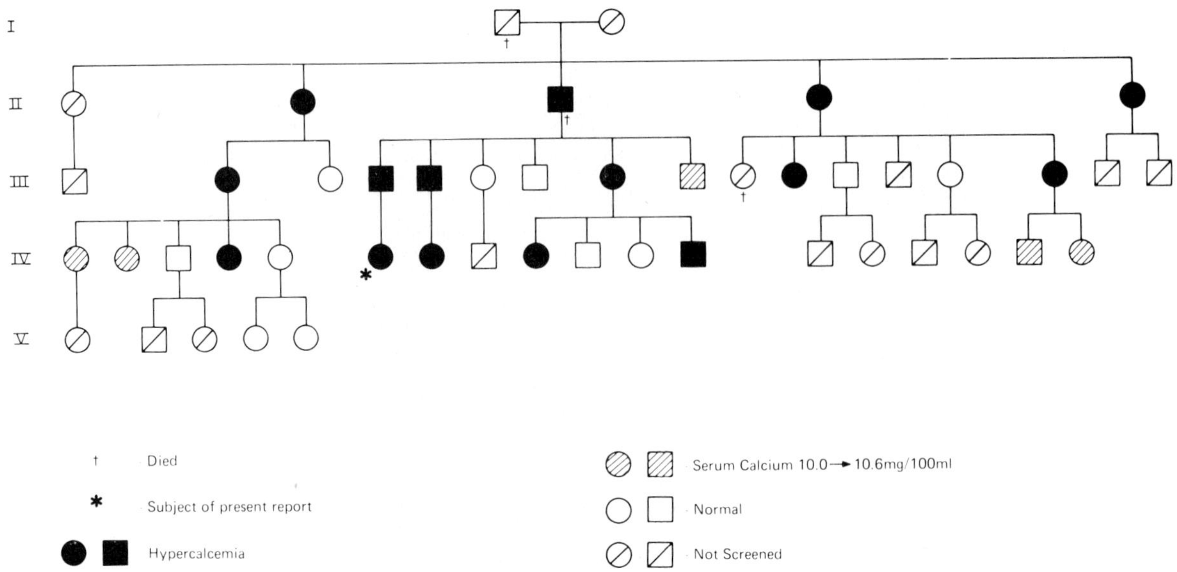

Figure 4–11. Pedigree of kindred of primary hyperparathyroidism with autosomal dominant inheritance. *Proband with congenital hyperparathyroidism. (From Spiegel, A. M., Harrison, H. E. Marx, S. J., Brown, E. M. and Aurbach, G. D.: Neonatal primary hyperparathyroidism with autosomal dominant inheritance. J. Pediatr. *90*:269–272, 1977. Reprinted with permission.)

Table 4–3. Summary of Clinical and Laboratory Findings in Hypercalcemic Subjects of Kindred with Familial Hyperparathyroidism

	Patient	Age at Diagnosis (yrs)	Serum *Ca* mg/dl	Serum *P* mg/dl	Symptoms	Parathyroid Pathology
	M.M.	64	12.2	2.2	Nephrolithiasis	N.E.*
II	J.D.	61	12.0	1.9	None	N.E.
	A.P.	54	13.7	2.1	Pseudogout	Chief cell hyperplasia
	I.C.	45	11.9	2.9	Peptic ulcer	N.E.
	B.N.	43	11.3	2.9	None	Chief cell hyperplasia
III	H.D.	22	12.5	1.8	None	N.E.
	R.D.	28	13.1	1.9	None	N.E.
	R.D.	30	11.9	3.0	Had infant with hypocalcemia	Chief cell hyperplasia
	W.B.	16	13.0	2.8	None	Chief cell hyperplasia
	B.S.	25	12.1	3.0	None	N.E.
IV	N.N.	15	12.1	3.1	None	N.E.
	C.D.	Newborn	13.9	3.4	Bone destruction	Chief cell hyperplasia
	D.D.	3	13.6	3.1	Growth retardation	Chief cell hyperplasia
	S.Di	10	12.6	3.1	None	N.E.
	W.Di	5	12.2	3.2	None	N.E.

*N.E. = not explored.

one of the females with hypercalcemia had had an infant with transient physiologic hypoparathyroidism during the newborn period. In this kindred, urine calcium excretion when determined has been consistently normal or low.

Although most of the reported cases of hyperparathyroidism presenting with symptoms at birth or soon thereafter have had a family pattern suggestive of autosomal recessive inheritance, familial hyperparathyroidism of the autosomal dominant variety may also present at birth, as the case of C.D. demonstrates.

Dominant Familial Hyperparathyroidism without Hypercalciuria or Skeletal Lesions

H.H. was a 7 year old girl brought to the outpatient department for evaluation. The only complaint was polydipsia and polyuria. It was found that in 1969 the mother had been evaluated for hypercalcemia, and on checking with the hospital where she had been studied it was learned that she had been diagnosed as having hyperparathyroidism. Serum studies on H.H. revealed the following concentrations: total calcium 12.3 mg/dl, ionized calcium 6.22 mg/dl, phosphate 3.7 mg/dl, alkaline phosphatase 4.7 Bessey-Lowry units, and urea nitrogen 14 mg/dl. Urine calcium excretion ranged from 50 to 105 mg/24 hrs or 0.14 to 0.25 mg per mg creatinine. Skeletal x-rays showed no evidences of bone abnormality. The child was admitted because of the polydipsia and polyuria which were thought to be symptoms of hypercalcemia. In the hospital these symptoms were not evident. Urine volume was never higher than 570 mg/24 hrs, and urine concentrating function was adequate with osmolalities up to 944 mOsm/l.

In view of the relatively benign manifestations of hyperparathyroidism in this child it was decided not to operate since it was probable that diffuse parathyroid hyperplasia was the underlying lesion. Analysis of the mother's serum concentrations showed calcium 11.95 mg/dl, ionized calcium 5.90 mg/dl, phosphate 3.4 mg/dl, alkaline phosphatase 2.7 Bessey-Lowry units, and urea nitrogen 15 mg/dl. The urine calcium excretion was only 0.09 mg/mg creatinine. These studies were made eight years after the original diagnosis of hyperparathyroidism in this woman and there had been no evidences of progressive renal injury during this time. It is again of importance that neither the child nor her mother had persistent hypercalciuria. The relatively benign course in the mother reinforced our decision not to do a parathyroidectomy on the child. The serum calcium and phosphate concentrations in the other three siblings were in the normal range. This child continues to do well with a serum calcium concentration of approximately 12.5 mg/dl. The polydipsia and polyuria were probably of psychogenic origin.

Familial Benign Hypercalcemia—a Variant of Mild Hyperparathyroidism

L.P., a 6 year old white boy, presented at the emergency room because of headache which on questioning had been occurring intermittently for 18 months. A channel 12 autoanalysis (SMA-12) was part of the routine study of this patient and a serum calcium concentration of 13.1 mg/dl was reported. Repeated serum calcium concentrations were 12.0 mg/dl or higher and serum phosphate concentrations were uniformly low, 2.7 and 3.1 mg/dl. No family history of hypercalcemia was obtained. Urinalysis, bone x-rays and IVP were negative. The corticosteroid suppression test showed no reduction of serum calcium with prednisone, 40 mg per day for ten days (Table 4–4). A tentative diagnosis of hyperparathyroidism was made and a neck and anterior mediastinal exploration was done. No parathyroid adenoma was found. Four parathyroid glands were identified and three were removed; the fourth was left in situ. None of the glands were thought to be grossly enlarged and the sections were reported to show normal parathyroid histology. Following the operation serum calcium concentrations

Table 4–4. Familial Benign Hyperparathyroidism (Proband L. P., Jr.); Refractoriness of Serum Calcium Concentration to Corticosteroids

DAY	SERUM *Ca* (*mg/dl*)	SERUM *P* (*mg/dl*)
Control	11.9	3.3
Control	12.0	4.0
Control	11.6	3.7
Control	11.5	4.5
Prednisone 40 mg/day		
8	13.3	3.7
10	12.2	3.7

remained elevated. At this time a careful family study was made despite absence of any complaints suggestive of hyperparathyroidism. The incidence of hypercalcemia in this kindred is diagrammed in Figure 4–12. The striking feature in this family is the low or normal urine calcium excretion of the hypercalcemic subjects. This finding has now been reported in other kindreds and is the basis of our separation of familial hyperparathyroidism into hypercalciuric and normocalciuric varieties, since in any one kindred patients appear to resemble each other with respect to urine calcium output.

Measurements of serum IPTH concentration in the patient and the other members of this family did not show an absolute increase, but serum IPTH was not suppressed by high calcium concentrations (Table 4–5). It is postulated that this kindred represents a mild and benign form of hyperparathyroidism, with hypercalcemia but without hypercalciuria, in which PTH output continues at an inappropriate rate despite hypercalcemia. Examination of the parathyroids was obtained only in the proband. The absence of histologic or gross evidence of hyperplasia is of interest but does not rule out hyperfunction of the parathyroid cells. None of the other members of the family were subjected to parathyroidectomy because of the benign nature of the disorder. Had the family study been done originally it is likely that the proband also would not have been parathyroidectomized.

Familial Hyperparathyroidism with Hypercalciuria and Nephrolithiasis

The patient was a 12 year old girl whose past history was unremarkable. The family history was important in that an older sibling had had hypocalcemia in the neonatal

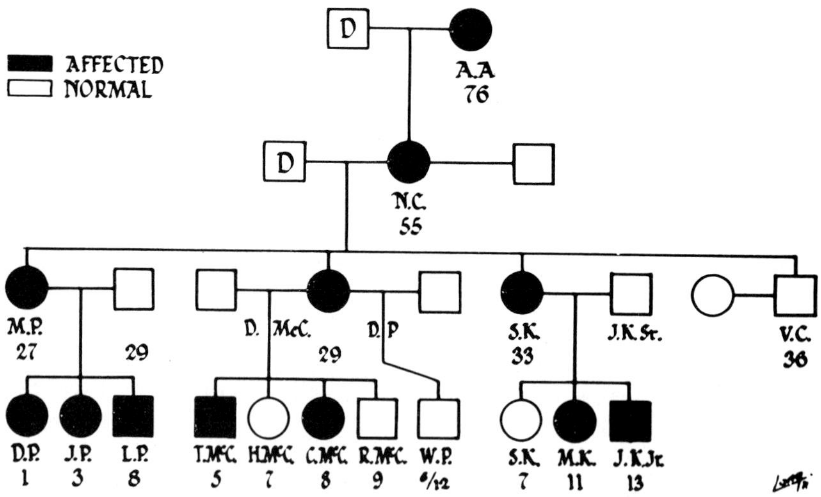

Figure 4–12. Diagram of kindred with benign variety of primary familial hyperparathyroidism, with autosomal dominant inheritance. (From Foley, T. P., Jr., Harrison, H. C., Arnaud, C. D. and Harrison, H. E.: Familial benign hypercalcemia. J. Pediatr. *81*:1060–1067, 1972. Reprinted with permission.)

Table 4–5. Manifestations of Benign Variety of Dominantly Inherited Hyperparathyroidism Without Hypercalciuria

PATIENT	AGE YRS.	SERUM Ca mg/dl	SERUM P mg/dl	SERUM IPTH μl/Eq/ml	URINE Ca mg/day
L.P., Sr.	29	10.1	3.0	20	187
M.P.	27	11.8	2.5	35	142
L.P., Jr.	7-8/12	12.2	3.8	40	36
J.P.	3	12.7	3.7	35	4
P.P.	1	12.3	5.3	30	10

period with complete recovery. The mother's serum calcium concentration had not been determined at that time. This previously healthy girl suddenly developed episodes of right flank pain radiating to the groin suggestive of renal colic. The urine showed microscopic hematuria and an IVP revealed dilatation of the right superior caliceal system. No stone was recovered. The serum calcium concentration was found to be 12.7 mg/dl and serum phosphate 3.1 mg/dl. At that time the mother's serum calcium concentration was found to be 11.8 mg/dl.

On referral to Baltimore City Hospital the girl was asymptomatic. Physical examination showed an early pubertal female without unusual findings. Urine analysis was negative except for a 24-hour excretion of 438 mg of calcium or 0.47 mg per mg creatinine, a value about three times the normal average for this age. Serum concentrations were calcium 12.9 mg/dl, phosphate 3.2 mg/dl, urea nitrogen 13 mg/dl, chloride 105 mEq/l, CO_2 29 mEq/l, sodium 143 mEq/l and potassium 5.0 mEq/l. The serum IPTH concentration was 46 μlEq/ml, with the upper limit of normal in this laboratory 40 μlEq. Skeletal x-ray examinations showed no abnormality. The studies on the patient's mother showed a serum calcium concentration of 12.0 mg/dl and phosphate of 2.6 mg/dl, whereas the father's serum calcium was 10.2 mg/dl and phosphate 3.1 mg/dl. The mother excreted 636 mg calcium per 24 hours, but despite this hypercalciuria had no story of kidney stone.

The diagnosis of dominant familial hyperparathyroidism with hypercalciuria was made. Skeletal x-rays of both patient and mother showed no evidences of bone demineralization. The possibility of multiple endocrine neoplasia type I was considered, but no clinical or laboratory evidences suggesting pituitary, thyroid, pancreatic or adrenal adenoma were found in either child or mother. It was thought, therefore, that diffuse chief cell hyperplasia of all parathyroids was most likely, and with this preoperative diagnosis and a documented episode of renal calculus the child was explored by Dr. J. Alex Haller at the Johns Hopkins Hospital. The parathyroids were not strikingly enlarged and only three parathyroids were identified. These were removed, but postoperatively the serum calcium remained elevated, rising as high as 14 mg/dl.

The child was then referred to the service of Dr. Gerald D. Aurbach at the National Institute of Arthritis, Metabolism and Digestive Diseases for parathyroid arteriography and catheterization of the neck veins for selective venous PTH determination. Arteriography revealed two areas of increased density, a 0.5-cm density below the left lobe of the thyroid and a smaller 0.2-cm density behind the right superior pole (Fig. 4–13). Parathyroid venography revealed "step-up" of IPTH in the area of the left upper pole and below the left inferior pole of the thyroid (Fig. 4–14). Peripheral IPTH values were also increased.

At reoperation an enlarged inferior parathyroid gland weighing 600 to 700 mg and a smaller right superior parathyroid gland were removed in toto. Histologically the glands showed chief cell hyperplasia. Since the child was considered to have had a total parathyroidectomy the Wells technique of parathyroid autotransplantation into the right forearm was employed. About one-half of the resected parathyroid tissue was cut into 20 separate sections and implanted in the right forearm. By eight hours postoperatively the serum calcium had fallen to 8.0 mg/dl. She was begun on 600 mg calcium lactate (78 mg calcium) orally four times per day, as well as 10 per cent calcium gluconate 1.5 ml/kg intravenously. The serum calcium concentration decreased to a low of 7.6 mg/dl on the second postoperative day. The patient had carpal spasm, a positive Chvos-

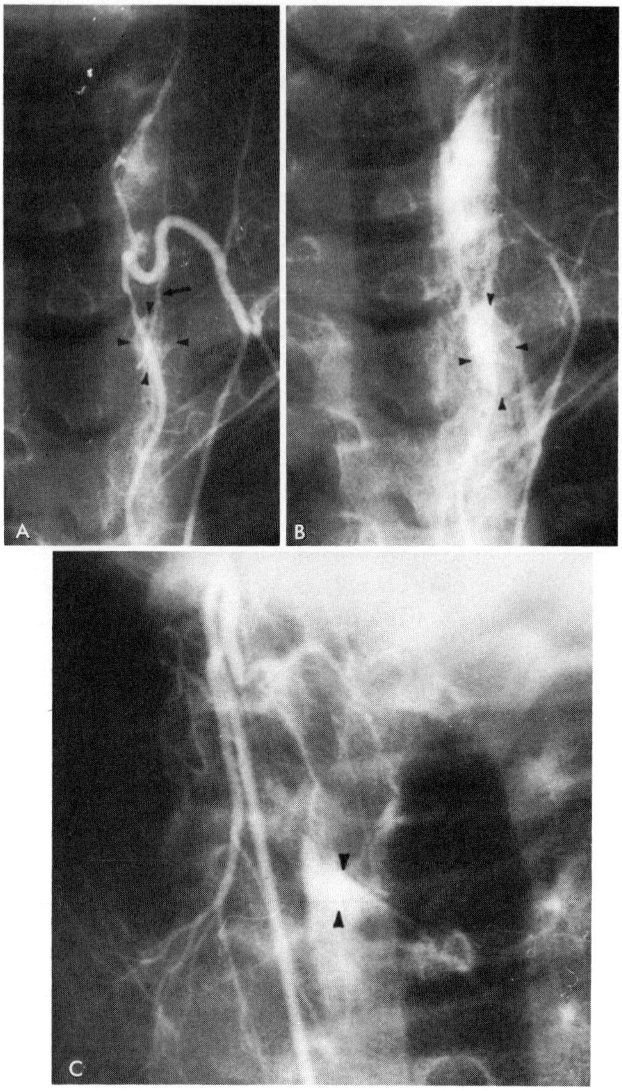

Figure 4–13. *a* and *b*, Arteriogram of neck, patient with familial parathyroid hyperplasia (N.P.). Suggestive densities indicating possible sites of hyperplastic parathyroids.

tek sign, and mild paresthesias of toes and fingers. The serum phosphate concentration rose from a preoperative value of 3.8 mg/dl to a high of 7.6 mg/dl. The patient was discharged on the 11th postoperative day on Neo-Calglucon, 5.0 ml (92 ml Ca) orally four times daily, and dihydrotachysterol, 0.25 mg per day. Subsequent follow-up studies indicated onset of function of the autotransplanted parathyroid tissue with return to a euparathyroid state, and supplemental calcium and dihydrotachysterol were discontinued. The child has remained well.

The mother was not operated on since she was asymptomatic and had no evidences of renal injury despite hypercalcemia and hypercalciuria. Close follow-up observation with serial renal function determinations is the current recommendation.

COMMENTS: These four families of dominantly transmitted familial hyperparathyroidism show the great variations in the severity and manifestations of this disease. One of the patients (N.P.) had marked hypercalciuria and presented with symptoms of renal stone. Only one subject (C.D.) had severe demineralizing bone disease. None of the patients had severe systemic manifestations of hypercalcemia, particularly anorexia, nausea and vomiting, constipation and neurologic symptoms. Such symptoms do occur, however, and only further follow-up studies of these patients will indicate whether generalized symptoms, renal in-

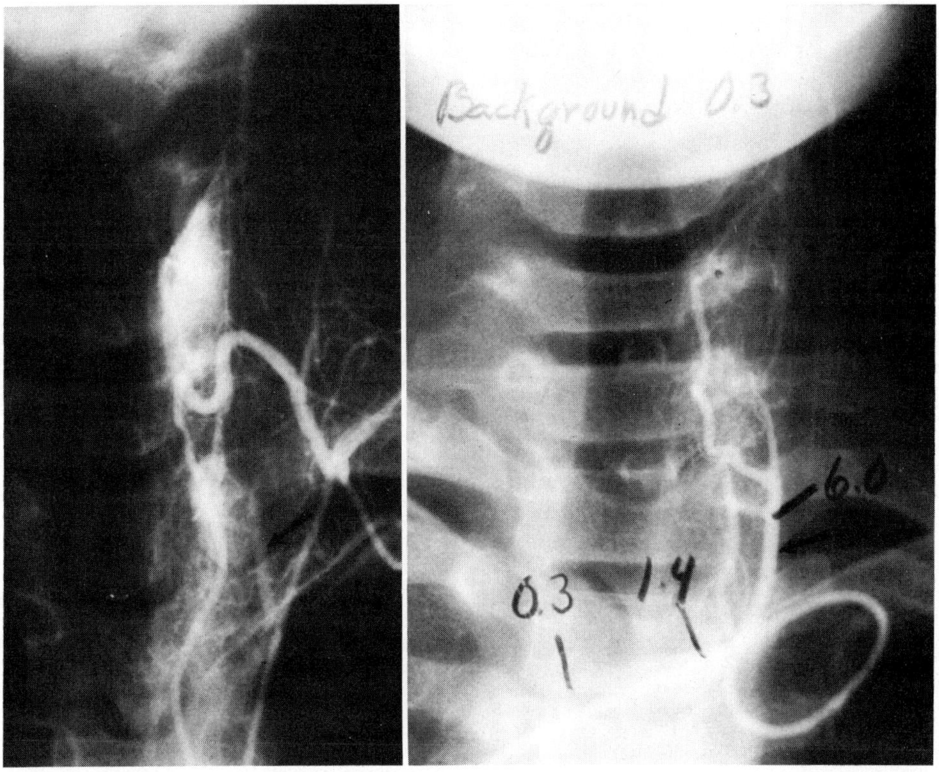

Figure 4–14. Selective catheterization of thyroid veins showing step-up of serum IPTH concentrations in blood draining left upper pole and left inferior pole of thyroid. (Courtesy of Dr. Allen M. Spiegel, NIH, Bethesda, Md.)

jury or bone demineralization may supervene after many years of relatively asymptomatic hyperparathyroidism.

The difference among kindreds in the severity of familial hyperparathyroidism is not a cause for surprise. What is difficult to explain is the lack of correlation of urine calcium excretion with serum calcium concentrations and the intrafamily consistency, i.e., no hypercalciuria in all members of a family or hypercalciuria in all members. This is an important finding since one of the major presenting manifestations of hyperparathyroidism is renal stone formation, which probably is dependent on the urine calcium concentration. The physiologic basis for the differences among families in the degree of hypercalciuria is unknown. Parathyroid hormone does influence various renal tubular functions, reducing reabsorption of phosphate, bicarbonate and amino acids but increasing reabsorption of magnesium and calcium. Increased tubular reabsorption of calcium operates to reduce urine output of calcium, except that the increased filtered load of calcium resulting from the higher concentration of ionized calcium is usually sufficient to outweigh the increased tubular reabsorption.

There may be differences among individuals in the renal tubular response to PTH with respect to calcium reabsorption, and if so such differences presumably are genetically determined because of the apparent segregation of families into hypercalciuric and nonhypercalciuric varieties of hyperparathyroidism. Even though patients with low or normal urine calcium excretion are not at risk for stone formation, prolonged hypercalcemia may cause renal calcinosis and renal insufficiency in the absence of hypercalciuria and stone formation. The calcinosis in this instance results from precipitation of calcium phosphate in the interstitial tissue of the kidney with fibrosis and destruction of nephrons. It is also suggested that hypercalcemia causes arterial constriction due to direct action on smooth muscle tone. Spasm of the different glomerular arterioles could cause increased renin output by the juxtaglomerular apparatus, and hyperreninemia has been reported in hyperparathyroidism. The resulting hypertension due to overactivity of the renin-angiotensin system further aggravates the renal ischemia, with progressive vascular injury and glomerular damage. Because of the possibility of insidious development of kidney damage, patients with asymptomatic hyperparathyroidism who are not subjected to operation must be followed closely with serial renal function studies as well as skeletal x-ray examination for progressive bone demineralization.

Transient Congenital Hyperparathyroidism in Infants of Mothers with Untreated Hypoparathyroidism

There have been several documented instances of infants with hyperparathyroidism born to women with hypocalcemia due to hypoparathyroidism. This variety of congenital hyperparathyroidism has been associated with skeletal lesions as well as elevated serum calcium concentrations. It is assumed that the maternal hypocalcemia during pregnancy prevented the placenta from maintaining the normally high serum calcium concentrations in the fetal plasma. As a result the fetal parathyroids were stimulated and became hyperplastic, with consequent demineralization of the fetal skeleton and postnatal hypercalcemia. The diagnosis is made by skeletal x-ray examinations and serum calcium determinations in infants whose mothers are found to be hypocalcemic. This form of hyperparathyroidism appears to be transient, and serum calcium concentrations return spontaneously to normal and the bone remineralizes without specific treatment.

Multiple Endocrine Adenomatosis — Type I

Discovery of hyperparathyroidism in more than one member of a family should stimulate search for multiple endocrine dysfunctions, since multiple endocrinopathies are genetic and transmitted as a dominant trait. The term *endocrine adenomatosis* may be somewhat of a misnomer since the parathyroid glands may show the histologic picture of chief cell hyperplasia rather than that of a circumscribed adenoma.

Multiple endocrine adenomatosis type 1 is the more common of the two constellations of involvement of multiple endocrine glands. This syndrome includes adenomata or hyperfunction of pituitary, parathyroid and pancreas with occasional involvement of thyroid and adrenal. The pancreatic lesion is usually a tumor of the gastrin-secreting cells of the pancreas resulting in the Zollinger-Ellison syndrome of hyperchlorhydria, peptic ulcers and diarrhea. Since calcium stimulates the release of gastrin, the association of hyperparathyroidism and a gastrin-secreting tumor causes massive secretion of gastric hydrochloric acid with severe intractable peptic ulcer formation and diarrhea. Although determined genetically with a dominant mode of inheritance, these manifestations of multiple endocrinopathy rarely appear in the first decade of life, and the disorder ordinarily is not recognized until adult life, with increasing frequency through the sixth decade. If adult members of a kindred are discovered to have Zollinger-Ellison syndrome and hypercalcemia the children should be studied for hypercalcemia since asymptomatic hyperparathyroidism may be discovered, and such patients should be carefully observed for development of more serious manifestations.

The pituitary lesion may be a nonfunctioning chromophobe adenoma. Growth retardation may be the result of such a tumor, and the tumor may be associated with enlargement of the sella turcica. An ACTH-secreting tumor with Cushing's syndrome has also been reported. Thyroid and adrenal adenomas are less commonly found as part of the constellation of multiple endocrine adenomatosis type I. The thyroid follicular adenomas and the adrenal cortical nodules are usually nonfunctioning as endocrine tumors and only discovered as anatomic entities. Skin lipomas have also been found in association with this variety of multiple endocrine adenomas.

The diagnosis and treatment of the hyperparathyroidism is similar to that of other forms of familial hyperparathyroidism. Since multiple involvement of the parathyroids is so frequently found, removal of a single enlarged parathyroid may be ineffective and dissection and identification of all of the parathyroids should be made, with removal of three and one-half glands or all of the parathyroids with autotransplantation of one of the parathyroids, as described above. If hypersecretion of hydrochloric acid and severe peptic ulceration continue despite restoration of serum calcium concentrations to normal, determinations of serum gastrin concentrations should be made. If these are elevated in the face of gastric hyperchlorhydria a gastrin-secreting tumor must be assumed to be present. The location and total removal of such a tumor are usually very difficult. The tumors may be multiple in the pancreas, and it has also been suggested that hyperplasia of gastrin-secreting cells of the gastric mucosa may be present. For this reason the recommended treatment for Zollinger-Ellison syndrome has usually been total gastrectomy.

MULTIPLE ENDOCRINE ADENOMATOSIS — TYPE II

Multiple endocrine adenomatosis type II is also genetic with a dominant mode of inheritance. This term is used for the constellation of calcitonin-secreting medullary carcinoma of the thyroid, pheochromocytoma and hyperparathyroidism. In some kindreds neuromas and neurofibromas have also been found. Hypercalcemia in a patient with hypertension, either paroxysmal or persistent, should suggest this syndrome. The diagnosis of pheochromocytoma is confirmed by determinations of catecholamines and metabolites such as vanillylmandelic acid in the urine. Hypercalcemia may result from excessive β-adrenergic stimulation, so that the pheochromocytoma should be treated first by surgical removal of the catecholamine-secreting tumor or tumors. If hypercalcemia persists after successful treatment of the pheochromocytoma the diagnosis of hyperparathyroidism must then be pursued, and the treatment again, as in patients with other forms of familial hyperparathyroidism, is parathyroidectomy with removal of three and one-half parathyroids or total parathyroidectomy with autotransplantation of parathyroid tissue. Obviously, as discussed above, the decision to remove the parathyroids depends on evidence that the hypercalcemia is interfering with the patient's normal function or is threatening renal injury. If a calcitonin-secreting tumor of the thyroid is also present this should be resected. As in multiple endocrine adenomatosis type I the clinical manifestations of this genetically determined disorder do not ordinarily develop until after the first decade of life.

IMMOBILIZATION HYPERCALCEMIA

It has long been known that lack of muscle pull on a bone and cessation of weight-bearing results in demineralization of the bone. A current theory is that deformation of bone mineral crystals by compression or tension causes a piezoelectric effect, i.e., development of an electrical current by the crystal. The electrical field so produced stimulates osteoblastic activity, with resultant increased formation of bone matrix and deposition of mineral in this matrix. The bone thus becomes thicker and more dense in response to the compressive forces of weight-bearing or of bending by muscle tension. Removal of these forces by paralysis of muscles or immobilization in a cast causes an abrupt decrease in the rate of bone mineral deposition without comparable reduction in bone resorption. Consequently, resorption and solubilization of bone mineral exceeds desposition of mineral so that there is a net flow of calcium and phosphate from the skeleton into the extracellular fluid. There is also evidence that immobilization actually increases the rate of bone resorption, possibly through increased PTH secretion. Increased concentrations of serum IPTH have been found in patients with immobilization hypercalcemia. The stimulus for the parathyroid secretory activity in the face of hypercalcemia is unknown. The early manifestation of the imbalance between bone mineral resorption and deposition is increased urinary calcium excretion, and hypercalciuria with renal calculus formation has been a complication of sudden immobilization as in patients with extensive paralytic poliomyelitis. The loss of bone mineral, however, may be so great that it cannot be compensated for by increased urine excretion of calcium, and hypercalcemia then occurs.

We have encountered the problem of immobilization hypercalcemia, particularly in active adolescent boys who have suffered a fracture of the femur. The sudden transition from the active physical life of an athletic boy to complete immobilization of the lower extremities because of a fractured femur can cause severe hypercalcemia, especially if both lower extremities are immobilized. However, even immobilization of a single lower extremity by traction has caused hypercalcemia. Rosen et al during a two-year period found six patients with immobilization hypercalcemia following a single limb fracture to a weight-bearing bone, out of a total of 12 patients with such fractures studied during this time. The six patients ranged from 4.3 to 15.8 years and had been in good health before the injury. All six were male.

The typical story in an immobilized patient with hypercalcemia is gradual onset of the rather nonspecific symptoms, namely, constipation, anorexia, nausea with occasional vomiting and weight loss. Lethargy and depression also result and polyuria may be present. These symptoms on occasion have been ascribed to the psychosomatic effects of the injury and hospitalization, and some patients have been allowed to go for weeks with considerable cachexia before the diagnosis of hypercalcemia was entertained. Acute onset of hypertensive encephalopathy has been reported in a patient with immobilization hypercalcemia, along with evidence of impairment of renal function. Serum calcium concentrations of 14 to 16 mg/dl have been found in patients who have had symptoms of hypercalcemia for several weeks. In our experience urine calcium excretion as high as 400 mg per day has been found. This is not a diagnostic problem once it has been recognized because hypercalcemia and hypercalciuria in a previously active child immobilized for several weeks are the definitive findings. Close monitoring of immobilized children and adolescents by urine calcium determination can detect incipient cases of hypercalcemia.

TREATMENT

The treatment of choice would be rapid mobilization of the patient but this usually is not possible. A high sodium and water intake, by the intravenous route if necessary, to increase urine excretion of calcium is of value to prevent the hypercalcemic crisis that might result from dehydration in these patients. Corticosteroid therapy has been successful in reducing the serum calcium concentration (Fig. 4–15), but if continued for long periods it would aggravate the already existing osteopenia. The physiologically appropriate treatment is injection of calcitonin, which reduces serum calcium concentration by inhibition of bone resorption. The preparation of salmon calcitonin now available, Calcimar, has been used with good results in several patients. This is injected intramuscularly in divided doses at six to eight hour intervals in a dosage of 5 to 8 MRC units per kg per dose. This may not necessarily reduce the serum calcium to normal but to concentrations which are in the range of physiologic tolerance, with abatement of symptoms of toxicity. The treatment can be supplemented by oral phosphate loading if desired in the form of Neutra-Phos or Neutra-Phos K in amounts providing 1 to 2 gm phosphorus per day. Treatment is continued until the serum calcium concentration does not rebound when the calcitonin is discontinued, and the period of treatment is shortened by as much muscular activity and weight-bearing as permitted by the patient's skeletal or muscular condition. This is a self-limited problem and permanent residua are not likely to occur unless cerebral injury has resulted, with coma and convulsions associated with hypercalcemia. Probably because of the relatively short duration of the hypercalcemia and hypercalciuria, permanent renal injury has not been a feature of this disorder.

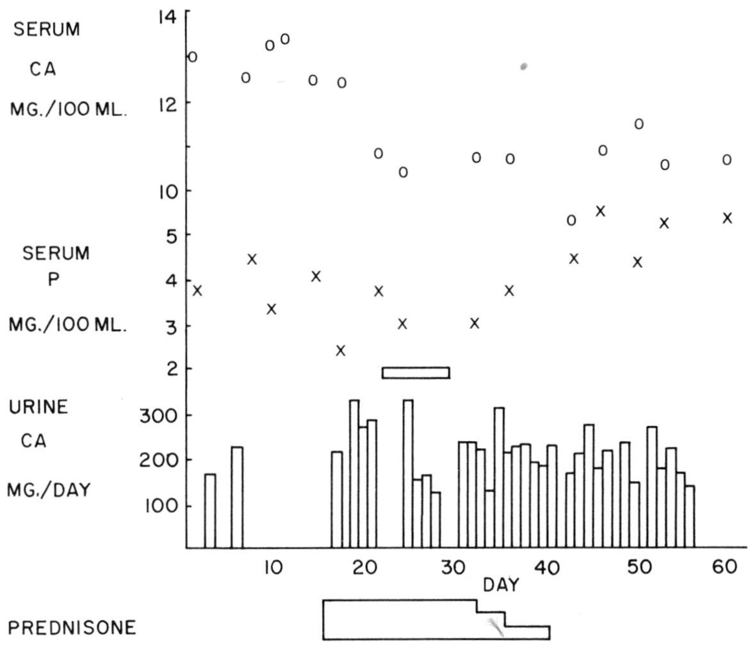

Figure 4–15. Immobilization hypercalcemia. Response of serum calcium concentration to corticosteroid treatment.

CASE REPORTS

Immobilization Hypercalcemia

K. Z. was a 16 year old boy who sustained a fracture of the femur while playing football. The fracture was reduced, he was placed in a double hip plaster spica and hospitalized on the orthopedic surgical service. During the second week of hospitalization he became anorectic and complained of nausea. He vomited occasionally. These symptoms were progressive, and weight loss was marked as indicated by the plaster cast becoming too loose. He was lethargic and depressed so that a psychiatric evaluation was requested in an attempt to explain the symptoms on a psychosomatic basis. All symptoms became progressively worse, and in the sixth week of his immobilization the serum calcium concentration was found to be 15.7 mg/dl, serum phosphate 2.2 mg/dl, creatinine 1.3 mg/dl and creatinine clearance 89 ml/min/1.73M^2. Urine calcium excretion was 403 mg/24 hrs. X-ray examination of the skeleton showed marked demineralization of all of the bones although the osteopenia was most marked in the bones of the lower extremities.

Treatment with increased sodium and water intake was instituted and urine volume increased to 1500 to 1900 ml/day. The serum calcium concentration varied between 13.0 and 15.4 mg/dl. After three weeks of this therapy a sodium and potassium citrate preparation was given with the purpose both of increasing urine citrate excretion to prevent nephrocalcinosis and possibly of stimulating bone mineral deposition. Urine citrate excretion was doubled and serum calcium concentration slowly decreased to values between 12 and 13 mg/dl in the next two weeks, while serum phosphorus rose to 3.5 mg/dl. Two months after onset of treatment serum calcium was 10.8 mg/dl, serum phosphate 4.0 mg/dl and BUN 15 mg/dl. During this time, however, partial mobilization was also achieved and eventually there was complete recovery without residual renal damage. Recovery in this case was slow and would probably have been expedited by steroid treatment, as indicated in the next case report. Addition of phosphate salts to the regime likely would also have been beneficial.

J. S. was a 10 year old boy admitted because of fracture and posterior dislocation of distal femoral epiphysis that resulted from his being caught in a landslide of sand and rocks. He was treated by closed reduction and immobilization in a double hip spica cast. He was discharged after six weeks still immobilized. He had been constipated during this hospitalization but no other symptoms were recorded. Following discharge from the hospital constipation became more troublesome, and he also complained of epigastric pain, nausea and vomiting; headache was also prominent. He was readmitted after 13 days at home because of these symptoms. At this time he showed evidence of considerable weight loss, and blood pressure was 120/80. Serum calcium concentration was 13.0 mg/dl and serum phosphate 3.7 mg/dl. He was treated symptomatically with a low calcium diet and high fluid intake but after two weeks serum calcium was still elevated, 12.4 mg/dl, and serum phosphate was 3.0 mg/dl. Prednisone was then given, 40 mg/day (2 mg/kg). After one week of this therapy the concentration of serum calcium was 10.8 mg/dl and that of serum phosphate 3.7 mg/dl. The serum calcium remained below 11.0 mg/dl; after 15 days the dose was quickly tapered and it was discontinued after 25 days. There was no rebound of the hypercalcemia and further convalescence was uneventful. The blood urea nitrogen had risen as high as 26 mg/dl during the course of treatment, but by eight weeks after the diagnosis of hypercalcemia was again within the normal range of less than 15 mg/dl.

A.B., a 14 year old white male, was transferred to the Montefiore Hospital for evaluation of persistent hypercalcemia. He had suffered a subtrochanteric fracture of the left femur three months earlier. Treatment had consisted of traction for seven weeks followed by a single leg cast. When treatment was initiated serum calcium concentration was 10.1 mg/dl. During the eighth week of immobilization serum calcium was found to be 13.8 mg/dl. When he was admitted to the Montefiore Hospital total and ionized calcium concentrations in serum were 13.2 and 5.9 mg/dl respectively. Urine calcium excretion was 420 mg per gm creatinine ($Ca/_{Cr} = 0.42$). Twenty-four hours after intravenous hydration serum total and ionized calcium concentrations were 13.3 and 6.4 mg/dl respectively. Salmon calcitonin treatment was instituted, 200 MRC units per day divided into two doses at 12-hour intervals injected intramuscularly. This sufficed to reduce serum ionized calcium concentration to 4.6 mg/dl and urine calcium excretion to 240 mg per gm creatinine within two days. Calcitonin treatment was continued for four weeks until complete ambulation was possible, and serum calcium concentrations and urine calcium excretion remained normal. Ten months later the patient continued to be asymptomatic with normal renal function.

COMMENTS: This patient was hypercalcemic for one month before treatment was instituted. Calcitonin treatment was begun after 24 hours of intravenous hydration in order to reduce the serum calcium concentration and urine calcium excretion as rapidly as possible. Initial treatment with a natriuretic and calciuretic drug such as furosemide might have reduced serum calcium concentration, but for prolonged hypercalcemia calcitonin is probably the treatment of choice. In some patients calcitonin may produce nausea and vomiting, which requires reduction of the dose. This patient showed no side-effects, and treatment was continued until he could be ambulated so that no rebound of serum calcium concentration occurred. (We are indebted to Dr. Laurence Finberg for the details of this case and permission to include them.)

HYPERCALCEMIA ASSOCIATED WITH TUMORS OTHER THAN OF THE PARATHYROID

There are two broad types of hypercalcemia associated with nonparathyroid tumors: (1) that in which there is evidence of primary or metastatic growth of the tumor in bone such as myeloma, leukemia or metastatic neuroblastoma; and (2) that in which there is no radiographic or biopsy evidence of invasion of bone by tumor. In the first type the hypercalcemia is thought to result from the destruction of bone with solubilization of bone mineral, either by the direct effect of proliferating tumor cells or by the secretion of tumor cells of an osteoclast-stimulating factor which causes secondary proliferation of osteoclasts that actively resorb bone.

In the second variety the secretion of a humoral factor that affects the metabolism of bone cells resembling the action of parathyroid hormone has been postulated. Some tumors, including both myeloid and lymphatic leukemias, have been found to secrete a peptide immunologically as well as physiologically similar to PTH. In such patients circulating PTH concentrations determined by radio immunoassay should be elevated, as in those with hyperparathyroidism. In other instances no such peptide has been found by immunoassays either of serum or of tumor extracts. Other compounds are presumably secreted which have a PTH- or vitamin D-like action. The prostaglandins are one such group of compounds, and there is experimental evidence that prostaglandin-secreting tumors in rats are associated with hypercalcemia, which is ameliorated when the tumor is removed or if prostaglandin secretion is lessened by indomethacin treatment.

Hypercalcemia has also been seen in patients with pheochromocytoma. Since pheochromocytoma and hyperparathyroidism can coexist in the syndrome multiple endocrine neoplasia type II, the hypercalcemia in subjects with pheochromocytoma has been thought to be evidence of associated parathyroid adenoma or hyperplasia. However, in at least two such patients the serum calcium concentrations returned to normal following removal of the pheochromocytoma without parathyroidectomy. In one of these patients, circulating IPTH concentrations were elevated during the period of hypercalcemia and decreased following removal of the adrenal tumor coincident with decrease in serum calcium concentration. This could be due to secretion of a peptide immunologically and physiologically similar to PTH by the adrenal medullary tumor, but a more probable explanation is that the high epinephrine concentrations stimulated PTH release, possibly through activation of adenyl cyclase in the parathyroid glands with increased concentrations of cyclic AMP.

The differential diagnosis of tumor-associated hypercalcemia depends of course mainly on the nature of the primary tumor. If there is invasion of bone by tumor, roentgenographic and biopsy findings establish the diagnosis. If there is no skeletal involvement the important question is whether hyperparathyroidism is present. The serum phosphate concentrations are not helpful since hypophosphatemia can occur in hypercalcemia without excess PTH. It may be that the nonparathyroid humoral factors that increase bone resorption also inhibit tubular reabsorption of phosphate. Circulating IPTH concentrations are important in the differentiation; these are usually undetectable in the peripheral blood of patients with tumor-induced hypercalcemia, so-called pseudohyperparathyroidism, whereas they are elevated or in the high normal range in the blood of those with true hyperparathyroidism.

At the moment determinations of prostaglandins in blood and urine are not clinically feasible, but conceivably the response of the serum calcium concentration to drugs that inhibit prostaglandin secretion such as acetylsalicylic acid, indomethacin and ibuprofen might provide indirect evidence of prostaglandin-induced hypercalcemia. The serum bicarbonate concentration is usually somewhat reduced in hyperparathyroidism, with reciprocal increase in serum chloride concentration. This is due to the action of PTH in reducing renal tubular reabsorption of bicarbonate as well as phosphate in the proximal tubule. This mild renal tubular acidosis is not usually found in tumor-associated hypercalcemia. Urinary calcium excretion can be increased, both in hyperparathyroidism and in hypercalcemia of nonparathyroid origin, but is usually higher in the latter at comparable concentrations of calcium in serum. This is explained by another action of PTH, i.e. its effect in increasing renal tubular reabsorption of calcium. The effect of corticosteroid treatment can also be of value since the hypercalcemia due to excess of PTH usually is not modified by corticosteroids, whereas that due to nonparathyroid tumors often is responsive.

The treatment of tumor-induced hypercalcemia includes both the primary treatment of the tumor and symptomatic therapy. The latter consists of all the measures discussed in the section on hypervitaminosis D. Calciuresis is induced by high sodium and water intake plus IV furosemide if the serum calcium concentration is rising rapidly or if symptoms are present. Steroid treatment such as prednisone, 2 mg/kg up to 60 mg per day, should also be instituted. The use of inhibitors of prostaglandin synthesis may be of therapeutic as well as diagnostic value, but there is insufficient knowledge of this problem to warrant recommendations for treatment. High phosphate intake orally can be instituted as an additional measure if renal function is not impaired. If pheochromocytoma is diagnosed, removal of the tumor or tumors with restoration of catecholamine excretion to normal should be done before parathyroid exploration, even if serum IPTH concentrations are increased, since (as mentioned) PTH secretion may be stimulated by high epinephrine concentrations.

REFERENCES

DIAGNOSIS

Drach, G. W. and King, J. S., Jr.: Estimating aberrant homeostasis: variance in serum calcium concentration as an aid in diagnosis of hyperparathyroidism. Clin. Chem. *16*:792–796, 1970.

Edmondson, J. W. and Li, T. K.: The relationship of serum ionized and total calcium in primary hyperparathyroidism. J. Lab. Clin. Med. *87*:624–629, 1976.

Sinha, T. K., Miller, S., Fleming, J. et al: Demonstration of a diurnal variation in serum parathyroid hormone in primary and secondary hyperparathyroidism. J. Clin. Endocrinol. Metab. *41*:1009–1013, 1975.

Stentstrom, G. and Heedman, P. A.: Clinical findings in patients with hypercalcemia. Acta Med. Scand. *195*:473–477, 1974.

HYPERVITAMINOSIS D

Counts, S. J., Baylink, D. J., Shen, F.-H. et al: Vitamin D intoxication in an anephric child. Ann. Intern. Med. *82*:196–200, 1975.

Holman, C. B.: Roentgenologic manifestations of vitamin D intoxication. Radiology *59*:805–816, 1952.

Hughes, M. R., Baylink, D. J., Jones, P. G. et al: Radioligand receptor assay for 25-hydroxyvitamin D_2/D_3, and 1α-25-dihydroxyvitamin D_2/D_3. Application to hypervitaminosis D. J. Clin. Invest. *58*:61–70, 1976.

Hunt, R. D., Garcia, F. G. and Hegsted, D. M.: Hypervitaminosis D in the New World monkeys. Am. J. Clin. Nutr. *22*:358–366, 1969.

Kanis, J. A. and Russell, R. G. G.: Rate of reversal of hypercalcemia and hypercalciuria induced by vitamin D and its 1α-hydroxylated derivatives. Br. Med. J. *1*:78–81, 1977.

Leeson, P. M. and Fourman, P.: Increased sensitivity to vitamin D after vitamin D poisoning. Lancet *1*:1182–1185, 1966.

Report of Committee on Nutrition, American Academy of Pediatrics: The prophylactic requirement and the toxicity of vitamin D. Pediatrics *31*:512–525, 1963.

Verner, J. V., Engel, F. L. and McPherson, H. T.: Vitamin D intoxication: report of two cases treated with cortisone. Ann. Intern. Med. *48*:765–773, 1958.

Wasserman, R. H., Henion, J. D., Haussler, M. R. et al: Calcinogenic factor in *Solanum malacoxylon:* evidence that it is 1,25-dihydroxyvitamin D_3-glycoside. Science *194*:853–855, 1976.

Ziegler, R., Minne, H., Rane, F. et al: Beobachtungen zur Vitamin D und Dihydrotachysterin Vergiftung. Dtsch. Med. Wochenschr. *100*:415–423, 1975.

IDIOPATHIC HYPERCALCEMIA

Barr, D. G. D. and Forfar, J. O.: Oral calcium-loading test in infancy with particular reference to idiopathic hypercalcemia. Br. Med. J. *1*:477–480, 1969.

Beuren, A. J., Schulze, C., Eberle, P. et al: The syndrome of supravalvular aortic stenosis, mental retardation and similar facial appearance. Am. J. Cardiol. *13*:471–483, 1964.

Fanconi, V. G. and Girardet, P.: Chronische Hypercalcäemie, kombiniert mit Osteosklerose, Hyperazotämie, Minderwuchs und kongenitalen Missbildungen. Helv. Paediatr. Acta *7*:314–349, 1952.

Friedman, W. F. and Mills, L. F.: The relationship between vitamin D and the craniofacial and dental anomalies of the supravalvular aortic stenosis syndrome. Pediatrics *43*:12–18, 1969.

Garica, R. E., Friedman, W. F., Kaback, M. M. et al: Idiopathic hypercalcemia and supravalvular aortic stenosis. N. Engl. J. Med. *271*:117–120, 1964.

Kenny, F. M., Aceto, T. Jr., Purisch, M. et al: Metabolic studies in a patient with idiopathic hypercalcemia of infancy. J. Pediatr. *62*:531–537, 1963.

Lightwood, R. and Stapleton, T.: Idiopathic hypercalcemia in infants. Lancet *2*:255–256, 1953.

Lindquist, B.: Idiopathic hypercalcemia. A disorder

of the calcium homeostatic mechanism. Acta Paediatr. *51*(Suppl. 135)144–152, 1962.

Smith, D. W., Blizzard, R. M. and Harrison, H. E.: Idiopathic hypercalcemia. A case report with assays of vitamin D in the serum. Pediatrics *24*: 258–269, 1959.

White, R. A., Preus, M., Watters, G. V. et al: Familial occurrence of the Williams syndrome. J. Pediatr. *91*:614–616, 1977.

HYPERCALCEMIC SARCOIDOSIS

Cushard, W. G., Simon, A. B., Canterbury, J. M. et al: Parathyroid function in sarcoidosis. New Engl. J. Med. *286*:395–398, 1972.

Hendrix, J. Z.: The remission of hypercalcemia and hypercalciuria in sarcoidosis by vitamin D depletion. Clin. Res. *11*:220, 1963.

Scholz, D. A.: Effect of steroid therapy on hypercalcemia and renal insufficiency in sarcoidosis. J.A.M.A. *169*:682–687, 1959.

HYPO- AND HYPERTHYROIDISM

Burman, K. D., Monchik, J. M., Earll, J. M. et al: Ionized and total serum calcium and parathyroid hormone in hyperthyroidism. Ann. Intern. Med. *84*:668–671, 1976.

Howard, J. E. and Thomas, W. C., Jr.: Clinical disorders of calcium homeostasis. Medicine *42*:25–45, 1963.

Lowe, C. E., Bird, E. D. and Thomas, W. C., Jr.: Hypercalcemia in myxedema. J. Clin. Endocrinol. Metab. *22*:261–267, 1962.

Parfitt, A. M. and Dent, C. E.: Hyperthyroidism and hypercalcemia. Q. J. Med. *39*:171–187, 1970.

Wilkins, L.: The Diagnosis and Treatment of Endocrine Disorders in Childhood and Adolescence, 3rd Ed. Charles C Thomas, Springfield, Ill., 1966 p. 488–489.

PHOSPHATE DEPLETION

Coburn, J. W. and Massry, S. G.: Changes in serum and urinary calcium during phosphate depletion: studies on mechanism. J. Clin. Invest. *49*:1073–1087, 1970.

Dent, C. E. and Winter, C. S.: Osteomalacia due to phosphate depletion from excessive aluminum hydroxide ingestion. Br. Med. J. *1*:551–552, 1974.

Freeman, S. and McLean, F. C.: Experimental rickets: blood and tissue changes in puppies receiving diet very low in phosphorus without vitamin D. Arch. Pathol. *32*:387–408, 1941.

Lotz, M., Zisman, E. and Bartter, F. C.: Evidence for a phosphorus-depletion syndrome in man. N. Engl. J. Med. *278*:409–415, 1968.

HYPERPARATHYROIDISM

Anspach, W. E. and Clifton, W. M.: Hyperparathyroidism in children. Am. J. Dis. Child. *58*:540–557, 1939.

Aurbach, G. D., Mallette, L. E., Patten, B. M. et al: Hyperparathyroidism: recent studies. Ann. Intern. Med. *79*:566–581, 1973.

Biernulf, A., Hall, K., Sjogren, I. et al: Primary hyperparathyroidism in children. Brief review of the literature and case report. Acta Paediatr. Scand. *59*:249–258, 1970.

Bouillon, R. and DeMoor, P.: Parathyroid hormone secretion in primary hyperparathyroidism. J. Clin. Endocrinol. Metab. *45*:261–269, 1977.

Christensson, T.: Familial hyperparathyroidism. Ann. Intern. Med. *85*:614–615, 1976.

Foley, T. P., Harrison, H. C., Arnaud, C. D. et al: Familial benign hypercalcemia. J. Pediatr. *81*:1060–1067, 1972.

Garcia-Bunuel, R., Kutchemeshgi, A. and Brandes, D.: Hereditary hyperparathyroidism. The fine structure of the parathyroid gland. Arch. Pathol. *97*:399–403, 1974.

Goldberg, E., Winter, S. T., Better, O. S. et al: Transient neonatal hyperparathyroidism associated with maternal hypoparathyroidism. Isr. J. Med. Sci. *12*:199–201, 1976.

Goldbloom, R. B., Gillis, D. A. and Prasad, M.: Hereditary parathyroid hyperplasia: a surgical emergency of early infancy. Pediatrics *49*:514–523, 1972.

Goldsmith, R. E., Sizemore, G. W., Chen, I. W. et al: Familial hyperparathyroidism. Description of a large kindred with physiologic observations and a review of the literature. Ann. Intern. Med. *84*: 36–43, 1976.

Gray, R. S. and Gillon, J.: Normotensive phaeochromocytoma with hypercalcemia: correction after adrenalectomy. Br. Med. J. *1*:378, 1976.

Hillman, D. A., Scriver, C. R., Pedvis, S. et al: Neonatal familial primary hyperparathyroidism. N. Engl. J. Med. *270*:483–490, 1964.

Jackson, C. E. and Boonstra, C. E.: The relationship of hereditary hyperparathyroidism to endocrine adenomatosis. Am. J. Med. *43*:727–734, 1967.

Kukreja, S., Hargis, G. K., Rosenthal, I. M. et al: Pheochromocytoma causing excessive parathyroid hormone production and hypercalcemia. Ann. Intern. Med. *79*:838–840, 1973.

Marsden, P., Anderson, J., Doyle, D. et al: Familial hyperparathyroidism. Br. Med. J. *3*:87–90, 1971.

Marx, S. J., Powell, D., Shimkin, P. M. et al: Familial hyperparathyroidism: mild hypercalcemia in at least nine members of a kindred. Ann. Intern. Med. *78*:371–377, 1973.

Marx, S. J., Spiegel, A. M., Brown, E. M. et al: Family studies in patients with primary parathyroid hyperplasia. Am. J. Med. *62*:698–706, 1977.

Nguyen, V. C., Sennott, W. M. and Knox, G. S.: Neonatal hyperparathyroidism. Radiology *112*:175–176, 1974.

Rhone, D. P.: Primary neonatal hyperparathyroidism. Am. J. Clin. Pathol. *64*:488–499, 1975.

Sann, L., David, L., Thomas, A. et al: Congenital hyperparathyroidism and vitamin D deficiency secondary to maternal hypoparathyroidism. Acta Paediatr. Scand. *65*:381–385, 1976.

Snyder, N., Scurry, M. T. and Deiss, W. P.: Five families with multiple endocrine adenomatosis. Ann. Intern. Med. *76*:53–58, 1972.

Spiegel, A. M., Harrison, H. E., Marx, S. J. et al: Neonatal primary hyperparathyroidism with autosomal dominant inheritance. J. Pediatr. *90*:269–272, 1977.

Weichert, R. F., III: The neural ectodermal origin of the peptide-secreting endocrine glands. Am. J. Med. *49*:232–241, 1970.

TUMORAL HYPERCALCEMIA

Berson, S. A. and Yalow, R. S.: Parathyroid hormone in plasma in adenomatous hyperparathyroidism, uremia, and bronchogenic carcinoma. Science *154*:907–909, 1966.

Cornbleet, M., Bondy, P. K. and Powles, T. J.: Fatal irreversible hypercalcemia in breast cancer. Br. Med. J. *1*:145, 1977.

Jayaraman, J. and David, R.: Hypercalcemia as a presenting manifestation of leukemia: evidence of excessive PTH secretion. J. Pediatr. *90*:609–610, 1977.

Powell, D., Singer, F. R., Murray, T. M. et al: Nonparathyroid humoral hypercalcemia in patients with neoplastic diseases. N. Engl. J. Med. *289*: 176–181, 1973.

Seyberth, H. W., Segre, G. V., Morgan, J. M. et al: Prostaglandins as mediators of hypercalcemia associated with certain types of cancer. N. Engl. J. Med. *293*:1278–1283, 1975.

Tashjian, A. H., Jr.: Prostaglandins, hypercalcemia and cancer. N. Engl. J. Med. *293*:1317–1318, 1975.

Tashjian, A. H., Jr., Tice, J. E., Sides, K.: et al: Biological activities of prostaglandin analogues and metabolites on bone in organ culture. Nature *266*:645–647, 1977.

Trummel, C. L., Mundy, G. R., Raisz, L. G. et al: Release of osteoclast activating factor by normal human peripheral blood leukocytes. J. Lab. Clin. Med. *85*:1001–1007, 1975.

Zidar, B. L., Shadduck, R. K., Winklestein, A. et al: Acute myeloblastic leukemia and hypercalcemia. A case of probable ectopic parathyroid hormone production. N. Engl. J. Med. *295*:692–694, 1976.

IMMOBILIZATION HYPERCALCEMIA

Bassett, C. A. L.: Biologic significance of piezoelectricity. Calcif. Tissue Res. *1*:252–272, 1968.

Bijvoet, O. L. M., Van der Sluys, V. J. and Jansen, A. P.: Effects of calcitonin on patients with Paget's disease, thyrotoxicosis, or hypercalcemia. Lancet *1*:876–881, 1968.

Dodd, K., Graubarth, H. and Rapoport, S.: Hypercalcemia, nephropathy and encephalopathy following immobilization. A case report. Pediatrics *6*:124–130, 1950.

Donaldson, C., Hulley, S. B., Vogel, J. M. et al: Effect of prolonged bed rest on bone mineral. Metabolism *19*:1071–1084, 1970.

Haymovits, A. and Rosen, J.: Calcitonin in metabolic disorders. *In* Advances in Metabolic Disorders (Eds.: R. Levin and R. Luft). Academic Press, New York and London, 1972, pp. 177–212.

Lawrence, G. D., Loeffler, R. G., Martin, L. G. et al: Immobilization hypercalcemia. J. Bone Joint Surg. *55-A*:87–94, 1973.

Lerman, S., Canterbury, J. M. and Reiss, E.: Parathyroid hormone and the hypercalcemia of immobilization. J. Clin. Endocrinol. Metab. *45*:425–428, 1977.

Rosen, J. F., Wolin, D. A. and Finberg, L.: Immobilization hypercalcemia after single limb fractures in children and adolescents. Am. J. Dis. Child. *132*:560–564, 1978.

Wolf, A., Chuinard, R. G., Riggins, R. S. et al: Immobilization hypercalcemia: a case report and review of the literature. Clin. Orthop. *118*:124–129, 1976.

HYPERCALCEMIA TREATMENT

Behn, A. R. and West, T. E. T.: Emergency treatment with calcitonin of hypercalcemia associated with multiple myeloma. Br. Med. J. *1*:755–756, 1977.

Najjar, S. S., Aftimos, S. R. and Kurani, R. F.: Furosemide therapy for hypercalcemia in infants. J. Pediatr. *81*:1171–1174, 1972.

Root, A. W. and Harrison, H. E.: Recent advances in calcium metabolism. II. Disorders of calcium homeostasis. J. Pediatr. *88*:177–199, 1976.

Schneider, A. B. and Sherwood, L. M.: Progress in endocrinology and metabolism. Calcium homeostasis and the pathogenesis and management of hypercalcemic disorders. Metabolism *23*:975–1007, 1974.

Silva, O. L. and Becker, K. L.: Salmon calcitonin in the treatment of hypercalcemia. Arch. Intern. Med. *132*:337–339, 1973.

Stapleton, F. B., Lukert, B. P. and Linshaw, M. A.: Treatment of hypercalcemia associated with osseous metastases. J. Pediatr. *89*:1029–1030, 1976.

West, T. E. T., Joffe, M., Sinclair, L. et al: Treatment of hypercalcaemia with calcitonin. Lancet *1*:675–678, 1971.

MISCELLANEOUS

Downie, W. W., Gunn, A., Patterson, C. R. et al: Hypercalcaemic crisis as presentation of Addison's disease. Br. Med. J. *1*:145–146, 1977.

Drummond, K. N., Michael, A. F., Ulstrom, R. A. et al: The blue diaper syndrome: familial hypercalcemia with nephrocalcinosis and indicinuria. Am. J. Med. *37*:928–948, 1964.

Duarte, C. G., Winnacker, J. L., Becker, K. L. et al: Thiazide induced hypercalcemia. N. Engl. J. Med. *284*:828–830, 1971.

Lee, J. C., Cantanzaro, A., Parthemore, J. G. et al: Hypercalcemia in disseminated coccidioidomycosis. N. Engl. J. Med. *297*:431–433, 1977.

Swinton, N. W., Clerkin, E. P. and Flint, L. D.: Hypercalcemia and familial pheochromocytoma—correction, after adrenalectomy. Ann. Intern. Med. *76*:455–457, 1972.

Chapter Five

RICKETS AND OSTEOMALACIA

Introduction

The name rickets according to one explanation has its origin in the Old English word *wrickken*, meaning to twist, and the application of this term to the skeletal deformities of rickets is obvious. Glisson suggested the similar-sounding but more elegant name of classical origin *rhachitis*, derived from the Greek *rhaxis*, spine, and both terms have been used interchangeably. The definition of rickets has been broadened to include more than the specific entity of the skeletal disorder of vitamin D deficiency. The essential morphologic lesion of rickets and osteomalacia, whatever the causation, is an excess of osteoid tissue, i.e., organic matrix which is produced in normal fashion in the growing cartilage and bone but remains unmineralized. The term *rickets*, therefore, includes all the disturbances of *growing* bone in which there is a lag in the mineralization of matrix so that uncalcified cartilage matrix and bone matrix, osteoid, accumulate to an abnormal extent. The delay of crystallization of the bone salt in the matrix in true rickets is secondary to abnormalities of the mechanisms that control the concentrations of the essential ions, inorganic phosphate and calcium, in the extracellular fluid phase of bone. As a result the concentrations of these ions, particularly that of phosphate, are too low to permit formation of the initial calcium phosphate precipitate even though cellular mechanisms and matrix composition are normal. Certain rare disturbances of bone mineralization that resemble rickets have been described in which the concentrations of phosphate and calcium ions in serum are entirely normal, and it is assumed that some local factor in the bone matrix necessary for mineralization is deficient, or that some inhibiting agent prevents the formation of bone salt crystals despite the normal concentrations of calcium and phosphate in body fluids.

Osteomalacia is simply rickets occurring after the linear growth of the skeleton has stopped, but the characteristic biochemical lesion, i.e., a deficiency of phosphate ion and possibly of calcium ion in extracellular fluid, is the same. The histologic lesion, an excess of osteoid, is also present since remodeling of adult bone by resorption and formation of new matrix and its mineralization continue in adult life even though linear growth ceases.

The diagnosis of rickets and osteomalacia, whatever the cause, is based on clinical signs and symptoms; analysis of serum for calcium, phosphate and alkaline phosphatase; roentgenographic examination of the bones; and histologic examination of the bones. The specific findings will be discussed in the sections on the various forms of rickets and osteomalacia.

CLASSIFICATION

Rickets and osteomalacia can be divided into two main categories, as listed in Table 5-1. In type I rickets a deficiency of the vitamin D hormone, 1,25-dihydroxyvitamin D, is the causative factor. This may occur as the result of lack of vitamin D itself or may be due to various abnormalities in the chain of metabolic events leading to the formation of the final active metabolite of vitamin D. Because calcium malabsorption and resistance to the action of parathyroid hormone are dominant features of 1,25-dihydroxyvitamin D deficiency, secondary hyperparathyroidism results, and in the more chronic forms of such deficiency the manifestations of secondary hyperparathyroidism may be predominant as in renal osteodystrophy. Type II rickets results from a primary reduction of inorganic phosphate concentration in extracellular fluid, most commonly owing to impairment of renal tubular reabsorption of phosphate. A variety of renal tubular disorders, congenital and acquired, can result in deficiency of the renal tubule phosphate transport system and hypophosphatemia, as shown in Table 5-1. In type II rickets deficiency of 1,25-dihydroxyvitamin D and secondary hyperparathyroidism are not major features although the former may be a complicating feature of renal tubular injury or dysfunction and the latter may occur as a consequence of treatment in genetic or acquired hypophosphatemia.

Table 5–1. Classification of Rickets

TYPE I *(Deficiency of 1,25-diOH Vitamin D)*	TYPE II *(Primary Phosphate Deficiency)*
Vitamin D lack	Disorders of renal tubule reabsorption of phosphate
Absence of sunshine	Fanconi syndrome
Dietary deficiency	Cystinosis
Vitamin D malabsorption	Tyrosinosis
Liver disease	Lowe's syndrome
Anticonvulsant drugs	Other causes
Renal disease	Renal tubular acidosis
Vitamin D dependency (abnormality of renal 1-OHase)	Genetic primary hypophosphatemia
	Hypophosphatemia with "nonendocrine tumors"
	Phosphate deficiency or malabsorption
	Parenteral hyperalimentation
	Low phosphate feeding in prematurely born infants
	$Al(OH)_2$ intake

TYPE I RICKETS

VITAMIN D DEFICIENCY RICKETS

History and Epidemiology

This classic disease of infancy and early childhood is fundamentally the result of climatic and environmental factors. The needed amounts of vitamin D are not available in the usual foods taken by the infant and young child. Human milk, like cow's milk, does not usually provide the minimal requirements of vitamin D even though it be the sole source of calories for the young infant (Table 5-2). The vitamin D content of human milk, to be sure, can be increased by feeding vitamin concentrates to the lactating woman or by her exposure to ultraviolet irradiation. In similar fashion the vitamin D content of cow's milk can be raised by adding vitamin D concentrates to the fodder.

The natural source of vitamin D for the infant and child, however, is the endogenous vitamin resulting from the irradiation of the 7-dehydrocholesterol of the skin by ultraviolet radiation. The activating radiation lies in the band between 230 and 313 mμ. During the winter months in the temperate zones the atmosphere filters out most of the radiation in the bands shorter than 306 mμ because of the oblique incidence of the sun's rays at this time, which increases the distance which they travel through the atmosphere. Even in the summer months infants living in crowded industrial cities are deprived of the necessary active radiation because much of the short wave ultraviolet radiation is filtered out by the soot and dust in the atmosphere.

Table 5–2. Vitamin D Content of Foods

Human milk	0 to 10 U/dl	Herring	1500 U/100 gm
Cow's milk	0.3 to 4 U/dl	Mackerel	1800 U/100 gm
Butter	35 U/100 gm	Canned salmon	300 U/100 gm
Egg yolk	25 U/average yolk	Canned tuna	250 U/100 gm
Calf liver	15 U/100 gm	Canned sardines	600 U/100 gm
	Cod liver oil	175 U/gm	

Rickets was thus the first important disease of atmospheric pollution. Its prevalence in the cities of northern Europe and North America is not surprising as industrialization of cities and increased crowding in the slums progressively reduced the possibility of exposing young infants to the ultraviolet radiation of sunlight. In 1917 Hess and Unger reported that almost all the infants in the Negro community of New York City had rickets, whether they were breast- or bottle-fed, and that the same was true to a slightly less extent in the Italian districts of the city. Similar experiences were collected in every large city of the Northern United States. The incidence of rickets in England and Europe was equally great in the large industrial communities. Before the discovery of vitamin D, Hess described the situation accurately by saying that at this time a map of the incidence of rickets was the "practical equivalent of a map of deficiency of sunlight."

This high incidence of rickets persisted despite the availability of cod liver oil. Although the latter was used as a medication as early as the 18th century, its value in the treatment or prevention of rickets continued to be doubted by the most eminent authorities. Only when rickets was produced experimentally in animals, and the effect of cod liver oil studied systematically, was its antirachitic action proved and accepted. A study of the history of cod liver oil should make any physician feel humble about the accuracy of clinical observations in the assessment of a therapeutic agent. The discovery of vitamin D as a specific agent that prevented rickets followed the development of diets that could produce experimental rickets in dogs and rats. Very quickly the role of sunlight in the story of human rickets was elucidated by the discovery that rickets in infants and experimental animals could be healed by exposure to ultraviolet radiation, and that some foods irradiated with short wave ultraviolet rays developed antirachitic potency. This started a series of brilliant investigations that culminated in the isolation of vitamin D and solved the riddle of the relationship of the energy of ultraviolet radiation to this accessory food substance.

A number of forms of vitamin D are known to exist, if vitamin D is used as a generic name for all sterols active in the prevention and cure of rickets. Only two of these compounds are of importance in human nutrition: (1) ergocalciferol, vitamin D_2, the reaction product resulting from irradiation of the plant sterol ergosterol; and (2) cholecalciferol, vitamin D_3, produced by the irradiation of 7-dehydrocholesterol, the sterol found in animal skin. Both of these sterols are equally potent as antirachitic agents for the rat and the human, although in the chick and the New World monkey cholecalciferol is much more active than ergocalciferol.

The role of dietary composition other than vitamin D content is of relatively little importance in human rickets. The albino rat, the subject of much of our nutritional information, does not develop hypophosphatemia with simple deficiency of vitamin D as does man, so that in this species rickets is not produced by vitamin D deficiency alone. In order to develop rickets in the rat, diets disproportionately low in available inorganic phosphate must be used. The rachitogenic effect of cereals, which was emphasized in early studies in the puppy, is almost certainly due to the fact that the phosphorus in cereal is present as phytin, inositol phosphate, which binds calcium and which is not a source of inorganic phosphate until it is hydrolyzed by the enzyme phytase. Since rickets in infants occurs during the period when milk is the major food source, the nutritional effect of excess phytin in the absence of vitamin D is mainly a problem of rickets in the older child and osteomalacia in the adult.

The incidence of rickets was progressively reduced by the discovery of an antirachitic substance, vitamin D, in cod liver oil, egg yolk and butter fat, and the subsequent development of vitamin D concentrates by the irradiation of the provitamin D sterols, ergosterol and 7-dehydrocholesterol. The present minimal occurrence of this disease is the result of the enrichment of evaporated milk, homogenized fresh cow's milk and special infant feeding preparations with such vitamin D concentrates. As will be discussed later, the daily intake of 400 units of vitamin D provided by one can of evaporated milk or one quart of enriched fresh milk is more than adequate to prevent rickets in the vast majority of infants. Since most infants are given vitamin D-enriched milk feedings as specially modified, commercially prepared cow's milk mixtures or homogenized pasteurized milk, few cow's milk-fed infants now develop rickets. Occasional cases of rickets are still seen in Baltimore and other large Northern American cities. Some of these infants are breast-fed and have received no vitamin D supplements because of the mistaken notion of the mother that breast-feeding obviates the need for vitamin supplements.

As has been mentioned, human milk is often as deficient in vitamin D as is cow's milk. The vitamin D content of human milk can be increased by exposure of the mother to ultraviolet radiation or feeding of vitamin D concentrates, but the ordinary dietary intake of this vitamin or the usual environmental exposure (or rather lack of exposure) to sunshine does not permit any appreciable amount of vitamin D to be transferred to the milk. In recent years the emergence of strict vegetarianism as a form of dietary faddism in which no food of animal origin is permitted except breast milk has been responsible for reappearance of rickets. Recently we were informed of a case of rickets developing in an infant who was thought to be sensitive to cow's milk protein and was fed on goat's milk obtained from a local farmer. This milk, of course, was not fortified with vitamin D, and since the routine use of vitamin D concentrates has been less common because of the usual feeding of vitamin D-enriched cow's milk preparation this infant received no source of vitamin D until florid rickets was present.

In pre-vitamin D days the unusually high incidence of rickets in black and other dark-skinned infants living in New York, Baltimore, Boston and other large northern cities was remarked. The possibility of a racial susceptibility to rickets was suggested with particular reference to the black infant. There is no evidence that the vitamin D requirement of the black is different from that of the white infant. Rickets has been effectively prevented in all groups by the same prophylactic measures. It is likely that the high incidence of rickets in the black infant was chiefly due to the more crowded housing conditions and almost complete deprivation of the sun's antirachitic rays. There is also evidence that the effective radiation of the sun is filtered out by skin pigment, so that darker-skinned peoples need a greater intensity of radiation to convert equal amounts of skin 7-dehydrocholesterol to vitamin D_3.

Individual differences do exist, however, in the response to or need for vitamin D. Some infants without known supplements of vitamin D and no unusual opportunity for exposure to sunshine escape rickets and presumably have minimal need for this sterol. A rare infant may develop rickets despite an authentic story of intake of vitamin D ordinarily adequate to prevent vitamin D deficiency. These infants may then respond to larger doses of vitamin D, suggesting that they need a greater than usual amount of this vitamin. Because of the previous lack of convenient methods for studying plasma levels of vitamin D and the absorption of this sterol, it was not possible to say whether this greater need was due to poor intestinal absorption or some alteration of the biochemical processes in which vitamin D interacts. This diagnostic handicap no longer obtains. The determination of 25-hydroxyvitamin D and 1,25-dihydroxyvitamin D in serum is now possible so that a more accurate classification of the causes of increased vitamin D requirement (vitamin dependency) can be made.

Finally, there is the group of infants who show hypercalcemia and the skeletal and renal changes suggestive of hypervitaminosis D despite doses of vitamin D not greater than those tolerated by the vast majority of infants. It has not been proved that "idiopathic hypercalcemia" is the result of hypersensitivity to vitamin D, but there is suggestive evidence that some of these patients have a defect in the mechanism for inactivation of this vitamin so that excessive concentrations of vitamin D persist in the serum.

Pathology of Rickets

The essential anatomic lesion of rickets is the result of the failure of crystallization of bone salt to keep pace with the formation of cartilage matrix and osteoid, i.e., bone matrix consisting of collagen and mucopolysaccharide. As a result, uncalcified matrix accumulates and the regular morphologic pattern of bone mineral deposition is distorted.

In the shafts of the long bones, layers of osteoid are formed by the osteoblasts at the surface of the trabeculae. In the rachitic bone this osteoid remains unmineralized so that excessively wide borders of osteoid are found at the periphery of the trabeculae (Fig. 5–1). The resorption of old bone is a physiologic process in the growth and remodeling of bone that continues at the usual rate in rachitic bone but is not accompanied by a parallel deposition of bone mineral in the newly formed matrix. As a result the bony trabeculae and the bone cortex become demineralized and an excessively large proportion of the volume of bone is osteoid, so that the shaft of the bone is less rigid and bends or twists under the weight of the body or the pull of muscles. The loss of bone strength is greatest at the sites of most rapid bone growth. The long thin bones of the extremities and the ribs are the most severely affected and the earliest to be deformed.

Further bending of the long bones results from displacement of the epiphysis owing to the rachitic changes at the cartilage-shaft junction described below. The epiphysis can be bent on the diaphysis in a backward, forward or lateral direction, and the shaft being formed is angulated with respect to the old. Angulation and bending of the bone may also result from fractures of the osteomalacic bone. Although the rachitic bone can be extremely thin and fragile it can also be excessively thick owing to excessive proliferation of osteoid. This is usually the result of long-standing rickets in which periods of activity alternate with periods of healing so that a series of new cortices are formed from the periosteum, one enveloping the other.

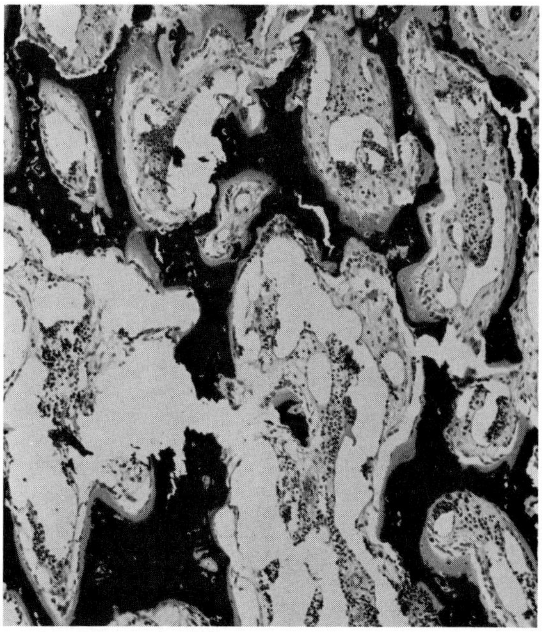

Figure 5–1. Section of human bone from subject with rickets showing broad layers of light staining osteoid (unmineralized matrix) overlying dark staining mineralized trabecular bone.

The histologic changes at the cartilage-shaft junction have been the subject of the most careful studies by investigators interested in the pathology of rickets in man and experimental animals. In the normal state the cartilage cells of the epiphyseal cartilage are invaded in an orderly fashion by the capillaries from the shaft side. The columns of cartilage cells lie in parallel rows in the lengthwise direction of the bone, embedded in the cartilage matrix consisting of collagen fibrils and mucopolysaccharide ground substance. The capillaries tunnel through the columns of cartilage cells, destroying them as they advance. Bone salt crystals are deposited in the matrix in close relation to these invading capillaries, which results in calcification of the cartilage at the cartilage-bone junction. Osteoblasts or their precursor cells are carried along with the capillaries; these cells form osteoid on the framework of the calcified cartilage matrix, and as the osteoid is mineralized a layer of new bone is formed. Meanwhile the bone grows in length by growth of the epiphyseal cartilage, so that a layer of fresh cartilage forms on the distal side as capillaries invade the proximal layer of the cartilage. The formation of bone from cartilage is so regular and orderly that the advancing border of bone remains sharp and distinct and the cartilaginous plate remains uniform in thickness.

In the rachitic subject the coordinated orderly transformation of cartilage to bone fails because of the lag in the deposition of bone salt in the cartilage matrix in proper relation to the proliferating capillaries. As a consequence of this defect of calcification of the cartilage matrix the zone of mature cartilage cells increases in width. The cartilage cells in the proliferating uncalcified cartilage appear to be more resistant to the destructive changes seen in the normally growing cartilage, and the ingrowth of capillaries is impeded since these invade the cell columns only as the cells undergo the destructive changes associated with maturation. The exact mechanism underlying the failure of the cartilage cells to undergo the usual sequence of changes is not known. The proliferative cartilage is penetrated by capillaries in an irregular fashion and the uncalcified cartilage matrix persists to a considerable extent. Osteoblasts accompanying the capillaries cover the surface of the matrix, and osteoid is produced. The epiphyseal cartilage of the rachitic bone is thus transformed into uncalcified chondro-osteoid trabeculae rather than into calcified bone.

This irregular mass of cartilage, blood vessels and chondro-osteoid trabeculae constitutes the rachitic intermediate zone or rachitic metaphysis. Figure 5–2 shows the histologic appearance of this zone in experimentally-induced rickets in rat bone in comparison with normal cartilage and bone. The lesion in experimental rickets is similar to that of the disease occurring in man, shown in Figure 5–3. Since all gradations of severity of rickets occur and healing phases may interrupt the progress of the pathologic process, islands and spurs of calcified matrix and bone may be present in this intermediate zone.

Healing of rickets following vitamin D therapy is marked by the deposition of bone mineral in the layer of epiphyseal cartilage adjacent to the distal boundary of the rachitic intermediate zone, and this precipitate shows up in the histologic specimen (Fig. 5–4) or x-ray film (Fig. 5–5) as a transverse line of increased density (Müller's line). Deposition of bone salt takes place also at various other points in the intermediate zone, which eventually is entirely transformed into bone. The fine organization of this bone remains abnormal, so that long after healing has occurred the abnormal structure of the rachitic bone is visible by x-ray examination. During healing the excessive osteoid laid down beneath the periosteum by subperiosteal osteoblasts is calcified and a layer of new bone forms, enveloping the old shaft. Both new and old bone are remodeled by resorption and redeposition of bone and the formation of haversian systems, so that during growth the abnormalities of the rachitic bone structure tend to disappear. Serious bends and twists persist, however, particularly if accentuated by abnormalities of muscle pull and weight-bearing resulting from the distorted shape of the bone. An understanding of the anatomic abnormalities of bone caused by rickets permits the interpretation of the radiographic changes of the bones which are described later.

Biochemical Abnormalities in Vitamin D Deficiency

The abnormalities of serum and urine constituents in rickets can be understood from the physiology of vitamin D, which has been discussed in detail in Chapter 2. The earliest manifestation is a reduction of serum calcium concentration secondary

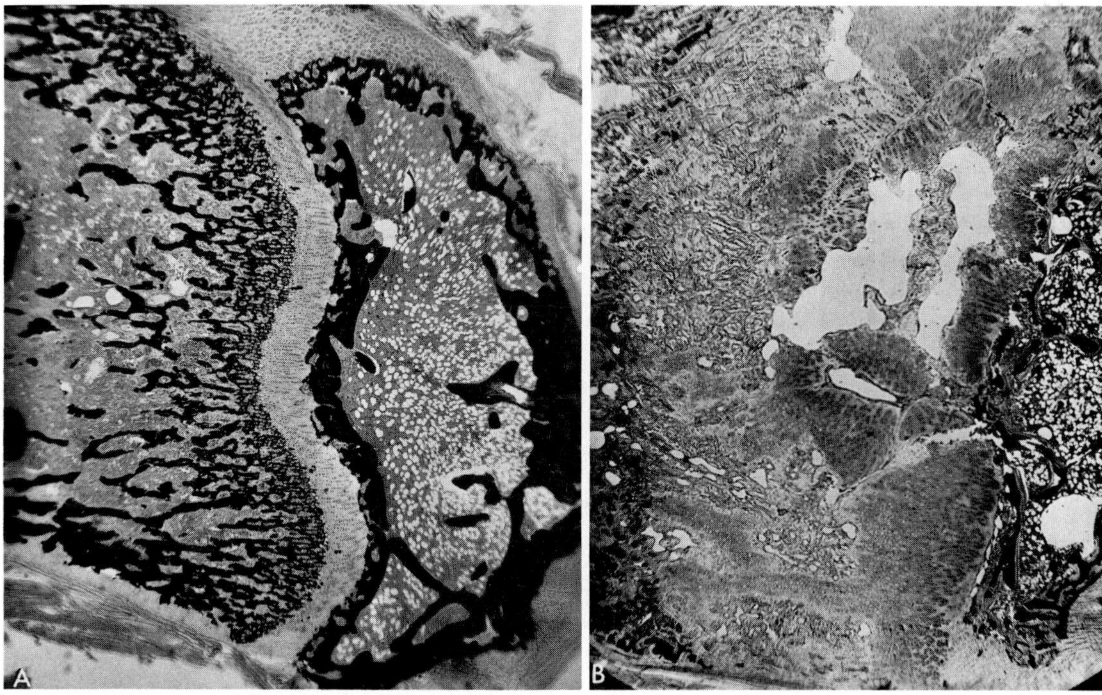

Figure 5–2. *a,* Section of tibia of normal rat showing thin uniform layer of unmineralized cartilage between metaphyseal plate and center of ossification in epiphyseal cartilage. *b,* Section of tibia of rachitic rat showing failure of mineralization of metaphyseal plate with broad irregular zone of chondro-osteoid separating metaphysis from center of ossification. (From Barnett, H. L.: Pediatrics, 15th Ed. Appleton-Century-Crofts, New York, 1972, p. 198. Reprinted with permission.)

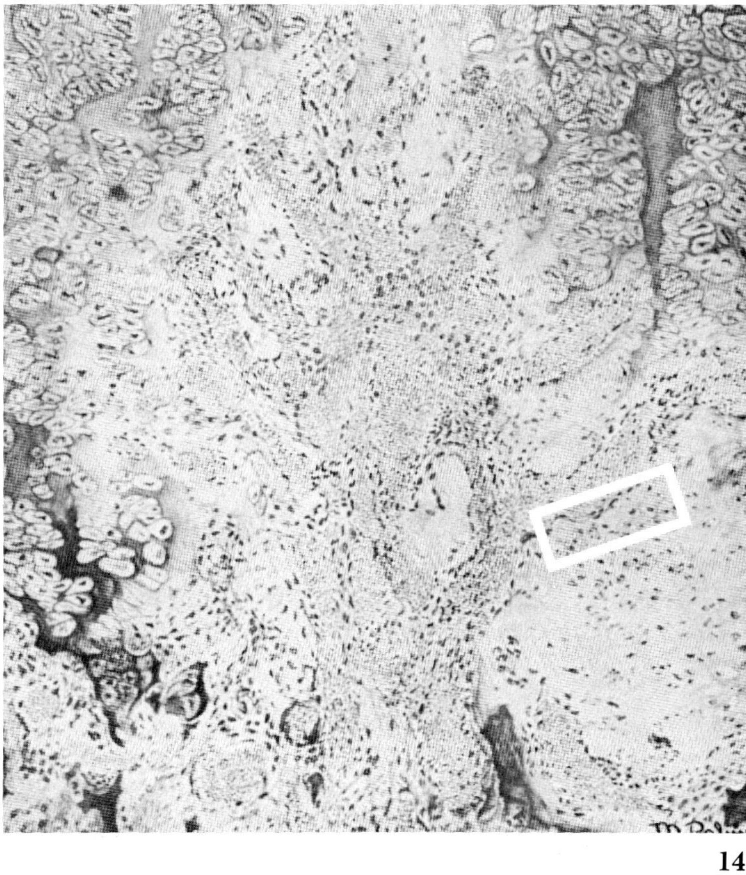

Figure 5–3. Histologic section of rachitic bone showing broad zone of unmineralized and disorganized chondro-osteoid at metaphysis.

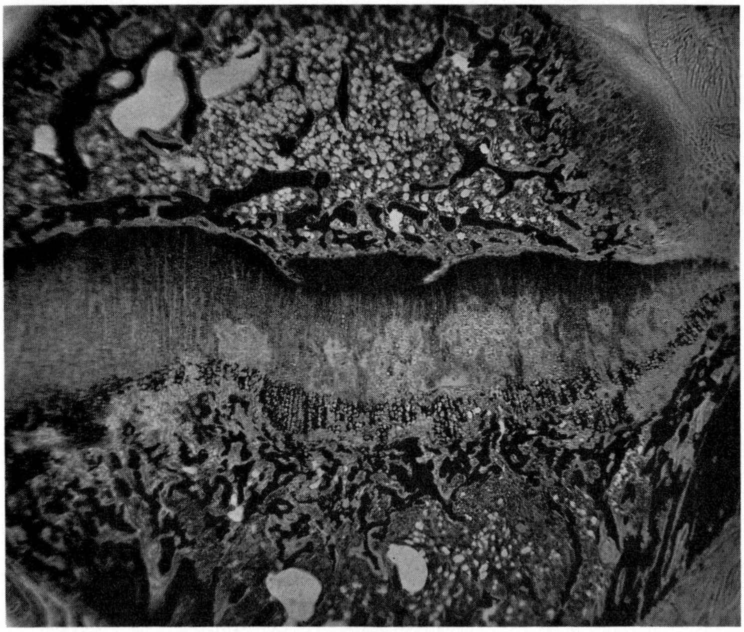

Figure 5–4. Healing of experimental rickets of rat, showing line of new mineral deposition in chondro-osteoid zone (Müller's line).

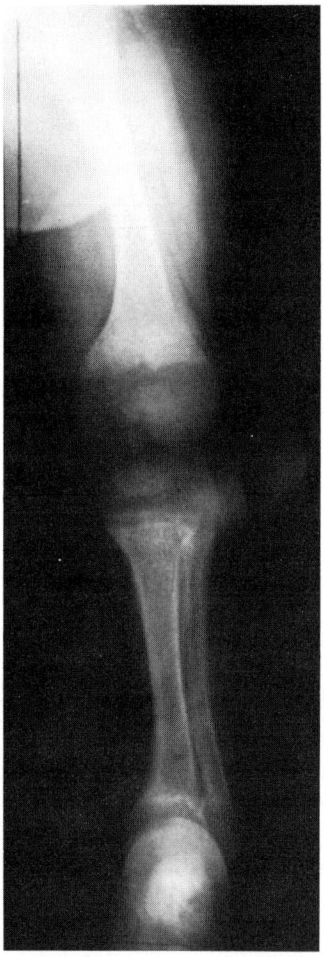

Figure 5–5. Radiograph of healing rickets in infant with zone of new mineral deposition separated from metaphyseal bone.

both to calcium malabsorption and to the refractoriness to PTH characteristic of the vitamin D-depleted subject. Such hypocalcemia can develop within two months in prematurely born infants who are not given vitamin D supplements (Fig. 5–6). Reduction of serum phosphate concentrations may be a slightly later phenomenon. It must be remembered that normal serum phosphate concentrations in the infant and child are considerably higher than those in the adult (Table 5–3), so that a serum concentration of 4 to 5 mg/dl may actually represent low values in the first two to three months of life, particularly in the prematurely born infant. Because of the hypocalcemia, hyperplasia of the parathyroids results, with excessive secretion of PTH. The hypophosphatemia is in part due to inhibition of renal tubular reabsorption of phosphate by PTH. Children with vitamin D deficiency excrete increased amounts of cyclic AMP in the urine, which indicates that this action of PTH on renal tubule cells is present. In addition there may be a direct effect of a vitamin D metabolite, 25-hydroxyvitamin D, on tubular phosphate transport, so that the high phosphate clearances of vitamin D deficiency can be the result of both lack of this vitamin D metabolite and excessive PTH.

Another manifestation of abnormality of renal tubule function in the vitamin D-

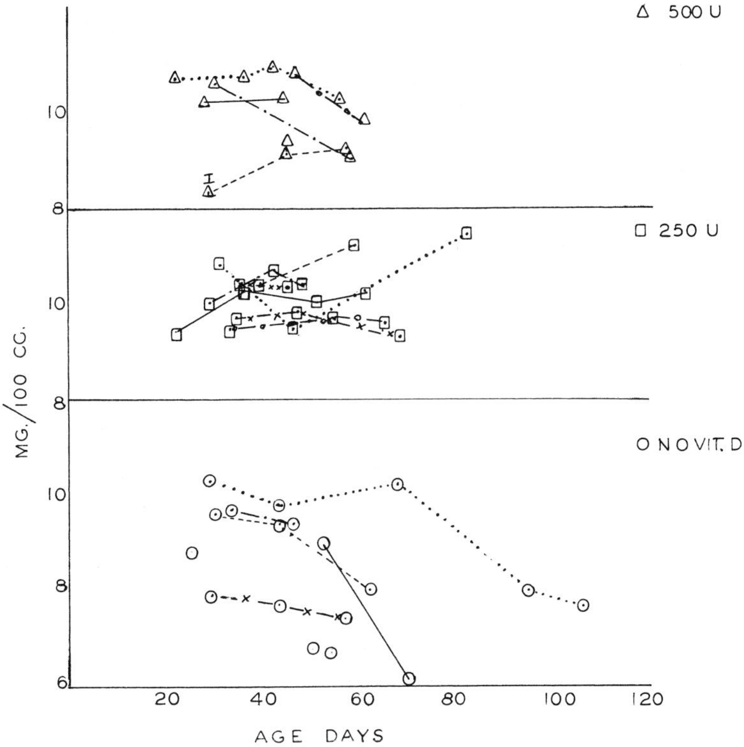

Figure 5–6. Serum calcium concentrations of small, prematurely born infants weighing 1500 grams or less at birth. All infants were given a vitamin D-free diet based on cow's milk. One group of infants was given no vitamin D supplement, a second group was fed 250 units of vitamin D daily and a third group 500 units daily. The infants receiving either 250 or 500 units of vitamin D from the first week of life showed no decrease of serum calcium concentration, whereas those not receiving vitamin D showed a dropping concentration of serum calcium at 6 to 8 weeks of age.

deficient patient is generalized aminoaciduria. This appears to be associated with the secondary hyperparathyroidism, and there is experimental evidence of an inhibitory effect of PTH on amino acid reabsorption by renal tubule. Many vitamin D-deficient infants show a slight reduction of bicarbonate concentration in serum. Since tubular reabsorption of bicarbonate is depressed by excess PTH this mild metabolic acidosis may also be a manifestation of secondary hyperparathyroidism.

Another of the striking and consistent biochemical abnormalities in vitamin D deficiency is the elevation of alkaline phosphatase activity in serum. This is one of the early changes resulting from deficit of vitamin D, and may appear during the first few months of life in infants on vitamin D-deficient feedings. This rise in serum alkaline phosphatase is associated with the overproliferation of osteoblasts in the rachitic intermediate zone along with proliferating capillaries and osteoid. These cells are rich in alkaline phosphatase, and the rise in serum enzyme is presumed to be the result of its discharge from the rapidly growing osteoblasts. Elevation of serum alkaline phosphatase activity is seen in other diseases in which osteoblastic activity is increased in the repair or attempted repair of bone, as in hyperparathyroidism with osteitis fibrosa cystica, and in Paget's disease. Because of the high concentration of alkaline phosphatase in the osteoblasts, attention has been given to the role of this enzyme in the process of mineralization of bone, as discussed in Chapter 1. The elevation of serum alkaline phosphatase activity in vitamin D-deficient states is of some value as an additional diagnostic test. Furthermore, it affords another measure of healing of rickets, since return of the al-

Table 5–3. Variation of Serum Inorganic Phosphate Concentrations with Age

AGE – GROUP	AVERAGE SERUM P (*mg/dl*)
Prematurely born infants	7.9±0.28
Full term newborns	6.1±0.33
1 to 10 years	4.6±0.16
Adults	3.5±0.19

All values are means ± SE of mean.

kaline phosphatase levels to normal usually occurs only when healing is complete, as indicated by roentgenographic evidences of restoration of bone mineralization to normal.

Clinical Manifestations

The availability of vitamin D concentrates of high potency, and the enrichment of milk and proprietary infant feeding preparations with vitamin D, have effectively eliminated rickets as an important public health problem. Most of the present cases of nutritional vitamin D deficiency in the Baltimore area occur in breast-fed infants whose mothers have not given them vitamin D concentrates. This failure has usually been due to the mistaken notion that breast milk supplies all essential nutrients for the infant and that vitamin supplements are not needed.

Even though vitamin D deficiency rickets has been abolished as a public health problem occasional cases will be seen in any large clinic. The early manifestations of such cases will not be the clinically recognizable deformities or even the characteristic radiographic signs, but rather the biochemical evidences of vitamin D deficiency. Presumably histologic examination of the bones in such infants would show early rachitic changes, since Park and Follis found a considerable incidence of histologic rickets in infants dying of a variety of diseases at the Harriet Lane Home of the Johns Hopkins Hospital. Vitamin D deficiency may be apparent by 2 to 3 months of age, particularly in prematurely born infants. The important finding at this stage is often hypocalcemia with only moderate reduction of serum phosphate concentrations (Table 5–4). Since the normal serum phosphate in the premature infant of this age is so high a level of 5 mg/dl may represent a relative hypophosphatemia. Although the value of 4 mg/dl is usually given as the dividing line between the normal concentration and the hypophosphatemia of rickets, the serum phosphate concentrations must be interpreted in terms of the normal values at different ages, as given in Table 5–3. Elevation of serum alkaline phosphatase levels also is usually present. Examination of the urine would show a generalized aminoaciduria. The precipitating factor in the patient's illness may be an acute infection such as pneumonia, and the serum calcium ion concentration may drop sufficiently to produce the symptoms of tetany, as in the case report. In other instances the hypocalcemia and other biochemical manifestations of vitamin D deficiency may be detected only by studies of serum calcium, phosphate and alkaline phosphatase concentrations in an infant with acute infection.

CASE REPORT

R. T., age 7 months, had been breastfed; he had been given cereal, strained vegetables, and fruits since the age of 3 months, but no eggs. Orange juice and cod liver oil supposedly had also been given starting at 3 months. Four days before admission the child had been vaccinated and given a DPT immunization. The day before admission he was feverish and irritable, and on admission the fever was more marked and the infant had a generalized convulsion.

Table 5–4. Serum Calcium and Phosphate Concentrations in Prematurely Born Infant with Early Rickets (B.G.E. — Age 2 Months)

Hospital Day	Serum *P* (*mg/dl*)	Serum *Ca* (*mg/dl*)	*Alk. P'tase Bodansky Units*
1	5.4	8.7	23
3	Vitamin D_2–600,000 units p.o.		
8	7.4	8.5	—
16	7.5	10.5	—
18	7.7	10.7	16

The significant findings on examination were a full tense fontanelle, craniotabes, and signs of meningeal irritation. The cerebrospinal fluid showed 100 lymphocytes per cu mm and an increased concentration of protein. Cultures were sterile. A diagnosis of meningoencephalitis was suggested. Serum analyses revealed the following concentrations: calcium 7.7 mg/dl; phosphate 3.2 mg/dl; alkaline phosphatase 37 Bodansky Units. The diagnosis of rickets was made and confirmed by radiographic findings in the long bones and a typical response to vitamin D therapy.

COMMENTS: Inasmuch as severe acute bacterial infections are frequently associated with reduction of serum phosphate levels and sometimes of serum calcium levels, the diagnosis of vitamin D deficiency cannot be made with assurance until it is determined that the hypocalcemia or hypophosphatemia persists after the infection has cleared, and that vitamin D treatment effects a return to normal levels. The roentgenographic signs of acute early rickets may be difficult to interpret and may be diagnostic only when healing is induced by vitamin D (Fig. 5–7).

The defect of bone mineralization reaches the stage of clinically evident skeletal deformities only after many months of vitamin D deficiency. These deformities are the end-stage of the various anatomic lesions of the bone (described earlier), individually or in combination: poor mineralization of membranous and tubular bones, with thinning of the cortices and trabeculae and consequent bending and twisting of the bone owing to muscle pull or weight-bearing; greenstick fractures of the long bones with angulation; thickening of the cortices owing to compensatory proliferation of osteoid tissue, with partial mineralization as the process becomes more chronic; and widening of the ends of the long bones as a result of lateral spread of the rachitic intermediate zone, the broad zone of unmineralized cartilage and osteoid between the shaft and the epiphysis at this weakened rachitic intermediate zone. The earliest clinical signs may be poor mineralization of the skull bones so that the bones are thin and easily indented, so-called craniotabes. During the first two or three months of life the skull bones normally are thin, particularly along the lambdoidal suture, so that physiologic craniotabes in this area may be found in otherwise normal infants. Rachitic craniotabes continues beyond 3 months of age and is more extensive, being especially noticeable in the posterolateral portion of the skull owing to the superimposed effect of the pressure of the intracranial contents in the recumbent infant, which increases bone resorption. After 6 months of age persistence of rickets usually results in compensatory thickening of the skull. The closing of the anterior fontanelle and the sutures is delayed and the frontal and parietal eminences of the skull become thickened, leading to the recognizable bossing of classic rickets.

At this period the chest deformity may also develop. The earliest changes are the thickening of the costochondral junctions of the ribs owing to overproduction and lateral spread of unmineralized chondro-osteoid. The expanded costochondral junctions can be seen as well as felt, forming the rachitic rosary. As the process becomes more severe the costal cartilages are pulled in by muscle pull, causing protrusion of the sternum, the pigeon-breast deformity. The lower portion of the rib cage is flared out and a depression is produced at the sites of the insertion of the diaphragm as a result of lessened rigidity of the ribs (Harrison's groove). Progression of the rickets into the second or third year of life is associated with deformity of the vertebral bodies and kyphoscoliosis. Such severe deformities of the chest are rarely seen now except in cases of extreme neglect.

The deformities of the long bones are at first the result of thickening and lateral spread of the structurally weak rachitic intermediate zone producing knobby prominences of the epiphyses at the wrists and ankles. At the ankles the increased width of the proliferative cartilage separating the epiphyses from the shaft of the bone may give the impression of a double epiphysis on palpation of the malleoli. Even before weight-bearing occurs angulation of the lower end of the tibia can develop owing to posterior displacement of the distal epiphy-

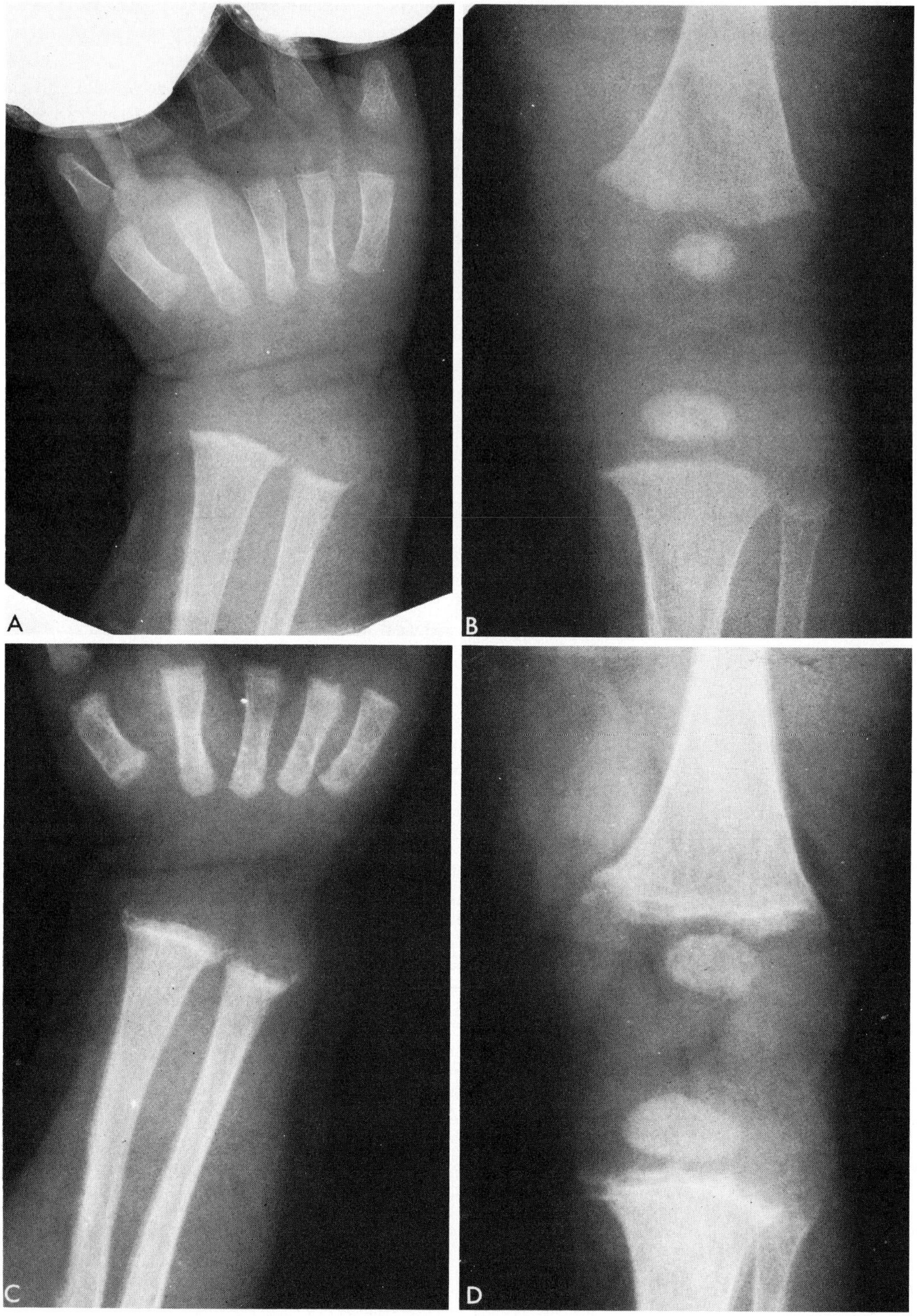

Figure 5–7. *a* and *b*, Radiographs of wrist and knee of young infant with rickets before receiving vitamin D. Although rickets is present the changes at the metaphyseal ends are minimal and diagnosis might be in doubt. *c* and *d*, Radiographs of wrist and knee of same infant after vitamin D, showing mineralization of rachitic chondro-osteoid, indicating the marked degree of rickets in this patient. (From Harrison, H. E.: *In* Holt, et al. Pediatrics. Appleton-Century-Crofts, New York, 1962, p. 279. Reprinted with permission.)

sis by muscle pull. The early deformity is therefore anteroposterior angulation and bowing of the tibias. When weight-bearing is superimposed, lateral bending of the femurs develops with either genu valgum or varum deformity.

In addition, medial torsion or twisting of the tibias can occur so that the lower ends of the tibias are inwardly rotated. In the effort to keep the feet from toeing in excessively the child walks with the femurs externally rotated. This causes instability of the hip joints and a waddling gait. Further interference with gait can result from bending of the femoral neck so that the normal obtuse angle between the neck and shaft of the femur is reduced to almost a right angle, coxa vara deformity. This causes the level of the trochanters to be raised with relation to the hip joint, which also interferes with the stabilization of the hip joint by the muscles inserted in the greater trochanter. The combination of twisting and bending of the weight-bearing bones, the femur and tibia, with angulation due to displacement of the epiphysis or to greenstick fractures, and compensatory deposition of subperiosteal osteoid with partial mineralization, may finally lead to the grotesque deformities of classic rickets illustrated in Figures 5–8 and 5–9. Fortunately such cases have virtually disappeared. Serious progressive rachitic deformities are still seen, however, in children with vitamin D-resistant rickets, discussed later.

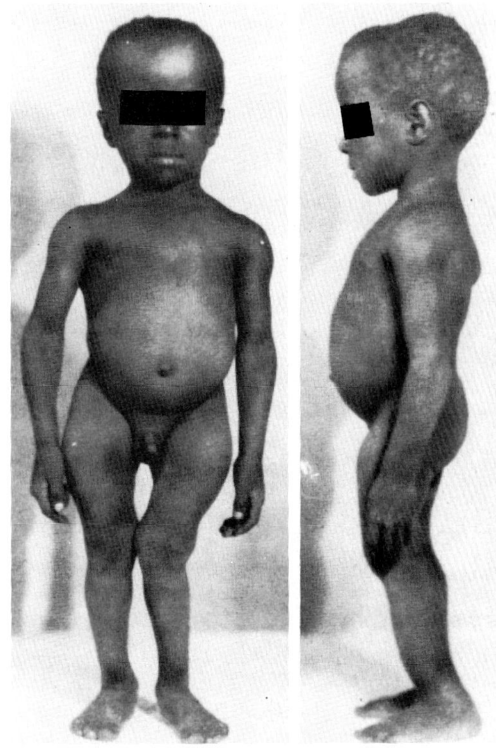

Figure 5–9. Severe rickets in child with skull bossing and marked bending and twisting deformities of lower extremities. (From Harrison, H. E.: *In* Holt, et al. Pediatrics. Appleton-Century-Crofts, New York, 1962, p. 295. Reprinted with permission.)

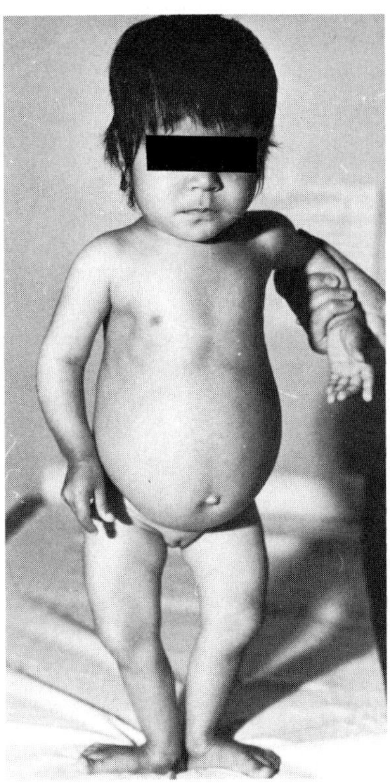

Figure 5–8. Rickets in young child. Deformities of chest and lower extremities are obvious.

Deformities of the pelvis due to rickets are of major historic interest since in the days when rickets was common such deformities, starting in childhood and persisting in the adult woman, were important causes of dystocia and fetal and maternal death. This is no longer of clinical significance.

Vitamin D deficiency can cause delayed eruption of the deciduous teeth. More important, however, is the hypoplasia of the

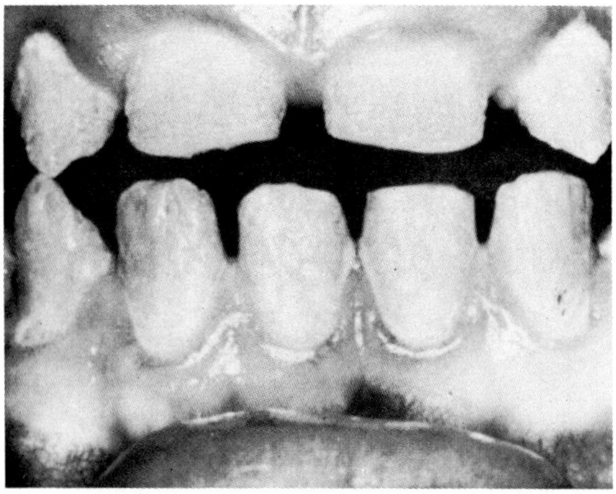

Figure 5–10. Enamel hypoplasia. (From Winter, G. B. and Brook, A. H.: Enamel hypoplasia and anomalies of the enamel. Dent. Clin. North Am. *19*:14, 1975. Reprinted with permission.)

enamel of the permanent dentition which can be seen in children who have had severe rickets in infancy. This failure of enamel formation is illustrated in Figure 5–10. A good deal of attention has also been paid to the role of vitamin D in the prevention of dental caries. Studies along this line have been confusing and contradictory. Although severe prolonged vitamin D deficiency does lead to defects of tooth development, there is no evidence that much of the dental caries seen at the present time is related to vitamin D lack. The suggestion that doses of vitamin D larger than those needed to prevent rickets will reduce the incidence of caries has not been consistently confirmed. At present, in view of the low incidence of true vitamin D deficiency, the cause of the prevalent dental caries which is still an important public health problem cannot be explained by lack of vitamin D, and successful methods of prevention must still be sought.

A variety of nonspecific symptoms were ascribed to rickets in the days when this was a common disease. Lethargy or irritability, retardation of motor development, profuse head sweats, head nodding and digestive disorders have all been considered signs of rickets, but there is little reason to believe that these were in any way specifically related to vitamin D deficiency. Muscular hypotonicity with laxity of abdominal musculature that results in the pot belly of rickets is, however, a real manifestation of vitamin D deficiency although its mechanism is obscure. The role of phosphate in the energy-producing reactions of the cell is well-known and it is tempting to attribute muscle weakness to deficiency of intracellular phosphate. The phosphate ester concentration of the red blood cell is reduced in vitamin D deficiency, but the uptake of phosphate by tissue cells continues even when extracellular phosphate is markedly depleted, so that no definite evidence exists that there is an important deficiency of available phosphate in the muscle.

There is also evidence that the calcium transport system of the muscle sarcoplasmic reticulum is influenced by vitamin D. Since skeletal muscle contraction and relaxation depends on fluctuations in calcium ion concentration in the fluid surrounding the muscle fibrils, this effect of vitamin D on the sarcoplasmic reticulum might be responsible for hypotonicity of muscle in severely rachitic children. In older children with apparently acquired primary hypophosphatemic rickets due to renal tubular phosphate transport defect, muscle weakness may be so severe as to suggest a primary muscular dystrophy. In these patients, phosphate depletion may be an important mechanism.

RACHITIC TETANY

In addition to the defect of mineralization of bone, an important physiologic disturbance results from deficit of calcium ion in extracellular fluid. This is tetany, namely, hyperexcitability of the peripheral motor nerves and of the motor nerve cells in the cortex. The instability of calcium ion concentrations in extracellular fluids of the rachitic subject has already been mentioned. The infant with rickets usually shows initially low serum calcium levels which then approach low normal values along with markedly reduced concentrations of serum phosphate. A sudden rise in extracellular phosphate due to high phosphorus intake, or liberation of phosphate from tissue cells because of fasting or increased metabolic requirements of fever, may cause the rapid crystallization of bone salt in the large mass

Table 5–5. Age Incidence of Rachitic Tetany*

AGE	PERCENTAGE
0 to 3 months	7.5
3 to 6 months	26
6 to 9 months	26
9 to 12 months	8
1 to 2 years	20
2 to 3 years	7
3 to 4 years	4

*Age incidence in 293 cases of rachitic tetany from the Johns Hopkins Hospital analyzed by Dr. Harriet Guild.

of unmineralized osteoid. When this occurs serum calcium levels drop rapidly and, as has been pointed out, the compensatory mobilization of calcium from bone is inhibited in the vitamin D-deficient subject.

Rachitic tetany, therefore, is often precipitated by acute febrile illnesses. It has also been mentioned that hypocalcemia is more likely to occur in the acute stages of rickets in the young infant than in the more chronic stages of rickets of later childhood. The maximum age incidence of rachitic tetany in Dr. Harriet Guild's analysis of cases at the Harriet Lane Home was between 5 and 7 months of age, and a significant number of cases appeared as early as 2 or 3 months of age (Table 5–5). Rare instances of tetany due to vitamin D deficiency have been reported in the first few weeks of life. Such cases of congenital rickets have occurred under conditions in which the mother had osteomalacia owing to vitamin D deficiency. This rare situation was observed in prerevolutionary China, where osteomalacia in pregnant women occurred because of the combination of lack of vitamin D in the diet and their seclusion and consequent little exposure to sunshine.

The symptoms of tetany are the manifestations of peripheral nerve discharge with local muscle spasm, or central nervous system cell discharge with convulsive seizures. These have been described in detail in Chapter 3.

Roentgenographic Diagnosis of Rickets

Radiographs give more accurate information concerning the existence of active or healing rickets than physical examination, but still have limitations for early diagnosis because the disease may be well under way at the costochondral junctions and yet be so poorly developed at the wrists or knees as not to reveal itself. The lower ends of the radius and ulna are most useful for the study of rickets by x-ray because of their small size and the ease with which the bone structure can be visualized. The lower end of the femur is too thick and the junction of the epiphysis with the diaphysis is too uneven for slight changes to show distinctly. The upper end of the tibia, however, is suitable for the study of certain of the early changes of rickets.

When radiographs of the radius and ulna are taken, the head and arm should be in complete supination. Even when pronation is slight, the lower end of the ulna appears cupped and sometimes lipped like a beaker. Evidence of rickets may often be detected at the lower end of the ulna because of its small size and almost flat terminal surface, when no sign is visible in the radius. Theoretically, rickets should develop earlier at the lower end of the ulna because of its more rapid growth, but histologically both bones show rickets at about the same time.

One of the very early roentgenographic evidences of rickets is widening of the space between the calcified plate at the end of the metaphysis and the center of ossification in the epiphysis owing to the proliferation of uncalcified cartilage and osteoid. This can be detected at the upper end of the tibia, since the epiphyseal center of ossification is well-developed and the width of the uncalcified zone can be readily estimated (Fig. 5–7*b*). The excess width of this rachitic intermediate zone is best appreciated when healing begins and calcification of the rachitic zone can be seen (Fig. 5–7*d*).

The changes at the cartilage-shaft junction are further characterized in the roentgenogram by cupping, spreading, spur formation, fringing and stippling. None of these signs by itself is pathognomonic but they produce a conglomerate picture which is characteristic of rickets.

"Cupping." The epiphyseal cartilage is virtually transparent under the x-ray, permitting the end of the shaft to be seen. Cup-

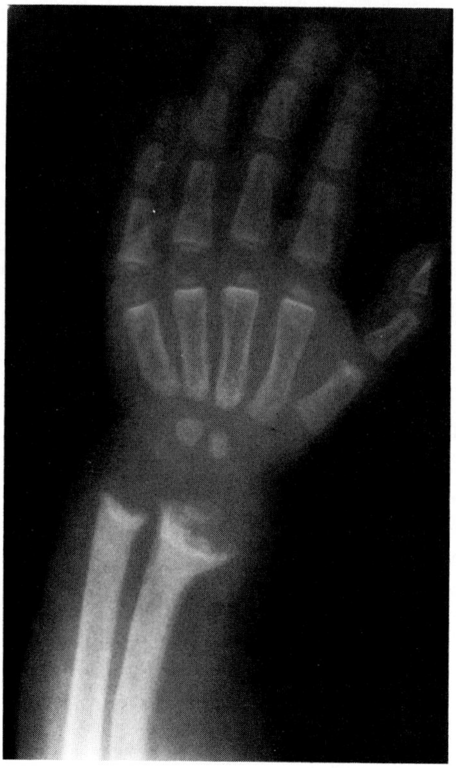

Figure 5–11. Radiograph of rickets showing cupping at distal end of ulna and radius.

ping is present when the end of the shaft is concave. It occurs only at those shaft ends whose normal configuration permits its development (Fig. 5–11). It is, therefore, a very early sign of rickets at certain cartilage-shaft junctions, but develops at others only after the disease has become advanced, and at some it never develops. The upper ends of the shaft of the humerus and femur are rounded and do not cup even in the most severe rickets, although bone formation may extend around the sides of the proliferative cartilage, which is the essential process in cup formation. The lower end of the radius, both ends of the tibia, and the lower end of the femur are slightly convex and do not show cupping until the rickets is advanced. Both ends of the fibula and the lower end of the ulna, however, are flat or slightly hollowed and consequently develop cupping soon after the rachitic process starts. Cupping at the anterior ends of the middle ribs is a well-marked normal phenomenon and rapidly increases as rickets develops.

Cupping may be seen in scurvy and in slight degree in the ulna of young infants, especially the prematurely born, whose bones are growing rapidly. Cupping may be very pronounced in chondrodystrophy, but the irregular craters characteristic of that disease need not be confused with rickets. In some of the most severe cases of rickets the distal ends of the radius and ulna appear to be flat and there is no sign of cupping. In such cases the cup is actually there but does not show because of the lack of lime salts in the organic tissue that forms the cup. When treatment is begun, cup formation comes into view. "Spreading" of the end of the shaft may accompany "cupping" or it may be present alone at the ends of bones that do not cup (Fig. 5–12). Spreading is not, however, a sufficiently well-defined sign to be valuable by itself.

"Cortical spurs." These are linear shadows that extend as prolongations of the shadows of the cortex along the sides of the proliferative cartilage. They are not always in the direct line of the cortex, but are exter-

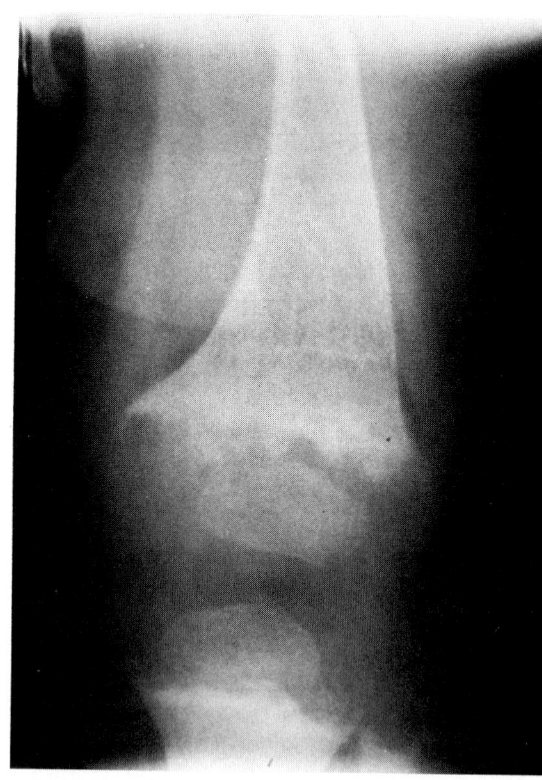

Figure 5–12. Radiograph of rickets demonstrating spreading of end of shaft.

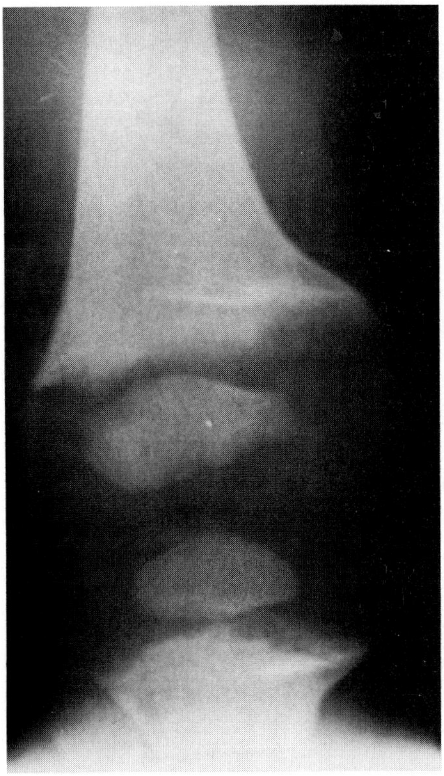

Figure 5–13. Characteristic radiographic appearance of rickets with fraying of metaphyseal end of bone.

nal to it since they lie in the perichondrial-periosteal layer and this envelops the cortex. Such shadows may be found on one or both sides; they may be straight and in the line of the cortex or they may arch outward. The spurs are composed of unmineralized matrix. They hug the proliferative cartilage. Their shape and direction, therefore, are determined by the configuration of the latter. If the proliferative cartilage bulges, the spurs curve outward; if the cartilage is no broader than the end of the shaft, as is frequently the case in congenital syphilis, they are straight. In advanced rickets and in congenital syphilis these spurs may be long and extend even to the resting cartilage. Very frequently the spurs are short. In the x-ray film it can often be seen that the spurs lie external to the cortex, overlapping and seeming to splint the cartilage-shaft junction. In that case they represent a new cortical layer forming outside the old.

"Fraying." This is seen in threadlike shadows extending from the end of the shaft into the transparent cartilage. When fully developed, they cause the end of the shaft to exhibit a "raveled sleeve" appearance. In early rickets the individual "threads" are short, thin and difficult to see, and give the end of the shaft a fuzzy outline. In moderate and advanced rickets the threads may be long and coarse; they are neither straight nor parallel but extend in various directions, exactly as would be expected from the disorder in the underlying pathologic condition. Fraying is a characteristic sign of rickets (Fig. 5–13). It may be lacking in the early stages of the disease. When judged by microscopic standards, rickets must be at least moderately developed for fraying to be present, even though seeming by x-ray examination to be slight. The lower end of the ulna usually "cups" before it "frays," but the radius "frays" before it "cups."

Fraying is the most reliable of the radiographic signs of rickets. However, it is found also in congenital syphilis, in which extensions of the calcification often take place along and around the cartilage canals. These formations, relatively speaking, are thick and appear as serrations at the end of the shaft, in contrast to the more delicate, fringelike formations of rickets. "Stippling" is present when the outline of the end of the shaft has an irregular dotted appearance (Fig. 5–14). It may be a sign of early rickets, early even according to microscopic standards, but is more often found in chronic forms of the disease in which bone salt deposits in some places but not in others. The dots correspond to deposits of mineral separated by uncalcified chondro-osteoid. Stippling is best seen when the margin of the end of the bone is thrown into sharp relief.

The changes at the cartilage-shaft junction in moderate or advanced rickets are easily recognized. The characteristic enlargements are clearly outlined. The distance from the ends of the shaft to the centers of ossification of the epiphyses or to other bony landmarks is abnormally great. The ends of the bones are usually spread or cupped, or both. The cupping may be so extreme that the shafts end in craters. Fraying may be so marked that the end of the bone looks fringed. At the extreme end of

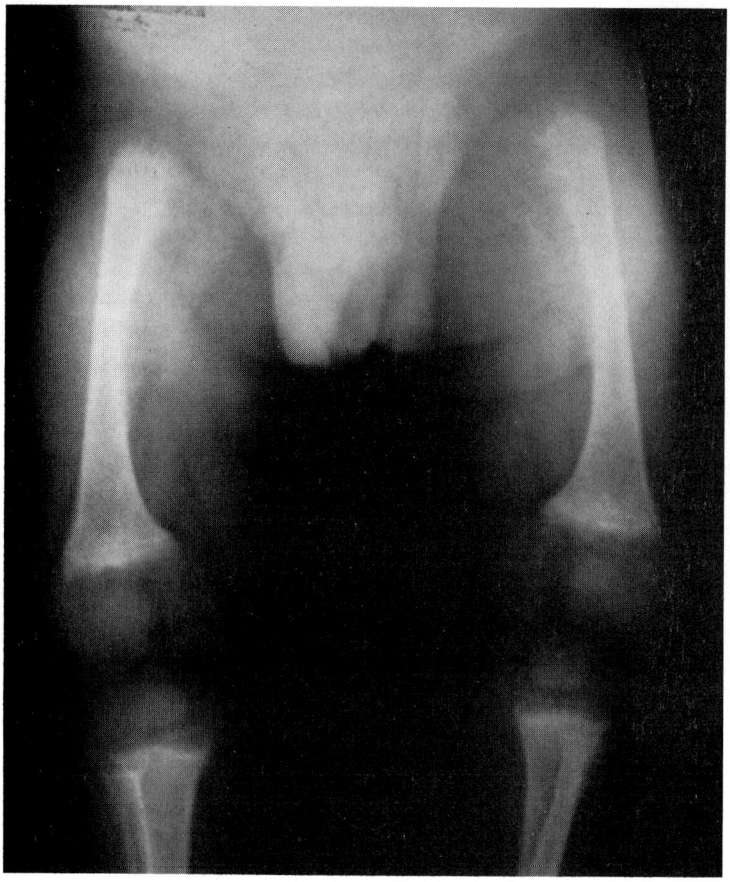

Figure 5–14. Radiographic appearance of rickets with stippling at end of shaft.

the shaft, just proximal to the point where the rachitic intermediate zone begins, the trabeculae often are transversely disposed and may look compressed and packed together; sometimes these transversely disposed trabeculae are only on one side.

The changes just enumerated are the ordinary ones at the cartilage-shaft junction in advanced rickets. In the most extreme rickets, however, only the end of the shaft, as it was before the rickets began, may be visible, and this is rounded or wedge-shaped, looking as if it had been lopped off with shears. The distance from the end of the shaft to the bony landmarks beyond is so great as to permit the deduction that the true end must be devoid of lime salts, and hence invisible. Under treatment the true end comes into view, showing the most extreme cupping or deformity.

The changes in the shaft vary with the disease: in the very early stages, no change is noted; in severe rickets the changes are characteristic. In the "atrophic" type (Fig. 5–15) perhaps the most striking change is the very marked diminution in the intensity of the shadow of the bone as a whole. In hypertrophic rickets (Fig. 5–16) there is diminution in intensity of the shadow but it is never so great as in the atrophic type, in which the bones may look like ghosts of themselves. The trabecular meshwork is often very coarse and irregular, varying in design in different parts of the bone. In the premature infant the diagnosis of rickets has been made from the very marked changes in the finer structure of the shaft, although the ends of the bones appeared normal.

In the atrophic form of rickets the cortex appears thin. Often it is invisible in places; then the border of the bone may appear ragged. It may be entirely invisible. Anomalous as it may seem, strands of osteoid extending from it to the also invisible periosteum may contain enough lime salts to cast shadows. In that case hairlike shadows stick

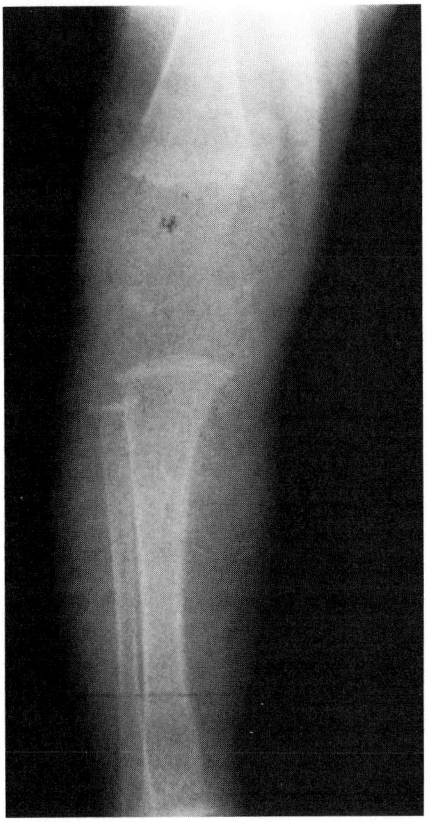

Figure 5–15. Atrophic rickets with markedly diminished density of bone.

out from the sides of the bone. Encasements of calcified subperiosteal layers of osteoid will not be seen if the cortex does not show plainly. When they are present they extend the entire or almost the entire length of the shaft and often form complete cylinders around it.

In the hypertrophic type the cortex appears thickened and composed of longitudinally disposed, slightly curved interlacing lamellae. Thickening of the cortex is usually greater on one side. The marrow cavity may be eccentric and its width greatly reduced. In this type periosteal encasements are usually visible. Several layers sometimes may be seen, appearing as fine lines parallel to the cortex.

Other changes in the shaft visible by x-ray examination are complete and partial fracture, callus formation, curvatures of the shaft with great thickening on the concave side, and displacement of the epiphysis on the diaphysis, due (as has been described) to bending at the rachitic intermediate zone.

When healing begins in earnest, lime salts deposit in great profusion at the cartilage-shaft junction and in the substance of the shaft. At the cartilage-shaft junction, the place where such deposits first occur is the zone in the proliferative cartilage that would have been the seat of bone salt deposition if no rickets had ever existed, and this zone forms the distal boundary of the rachitic intermediate zone. Treatment reveals its situation, because the depositing bone mineral in this region renders it visible. The first evidence of recovery is, therefore, the formation of a linear shadow, very faint and fragmentary at first, beyond and parallel to the end of the shaft (Fig. 5–17). Its distance from the end of the shaft depends on the depth of the rachitic intermediate zone. If the latter is deep, the line of shadow may be 0.5 to 1 cm (0.2 to 0.4 in) away from the end

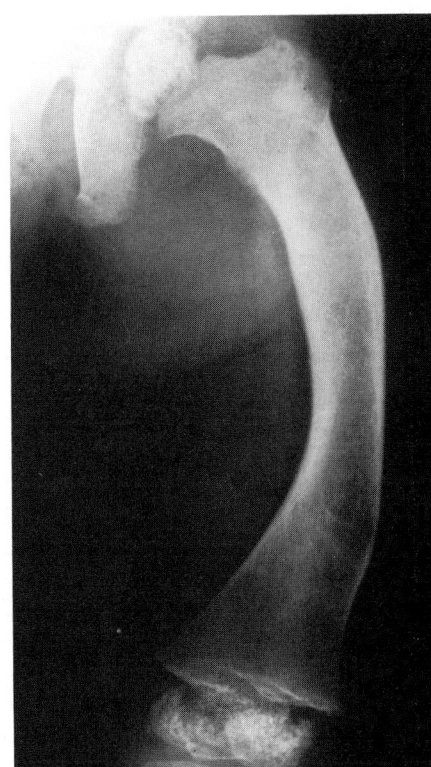

Figure 5–16. Chronic hypertrophic rickets with thickening of bone, coarse trabecular structure and over-all reduction of mineral density.

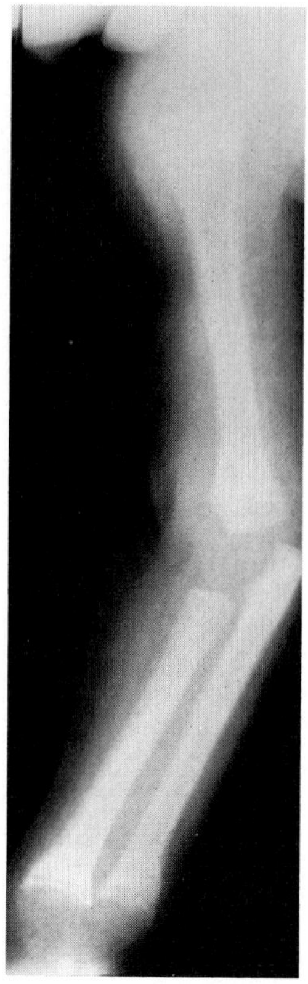

Figure 5-17. Radiographic signs of healing rickets with beginning deposition of mineral in epiphyseal cartilage and beneath periosteum.

of the shaft. If lime salts precipitate freely and growth continues, this transverse line of lime salt deposition increases in thickness. It may seem to double. The increase in thickness always occurs on the epiphyseal side. At the same time, the intermediate zone, which has cast little or no shadow before, develops a splotchy appearance. The splotches have mainly a longitudinal, but in some places a transverse, direction. The longitudinal splotches are caused by bone salt deposits in the cartilage bordering on the cartilage canals. As this reparative process continues, the intermediate zone acquires a ground-glass appearance because of the great number of interlacing osteoid trabeculae, all of which gradually take up bone mineral and begin to cast shadows. Bone salt is also deposited profusely in the periosteal encasements of the shaft. In consequence these become visible and often surprise the observer who has not suspected their presence (Fig. 5-17).

As soon as bone mineralization is restored, the process of new bone formation becomes entirely normal. With increasing growth of the bone, the rachitic intermediate zone is left farther and farther behind in the shaft, and gradually the shadow that it casts becomes changed to normal. For a long time, however, perhaps two years or more, the site of the rachitic intermediate zone may be recognized by a bulge in the outline of the bone, by differences in the architecture, by areas of increased density or by transversely arranged trabeculae. The zone of coarse, transverely or irregularly disposed trabeculae that marked the end of the shaft at the time the rickets began may remain as a "scar" visible for a year or several years.

Prevention of Rickets

The tool for successful prevention of vitamin D deficiency rickets was in hand long before it was actually used to proper advantage. Cod liver oil has a long history as an agent for the prevention and cure of rickets but a study of this history is a humiliating one for the physician. As late as 1922 a scholarly monograph on rickets published in England by Dick mentions cod liver oil with little enthusiasm, dismisses it as a nonspecific antirachitic agent and classes it with other food fats to its obvious disadvantage. To quote: "The scope for the medicinal treatment of rickets is comparatively limited. The administration of cod liver oil has already been dealt with as this is more a food than a drug. As a rule the tendency is to give the oil in too large doses. For an infant it is well to begin with small doses of ten drops three times a day."

It is easy to see why there was so much reluctance to accept cod liver as a specific antirachitic medication. The oil available was almost certainly of variable potency before methods of assay of vitamin D were known. The curative effect of cod liver oil on rickets was difficult to determine because of the relatively low and variable doses of vitamin D given, and because the clinical response of the infants could not be evaluated accurately until roentgenographic and biochemical methods for study of rickets were perfected. It would have been possible, of course, to determine the prophylactic value of cod liver oil if controlled studies on large series of infants were made, but medical research and preventive medicine had not reached this level of sophistication. Following the production of experimental rickets in animals, methods of bioassay of cod liver oil and other vitamin D preparations were available. Within a relatively few years the availability of fish liver oils much more potent than that from cod liver (percomorph liver oil) and concentrates prepared from irradiated ergosterol made cod liver oil obsolete.

The cumulative experience of a number of studies in infants, which utilized bone x-rays and serum calcium, phosphate and alkaline phosphatase determinations as indices of rickets, indicated that the daily intake of 400 units (10 μg) of vitamin D was adequate to prevent rickets in almost all infants. For many much less than this is required and 100 units (2.5 μg) per day may be sufficient. Four hundred units of vitamin D could be supplied by one teaspoonful of standardized cod liver oil, but can be much more conveniently given as a few drops of calciferol in propylene glycol (Drisdol, 200 units per drop). A variety of multivitamin preparations are also available supplying 400 units of vitamin D in a standard amount of 0.6 ml given by calibrated dropper. The most important step in the prevention of vitamin D deficiency rickets was the enrichment of evaporated milk and homogenized, pasteurized milk with vitamin D, 400 units per can of evaporated milk or per quart of homogenized milk. The various proprietary infant feeding preparations are also supplemented with comparable amounts of vitamin D. Infants receiving vitamin D-enriched milks can obtain their vitamin D needs in their milk feedings and therefore additional concentrates or fish liver oils are not needed. Aside from the ease and certainty of vitamin D administration by its incorporation in milk, there is evidence that it is more efficiently utilized when given with milk than as an oily concentrate. The breast-fed baby of course should receive vitamin D as a supplement since human milk is not a dependable source of this vitamin.

Rickets in the Prematurely Born Infant

The great susceptibility to rickets of the prematurely born infant has been noted repeatedly. Biochemical evidences of rickets can appear within two months in small premature infants (less than 1500 gm at birth) if vitamin D is withheld. The high incidence of rickets in premature infants in the days before suitable concentrates of vitamin D were available suggested that they needed larger amounts of vitamin D than full term infants, and in the past daily intakes of 5000 units or more were recommended for this group. Present evidence indicates that the prematurely born infant does not require more than 400 units of vitamin D daily to prevent vitamin D deficit, but that this amount should be supplied as soon as he begins to take food, i.e., by the second week of life.

Figures 5–6 and 5–18 summarize a study made in prematurely born infants weighing 1500 gm or less at birth without complications other than prematurity. All infants were given a cow's milk-based feeding containing no added vitamin D. One group was given no vitamin D, a second 250 units of vitamin D daily starting as soon as milk feeds were well taken by the end of the first week of life, and a third group 500 units daily. Serum calcium, phosphate and alkaline phosphatase determinations were made weekly. As the data in these figures indicate, by 6 to 8 weeks of age the infants in the no vitamin D group showed a decrease in serum calcium and phosphate and an increase in serum alkaline phosphatase, wherease those given 250 and 500 units of vitamin D daily showed no appreciable changes in these measurements. When the infants of the 250-unit group were given extra vitamin D at the end of the initial study no increase of serum calcium resulted, suggesting that this dose had maintained adequate vitamin D concentrations.

More recent studies of serum 25-hydroxyvitamin D concentrations in prematurely born infants of mothers living in New York City, and who presumably had re-

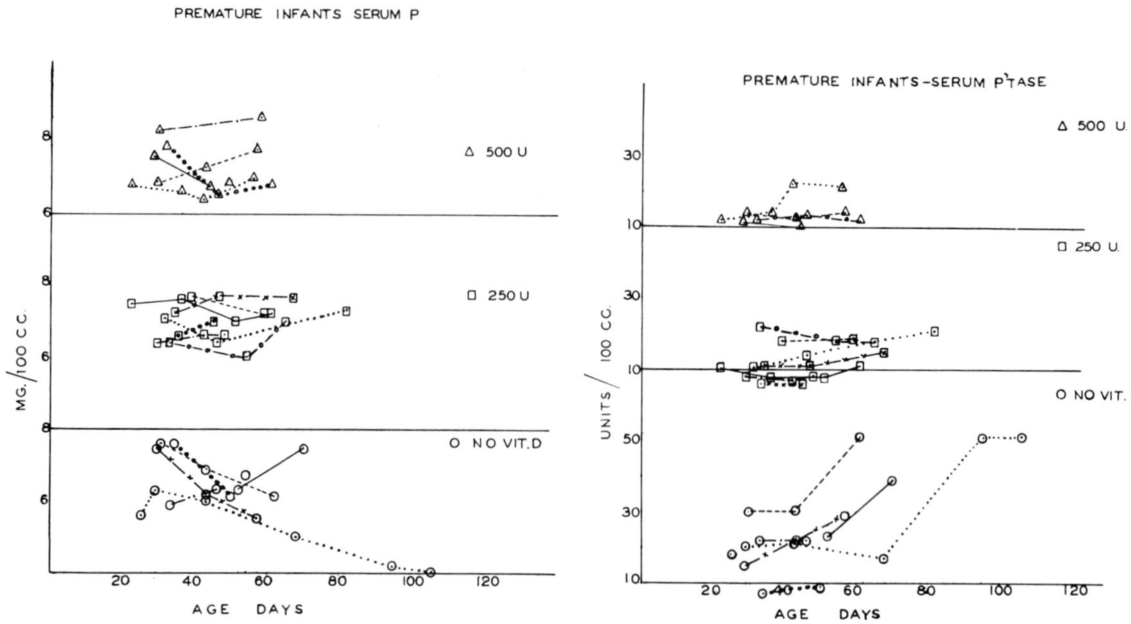

Figure 5–18. Serum phosphate and alkaline phosphatase concentrations of prematurely born infants fed a vitamin D-free milk, either without vitamin D addition or with 250 or 500 units daily.

ceived little prenatal care, showed low concentrations of this vitamin D metabolite in both maternal and infant blood. Since the incidence of prematurity is greater in women of low socioeconomic status who have had minimal prenatal medical care, it may be expected that many prematurely born infants will have low stores of vitamin D at birth, and with their rapid growth the need for starting vitamin D as soon as possible may be explained. It is not necessary, however, to give unusually large amounts.

The vitamin D-enriched milks are not useful as a sole source of this vitamin for the premature infant, since the milk intake of these small infants is much below the quart a day or its equivalent needed to provide 400 units. It is, therefore, necessary to supply a daily supplement of vitamin D to these infants in addition to any contained in the milk feeding. Since they also require vitamin C and tocopherol, preparations supplying vitamins C and D and tocopherol can well be employed.

Prevention of Rickets in the Older Child

It has been easy enough to determine the vitamin D needs of the infant during the first year of life, but the requirement of the older infant and child is more difficult to assess. Rickets has its onset during the first year of life and, although it may progress during the second year, spontaneous healing often occurs during this period if the child is allowed outdoors and is exposed to sunshine. If vitamin D deficiency is not present by the end of the first year of life it rarely appears at a later date except under the unfavorable climatic and housing conditions that cause deprivation of ultraviolet radiation. The question arises whether vitamin D is needed in significant amounts after the period of rapid growth in early infancy. The older child can obtain some vitamin D from the addition of fish such as herring, sardines, salmon and tuna to the diet as well as egg yolk. This, plus the greater exposure to sunlight of the more mobile toddler and young child, may provide the minimum requirements of vitamin D and explain the "spontaneous" healing of rickets during the second year of life.

Osteomalacia can occur in older children and adults who for some reason are not permitted normal exposure to sunshine. It has been seen, for example, in adult women who live under conditions of purdah as in

Pakistan and India, or comparable seclusion and almost complete indoor existence as in China in the days of the old regime. It is true that such osteomalacia is much more likely to occur following pregnancy and lactation, which may increase vitamin D needs in the woman. Rickets and osteomalacia may also occur in older children and adults with sprue or steatorrhea due to other causes, and this can be overcome by increased amounts of vitamin D. Such evidence indicates that some vitamin D is needed by the older child and adult but that manifest deficiency results only from the most extreme deprivation. It seems justifiable to provide a guaranteed amount of vitamin D to the older child and it may be presumed that 400 units daily should be sufficient for this purpose.

Some question has been raised about the vitamin D requirement during the adolescent growth spurt. Johnston has suggested that calcium retention can be increased in the adolescent period by vitamin D intakes several times the suggested 400 unit allowance. Intakes of 1000 to 1500 units daily would not be excessive but there is no general agreement concerning the need for this dose. It is suggested that vitamin D is needed throughout the growing period and probably also in adult life, but that the requirement is not greater than that of the infant except under conditions of impaired absorption of fat-soluble vitamins. On ordinary mixed diets and with usual exposure to sunshine the older child and adult probably obtain adequate amounts of vitamin D without specific supplements. In the adult 100 units/day has been found adequate to prevent osteomalacia under conditions of limited exposure to sunshine. The use of vitamin D-enriched milk is, however, a perfectly satisfactory method of providing a known amount of vitamin D in the diet, and there is certainly no contraindication to the addition of this amount (200 to 400 units per day) even to already adequate diets.

Treatment of Vitamin D Deficiency Rickets

The definitive treatment of vitamin D deficiency rickets is of course the administration of vitamin D. The amount to be given and for how long is a matter for discussion. Since vitamin D can be efficiently retained in the body tissues, doses only several times the minimum maintenance needs will result in complete healing of rickets if given over a sufficiently long time. Rickets can then be healed by a few thousand units of vitamin D given daily over a period of several months. More rapid healing can be induced by much larger amounts and one suggested method is the so-called "stoss" treatment. This was made possible when purified or semipurified preparations of irradiated ergosterol or 7-dehydrocholesterol of high potency became available. In this form of therapy a single large dose of vitamin D, 600,000 units, is given orally or in some instances intramuscularly. The IM injection of vitamin D, however, has no advantages over its oral feeding, and in the United States IM preparations are no longer recommended or readily available. In order to obtain maximum absorption of the vitamin D we have divided the 600,000 units into six doses of 100,000 units each given at two-hourly intervals. Vitamin D_2 is usually given since there is no apparent advantage of vitamin D_3 in the rachitic infant, and irradiated ergosterol preparations are less expensive than irradiated 7-dehydrocholesterol.

There are several advantages to the provision of this large load of vitamin D in a single day. The biochemical evidences of vitamin D effect appear within a few days, so that by the end of a week it is possible to determine whether a characteristic response will occur, i.e., a rise of serum phosphate levels to values above 5 mg/dl and increase of serum calcium to above 10 mg/dl. Healing by x-ray can be detected ten days to two weeks after the vitamin D is given. This rapid response permits the differentiation of true vitamin D deficiency rickets from other varieties that are relatively resistant to vitamin D treatment. At present the comparative infrequency of true vitamin D deficiency rickets, and the increasing recognition of the various forms of vitamin D-refractory rickets, make this rapid differentiation of practical value. Examples of the typical biochemical response

Table 5–6. Typical Response to Vitamin D Loading Test of Infants with Uncomplicated Vitamin D Deficiency Rickets

DAY	SERUM P (mg/dl)	SERUM Ca (mg/dl)	P'tase Bodansky Units
	A.S.–Age 6 Months		
−8	3.0	8.9	
−1	3.0	9.9	
0	Vitamin D–600,000 units		
7	4.6	10.6	
11	6.9	11.2	
	R.T.–Age 7 Months		
−4	3.2	7.7	27
−2	3.2	7.2	27
0	Vitamin D–600,000 units		
3	4.1	9.3	16
6	4.7	9.0	16
13	5.5	10.8	21

to 600,000 units of vitamin D given as described above are shown in Table 5–6.

Failure of complete response to 600,000 units of vitamin D given orally may result from poor absorption of vitamin D owing to steatorrhea, as indicated in Table 5–7. Severe infections may also interfere with response to vitamin D by a mechanism as yet unknown, as shown in Table 5–8. If serum phosphate and calcium concentrations do not return to normal by two weeks after a course of 600,000 units of vitamin D, a second dose of the same amount can be given. In the patients with steatorrhea and acute infection the second course was followed by complete healing, whereas only a partial response was obtained with the first dose. We have not given more than a total of 1,200,000 units for the treatment of infants with vitamin D deficiency rickets, and no toxic manifestations have resulted from this type of management. The characteristic rise of serum phosphate concentrations to high normal values does not occur when patients with primary hypophosphatemic rickets are challenged with doses of 600,000 or even 1,200,000 units (Table 5–9), so that a rapid differentiation of this form of rickets is possible.

Apart from the advantage of rapid differentiation of the various types of rickets, the large dose method of treatment pro-

Table 5–7. Delayed Response to Vitamin D Loading Test of Child with Steatorrhea and Vitamin D Deficiency Rickets (J.W. — Age 1 Year)

DAY	SERUM P (mg/dl)	SERUM Ca (mg/dl)	P'tase Bodansky Units
−12	2.6	8.6	80
−6	2.5	8.5	–
0	Vitamin D–600,000 units		
6	3.5	9.7	103
10	2.6	8.8	–
11	Vitamin D–600,000 units		
15	2.2	8.8	
20	3.0	9.1	
34	4.6	9.8	
90	5.7	10.2	
190	6.4	11.4	34

This case illustrates the slow but eventually adequate response to vitamin D therapy of a child with vitamin D deficiency due to steatorrhea. The final increase of serum phosphate concentrations to normal differentiates this type of patient from the child with primary hypophosphatemic rickets.

Table 5–8. Delayed Response to Vitamin D Load of Rachitic Infant with Active Infection (Pneumonia) (C.F.E. — Age 6 Months)

DAY	SERUM P *mg/dl*	SERUM Ca *mg/dl*	*P'tase Bodansky Units*
−6	3.6	7.4	17
−1	4.4	6.8	24
0	Vitamin D – 600,000 units		
4	4.5	6.4	–
10	4.8	6.6	26
11	Vitamin D – 600,000 units		
16	5.1	7.1	30
22	5.5	8.3	–
31	6.4	10.8	–

Following the first vitamin D dose the serum phosphate concentration rose slightly but the serum calcium levels remained low. With the repeat dose of vitamin D there was a rapid rise of serum phosphate and a slower return of serum calcium to normal.

duces more rapid healing and particularly reduces the possibility of low calcium tetany occurring during the period of treatment. If very small doses of vitamin D are given to rachitic infants reduction of serum calcium may occur in the initial phases of healing, because of rise of inorganic phosphate in extracellular fluid and deposition of calcium in osteoid before the complete effect of vitamin D on mobilization of calcium is obtained. When large doses of vitamin D are given the vitamin D effect on calcium homeostasis is obtained early and the danger of hypocalcemia occurring after this time is lessened.

The possible objection to the massive dose or "stoss" therapy is the danger of hypervitaminosis D. No evidences of toxic effect have been seen in the infants given one or two courses of 600,000 units each. Of course, repeated administration of this dose would produce cumulative toxic effects, but not if treatment is under the physician's control. Following the curative dose of vitamin D no further vitamin D need be given for three months or so, and then the usual daily prophylactic amount of vitamin D can be started and continued as in any other child.

Unfortunately the available pharmaceutic preparations of vitamin D make the "stoss" treatment of rickets difficult. A concentrated preparation of vitamin D in oil is needed for this therapy. The dilute solution of ergocalciferol (vitamin D_2) in propylene glycol, Drisdol, commonly used in the prophylaxis

Table 5–9. Abnormal Response to Vitamin D Loading Test of Child with Genetic Primary Hypophosphatemic Rickets (D.M. — Age 18 Months)

DAY	SERUM P *mg/dl*	SERUM Ca *mg/dl*	*P'tase Bodansky Units*
−2	2.4	10.6	24
0	Vitamin D – 600,000 units/day		
8	2.0	10.1	42
9	Vitamin D – 50,000 units/day		
16	Vitamin D – 100,000 units/day		
22	2.3	9.8	48
23	Vitamin D – 150,000 units/day		
29	2.2	9.2	74
*45	3.2	10.8	58
*57	3.3	10.1	57

*Early roentgenographic signs of healing of rickets noted at this time despite failure of serum phosphate to rise to normal range. Serum phosphate levels remained low in this child until toxic effects of excess vitamin D developed. Note that the serum calcium concentrations were never low.

of rickets, provides 8000 units per ml and cannot be used for massive dose treatment of rickets. The quantity of propylene glycol involved would cause intoxication with prolonged stupor since this liquid, although metabolized, has the toxic effects of an alcohol. The vitamin D-concentrated oil preparations are formulated for adults and are put up in capsules containing 50,000 units each. It is possible to give the contents of the capsules to infants but this is a clumsy procedure, and it is simpler to treat vitamin D deficiency rickets by giving smaller doses of vitamin D daily over a period of several weeks. Twelve-thousand to 16,000 units of vitamin D (1.5 to 2.0 ml Drisdol) can be given daily for three weeks, at which time there should be clearly demonstrable biochemical and roentgenologic evidences of healing if the rickets is due to uncomplicated vitamin D deficiency. Following this course the dose can be reduced to the usual prophylactic amount, 400 units per day.

The treatment of rachitic tetany is partly of course the treatment of rickets. These patients should receive the massive dose of vitamin D described above to produce the calcium-mobilizing effects of vitamin D as rapidly as possible (Table 5–6, R.T.). In addition, however, the administration of calcium salts is necessary to stop convulsions or to prevent the possibility of convulsions in infants with latent tetany. The convulsive seizures of hypocalcemia are not controlled satisfactorily by barbiturates. If the infant is convulsing or has convulsed, the initial treatment should be the intravenous injection of 10 percent calcium gluconate solution in a dose of 1 to 2 ml/kg up to 10 ml, given slowly enough so that bradycardia is not induced (see Chapter 3, *Treatment of Hypocalcemia*).

In infants with severe infections or other complications preventing feeding for several days, 10 percent calcium gluconate can be given by slow infusion after the initial "stat." amount in a dose of 20 to 30 ml per day diluted with the usual parenteral fluids. If the infant is taking feeds the calcium can be given orally as calcium lactate added to the milk feedings. An additional 1 to 1.5 gm of calcium per day should be given in this way. Since calcium lactate powder is only about 13 percent calcium by weight, this means that about 8 to 12 gm of the calcium lactate powder needs be added to the daily intake of milk. This amount of calcium lactate dissolved in the milk mixture is accepted by the infant and does not produce gastrointestinal disturbances. After a few days the calcium supplement can be progressively diminished and discontinued since the vitamin D effect should then be adequate to maintain calcium homeostasis.

Vitamin D Malabsorption

Under conditions in which the major source of vitamin D is the diet rather than skin exposure to ultraviolet light, fat malabsorption can result in rickets and osteomalacia owing to failure of vitamin D absorption. Hypovitaminosis D may occur in such patients, along with deficiencies of other fat-soluble vitamins, particularly α-tocopherol, carotene and vitamin K. Deficiency of both vitamins A and D may be present, but overt rickets is not likely to occur in such patients because of the failure of the growth of the vitamin A-deficient subject. Fat malabsorption due to intestinal disease, but with normal pancreatic lipase, may produce the most severe degree of calcium malabsorption because of the precipitation of calcium in the gut as insoluble, and therefore unabsorbable, calcium soaps of long chain fatty acids. It is of great interest that rickets is almost never seen in infants and children with cystic fibrosis despite the presence of pancreatic steatorrhea. It was once held that this was due to the growth failure of such subjects. However, with current methods of control of infection of the respiratory passages and use of pancreatic enzyme therapy, reasonably good growth may be seen in these young patients even though fat malabsorption is not completely corrected by pancreatic enzyme treatment. Nevertheless, rickets does not occur, and vitamin D intake in these infants and children does not have to be unusually large. It is possible that the vitamin D absorption is more efficiently stimulated by the pancreatic enzyme replacement therapy than that of α-tocopherol or carotene, since some of these subjects may have no evidences of vitamin D deficiency despite low serum levels of tocopherol and carotene.

Fat malabsorption should be considered as one of the possible mechanisms of rickets when unexplained evidences of vitamin D deficiency are noted in infants and children with known intake of vitamin D. We have seen two siblings who developed rickets, each at approximately 1 year of age. Vitamin D intake was adequate by history. Neither infant was grossly malnourished but one of the siblings presented with bleeding consequent upon vitamin K malabsorption and hypoprothrombinemia. The mechanism of fat malabsorption in these infants was not completely elucidated but presumably was genetically determined. The mode of transmission may have been as a sex-linked recessive since both patients were males, but there was no history of defect of intestinal fat absorption in the mother's family. Both patients responded satisfactorily to treatment with increased amounts of vitamin D and reduced dietary intake of fat.

The vitamin D deficiency of fat malabsorption may present with symptoms of hypocalcemia, as illustrated by the following case.

CASE REPORT

This 2½ months old infant was hospitalized because of sudden onset of convulsive seizures. He was full term at birth and there had been no unusual birth trauma or neonatal problems. He had been given a vitamin D-fortified commercial infant feeding preparation. The initial serum calcium concentration was 5.6 mg/dl and phosphate 8.0 mg/dl. The diagnosis of either late transitory physiologic hypoparathyroidism or congenital hypoplasia of the parathyroids was entertained because of the hypocalcemia and high normal serum phosphate concentrations. As shown in Table 5–10, addition of calcium to the diet as calcium lactate, with a final Ca:P ratio of 4, was completely ineffective in raising serum calcium concentration, although the addition of calcium decreased the serum phosphate. Serum carotene concentration was found to be low and quantitative stool examination revealed an excessive excretion of fat. Administration of a large vitamin D dose resulted in a rapid increase of serum calcium concentration.

COMMENTS: The marked hypocalcemia that was the presenting complaint in this infant might have been due to severe calcium malabsorption resulting from calcium soap formation as well as vitamin D deficiency. The diagnosis in this patient was suggested by a low serum carotene concentration that led to study of fecal fat excretion. We can now determine the serum 25-hydroxyvitamin D concentration and this makes possible a definitive diagnosis, since this would not be reduced in hypoparathyroidism but would be extremely low in vitamin D deficiency due to malabsorption of fat-soluble vitamins. Treatment in such patients is oral administration of vitamin D in doses of 10 to 20 times the usual daily allowance of 10 μg (400 units). With such large doses sufficient vitamin D is absorbed de-

Table 5–10. Vitamin D Deficiency Secondary to Fat Malabsorption

AGE DAYS	SERUM Ca (mg/dl)	SERUM P (mg/dl)	*Alk. P'tase Bodansky Units*
75	5.6	8.0	24.7
76	Calcium lactate supplement	Ca:P = 4	
77	6.9	4.8	–
79	5.0	6.5	18.0
80	5.2	6.5	–
81	Vitamin D_2 – 600,000 units p.o.		
84	6.7	7.7	20.0
88	8.6	7.1	–
108	10.5	6.0	–
151	9.8	4.8	14.7

spite continued fat malabsorption. If the basis of the fat malabsorption can be diagnosed and specific treatment given, as in pancreatic lipase deficiency or gluten-induced enteropathy, reduction of stool fat loss can lessen the dose of vitamin D needed. Serial measurements of serum 25-hydroxyvitamin D could provide the most definitive method of determining the vitamin D dose needed to maintain normal vitamin D status.

HEPATIC RICKETS

The role of liver disease, especially that associated with atresia of the external biliary tract, in causing rickets and osteomalacia is well-known. It was once thought that the vitamin D deficiency in these patients was due to vitamin D malabsorption. In complete obstruction of the biliary tract this may well be the case since there is good experimental evidence that vitamin D is not absorbed in the absence of bile salts in the intestinal lumen. However, the occurrence of rickets in infants with neonatal hepatitis who had only transitory acholic stools was difficult to explain on this basis alone, unless it was postulated that bile salt secretion was reduced because of hepatocellular damage. With our current knowledge of the vitamin D cycle it seems likely that failure of vitamin D 25-hydroxylation by microsomal or mitochondrial liver cell systems may be the causative factor in some patients, since 25-hydroxylation is the necessary preliminary reaction before 1-hydroxylation by the kidney mitochondrial system can occur. Patients with liver disease, therefore, can have reduced concentrations of serum 25-hydroxyvitamin D and serum 1,25-dihydroxyvitamin D.

The mode of presentation of the deficiency of the vitamin D hormone can be early hypocalcemia before radiologic signs of rickets are present. In the more chronic cases severe skeletal lesions of rickets and osteomalacia develop. Because of the significant incidence of manifestations of vitamin D deficiency in infants and children with liver disease these patients routinely should be given increased amounts of vitamin D. It would be more efficient to treat them with 25-hydroxyvitamin D, and this vitamin D metabolite, as expected, has been shown to be effective in small doses in the treatment of the evidences of vitamin D deficiency in patients with liver injury, including those with biliary tract atresia. However, 25-hydroxyvitamin D has not yet been approved by the Food and Drug Administration and is, therefore, still an investigational drug.

In the absence of this metabolite, which would be the physiologic replacement form of vitamin D in hepatic disease, large doses of vitamin D, 0.1 to 0.2 mg per day (4000 to 8000 units), should be given, and in some children with atresia of the bile ducts even doses of this magnitude are not sufficient. Again, if it is possible to monitor treatment with measurements of 25-hydroxyvitamin D this is the most efficient method of determining the actual required dose. In the absence of this measurement the treatment can be controlled by determinations of serum calcium and phosphate concentrations and bone x-rays. The serum alkaline phosphatase is, of course, of no value in determining healing of the rickets in these subjects since it can be elevated by the hepatic disease, and separation of the liver and bone forms of the enzyme is not usually done.

CASE REPORTS

J.F. was a male infant admitted at 3 months of age with jaundice of two weeks' duration associated with vomiting and epistaxis. The liver and spleen were both enlarged; stools were acholic and the serum bilirubin concentration was 8.5 mg/dl total and 5.8 mg/dl direct. The serum glutamic pyruvate transaminase concentration was 950 units. The concentrations of serum carotene and α-tocopherol in serum were both low. Serum calcium concentration was normal but serum phosphate was reduced, 3.6 mg/dl. The roentgenogram of the long bones showed suggestive evidences of rickets (Fig. 5–7*a* and *b*). More conclusive radiographic proof of rickets appeared when healing occurred following oral administration of 600,000 units of vitamin D (Fig. 5–7*c* and *d*).

COMMENTS: This infant had been fed a vitamin D-fortified milk since the first days of life so that the vitamin D deficiency must be ascribed to the liver disease. In this instance some degree of vitamin D malabsorption may have been a factor since concentrations in serum of other fat-soluble substances, carotene and α-tocopherol, were also low. However, a defect of 25-hydroxylation of vitamin D may have been an important contributory factor. The diagnosis of hepatitis, type undetermined, was made in this infant and there was apparent complete recovery over a period of several weeks. The initial bone x-rays were thought to show only minimal evidences of rickets, but when healing was induced by a large vitamin D load the healing pattern showed that the rickets was quite pronounced and that a broad layer of chondro-osteoid was present at the metaphyseal end of the bone in the initial study, which could only be demonstrated when deposition of bone mineral in this matrix occurred after the vitamin D treatment.

Infant B.Z. was full term and apparently normal at birth. There was no history of maternal illness during pregnancy or postpartum. However, he was noted to have prolonged jaundice for the first few weeks of life with an increased concentration of direct reacting bilirubin. The liver was thought to be enlarged and the activity of serum glutamic pyruvate transaminase was high. The stools were never acholic. A diagnosis of neonatal hepatitis of unknown pathogenesis was made. On follow-up the baby appeared to be clinically well and the hyperbilirubinemia gradually diminished. Because of the apparent mildness of the liver injury extra vitamin D was not given and the intake was that provided in a fortified infant feeding preparation. At 2½ months of age, however, the patient's serum calcium concentration was found to be 7.0 mg/dl. Roentgenograms of the bones did not show clear-cut rickets although the bones were poorly mineralized. Further studies are summarized in Table 5–11.

COMMENTS: The hypocalcemia in this infant was presumably due to failure of 25-hydroxylation of vitamin D as the result of hepatocellular injury. The noteworthy fact is the mildness of the hepatitis by usual clinical criteria, and yet the liver cell damage was sufficient to interfere with vitamin D metabolism. Serum 25-hydroxyvitamin D concentrations unfortunately were not determined in this patient but a low concentration would have been the specific diagnostic feature. This experience indicates that all infants with neonatal hepatitis, even though apparently self-limited with spontaneous recovery, should receive an increased allowance of vitamin D, and their vitamin D status should be monitored by serum 25-hydroxyvitamin D concentrations if possible and by serum calcium determinations.

Patients S.K. and P.K. were monoamniotic uniovular twin girls delivered by cesarean section supposedly post term. S.K. weighed 1400 gm and P.K. 1670 gm at birth. Apgar scores were 8. S.K. fed well and gained weight until 7 weeks of age when her stools were noted to be acholic. At that

Table 5–11. Hypocalcemia and Rickets as a Complication of Neonatal Hepatitis

	Serum			
Age Days	*Bilirubin*	*Ca mg/dl*	*P*	*Alk. P'tase Bodansky Units*
3	2.0/11.4	–	–	–
69	3.3/ 5.0	7.2	6.7	29.8
76	2.6/ 3.9	7.2	5.9	37.3
81	1.8/ 3.1	7.0	5.6	41.2
		Vitamin D treatment started		
92	1.0/ 1.3	10.0	5.9	45.3
124	0/ 1.6	11.5	7.3	17.7

time examination showed mild jaundice and marked enlargement of liver and spleen. Biopsy of the liver showed mild hepatitis and cytomegalovirus was cultured from the biopsy and urine. X-ray examinations at 9 weeks of age showed severe osteomalacia. P.K. had an episode of fever and vomiting at 8 days of age but cultures were negative for bacterial infection She apparently did well thereafter but at 8 weeks of age was reexamined because of the problems found in her twin sister. At that time she was mildly jaundiced and liver and spleen were enlarged. At 11 weeks of age a deformity of the left wrist was noted and radiographs showed fractures of both distal radii and left distal femur with severe generalized osteomalacia. Serum Ca was 7.2 mg/dl, serum P 5.3 mg/dl, and alkaline phosphatase 15 K.A. units. Both were treated with 10,000 units vitamin D daily. Serum calcium concentration quickly rose to normal and four months later radiographs showed good mineralization of the bones.

COMMENTS: In these infants with severe liver injury due to cytomegalic inclusion body disease, serious bone disease developed with spontaneous fractures in one infant before it was recognized that a defect of vitamin D metabolism was present. This again reinforces the recommendation that all infants and children with liver disease be considered at risk for abnormality of vitamin D metabolism; that they receive an increased allowance of vitamin D, 10 to 20 times the normal prescribed daily intake; and that they be followed carefully for biochemical evidences of lack of the vitamin D hormone. Despite acholic stools and steatorrhea large doses of vitamin D given orally are effective in both prevention and treatment of the vitamin D-deficient state.

(Cases S.K. and P.K. were reported by Kopelman, A. E. et al: J. Pediatr. *81*:101–105, 1972.)

RICKETS ASSOCIATED WITH ANTICONVULSANT DRUG TREATMENT

Instances of rickets in older children and osteomalacia in adults on long-term treatment with anticonvulsant drugs, particularly combinations of drugs including phenobarbital, phenytoin and primidone, have been reported repeatedly following the original reports of Schmid and of Kruse in 1967 and 1968. We observed at that time two cases of rickets in older children with seizure disorder who had severe motor difficulty and were essentially bed-ridden. It was our thought that these children had not been given any outdoor exposure to sunshine and that their dietary intake of vitamin D had been inadequate. Many of the cases of rickets and osteomalacia in subjects treated with anticonvulsant drugs arose in institutionalized patients who also might have suffered from lack of sunshine exposure. As knowledge of vitamin D metabolism developed, however, the possibility that anticonvulsant drugs imposed an increased requirement of vitamin D through an effect on vitamin D metabolism became a more attractive hypothesis. Phenobarbital and phenytoin are known to be enzyme-inducing drugs. In the liver these drugs increase the formation of mixed oxidases of the cytochrome P-450 system. These enzymes hydroxylate cyclic compounds including vitamin D, and these more polar hydroxylated compounds are then glycuronidated and excreted as inactive glycuronides in the bile. The action of phenobarbital, for example, has been used to increase the glycuronidation and excretion of bilirubin glycuronide in patients with a deficiency of glycuronyl transferase (Crigler-Najjar syndrome).

Haddad and others showed that serum 25-hydroxyvitamin D concentrations were lower in patients being treated with phenobarbital and phenytoin than in control subjects in the same environment and the same season of the year with approximately the same dietary intake of vitamin D. This reinforced the concept that inactivation of vitamin D by further hydroxylation in the liver beyond the 25-hydroxy stage with glycuronidation and excretion of the vitamin D glycuronides in the bile was an important mechanism explaining the increased risk of rickets and osteomalacia in subjects treated with anticonvulsant drugs. On the other hand, determination of 1,25-dihydroxyvitamin D in these patients did not indicate a deficiency of the vitamin D hormone, so that the reduced concentration of the substrate, 25-hydroxyvitamin D, did not appear to be sufficient cause for the increased incidence of the hypocalcemia, hypophosphatemia, rickets and secondary hyperparathyroidism of the patients receiving enzyme-inducing anticonvulsant drugs such as phenobarbital, phenytoin and primidone.

Experimental evidence in animals indicates that phenytoin or phenobarbital and phenytoin in combination inhibit calcium transport by the intestinal mucosa, thus preventing the vitamin D enhancement of calcium absorption. This direct effect on calcium transport may be analogous to the role of phenytoin in inhibiting calcium uptake by synaptasomes, which has been thought to be a possible mechanism for its anticonvulsant effect. The increased vitamin D requirement of patients receiving anticonvulsant drugs, therefore may be the integrated effect of an action on the intestinal mucosal transport of calcium and hepatic inactivation of vitamin D by hydroxylation and glycuronidation.

There has been sufficient experience with many thousands of patients with seizure disorders receiving anticonvulsant drugs to enable some conclusion to be reached concerning their need for additional vitamin D beyond that ordinarily available in the diet. One of the major conclusions is that the active child and adult living in the temperate zone receives sufficient sunshine, particularly in the summer months to supply most of his vitamin D requirements and that dietary intake plays a less important role. This also seems clear in the patient receiving anticonvulsant drugs. Most of the instances of hypocalcemia and rickets and osteomalacia have been in institutionalized patients or those with severe motor handicaps preventing them from outdoor activity. There are some exceptions to this, and it may be that occasional patients have vitamin D requirements considerably greater than the average (this has been discussed in the section on normal vitamin D requirements), and this metabolic difference, plus the increased rate of vitamin D inactivation secondary to anticonvulsant drugs, results in a much higher vitamin D need than can be met by usual skin exposure to the ultraviolet radiation of sunshine. Such patients need considerably higher oral intake of vitamin D to maintain normal serum calcium concentrations, as illustrated in the case reports.

It seems logical that all patients being treated with combinations of anticonvulsant drugs including phenobarbital, phenytoin or primidone should have as part of their follow-up studies measurements of serum calcium and phosphate concentrations. Determinations of serum alkaline phosphatase activity are less useful since the enzyme-inducing drugs increase hepatic cell output of alkaline phosphatase, so that the determined serum alkaline phosphatase may be high as a result of hepatic enzyme without any abnormality of bone mineralization. If there is any limitation of exposure to sunshine because of motor handicaps, severe mental retardation with limited activity, or environmental factors, patients receiving anticonvulsant drugs should be given extra vitamin D routinely as an oral supplement. This need not be a very large supplement and it has been proposed that an additional 3000 units per week may be sufficient.

CASE REPORTS

C.V., a 43 year old woman, had been diagnosed as having cerebral palsy with convulsive seizures since childhood. She was markedly retarded and had been cared for at home by her mother. Her anticonvulsant medications were phenobarbital, phenytoin and primidone. In 1974 she sustained a fracture of the tibia following a fall, and x-ray examinations indicated that her bones were poorly mineralized (Fig. 5–19*a* and *b*). When referred for further study it was found that the serum ionized calcium concentration was consistently less than 4 mg/dl. Total serum calcium concentration was reduced and serum phosphate was also borderline low, as shown in Table 5–12. The serum 25-hydroxyvitamin D concentration was below the normal range. At this time she was unable to walk because of pains in both extremities presumably due to her os-

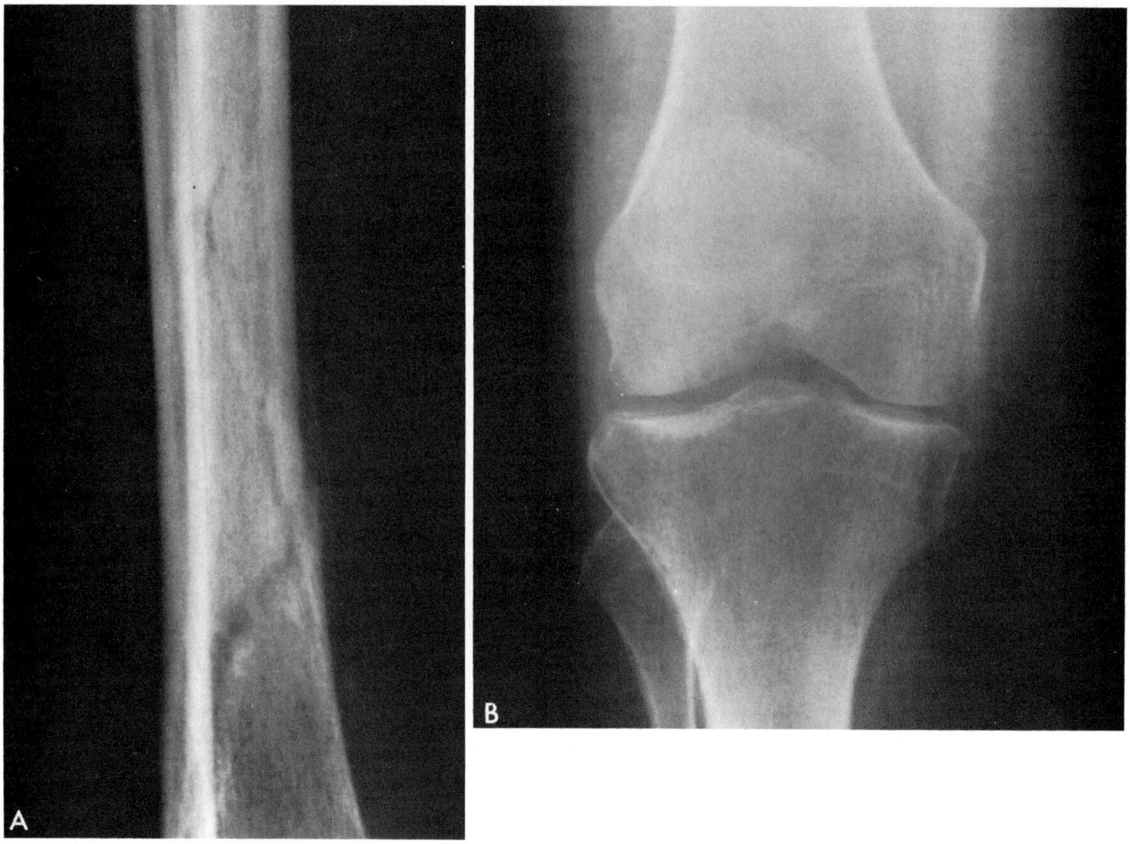

Figure 5–19. Osteopenia in patient (C.V.) who had been receiving long-term anticonvulsant medication: *a*, tibial fracture; *b*, poor mineralization of bones of knee joint.

teopenic state. She was started on vitamin D, 16,000 units per day, with subsequent rise in serum ionized calcium concentration and increase of serum 25-hydroxyvitamin D concentrations to levels considerably greater than normal (Table 5–12). On this regime the patient showed marked improvement of bone mineralization, the bone pain disappeared and she became mobile once more.

COMMENTS: This retarded woman with limited activity developed hypocalcemia and osteopenia with fracture while on anticonvulsant drug treatment without supplementary vitamin D. These manifestations of lack of physiologic effects of vitamin D were reversed when large doses of vitamin D were given. This patient may have had a markedly increased requirement of vitamin D resulting from anticonvulsant medication, and her limited outdoor physical activity may also have been an important factor in her development of the vitamin D-deficient state.

D.W. was a 29 year old white male who had been on anticonvulsant medication from 1964 until 1976. His drug regimen included Tegretol, Mebroin and Clonopin. He was one of several hundred patients in a seizure clinic who had been routinely screened to determine the incidence of minimal evidences of abnormality of calcium metabolism associated with anticonvulsant drug treatment. The screening program included determinations of serum ionized calcium concentration, and D.W.'s was found to be 3.8 mg/dl, which was lower than the minimal value of 4.0 mg/dl found in our control group. This was repeated several months later without change. D.W. was an inactive, moderately retarded young adult who remained indoors most of the time so that exposure to sunshine was limited. Radiographs of the knee indicated reduced

Table 5–12. C.V.

DATE	Total Ca	Ionized Ca mg/dl	SERUM P	Alk. P'tase. Units*	25-OHD ng/ml
6/11/75	7.9	–	2.8	–	<5
		Vitamin D – 16,000 units per day			
6/18/75	8.3	–	3.6	–	180
7/16/75	8.8	–	2.9	2.6	255
12/10/75	9.4	–	2.9	–	240
4/14/76	8.7	3.78	2.5	2.4	270
10/20/76	8.6	4.00	2.2	2.5	147
3/9/77	8.9	4.30	2.7	1.6	–
6/8/77	8.9	4.50	2.8	–	150

*Bessey-Lowry Units.

bone mineralization (Fig. 5–20). Serum 25-hydroxyvitamin D concentration was too low to determine (less than 5 ng/ml), which was compatible with poor vitamin D intake plus the effect of anticonvulsant drugs on 25-hydroxyvitamin D metabolism. The course after vitamin D treatment is given in Table 5–13.

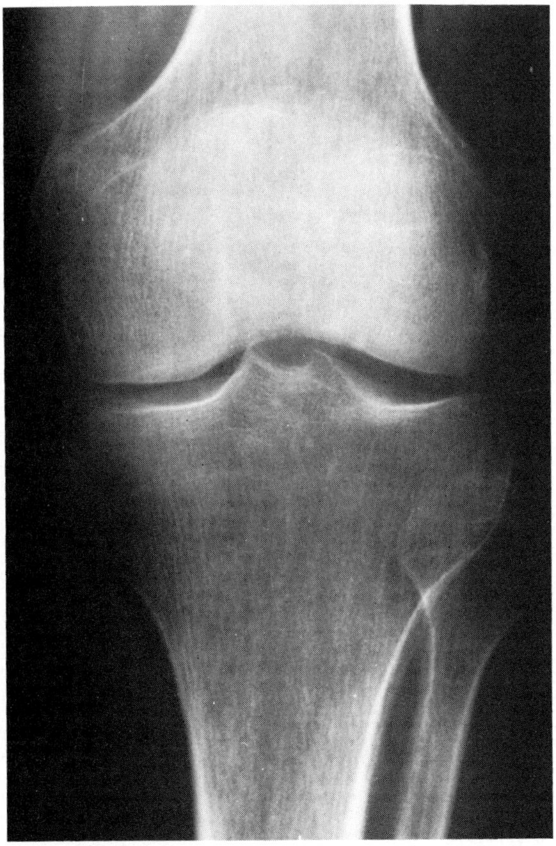

Figure 5–20. Osteopenia in patient (D.W.) who had been receiving anticonvulsant medication.

COMMENTS: This patient represents the more usual disorder of calcium metabolism related to anticonvulsant drug treatment, i.e., reduced vitamin D stores because of the combination of limited exposure to sunshine and the effect of enzyme-inducing anticonvulsant drugs in increasing the rate of 25-hydroxyvitamin D inactivation. There were no obvious clinical manifestations but there were biochemical evidences of abnormality of vitamin D metabolism, a very low concentration of serum 25-hydroxyvitamin D and a reduced concentration of ionized calcium in serum. There was no hypophosphatemia nor elevation of serum alkaline phosphatase, and bone x-rays showed some degree of osteopenia but no obvious osteomalacic change. This patient required an increased amount of vitamin D to maintain serum 25-hydroxyvitamin D above 30 ng/ml and ionized calcium above 4 mg/dl. Whether he will continue to need as much as 4000 units per day remains to be determined. If his outdoor exposure to sunshine were increased his oral intake of vitamin D could be less.

VITAMIN D-DEPENDENT RICKETS

Following the availability of high potency vitamin D preparations a category of rickets was described that resembled vitamin D deficiency rickets in its manifestations but which required very large doses of vitamin

Table 5–13. Anticonvulsant Drug Hypocalcemia and Osteopenia (D.W.)

DATE	*Total Ca*	*Ionized Ca* *mg/dl*	SERUM *P*	*Alk. P'tase.* *Units**	*25-OHD* *ng/ml*
3-27-76	8.8	3.80	4.0	2.8	<5
5-14-76	9.6	3.80	3.1	2.7	9
		Vitamin D – 4000 units/day			
7-18-76	9.2	4.02	3.2	2.5	35
2-15-77	9.7	4.48	3.8	2.1	54

*Bessey-Lowry Units.

D for its healing. When these large doses were discontinued the evidences of vitamin D deficiency recurred, hence the name "vitamin D-dependent rickets." Possibly a better term would be "increased requirement for vitamin D," since true vitamin D deficiency rickets is also vitamin D-dependent, but infants and children with vitamin D deficiency, once treated, do not relapse. Their vitamin D requirements are now met by skin exposure to sunshine plus the small amounts of vitamin D in a mixed diet. Vitamin D-dependent rickets can be familial and the studies of kindreds indicate that the genetic pattern is that of an autosomal recessive gene transmission.

The children with this disorder develop progressive manifestations of rickets despite proved intake of vitamin D in quantities sufficient to prevent or cure true vitamin D deficiency. Biochemically the findings are those of a vitamin D deficiency state, namely, reduced concentrations of calcium and of phosphate in serum, increased alkaline phosphatase concentration and generalized aminoaciduria. More recently, increased urinary excretion of cyclic AMP has also been found. The urinary findings are compatible with the secondary hyperparathyroidism that these patients manifest. The radiologic evidences are usually those of chronic progressive rickets, although in one patient summarized in the following case reports the lesions of hyperparathyroidism predominated.

The diagnosis of vitamin D-dependent rickets is made on the evidence of rickets associated with hypocalcemia and generalized aminoaciduria in a child who has been documented to have received sufficient vitamin D to prevent vitamin D deficiency. A family history of another sibling with the same problem, of course, would be confirmatory. Determinations of 25-hydroxyvitamin D in the serum show normal or high values, confirming the history of adequate vitamin D intake and ruling out both vitamin D malabsorption or hepatocellular injury as the basis for increased vitamin D needs. The latter would also be eliminated by the standard tests for liver function. Another form of rickets refractory to vitamin D, primary hypophosphatemic rickets, is eliminated by the hypocalcemia and generalized aminoaciduria which are not present in primary hypophosphatemia. Rickets due to the renal tubular injury of the Fanconi syndrome should not present with hypocalcemia early, as does vitamin D-dependent rickets. Although patients with the latter do have generalized aminoaciduria, as do those with the Fanconi syndrome, they do not show glycosuria or the other evidences of renal tubular dysfunction found in the Fanconi syndrome, namely, bicarbonate wasting, potassium wasting, polyuria with failure of concentrating ability, and proteinuria.

The biochemical basis for genetically transmitted vitamin D-dependent rickets has been elucidated by the recent studies of Fraser, Scriver and co-workers. They found that the biochemical and radiologic evidences of rickets in these patients could be corrected by physiologic quantities of 1,25-dihydroxyvitamin D. In other words, although amounts of vitamin D as high as 1000 μg per day were required to prevent relapse of rickets (100 times the physiologic requirement) the requirement of 1,25-dihydroxyvitamin D was normal, 0.5 to 1μg per day. This indicated an abnormality of the renal 25-hydroxyvitamin D 1-hydroxylase enzyme system which converts 25-hydroxyvitamin D to 1,25-dihydroxyvitamin D. The effect of large doses of vitamin D in these patients may be explained by the high serum concentrations of 25-hydroxyvitamin D that result from this treatment. Either increased concentrations of substrate may push the reaction in the renal tubules toward production of adequate amounts of 1,25-dihydroxyvitamin D if the enzyme is not completely lacking or inactive, or high concentrations of 25-hydroxyvitamin D may substitute for 1,25-dihydroxyvitamin D, since it is known that 25-hydroxyvitamin D in supraphysiologic amounts is effective in the nephrectomized subject who cannot form 1,25-dihydroxyvitamin D. Whether vitamin D-dependent rickets is a uniform entity with a single abnormality of the renal enzyme remains to be determined. It is possible that a number of disorders of the enzyme system for 1-hydroxylation of 25-hydroxyvitamin D may exist with increased requirement for vitamin D as the common final manifestation.

The present treatment of vitamin D-dependent rickets is the administration of large doses of vitamin D; the exact amounts must be determined individually since there are considerable differences among patients. As much as 50,000 units per day has been needed by some patients to produce complete healing of rickets, whereas others have responded to doses of 10,000 to 20,000 units per day. The required dose is determined by serial measurements of serum calcium concentration and urine calcium excretion, and should be sufficient to raise the serum calcium into the normal range —9 to 10.5 mg/dl — without producing hypercalciuria, i.e., urine calcium >0.25 mg per mg creatinine or >4 mg/kg/day. 1,25-dihydroxyvitamin D is now available although currently approved only for treatment of patients with renal insufficiency. This compound in doses of 0.5 to 1μg per day would be the most specific therapy. If it is too costly, however, vitamin D would still be the most convenient method of treatment.

CASE REPORTS

Increased Requirement of Vitamin D

E.McF., a girl aged 8 years, 10 months, came to the emergency room because of a fall and pain in one leg. Radiographs of the leg showed lesions suggestive of rickets at the metaphyseal ends, and she was admitted for further study. There was no history of bone pains or deformity prior to this fall. The diet had included vitamin D-fortified milk and the child had the usual outdoor activity, so that vitamin D deficiency was unlikely. There was no family history of rickets. The serum calcium, phosphate and alkaline phosphatase concentrations are given in Table 5–14. The initial calcium concentration was slightly low, 8.8 mg/dl. Following one month of hospitalization without added vitamin D except that provided by fortified milk and other foods, the serum calcium concentration was 10.0 mg/dl and that of phosphate 4.4 mg/dl. However, there was little radiographic evidence of healing. She was then given a single dose of 500,000 units of vitamin D with prompt rise in serum phosphate to 7.2 mg/dl. No further vitamin D treatment was given. Follow-up studies were inadequate because of lack of family cooperation. Two years later the serum calcium concentration was again found to be somewhat low, 8.6 mg/dl, but no other examinations were made.

She was seen again at 13 years, 2 months because of painful swelling of the lower third of the left tibia. At this time x-ray examination showed a cystic lesion suggestive of a nonossifying fibroma, and subsequently a pathologic fracture developed through this cyst (Fig. 5–21*a* and *b*). Biochemical evidences of vitamin D deficiency were more marked, with a serum calcium concentration of 7.5 mg/dl (Table 5–14) and alkaline phosphatase activity of 26.9 Bodansky Units. She was again treated with vitamin D, with return to normal of serum calcium, phosphate and alkaline phosphatase concentrations. With continued vitamin D treatment, 5000 units per day, she remained

Table 5–14. Serum Calcium and Phosphate Concentrations of Child with Increased Requirement of Vitamin D (E. McF.)

AGE YEARS	SERUM Ca mg/dl	SERUM P mg/dl	Alk. P'tase Bodansky Units
8-10/12	8.8	4.7	14.8
	Hospitalized for 1 month—rickets by x-ray		
	No extra vitamin D except that in diet		
8-11/12	10.0	4.4	17.3
* 8-11/12	10.1	7.2	–
10-7/12	8.6		
13-2/12	7.5	4.1	26.9
14-11/12	7.8	3.8	14.4
*14-11/12	9.9	5.0	14.0
	5000 units vitamin D/day		
15-9/12	10.0	3.8	7.9

*6 days following single dose of 500,000 units of vitamin D.

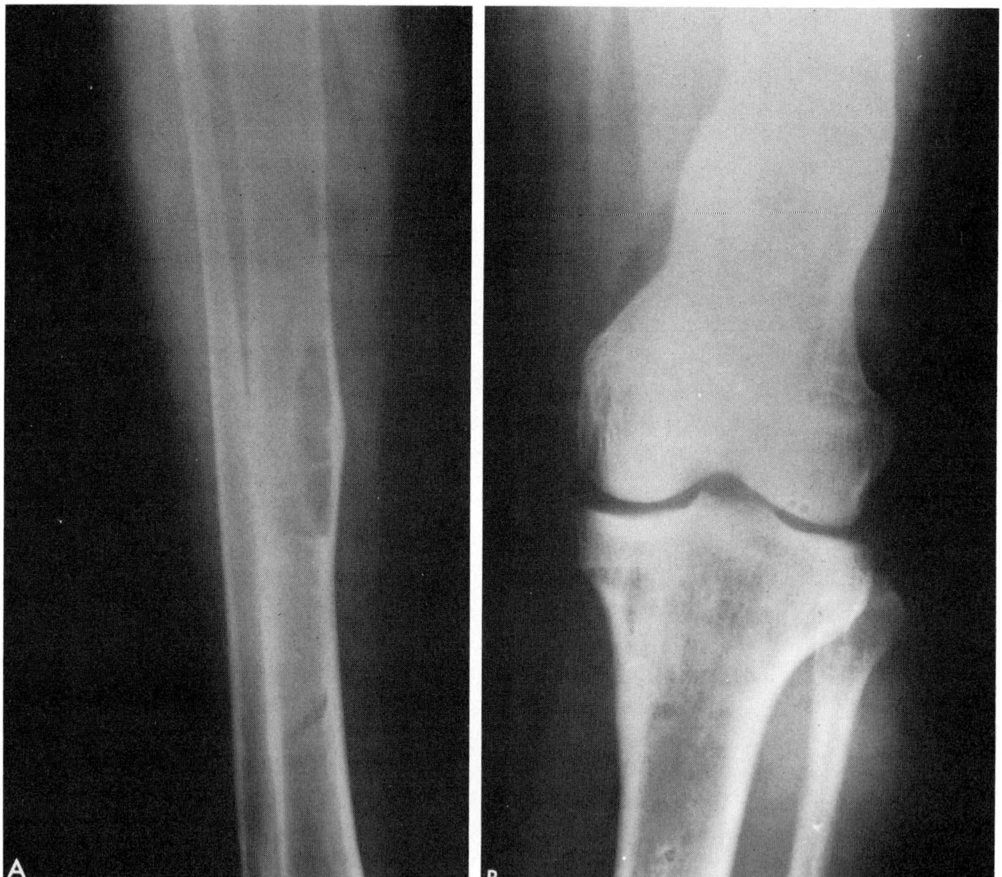

Figure 5–21. Radiographs of child (McF) with increased requirement of vitamin D and recurrent rickets: *a*, nonossifying fibroma of tibia forming radiolucent cystic structure; *b*, deformity following pathologic fracture through cyst, and subsequent slow healing.

normocalcemic and no further evidences of skeletal demineralization.

COMMENTS: This is an example of evident increased requirement for vitamin D with approximately a tenfold greater requirement than normal. Whether this represents a deficiency of 1-25-dihydroxyvitamin D is not known since this child was studied before knowledge of this vitamin D metabolite was available. Most patients with familial vitamin D-dependent rickets have had an earlier onset of the clinical, biochemical, and roentgenographic signs of rickets, and their vitamin D requirements have been greater than that of this girl. Presumably the deficiency of vitamin D metabolism was less marked in this patient than in the usual vitamin D-dependent rickets, accounting for the late onset of rickets and its slow progression.

J. McK., a white male, at age 9 or 10 years had painful swelling of the right knee that prevented him from walking. He was hospitalized for almost one year without diagnosis despite operation on the right knee. He presented to the orthopedic clinic at 13 years because of persistence of the painful knee swelling and limp. Examination showed limitation of motion of both knees and right wrist, with swelling of the joints. Radiographs showed demineralizing lesions of skeleton with radiolucent metaphyseal bands at ends of long bones and multiple cystic lesions. Serum Ca was 9.7, 10 mg/dl, serum P 5.7, 5.2 mg/dl, and serum alkaline phosphatase 42 I.U. There was generalized aminoaciduria. Urine calcium excretion was increased 234 and 228 mg/24 hrs. Urine hydroxyproline excretion was also increased, 186 mg/24 hrs. Synovial biopsy showed mild chronic inflammatory change, and bone biopsy was reported as possible "metabolic bone disorder in the category of rickets–osteomalacia." No diagnosis was made and the skeletal lesions progressed so that the boy was unable to walk because of severe knee pains; wrists and hands also had become swollen with limitation of motion. He had recurrent episodes of asthma starting at age 14.

At age 15, he was admitted to the hospital because of bilateral pneumonitis and pleural effusion. Reevaluation of the orthopedic problems showed a short, stocky adolescent boy with inability to walk because of pain in knees and wasting of muscles of lower extremities secondary to lack of activity. There was swelling of both knees and swelling and contractures of both wrists. Radiographs showed extreme osteopenia with rarefaction of metaphyseal ends of long bones and cystic lesions of all bones (Fig. 5–22*a* and *b*). There was absence of lamina dura and lytic lesions around roots of molar teeth (Fig. 5–22*c*). Pertinent laboratory studies showed serum Ca 7.7 mg/dl, serum P 2.4 mg/dl, alkaline phosphatase 57 I.U., and urine Ca 51 mg/24 hrs (Ca/Cr — 0.06). Bone biopsy was reported as showing histologic evidence of hyperparathyroidism with minimal evidences of osteomalacia.

It was postulated that this boy had a defect of vitamin D metabolism with secondary hyperparathyroidism which was so severe as to produce skeletal lesions of bone demineralization and osteitis fibrosa cystica. In order to delineate the problem the patient was given 25-OH vitamin D, 20 μg per day, to see if there was any improvement. The results of treatment are summarized in Table 5–15. The serum calcium and phosphate concentrations had risen somewhat before treatment owing to the control of infection. Following 25-OH vitamin D these concentrations returned to normal, and there was a temporary increased concentration of alkaline phosphatase, which can be seen in vitamin D deficiency states following institution of treatment. Following several weeks of treatment there was definite clinical improvement; pain in the knees was less and ambulation was begun. By six months of treatment clinical improvement was unquestioned but remineralization of bones by x-ray was slow, so that the dose of 25-OH vitamin D was increased to 40 μg/day. After 11 months of treatment with 25-OH vitamin D it was decided to change to vitamin D at a high dose to see if improvement was maintained. With daily intake of 400 μg vitamin D (16,000 units) serum calcium and phosphate concentrations remained normal, serum alkaline phosphatase activities returned to normal,

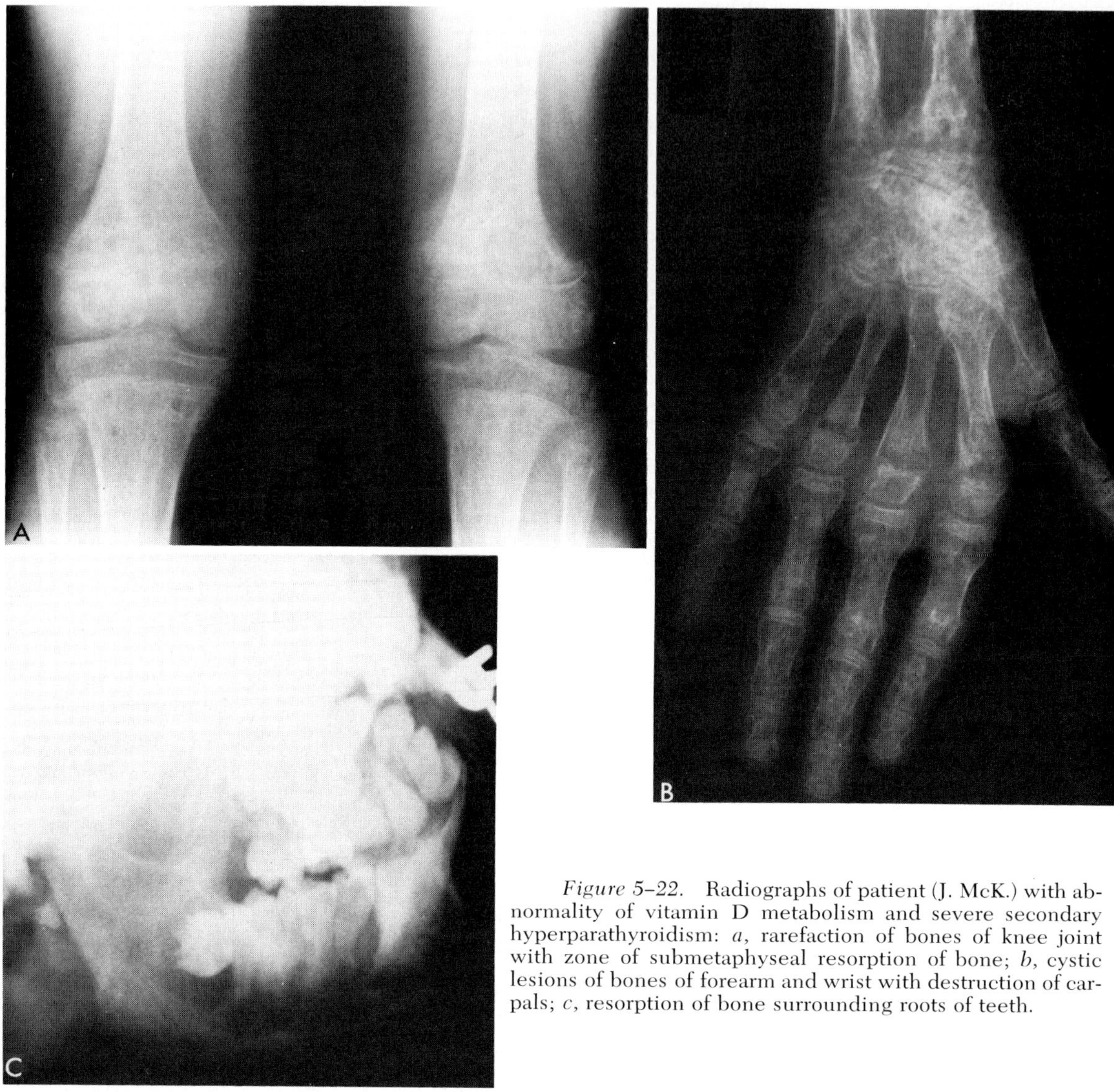

Figure 5–22. Radiographs of patient (J. McK.) with abnormality of vitamin D metabolism and severe secondary hyperparathyroidism: *a*, rarefaction of bones of knee joint with zone of submetaphyseal resorption of bone; *b*, cystic lesions of bones of forearm and wrist with destruction of carpals; *c*, resorption of bone surrounding roots of teeth.

Table 5–15. Vitamin D Dependency with Extreme Seconday Hyperparathyroidism (J. McK.)

	SERUM			
DATE	*Ca* *mg/dl*	*P*	*Alk. P'tase. Units*	*25-OHD ng/ml*
4-10-74	7.7	2.4	57*	
4-11	8.1	2.7	—	
4-16	7.8	3.5	59*	
4-18	8.2	4.2	—	
4-22	25-OH vitamin D—20 μg/day			
4-25	9.1	—	—	
5-1	9.2	5.0	—	
5-4	9.4	5.3	112*	
7-3	9.1	3.8	148*	
10-21	9.7	4.5	88*	
10-22	25-OH vitamin D—40 μg/day			
11-19	9.8	4.8	80*	
12-16	9.8	4.4	10.2†	
2-10-75	9.7	5.2	10.5†	
3-18	9.4	4.3	8.9†	
3-19	Vitamin D_2—400 μg (16,000 units/day)			
5-19	9.7	5.0	10.6†	
6-16	9.9	5.0	—	
12-15	10.6	5.1	5.7†	
3-16-76	9.9	3.2	3.4	135
9-20	10.4	4.0	3.1	197
4-13-77	10.1	3.8	3.1	110

*mIU (upper limit of normal for adult = 32).
†Bessey-Lowry units (upper limit of normal for adult = 3.5).

and clinical improvement continued. Following a total of two years of treatment the young man is normally active, has no pain and has been working in a job requiring considerable physical exertion. He is short in stature, 168 cm, and has considerable contracture deformity of one hand, with limitation of motion and ankylosis of radius and ulna of right arm. Radiographs show partial remineralization of the huge cystic lesions of the bones (Fig. 5–23).

COMMENTS: This boy was an extraordinarily puzzling case for a number of years during which he became progressively more crippled and finally completely bedridden and unable to walk. The radiographic findings were compatible with hyperparathyroidism with osteitis fibrosa cystica, and the bone biopsies also were consistent with hyperparathyroidism with that degree of osteomalacia that can be seen in primary hyperparathyroidism. The biochemical findings, however, were not those of primary hyperparathyroidism. Serum calcium concentrations were normal or low, particularly when he developed a severe infection. It was concluded that he must have a marked degree of secondary hyperparathyroidism, presumably resulting from a defect in vitamin D metabolism. The response to 25-OH vitamin D suggested a possible defect in 25-hydroxylation. However, when he was placed on large doses of vitamin D a serum 25-OH vitamin D concentration of almost 200 ng/ml was found, which is three to four times normal values. This indicated that 25-hydroxylation of vitamin D probably was not affected but that the increased requirement for vitamin D was at a later stage in the vitamin D cycle. This patient may also represent an example of a defect of 1-hydroxylation of 25-OH vitamin D, but serum 1,25-diOH vitamin D concentrations have not been determined. He has maintained remarkable clinical improvement on moderately large doses of vitamin D, 16,000

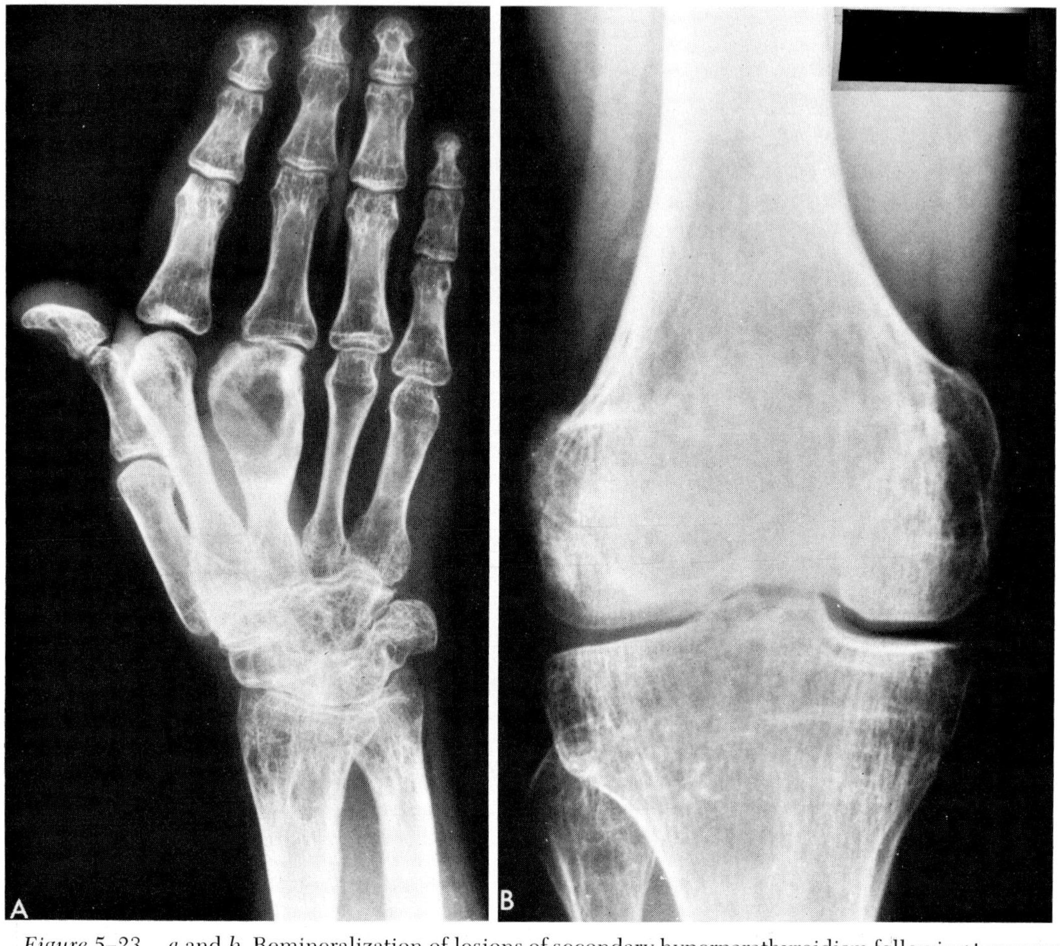

Figure 5–23. *a* and *b*, Remineralization of lesions of secondary hyperparathyroidism following two years of treatment, first with 25-hydroxyvitamin D and subsequently vitamin D (patient J. McK.).

units per day, but there is residual deformity of the small bones of the hand and forearm, which were the seat of extreme cystic change at the height of the reactive hyperparathyroidism, and ankylosis has occurred during the healing of the destructive bone lesions.

Following the discovery of 1,25-diOH vitamin D and the development of a method for determination of this vitamin D metabolite in serum, several clinical entities have been described in which rickets and secondary hyperparathyroidism have been associated with excessively high concentrations of 1,25-diOH vitamin D in serum, suggesting target cell resistance to the active vitamin D hormone. In one family studied by Finberg and Rosen two siblings were affected; both parents were free of disorders of calcium metabolism, indicating autosomal recessive inheritance. In addition to hypocalcemia, hypophosphatemia, generalized aminoaciduria, short stature and severe skeletal changes of rickets, both children had alopecia. The rickets was resistant to large doses of vitamin D, which raised serum 1,25-diOH vitamin D concentrations to levels greatly in excess of normal values. A young adult with chronic hypocalcemia, osteomalacia and secondary hyperparathyroidism has also been described in whom the concentrations of 1,25-diOH vitamin D were elevated above normal (Brooks et al). This patient showed biochemical evidences of healing when given 4000 units of vitamin D per day, with further increase of serum 1,25-diOH vitamin D concentration to 297 pg/ml (normal 20 to 55 pg/ml). It has been proposed that such patients be classified as a separate variety of vitamin D-dependent rickets, type II, resulting from abnormality of end-organ response to 1,25-diOH vitamin D. The possibility of a deficiency of cell cytosol receptors for the vitamin D hormone as a cause of refractoriness of target cell response is an attractive hypothesis but remains to be proved. As current methods for the measurement of vitamin D metabolites become more widely available it seems likely that new subgroups of rickets hitherto labeled as vitamin D-dependent or -resistant will be recognized.

DISORDERS OF CALCIUM AND PHOSPHATE METABOLISM SECONDARY TO RENAL INSUFFICIENCY

Pathophysiology

Bone disease, so-called renal osteodystrophy, is an important complication of chronic renal disease and can be a major disabling condition. As patients with progressive renal insufficiency have been given longer life expectancy by improved methods of management, including diet, hemodialysis and renal transplantation, the importance of understanding and preventing the disability of chronic bone disease in such patients has increased. This complication has a higher incidence in children and adolescents than in adults, and in several series 25 to 39 per cent of children with renal failure have had severe skeletal abnormalities. In addition to the specific skeletal lesions of renal insufficiency, retardation of skeletal growth and delay of skeletal maturation (retarded bone age) are characteristic of chronic renal disease. These latter changes may be the result not of the disorders of calcium and phosphate metabolism of renal insufficiency but of a separate problem, possibly reduced somatomedin production in the child with renal insufficiency. The total picture is complex, with multiple factors involved in the final abnormality of skeletal structure.

The lesions of the bone in renal insufficiency are the result of growth retardation, osteosclerosis, osteomalacia and secondary hyperparathyroidism. The fact that osteosclerosis can occur simultaneously with osteomalacia and secondary hyperparathyroidism may seem strange, but this lesion is seen predominantly in the vertebral column with increased density of the vertebral body end-plates. This, in association with diminished mineralization of the trabecular structure of the vertebral body, results in a striped appearance, the so-called rugger jersey spine. The mechanism of this osteosclerosis is uncertain. One explanation is that it results from the increased concentration of inorganic phosphate in extracellular fluid characteristic of the patient with reduced glomerular filtration rate. This osteosclerotic lesion is probably less pronounced in children than in the adult with renal disease.

The major lesion in the skeleton is the demineralization and destruction of bone structure resulting from secondary hyperparathyroidism. Patients with renal disease who have significant reduction of GFR have reduced concentrations of serum ionized calcium, which stimulates parathyroid hormone secretion and progressive hyperplasia of the parathyroid glands. The serum IPTH concentrations rise to values that may exceed those seen in primary hyperparathyroidism. As a result of this extreme secondary hyperparathyroidism severe destructive lesions of bone occur, and this is the dominant bone lesion of renal insufficiency. The basis of the intestinal calcium malabsorption and the reduced ionized calcium concentration in ECF is the abnormality of vitamin D metabolism associated with impairment of renal function.

As discussed in the section on vitamin D metabolism, the vitamin D metabolite most active in the stimulation of calcium transport by the intestine is 1,25-diOH vitamin D, which is the product of action of a renal tubule mitochondrial system on the substrate, 25-OH vitamin D. This system is controlled by PTH and inorganic phosphate concentrations in body fluids, and also of course requires intact functioning renal tubule cells. In renal insufficiency one of the early factors interfering with the production of 1,25-diOH vitamin D is hyperphosphatemia. As GFR decreases the balance between the quantity of phosphate filtered and that reabsorbed is altered so that serum phosphate concentration rises. This then reduces serum ionized calcium concentrations by formation of calcium phosphate associates in ECF as well as by inhibition of 1,25-diOH vitamin D formation. The resulting hypersecretion of PTH can cause a new equilibrium by reduction of renal tubular reabsorption of phosphate, but as GFR progressively decreases, depression of tubular reabsorption of phosphate by increased PTH concentrations cannot compensate for the reduction of filtered phosphate. With loss of functioning nephrons the mass of tubular cells also becomes insufficient to maintain 1,25-diOH vitamin D production, and progressive deficiency of this vitamin D metabolite results.

The net effect is a progressive downhill spiral of intestinal calcium malabsorption, reduced extracellular ionized calcium, parathyroid hyperplasia and hypersecretion, and demineralization and resorption of bone structure by the action of PTH, which at extraordinarily high concentrations is effective despite the deficiency of 1,25-diOH vitamin D.

Histologically the bones of patients with renal insufficiency also show some evidences of osteomalacia, i.e., accumulation of the unmineralized matrix, osteoid. The basis for the osteomalacia has been debated since mineralization of new matrix should be adequate in the presence of high concentrations of extracellular inorganic phosphate even if ionized calcium concentrations are reduced. It has been suggested that toxic factors accumulating in body fluids in renal insufficiency inhibit the mineralization process despite calcium and phosphate concentration products that should permit mineralization of matrix as it is formed. The chronic acidosis of patients with inadequately treated chronic renal disease may be one of the factors impeding mineralization, since the formation of bone mineral is an H^+ ion generating process, and H^+ ion acceptors are required to maintain a pH suitable for precipitation of bone mineral in the matrix.

Diagnosis

Although the osteodystrophy of chronic progressive renal disease is often a late manifestation it can develop with surprising rapidity in young infants with congenital malformations of the urinary tract, and may be full-blown by 1 year of age, as indicated in the case reports. The manifestations of renal bone disease are pain due to skeletal destruction and microfractures, and deformity resulting from lack of ability of the bone to resist the stresses of muscle pull and weight-bearing as well as destruction of the growing ends of the bone. If the renal osteodystrophy is detected early, or preferably prevented by proper treatment, there is little or no deformity, and the major skeletal change is retarded growth and delayed bone maturation. If neglected the common deformities are severe genu valgum, coxa vara, twisting and bending of the femurs and tibias, destruction of the distal end of the clavicles, and destruction of the small bones of the hands with ankylosis and wrist deformity. In extreme cases patients are bed-ridden and immobilized because of the bone pain and the marked twisting and bending deformities of the lower extremities.

The roentgenographic changes in the long bones can be confused with rickets because the destruction of the growing ends of the bones results in ragged unmineralized metaphyses that superficially resemble the irregular unmineralized or poorly mineralized chondro-osteoid of the growing ends of the bones in rickets (Fig. 5–24). In addition to the irregular destruction of the metaphyseal ends of the bones, there is irregularity of the cortex owing to subperiosteal resorption of bone. This is most striking at the proximal ends of the tibias along the medial surface (Fig. 5–25), in the phalanges and at the distal ends of the clavicles. The phalangeal destruction may result in marked irregularity and almost complete disappearance of the normal tufted ends of the terminal phalanges. As mentioned above, the vertebral bodies may be demineralized, but the terminal plates can be osteosclerotic, resulting in a striped appearance, the rugger jersey effect.

The biochemical abnormalities associated with renal bone disease are reduced serum calcium concentration, elevated serum phosphate and low serum bicarbonate. Since GFR is usually reduced to less than 50 per cent of normal before bone disease develops, the urea nitrogen and creatinine concentrations in serum are increased. Serum IPTH concentrations usually are markedly elevated. The serum alkaline phosphatase activity is increased, and there is a positive correlation between serum alkaline phosphatase and serum IPTH concentrations and radiologic and histologic evidence of bone resorption and osteitis fibrosa. Urine calcium excretion is reduced, correlating with poor intestinal calcium absorption. Serum 25-hydroxyvitamin D concentrations are variable depending on vitamin D intake, but concentrations of the renal metabolite, 1,25-dihydroxyvitamin D, are reduced.

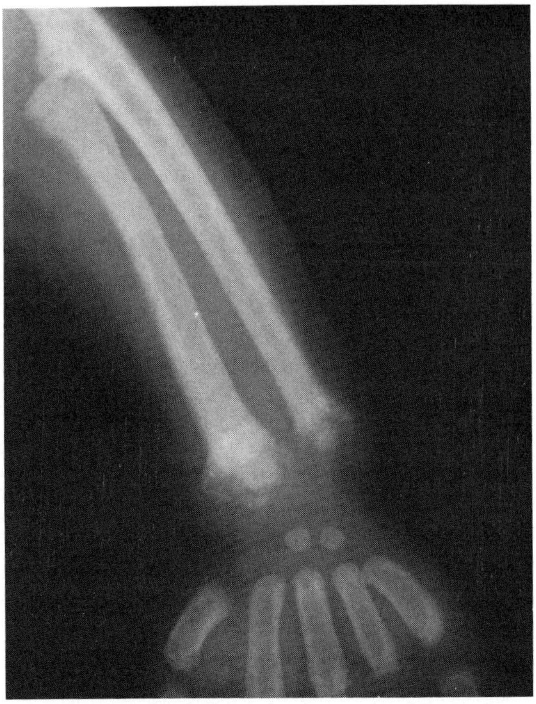

Figure 5–24. Lesions of hyperparathyroidism in growing bone, showing ragged unmineralized metaphyseal ends superficially resembling rickets.

Prevention and Treatment

Although not all the skeletal manifestations of uremic bone disease can be prevented or treated except via restoration of renal function by transplantation, the major bone lesions of secondary hyperparathyroidism can be prevented or reversed by proper management. Patients on hemodialysis particularly require careful management to minimize crippling renal osteodys-

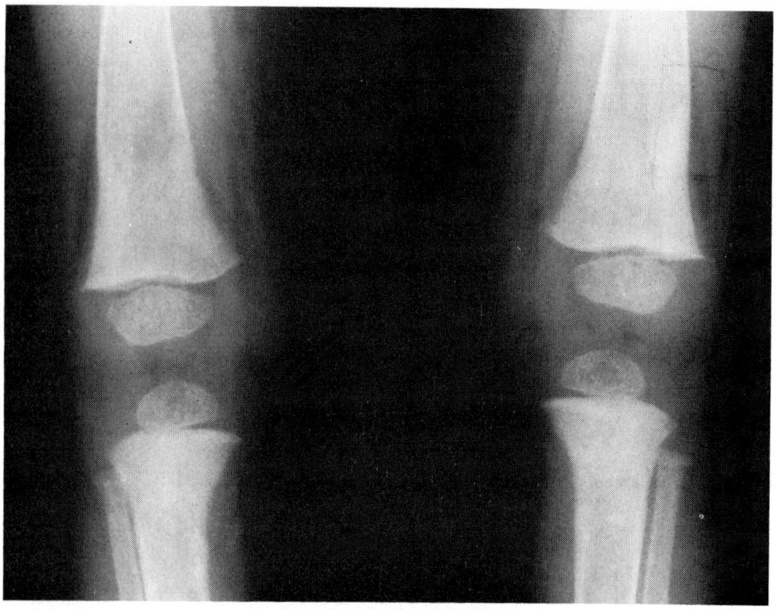

Figure 5–25. Subperiosteal erosions of medial surface of proximal tibia characteristic of hyperparathyroidism.

trophy, which might be the limiting factor in ambulatory and useful existence. Prevention of renal bone disease should begin as soon as there is evidence of chronic progressive renal disease with reduction of GFR below 50 per cent of the normal value (125 ml/mm per 1.73 m^2 surface area). This is particularly important in infants with congenital obstructive abnormalities of the urinary tract with hydronephrosis if both kidneys are involved, or congenital dysplasia of the kidneys such as renal hypoplasia, multicystic or polycystic kidneys, since as mentioned the skeletal lesions of secondary hyperparathyroidism can develop rapidly in such patients.

Dietary management is directed at prevention of chronic bicarbonate deficiency and hyperphosphatemia. The bicarbonate deficiency is the result of failure of excretion of H^+ ion, most of which is endogenous in origin as the result of oxidation of the sulfhydryl groups of amino acids to inorganic sulfate. A lesser amount is contributed by the hydrolysis of organic phosphate esters to inorganic phosphate. Reduction of protein intake serves to limit both endogenous H^+ ion production and phosphate load. However, an adequate protein intake is needed to maintain growth in infants and children. The protein intake can be reduced to provide about 7 to 8 per cent of the total calorie intake if the protein is of excellent biologic quality. In the infant this amounts to a minimum of 2 gm per kg, gradually decreasing in the older child to 1.5 to 1.0 gm protein per kg body weight. Although protein intakes of less than 1.0 gm/kg have been given to older children with far advanced renal insufficiency, it is questionable whether growth can be maintained with such low protein intakes. Egg white is an excellent protein because of its amino acid composition. The other animal proteins, milk, meat and fish, also have high biologic value but provide considerable amounts of sulfur-containing amino acids as well as phosphorylated amino acids, so that they must be given in limited amounts. In addition to reduction of endogenous H^+ ion production by low protein intake, additional buffer base is provided as sodium bicarbonate or sodium salts of organic acids which are metabolized to CO_2. We have used both sodium lactate and sodium citrate for this purpose. The quantity of sodium bicarbonate or sodium salts of organic acids must be titrated in each patient to cover the quantity of endogenous H^+ ion production that cannot be excreted because of failure of renal acidification of the urine. This may be in the range of 1 to 2 mEq/kg per day but obviously is a function of the diet.

The limitation of protein intake also reduces the phosphate load, but phosphate is present in all foods containing animal or vegetable cells, and of course is also found in high concentration in milk as a phosphoprotein or in egg yolk as phospholipids. It is not possible to limit phosphate intake severely without making the diet nutritionally inadequate or unpalatable. In order to prevent hyperphosphatemia, therefore, it is necessary to resort to phosphate sequestering agents that bind phosphate in the intestinal lumen and prevent its absorption. The phosphate binders that can be used are either calcium or aluminum salts, since both calcium phosphate and aluminum phosphate are insoluble at the pH of the small intestine. We prefer to use calcium as the initial binding cation since an excess of calcium promotes calcium absorption as well as binding phosphate.

In the young child a soluble calcium salt such as calcium lactate or calcium gluconate can be given dissolved in milk, fruit juice or flavored drink. Calcium lactate is only 13 per cent calcium so that 8 gm of this powder must be given for every gram of calcium to be supplied. Since it may be necessary to give 2 to 3 gm of calcium depending on the phosphate intake the quantity of calcium lactate powder needed is quite large, but can be tolerated by the child without gastrointestinal upset. Calcium gluconate powder is only 9 per cent calcium but a preparation of calcium gluconate is available, syrup of Neo-Calglucon, which supplies 92 mg calcium per teaspoonful. The needed dose, 10 to 20 teaspoonfuls, is not only expensive but may cause diarrhea, possibly because of the large quantity of sucrose in the syrup.

In the older child who will accept a flavored suspension, a 10 per cent suspension of calcium carbonate flavored with mint is an excellent vehicle for supplying the needed calcium. Calcium carbonate is 40 per cent calcium so that 1 gm of calcium is supplied by 2.5 gm of calcium carbonate or 25 ml of the suspension. In the older patient as much as 5 to 10 gm of calcium carbonate may be needed (2 to 4 gm of calcium) to bind the phosphate of even a restricted protein diet. The dosage is determined by the serum phosphate concentration, and the goal is to maintain this concentration in the range between 4.5 and 6 mg/dl in children under 1 year of age, with an upper limit of 5.5 mg/dl in older children. Concentrations much below 4.5 mg/dl are too low for growing children and excessive treatment with phosphate binding agents may superimpose hypophosphatemic bone disease, i.e., true rickets and osteomalacia, on the lesions of secondary hyperparathyroidism.

If the serum phosphate is not adequately controlled by calcium as a binding cation the addition of aluminum hydroxide or carbonate is indicated. Many use aluminum as the standard binding agent because it has been employed so widely and its effectiveness is demonstrated. However, in recent years some evidence has been presented that aluminum may be absorbed and may produce some toxic effects, and although serious toxicity seems unlikely in view of the widespread use of aluminum salts as an antacid as well as a phosphate binder, it seems wise to limit the quantity of aluminum if possible. The preparation may be either an aluminum hydroxide or aluminum carbonate suspension, but it should not be a magnesium-containing aluminum preparation since hypermagnesemia can result if one of the proprietary aluminum magnesium hydroxide preparations designed for antacid use is given to patients with renal insufficiency. The aluminum dose again is determined by the quantity required to maintain the serum phosphate concentration in the appropriate range for age. A dose of 135 ml aluminum hydroxide suspension or 18 300-mg tablets of aluminum hydroxide provides 100 mM aluminum. Ten gm of calcium carbonate supplies 4 gm calcium or 100 mM calcium. A phosphate intake of 600 mg phosphorus per day provides 20 mM of phosphate, and a several-fold excess of aluminum or calcium or both together is required to precipitate the phosphate and prevent its absorption. Because of the unpalatability of aluminum hydroxide it has been incorporated into cookies, which have been accepted by children.

Aluminum Hydroxide Cookies

1 cup salt-free butter
1½ cups sugar
1½ cups cake flour
¾ cup corn starch
1 egg
3 tsp vanilla or almond extract or 2 tsp cinnamon
100 tablets aluminum hydroxide tablets (Amphojel) powdered by grinding in blender
1 cookie = 2 aluminum hydroxide tablets

It is possible to make similar cookies with calcium carbonate powder as well, so that a suitable dose of calcium carbonate and aluminum hydroxide might be made available in this fashion to children who cannot be persuaded to take suspensions or tablets.

Even though serum bicarbonate concentrations are maintained in the normal range, and sporadically determined serum phosphate and calcium concentrations in the postabsorptive state are not outside the limits of normal, parathyroid hyperplasia and the bone lesions of excessive parathyroid secretion may develop in patients with moderate degrees of renal insufficiency with serum creatinine concentrations in the 3 to 5 mg/dl range.There are several possible explanations. Although fasting serum calcium and phosphate concentrations are not below or above the limits of normal respectively, the integrated serum phosphate concentration during the 24 hours may be increased because the rises following meals may be greater and more prolonged than in subjects with normal kidney function. Similarly, serum ionized calcium may remain at the lower limits of normal and not be increased following calcium intake, so that there is parathyroid gland stimulation because of failure of the small but significant increase of calcium ion concentrations owing to intestinal absorption of this ion following meals. Occasional determinations of total serum calcium also may not accurately reflect the integrated extracellular ionized calcium concentration to which the parathyroid gland cells are exposed during the whole day.

Another factor of importance in the control of parathyroid gland function is the concentration of the vitamin D metabolities which are the products of kidney cell function, 1,25-dihydroxyvitamin D and 24,25-dihydroxyvitamin D. Although the latter metabolite is considered to be the first step in the inactivation of vitamin D, there is evidence that both it and 1.25-dihydroxyvitamin D are taken up by the parathyroid gland cells and inhibit parathyroid gland hormone secretion directly, supplementing the action of 1,25-dihydroxyvitamin D through its effect on intestinal absorption of calcium and extracellular calcium ion concentration. There is, therefore, a specific role of the vitamin D metabolites that may explain the occurrence of secondary hyperparathyroidism in patients with renal insufficiency whose serum calcium concentrations have never been documented to be below 8.5 mg/dl.

The prevention and treatment of the secondary hyperparathyroidism of renal insufficiency requires, therefore, some way of compensating for the deficiency of the renal metabolites of vitamin D. Historically, the treatment of these patients with amounts of vitamin D greatly in excess of normal physiologic requirements has been successful in increasing serum calcium concentration and improving the radiologic evidences of secondary hyperparathyroidism. This treatment probably works by increasing the concentration of 25-hydroxyvitamin D in serum, since the hepatic metabolism of vitamin D is not closely controlled and concentrations of 25-hydroxyvitamin D five times or more greater than normal can be produced by high intake of vitamin D.

There is evidence that 25-hydroxyvitamin D is a physiologically active metabolite that can function in the anephric subject. It is possible, then, that in high concentrations it can suppress parathyroid gland activity even though the deficiency of the renal metabolites of vitamin D is not corrected. The problem with the use of the supraphysiologic amounts of vitamin D is the danger of excessive concentrations of 25-hydroxyvitamin D with hypercalcemia, hypercalciuria and metastatic calcification in blood vessels and the kidney, further accelerating the downward course of renal function. Much of the vitamin D when given in large doses is stored in adipose tissue and circulates in the plasma, acting as a continued substrate for increased 25-hydroxyvitamin D production. Once toxic effects are recognized discontinuance of further vitamin D dosage does not remove this store of vitamin D in the tissues, which may remain for months and cause continued toxicity for this prolonged period.

It would be desirable to provide the metabolites of vitamin D which have a much shorter biologic half-life than vitamin D and are active at much lower doses, rather than giving vitamin D itself. Studies have been made of 25-hydroxyvitamin D, 1α,25-dihydroxyvitamin D and 1α-hydroxyvitamin D for their effectiveness in treating renal bone disease. The last compound, 1α-hydroxyvitamin D, is a synthetic vitamin D which is hydroxylated by the liver to 1α,25-dihydroxyvitamin D so that it is essentially a source of the active renal metabolite. In a number of experimental studies, 25-hydroxyvitamin D in doses of 25 to 200 μg per day has been effective in raising serum 25-hydroxyvitamin D concentrations and in healing the bone lesions of secondary hyperparathyroidism. Overdosage with 25-hydroxyvitamin D also leads to hypercalcemia, but its concentration in serum is directly related to the dose, and its disappearance is rapid after treatment is discontinued.

Similar studies with 1,25-dihydroxyvitamin D and 1α-hydroxyvitamin D have shown that these compounds can reverse the lesions of secondary hyperparathyroidism in doses of 0.25 to 1.0 μg of 1,25-dihydroxyvitamin D daily and slightly greater amounts of the 1 α-hydroxyvitamin D. These compounds can also produce hypercalcemia in these doses, requiring close monitoring of patients, but the half-life of these compounds is even shorter than that of 25-hydroxyvitamin D so that the effects of excessive dosage are dissipated within a few days after treatment is stopped. 1,25-dihydroxyvitamin D is now approved and is available for the treatment of patients with renal insufficiency.

There is a vitamin D-related compound that can be used effectively as a substitute for pharmacologic doses of vitamin D in patients with renal insufficiency. This is dihydrotachysterol which, like vitamin D_2, ergocalciferol, is produced by the ultraviolet irradiation of ergosterol. The actual compound produced in the irradiation process is tachysterol, which is not stable enough to be used as a drug but is reduced to dihydrotachysterol, which is quite stable. Dihydrotachysterol differs from vitamin D in that the A ring is rotated 180° so that the 3-OH group is in the geometric position of the 1-OH group of 1,25-dihydroxyvitamin D (see Chapter 2). Dihydrotachysterol can be considered as a pseudo-1-hydroxy compound, and because of this stereochemical configuration it is not acted upon by the kidney and is active in the absence of the kidneys. Dihydrotachysterol is metabolized by the liver to 25-hydroxydihydrotachysterol and this is probably the active compound. Dihydrotachysterol is effective in increasing calcium absorption, raising serum calcium concentrations and suppressing secondary hyperparathyroidism in patients with renal insufficiency, in doses much less than those of vitamin D and probably not much greater than those of 25-hydroxyvitamin D. Like the latter its effects are dissipated much more rapidly than those of vitamin D when treatment is discontinued. We prefer dihydrotachysterol* to vitamin D because of the obvious advantage

*Dihydrotachysterol has not been used as consistently in the prevention and treatment of renal osteodystrophy as its advantages would suggest. One reason has been the high cost of the pharmaceutical preparation, Hytakerol. It is possible to purchase crystalline dihydrotachysterol at a relatively low cost. This material can be dissolved in oil such as sesame oil (which has the advantage that it does not become rancid) to make up a solution providing 1 mg/ml.

of shorter duration of activity; the necessary dose can be more readily determined, and if an excessive effect is obtained the toxicity is of short duration once the drug is discontinued.

The dosage of dihydrotachysterol in children with renal disease is not much greater than that of 25-hydroxyvitamin D. In infants we have used 0.05 mg, and the doses have been graded up to 0.5 mg per day in older children. These doses must be based on individual study of patients. One of the important indices is the urine calcium excretion. The low urine calcium excretion of patients with moderate renal insufficiency is increased when intestinal calcium absorption is enhanced by any of the vitamin D preparations. The urinary calcium should be maintained in the range of 25 to 150 mg/24 hours, but should not exceed 3 mg/kg/day or 0.18 mg calcium per mg creatinine excreted. These should be safe limits since the upper limit of normal urine calcium excretion in children with normal renal function is 4 mg/kg/day or 0.25 mg per mg creatinine.

The serum calcium and phosphate concentrations must also be monitored. It must be stressed that reduction of serum phosphate to the normal range must be achieved before treatment with vitamin D or vitamin D metabolites or analogues is begun. In the presence of hyperphosphatemia increased intestinal absorption of calcium due to activated sterol treatment can raise the Ca × P product to values at which precipitation of calcium in nonosseous tissue occurs. The upper limit of the safe Ca × P product is often given as 70, but this value may differ at different ages since this product in the rapidly growing, prematurely born infant may exceed this value without metastatic calcification. For practical purposes, in infants and children with renal insufficiency, this upper limit of 70 can be a useful guideline. Serum ionized calcium determinations would be more suitable as a measure of safe calcium concentrations than total serum calcium, since protein-bound calcium may be low in patients with renal disease, but in the absence of the ready availability of such determinations the total serum calcium must be evaluated in relation to serum protein concentration, and the dose of activated sterol adjusted so that the corrected serum calcium does not rise above the normal limits. By the use of both urine and serum calcium determinations the dose of dihydrotachysterol or of 1,25-dihydroxyvitamin D can be adjusted to maintain extracellular calcium concentrations within the range adequate to suppress parathyroid gland overactivity without producing extraskeletal calcification.

As renal insufficiency progresses and hemodialysis becomes necessary, the control of renal bone disease demands careful attention to serum calcium, since urine calcium excretion is no longer a measure of the increased extracellular concentration of calcium due to intestinal absorption of calcium. In such patients prevention of secondary hyperparathyroidism is of crucial importance, since bone pain and deformity may vitiate the goals of dialysis therapy, namely, comfortable and useful life. Moreover, if these patients eventually undergo renal transplantation, the parathyroid hyperplasia may be so great at the time of transplantation that suppression of parathyroid hypersecretion by the normal vitamin D metabolism of the functioning transplant may not be possible.

Under these conditions post-transplant hyperparathyroidism with marked hypercalcemia has occurred. This degree of parathyroid hyperplasia should have been prevented by early treatment. If, however, a patient has developed severe parathyroid hyperplasia with advanced bone disease and the biochemical manifestations of very high concentrations of serum IPTH and alkaline phosphatase, parathyroidectomy may be indicated whether the patient is to undergo renal transplantation or prolonged dialysis. In either case the parathyroid hyperplasia and bone disease may no longer be controllable by medical management, and hypercalcemia occurs when activated sterols are given without suppression of parathyroid activity. It must again be stressed that careful patient management in the early stages of renal insufficiency should prevent the complication of extreme parathyroid hyperplasia and avoid the need for subtotal parathyroidectomy in the control of renal osteodystrophy.

CASE REPORTS

K.H. was the 5¼-lb product of a full term pregnancy complicated at five months by urinary tract infection. At delivery polyhydramnios was noted. The infant gained weight poorly and was discharged at 3 weeks of age weighing 4½ lb. At 6 weeks she was admitted because of acidosis and dehydration and was found to be azotemic. Studies including intravenous pyelogram were compatible with a diagnosis of congenital renal hypoplasia. During the next year she had a stormy course and at least one episode of *E. coli* infection of the urinary tract. At 18 months of age she was a miserable child in severe pain, unable to sit or bear weight on her extremities because of severe bone disease. She was transferred at this time to the Johns Hopkins Hospital. The bone roentgenograms are shown in Figure 5–26*a* and *b*. She was treated systematically with a moderate restriction of protein intake, sodium and potassium citrate solution (Polycitra) to maintain serum bicarbonate concentrations in the normal range, calcium lactate or gluconate supplements, and low doses of dihydrotachysterol, 0.05 to 0.1 mg. It was possible to keep serum calcium, phosphate and bicarbonate concentrations in the normal range, and progressive improvement in the bone changes followed, as seen in Figure 5-27*a*, *b* and *c*. Subsequently the only treatment needed has been a moderately restricted protein intake (45 gm/day), Polycitra 30 ml per day, calcium gluconate equivalent to 1000 mg calcium per day, and low doses of dihydrotachysterol. At age 15 years she is an energetic but small pubertal girl who has some decrease of exercise tolerance but no bone pain or deformity. The only urinary symptom is polyuria. At a recent evaluation the following serum electrolyte concentrations were found: sodium 142 mEq/l, chloride 110 mEq/l, CO_2 20 mEq/l, potassium 4.0 mEq/l, SUN 36 mg/dl, creatinine 3.6 mg/dl. calcium 10.6 mg/dl, phosphorus 4.5 mg/dl and alkaline phosphatase 90 IU; urine calcium excretion 7.3 mg/24 hrs. Because of the low urine calcium output despite calcium supplement, dihydrotachysterol was prescribed, 0.1 mg/day.

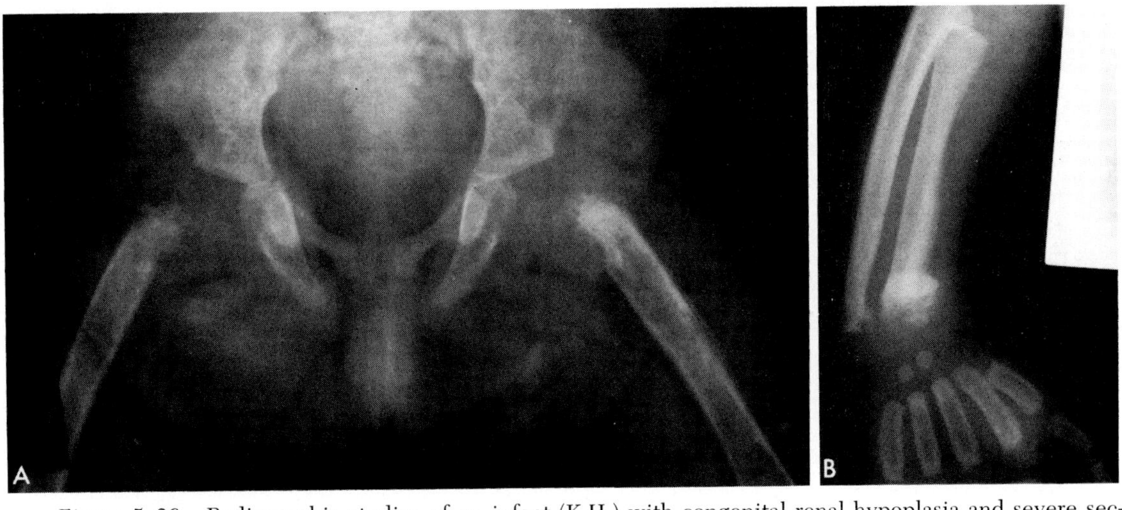

Figure 5–26. Radiographic studies of an infant (K.H.) with congenital renal hypoplasia and severe secondary hyperparathyroidism at age 18 months: *a*, destructive lesions of femoral necks; *b*, wrist showing "rickets-like" irregularity of metaphysis.

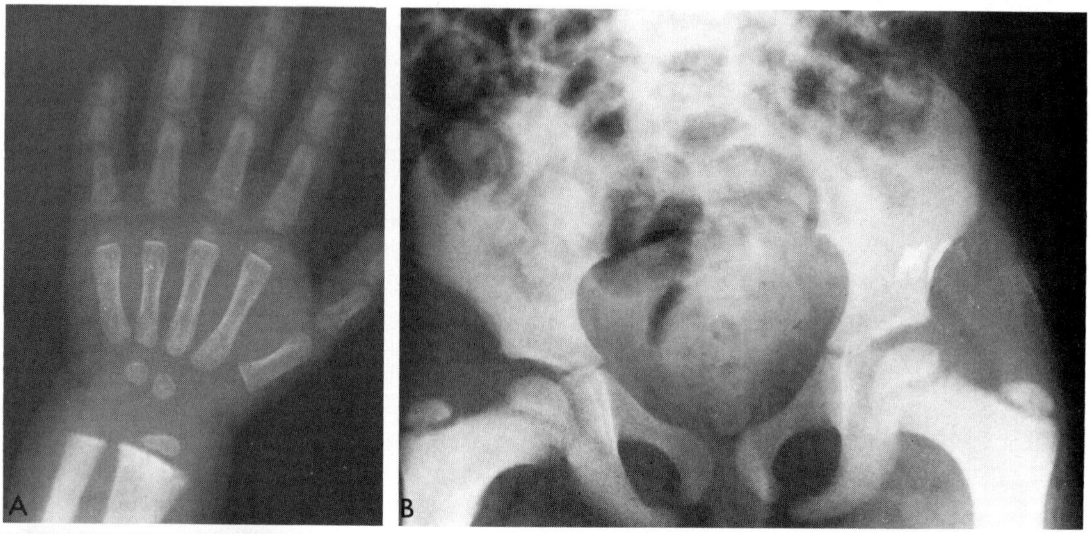

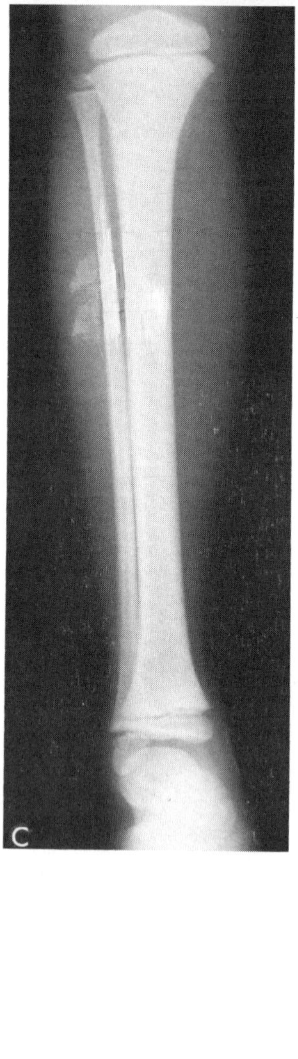

Figure 5–27. Radiographs of patient (K.H.) showing restoration of normal bone architecture following treatment directed at suppression of excessive parathyroid hormone secretion: *a*, wrist at 2½ years—normal bone architecture; *b*, age 6 years—hip joints showing remineralization of femoral necks; *c*, age 6 years—tibia—knee and ankle structure normal—no subperiosteal erosion.

COMMENTS: The skeletal lesions of secondary hyperparathyroidism in this child were evident at 10 months of age and were marked at 18 months, at which time there were severe symptoms of bone demineralization. The secondary hyperparathyroidism was successfully treated and there are no residual skeletal deformities at 15 years of age. GFR has remained stable at approximately 30 to 40 ml/min./1.73M^2. This case illustrates the great propensity of the infant and young child to develop the skeletal lesions of secondary hyperparathyroidism and the rapidity with which such lesions progress at this age. The successful treatment and prevention of recurrence of renal osteodystrophy illustrates the value of preventive treatment in the child with impairment of renal function.

T.T., a black male infant, was admitted at 14 months of age. During most of the first year of life he had had repeated episodes of fever and apparent dysuria. At 11 months he was found to have bladder neck obstruction with marked hydronephrosis, and was treated by a vesicostomy. He failed to do well, and when admitted to the Johns Hopkins Hospital had severe urinary tract infection with a variety of organisms. A diagnosis of posterior urethral valve was made, but because of the marked ureteral dilatation bilateral loop ureterostomies were done. Skeletal x-rays at this time were normal (Fig. 5–28). The immediate postoperative course was stormy, but on discharge at 17 months of age he was markedly improved. At this time SUN was 35 mg/dl, HCO_3^- 22 MEq/1, and creatinine clearance 19 ml/min/1.73 M^2. He seemed to be doing satisfactorily at home and had been given only ordinary prophylactic amounts of vitamin D plus sodium citrate for prevention of bicarbonate deficit. He began walking at 18 months of age but stopped at 22 months. Radiographs at that time showed marked secondary hyperparathyroidism (Fig. 5–29). Treatment with dihydrotachysterol, 0.2 mg/day, and calcium lactate supplement was started, but soon afterwards he fell from a chair, fractured the left femur and was readmitted.

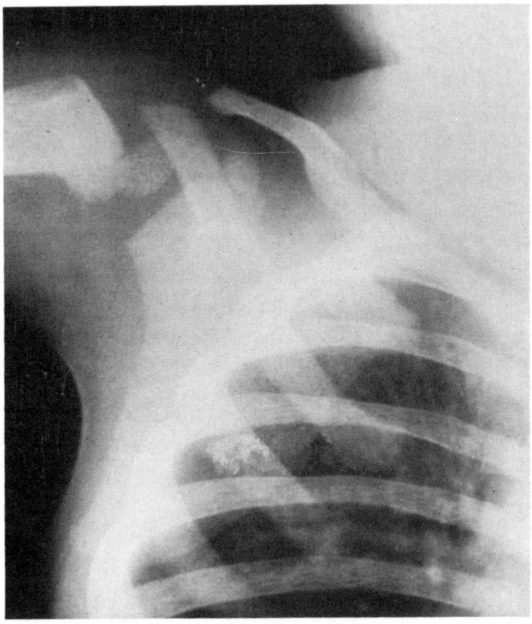

Figure 5–28. Radiographic studies of patient (T.T.) at 11 months of age, with posterior urethral valve and unilateral hydronephrosis. No evidence of secondary hyperparathyroidism present.

Examination showed a chronically ill, somewhat pale 22 month old child, weight 8.5 kg, with left leg in skeletal traction. Laboratory data: Hct. 29.6, serum electrolyte concentrations, Na 142, K 4.2, Cl 104, and HCO_3^- 22 mEq/1. Urea nitrogen 30 mg/dl, Ca 9.7 and P 5.2 mg/dl. Urine cultures — Klebsiella.

Treatment was continued with dihydrotachysterol, 0.2 mg per day, calcium lactate, 5.4 gm per day (equivalent to 700 mg calcium) and sodium citrate in amounts adequate to maintain normal serum bicarbonate. The urinary tract infection was treated with cephalosporin. The fracture healed satisfactorily, and after four weeks of traction a cast was applied and the patient was discharged on the same dose of dihydrotachysterol and calcium lactate. At discharge serum findings were: Ca 10.5 mg/dl, urea nitrogen 13 mg/dl, Na 145, K 4.8, Cl 106, and HCO_3^- 27 mEq/1.

COMMENTS: This child with congenital obstruction of the bladder outlet and se-

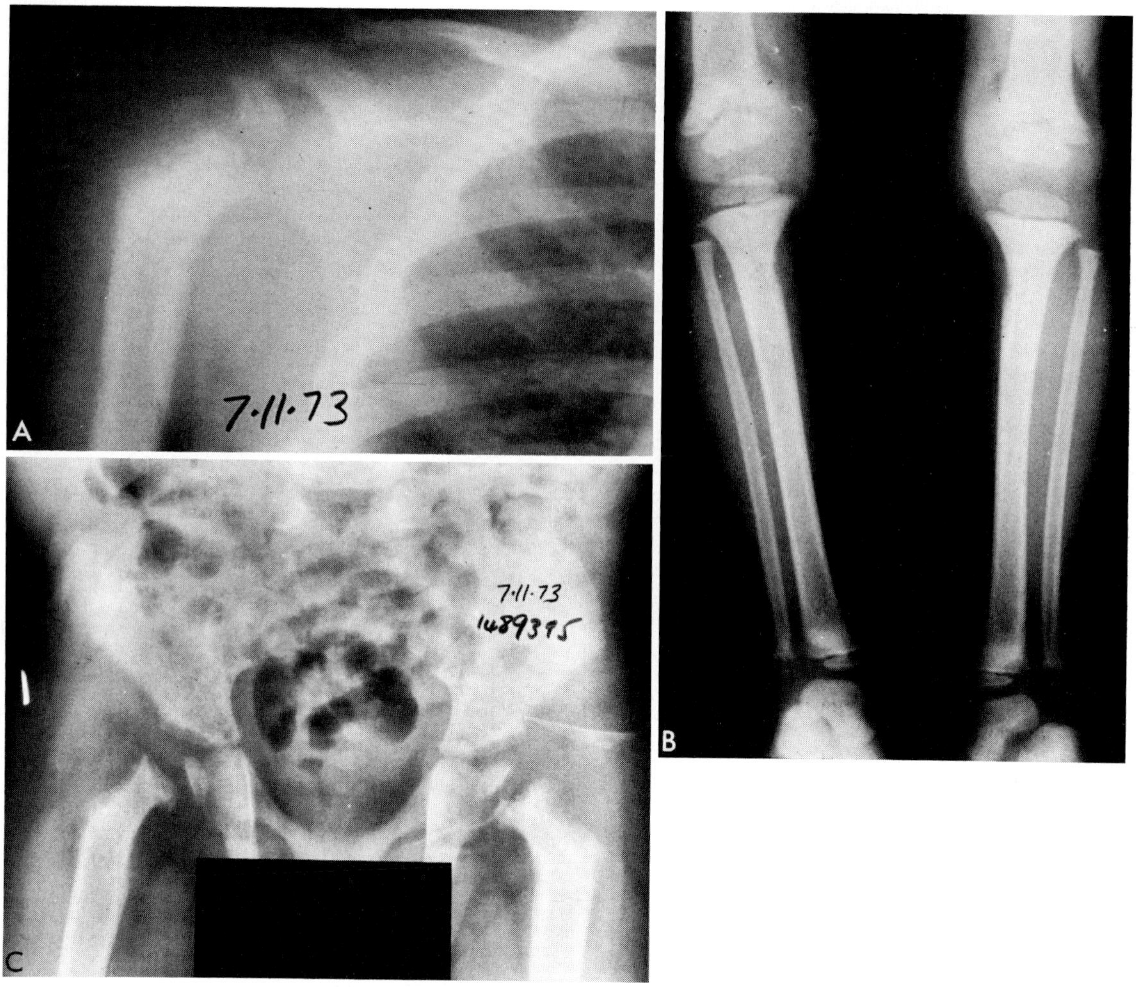

Figure 5–29. *a*, *b*, and *c*, Radiographs of patient (T.T.) at 22 months of age, showing marked evidences of secondary hyperparathyroidism.

vere bilateral hydroureter and hydronephrosis was given adequate drainage of the urinary tract and treated for urinary tract infection. However, no specific steps were taken to prevent the secondary hyperparathyroidism of major renal parenchymal injury and presumed failure of 1,25-diOH vitamin D production. As the radiographs dramatically demonstrate, the skeletal changes of secondary hypoparathyroidism were not present when he was studied at 11 months of age but were far advanced at 22 months. Although treatment was successful in reversing these changes, proper management should have prevented these destructive lesions of the skeleton. The physicians supervising his care may have been overly optimistic that recovery of renal function following ureteral drainage and antibiotic therapy of infection was sufficient to prevent the abnormality of vitamin D metabolism characteristic of renal injury.

CALCIUM DEFICIENCY RICKETS

This is a rare form of iatrogenic rickets resulting from the prescription of special milk-free diets because of lactose intolerance, milk protein sensitivity or chronic diarrhea of unknown cause in which milk has been eliminated. If these diets are formulated from strained meats as a source of protein and calcium salts are not added, it is possible to prepare a diet so low in calcium as to produce calcium deficiency. The consequence of this is stimulation of parathyroid hormone secretion and secondary hyperparathyroidism, with hypophosphatemia as well as hypocalcemia. The urine as expected shows a generalized aminoaciduria compatible with hyperparathyroidism. Radiographically the lesions are indistinguishable from rickets although presumably there is a considerable element of bone resorption. Serum 25-OH vitamin D concentrations in one such patient were normal, ruling out concomitant vitamin D deficiency. Treatment is directed at institution of a proper diet that provides normal calcium requirements.

There is also a puzzling form of rickets reported from South Africa. The patients are black children subsisting on diets of corn and leafy vegetables. The possibility of an anti-vitamin D factor blocking the normal metabolism of vitamin D has been proposed, but it seems more likely that this diet is severely deficient in available calcium. Treatment with increased calcium intake without supplementary vitamin D has produced healing of the rickets.

TYPE II RICKETS

HYPOPHOSPHATEMIA SECONDARY TO A PRIMARY DISORDER OF RENAL TUBULAR REABSORPTION OF PHOSPHATE

Hypophosphatemia, even in the presence of normal stores of vitamin D and normal concentrations of vitamin D metabolites including 1,25-dihydroxyvitamin D, can cause rickets and osteomalacia if the concentration of inorganic phosphate is too low to permit formation of the initial amorphous calcium phosphate precipitate, $Ca_3(PO_4)_2$. Such reduction of inorganic phosphate concentrations does result from abnormalities of renal tubule transport of phosphate in which there is inadequate tubular retrieval of filtered phosphate. There are a variety of renal tubular disorders in which renal tubular reabsorption of phosphate is deficient, and these can be listed under the headings of Fanconi Syndrome, Renal Tubular Acidosis and Primary Renal Tubular Hypophosphatemia.

FANCONI SYNDROME

This eponym includes a variety of disorders of renal tubular function in which multiple transport functions of the tubule are disturbed, but always including impaired tubular reabsorption of glucose, amino acids and phosphate, resulting in the basic triad of renal glycosuria, aminoaciduria and hypophosphatemia. The term *renal hypophosphatemia* rather than *renal phosphaturia* is used advisedly since, unlike glucose and amino acids which are almost completely reabsorbed by the renal tubule under normal physiologic conditions, tubular reabsorption of phosphate is not complete except when the serum concentration (and therefore the filtered load) is reduced. The actual quantity of phosphate excreted in the urine may be no greater in patients with renal tubular hypophosphatemia than in normals, but this same quantity of phosphate is excreted at much lower concentrations of serum phosphate, representing a marked reduction of tubular reabsorption of phosphate. Urine phosphate excretion must always be related to serum phosphate concentration if disturbances of tubular reabsorption of phosphate are to be diagnosed (see Chapter 2).

The Fanconi syndrome results from renal tubular injury due to a number of toxic agents both of endogenous and exogenous origin, as indicated in Table 5-16.

Nephropathic Cystinosis

This is probably the most important single metabolic derangement causing the Fanconi syndrome in childhood. It is an inheritable disorder transmitted as an autosomal recessive. The specific enzymatic abnormality is still unknown but the primary metabolic consequence of the abnormal gene is the accumulation of cystine in reti-

Table 5–16. Etiology of Fanconi Syndrome

A. ENDOGENOUS INJURIOUS FACTORS
  1. Inborn Errors of Metabolism
    (a) Cystinosis-nephropathic and intermediate
    (b) Tyrosinosis
    (c) Lowe's syndrome
    (d) Wilson's disease
    (e) Fructose intolerance
    (f) Idiopathic Fanconi syndrome
  2. Tumors
    Multiple myeloma
B. EXOGENOUS TOXIC AGENTS
  1. Heavy metals – Pb, Hg, Cd
  2. Organic phenolic compounds
  3. Tetracycline degradation products
  4. Maleic acid (experimental)

culoendothelial cells and granulocytes. This accumulation is thought to be within lysosomes since the cystine aggregates are found in membrane-bound cell organelles. The concentration of cystine in plasma is not increased, however, so that its source is of cell metabolic origin. The intracellular cystine is found early in bone marrow, the retina and cornea of the eye, the kidney and the liver. The retinal accumulation is associated with a mild retinopathy which is important only as a diagnostic clue but does not lead to serious visual disturbance. The corneal accumulation causes some photophobia, probably owing to the reflection of light by cystine crystals, but there is no severe keratitis. The bone marrow and liver accumulation of cystine is of diagnostic significance but does not cause serious functional derangements of these tissues. The high concentration of cystine in kidney, however, causes severe cellular injury, first functional injury to renal tubule cells manifest by disturbances of renal tubule transport functions, and then by disappearance of cells with interstitial fibrosis and eventual loss of functioning nephrons.

Symptoms and Signs. Infants with cystinosis are apparently normal at birth, although measurement of cystine in the separated white blood cell layer would show an increased concentration of this amino acid. However, histologic study does not reveal cystine crystals in the tissues. Renal function is apparently normal and glycosuria and aminoaciduria are not present in the first few months of life. Ophthalmoscopic examination may reveal a retinopathy with clumping of pigment. At several months of age the deposition of cystine crystals in the cornea can be detected by slit lamp microscopy although the cornea does not show any abnormalities on ordinary examination. Formation of pigment in skin and hair is deficient in this disorder and the children are often strikingly fair-skinned and blond. By 4 to 6 months of age renal tubular disorders are present. An early manifestation on ordinary urine examination often is renal glycosuria, and at this time studies of amino acid excretion reveal a generalized aminoacidura (Fig. 5-30). A mild proteinuria may be present. More detailed examination of the urine protein shows it to consist of low molecular weight proteins characteristic of tubular proteinuria. Serum phosphate concentrations are reduced but serum calcium is normal.

These are all manifestations of proximal tubular injury. However, disturbances of distal tubule function are also prominent. These may be secondary to failure of sodium, chloride, bicarbonate and water reabsorption in the proximal tubule or loop of Henle, so that excessive volume of glomerular filtrate is dumped into the distal tubule; this causes an abnormal rate of potassium exchange for sodium, with kaliuresis and hypokalemia plus failure of urine concentration with Pitressin-resistant diabetes insipidus. In some patients marked polyuria is the earliest obvious symptom of the renal tubular disorder. By the end of the first year of life the tubular transport abnormalities are well-developed; because of the electrolyte abnormalities growth is retarded and some patients may present to the physician because of dwarfism and polyuria. In others the hypophosphatemia is the important problem and the major presenting complaint can be the skeletal deformity of rickets. If the growth retardation is extreme, rickets may not be present, since growth of the skeleton is needed for the deficiency of inorganic phosphate to manifest itself by the characteristic failure of mineralization of the growing bone.

The serum electrolyte findings at this stage are usually a normal concentration of

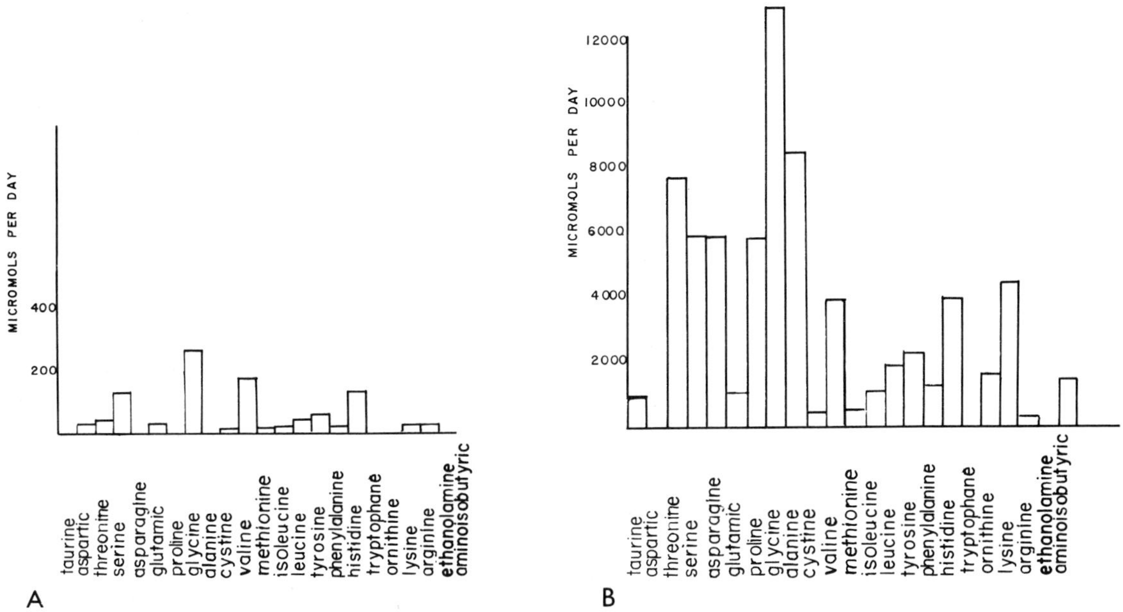

Figure 5–30. Aminoaciduria in Fanconi syndrome: *a*, pattern of amino acid excretion in urine of normal infant; *b*, pattern of amino acid excretion in urine of infant with Fanconi syndrome secondary to cystinosis. (From Harrison, H. E. and Harrison, H. C.: Aminoaciduria in relation to deficiency diseases and kidney function. J.A.M.A. *164*:1571, 1957. Reprinted with permission.)

serum sodium with reduced serum bicarbonate and elevated serum chloride. The serum phosphate concentration is low as is the potassium concentration. The serum urea nitrogen and creatinine concentrations are normal, consistent with a normal glomerular filtration rate.

Diagnosis. The diagnosis is suggested by the evidences of the Fanconi syndrome in a young child, i.e., renal glycosuria, aminoaciduria and hypophosphatemia in addition to the other evidences of tubular dysfunction: potassium wasting, bicarbonate loss and failure of concentration. Slit lamp microscopy of the cornea by an experienced ophthalmologist will almost always be diagnostic, showing the characteristic pattern of light scattering by cystine crystals in the cornea. Bone marrow aspirates also reveal clumps of cystine crystals (Fig. 5-31) which on careful examination are usually found to be within macrophages swollen with cystine. If renal biopsy is performed similar masses of cystine crystals are found. Additional pathologic findings in the kidney may be minimal initially, with interstitial fibrosis and tubular atrophy and dilatation at a later stage. Biochemical determinations of cystine can also be made in the separated WBC layer of blood and the concentration is markedly increased. The diagnosis of cystinosis clearly is not a difficult one.

Clinical Course. Once the renal tubular manifestations of cystinosis have appeared the picture is one of progressive renal injury, so that the initial manifestations of limited disorders of renal tubular transport systems are gradually replaced by progressive destruction of nephrons, and by 8 to 10 years of age almost all patients show severe renal insufficiency. As the reduction of GFR progresses, the disorder of calcium and phosphate metabolism changes from one of primary hypophosphatemia due to renal wastage of phosphate to one of phosphate retention and failure of 1,25-dihydroxyvitamin D production with hypocalcemia and secondary hyperparathyroidism. The picture of renal osteodystrophy replaces that of rickets and osteomalacia. Nephropathic cystinosis invariably leads to the terminal stages of renal insufficiency, and patients can only survive through hemodialysis or renal transplantation.

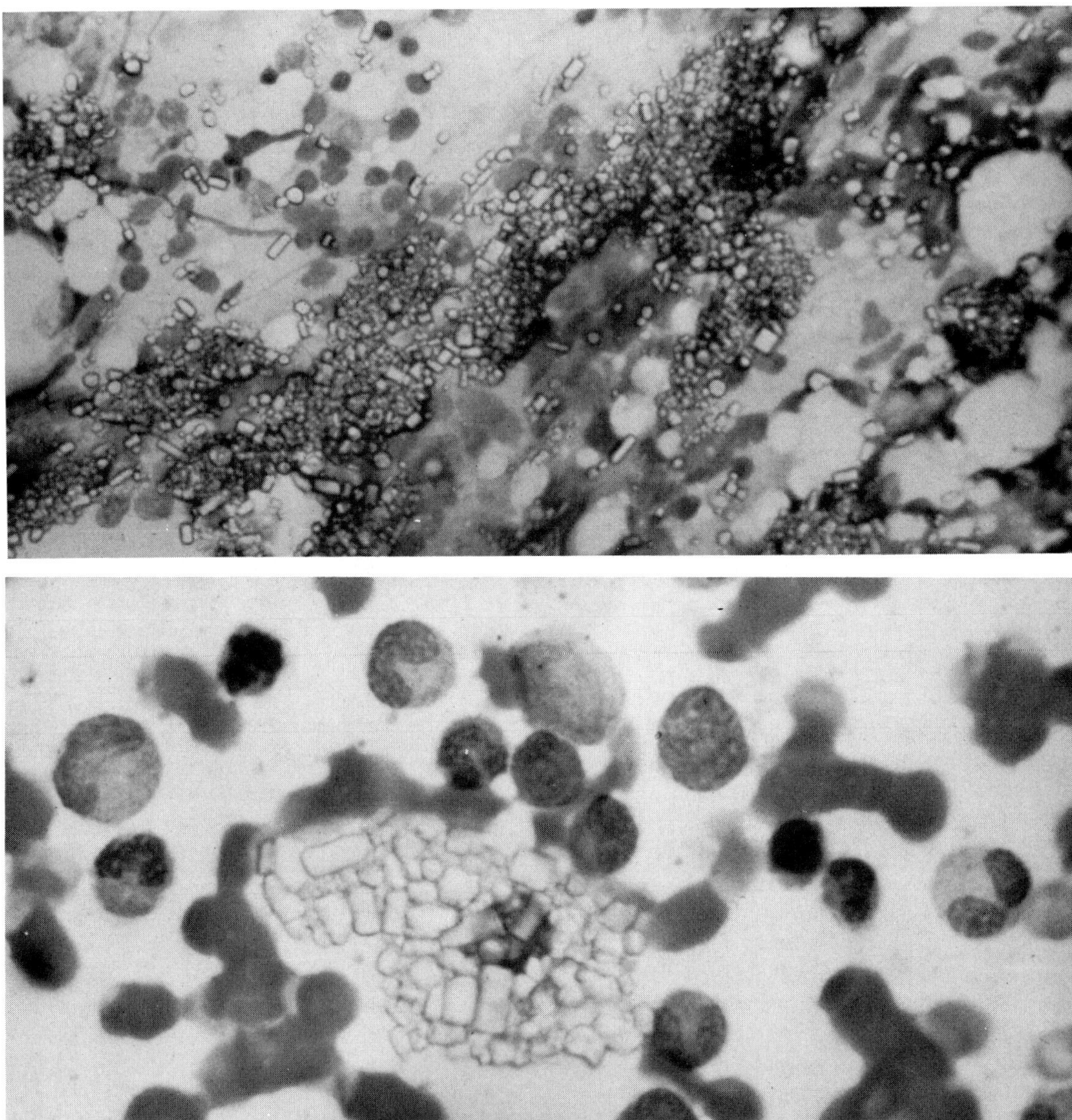

Figure 5–31. Bone marrow aspirates from children with cystinosis. Low and high power photomicrographs showing cystine crystals appearing as hexagonal plates within macrophages.

The extreme susceptibility of the kidney to the toxic effects of accumulated cystine poses somewhat of a mystery since bone marrow and liver in which cystine-laden macrophages also accumulate do not show major pathologic changes. It may be that this is simply a concentration difference with much higher concentrations of cystine occurring in kidney than in other tissues. There is also cystine accumulation in the thyroid gland and clinical hypothyroidism has been seen in cystinotic children. Determinations of serum thyroxine concentration should be made in these patients to detect impairment of thyroid gland function. An even more sensitive method of screening for such impairment is the determination of thyroid stimulating hormone (TSH) since this can be increased before circulating thyroxin concentrations are low enough to be diagnostic.

Treatment. There is as yet no specific treatment for cystinosis. As mentioned before, the specific metabolic or enzymatic defect responsible for accumulation of cystine in phagocytic cells is unknown. High concentrations of reducing substance such as ascorbic acid can remove cystine from cystinotic cells in vitro and an investigational protocol for the treatment of patients with nephropathic cystinosis with large amounts of ascorbic acid was tried, but there was no evidence of alteration of the course of the disease.

The hypophosphatemic rickets that may be the major presenting manifestation in some subjects can be ameliorated by treatment. As in other forms of renal acidosis, these patients require an excess of cation intake to provide buffer base lost because of the inability of the renal tubules to reabsorb bicarbonate or excrete the H^+ ion which is produced endogenously by metabolism of sulfur-containing amino acids. Since these patients also have marked potassium-losing renal tubular dysfunction a preparation providing both sodium and potassium bicarbonate is needed. We use for this purpose a solution of sodium citrate and potassium citrate (Polycitra) which provides 1 mEq/ml each of sodium and potassium. Since the citrate is rapidly metabolized to CO_2 this is essentially a source of sodium and potassium bicarbonate. The commercial preparation is flavored and the taste of the citrate salts can be disguised more effectively than that of the bicarbonate salts, so that the preparation is more acceptable to children and there is no gastric discomfort. The dose must be adjusted to the patient's loss of bicarbonate and is usually 1 to 3 mEq/kg each of Na and K given in divided doses during the day. If the serum bicarbonate concentration is corrected but serum potassium remains low, additional potassium can be given as potassium citrate, phosphate or chloride to maintain potassium balance.

The hypophosphatemic rickets of cystinosis can be healed or ameliorated by the combination of increased phosphate intake and pharmacologic doses of vitamin D of the order of 25,000 units per day, or dihydrotachysterol 0.1 to 0.2 mg/day. The therapeutic effect of large doses of vitamin D may seem puzzling in view of the presumption that vitamin D metabolism is not primarily altered in cystinosis, but that renal tubular reabsorption of phosphate is reduced despite normal concentration of vitamin D metabolites. There are several possible explanations: one being that suppression of normal PTH output by the increased intestinal absorption of calcium produced by high concentrations of 25-hydroxyvitamin D or by 25-hydroxydihydrotachysterol eliminates the inhibitory action of physiologic concentrations of PTH on one component of tubular transport of phosphate, and thereby permits increased renal tubular reabsorption of phosphate. It is also possible that the concentrations of 25-hydroxyvitamin D or 25-hydroxydihydrotachysterol attained by this treatment could directly stimulate phosphate reabsorption and overcome the inhibitory action of cystinosis. There is also some evidence that the activity of the renal enzyme system that metabolizes 25-hydroxyvitamin D to 1,25-dihydroxyvitamin D is reduced by the tubular injury of cystinosis, so that there is some deficiency of 1,25-dihydroxyvitamin D.

By combining the vitamin D or dihydrotachysterol treatment with increased phosphate intake, higher serum phosphate concentrations can be achieved with more complete healing of rickets and possibly some improvement of growth. Since these children require high potassium intake the phosphate load should be given as a solution of potassium phosphate such as Neutra-Phos K. Each 300 ml of a standard dilution of the Neutra-Phos K powder provides 1.0 gm of phosphorus and 57 mEq of potassium. Each capsule of Neutra-Phos provides 250 mg of phosphorus and can be dissolved in about 75 ml of water. In a child of 2 to 3 years the appropriate dose is about 1 gm of phosphorus per day, increasing to 1.5 to 2.0 gm per day in the older child.

As the renal injury progresses, however, and GFR as measured by serum creatinine diminishes, the additional phosphate is discontinued and the regimen then altered to that described in the section on renal osteodystrophy, namely, a restricted protein and phosphate intake with special attention to prevention of acidosis by increased intake of bicarbonate precursors. Polycitra can be continued for this purpose unless potassium concentrations begin to rise above normal, when sodium bicarbonate, sodium lactate or sodium citrate is substituted in amounts that maintain normal serum bicarbonate concentration. Eventually, calcium supplements or aluminum hydroxide will be needed to reduce intestinal absorption of phosphate to prevent hyperphosphatemia. The transition from hypokalemia, hypophosphatemia to hyperkalemia, hyperphosphatemia can occur within a relatively few months. The treatment with vitamin D or dihydrotachysterol is continued in order to inhibit excessive parathyroid gland function, and the dose is monitored by urine calcium excretion and serum calcium concentration.

When end-stage renal insufficiency is reached renal transplantation can be considered as in other forms of renal insufficiency. It is somewhat surprising that renal transplants in patients with cystinosis show the same survival time as in those with renal insufficiency of other etiologies. Since the patient has a genetic abnormality, the metabolic consequences of which has led to destruction of his own kidneys, it would be expected that the transplanted kidney would also be injured by cystine since the patient's cystine-laden leukocytes and macrophages do invade the transplant. Biopsy of the transplanted kidneys shows cystine-containing cells, but the kidney does not show renal tubular dysfunction characteristic of the Fanconi syndrome and no progressive disappearance of nephrons aside from the usual rejection problems of any kidney transplant. Why the donor kidney is not susceptible to cystine damage as were the patient's own kidneys is not clear. It may be a concentration effect and that the quantity of cystine brought in by invading leukocytes and cells of the reticuloendothelial system is not great enough to produce the toxic effect of cystine. This concentration effect will be mentioned again in subsequent discussions of intermediate (late onset juvenile) or benign (adult) cystinosis.

Prenatal Diagnosis and Genetic Counseling. Inasmuch as cystinosis is transmitted by an autosomal recessive gene and the consequences of this metabolic error are so dire, the possibility of prenatal diagnosis once a case has appeared in a family is of importance. Cultured fetal cells obtained by amniocentesis do accumulate high concentrations of cystine if the fetus is cystinotic. The differentiation between homozygous and heterozygous fetuses is possible on the basis of concentration of cystine. It is thus feasible to offer this service to families who have had a cystinotic child so that they may plan for others in the knowledge that a fetus with proved cystinosis can be aborted.

CASE REPORTS

L.H., a white female, was seen at 3 years of age because of growth failure and polyuria. She was the only child of apparently healthy parents. Increased urine output was noted during the first year of life, and by 1 year of age it was apparent that she was growing at a subnormal rate. Her appetite was poor but there were no other symptoms of illness. When first examined she was a tiny child well below the third percentile for height and weight. Her skin was fair and hair blonde. There were no obvious deformities. Serum electrolyte determinations showed: Na 136 mEq/l, Cl 93.0 mEq/l, HCO_3 25 MEq/l, K 2.0 mEq/l, urea N 18 mg/dl, Ca 11.4 mg/dl and P 3.3 mg/dl. Urine analysis: Sp.Gr. 1.006, pH 6, protein$^+$, glucose negative. X-ray examination revealed poorly mineralized bones but no evidence of rickets. The marked hypokalemia suggested Bartter's syndrome but the hypophosphatemia and relatively normal serum HCO_3 did not support this diagnosis. The possibility of cystinosis was considered, but absence of glycosuria and the normal serum HCO_3^- concentration made this diagnosis seem unlikely. However, slit lamp microscopy of the cornea was done and revealed the characteristic appearance of crystals compatible with cystinosis. Bone marrow aspirates were positive for cystine crystals within macrophages.

The child was treated with potassium phosphate (Neutra-Phos K) providing 1000 mg P/day and 8000 units of vitamin D per day. The serum P remained low, about 3.5 mg/dl, and serum K continued low, 2.6 mEq/l, despite high potassium intake. She was placed on the experimental protocol of the ascorbic acid study of the N.I.H. Serum HCO_3^- concentrations decreased and sodium and potassium citrate solution was added to the regime, providing 90 mEq K daily. The patient has continued to be an active and alert little girl but growth has remained markedly retarded. Kidney function as measured by creatinine clearance has not shown any change.

COMMENTS: This is an example of a child with cystinosis presenting primarily with dwarfism and a history of polyuria. No rickets has developed despite the hypophosphatemia, presumably because growth has been so slow. The marked hypokalemia is probably an important factor in the growth retardation. She shows the lack of skin and hair pigmentation frequently seen in cystinosis. The mother became pregnant following the diagnosis of cystinosis in this child; an amniocentesis was done and the fetal fibroblasts which were cultured showed no evidence of cystine accumulation. The pregnancy went to term, the infant was normal at birth and leukocyte cystine content was normal. Although treatment of L.H. has not been successful in overcoming the renal tubular disorders of cystinosis, the value of genetic counseling and prenatal diagnosis has been demonstrated.

C.M. was referred at 8 years of age because of growth retardation, glycosuria, proteinuria and possible rickets. Glycosuria and proteinuria were first noted at 6 years of age when he was admitted to a hospital for repair of hypospadias. A glucose tolerance test was normal. A year later glycosuria and proteinuria were still present. About one month before referral he had cervical lymphadenitis with fever and his physician noted that he was below the third percentile in height and weight (Fig. 5–32). Radiographs of the wrists showed suggestive evidences of rickets and he was referred to our clinic.

On examination he was a small thin boy who looked chronically ill. Skin and mucous membranes were pale. BP was 110/70. There was no hepatosplenomegaly and there were no striking deformities of the extremities.

Laboratory Data: Blood — Hct. 27.8, Hgb. 9.0, ESR 60 mm/hr. Urine — Sp.Gr. 1,020, pH 6, protein 2^+, glucose 3^+. Microscopic — numerous RBCs. Amino acids — generalized aminoaciduria. Serum — Na 137 mEq/l, Cl 106 mEq/l, HCO_3 17 mEq/l, K 2.1 mEq/l, urea N 56 mg/dl, creatinine 2.5 mg/dl, Ca 8.0 mg/dl, P 4.3 mg/dl, alkaline phosphatase 9.8 Bodansky Units/dl.

X-ray examination showed marked delay of bone maturation and slight evidences of rickets. There were no lesions suggestive of hyperparathyroidism (Fig. 5–33*a* and *b*). Slit lamp microscopy of the eyes showed diffusely distributed crystalline stromal infiltrate diagnostic of cystinosis. A bone marrow aspirate contained numerous cystine crystals. The diagnosis of nephropathic cystinosis with moderately advanced renal insufficiency was made. The child was placed on a low protein diet, 20 gm/day, sodium and potassium citrate solution, 30 ml/day providing 2 mEq/kg/day each of Na and K, and dihydrotachysterol 0.25 mg/day.

After six months of this treatment the patient was symptomatically better in that he

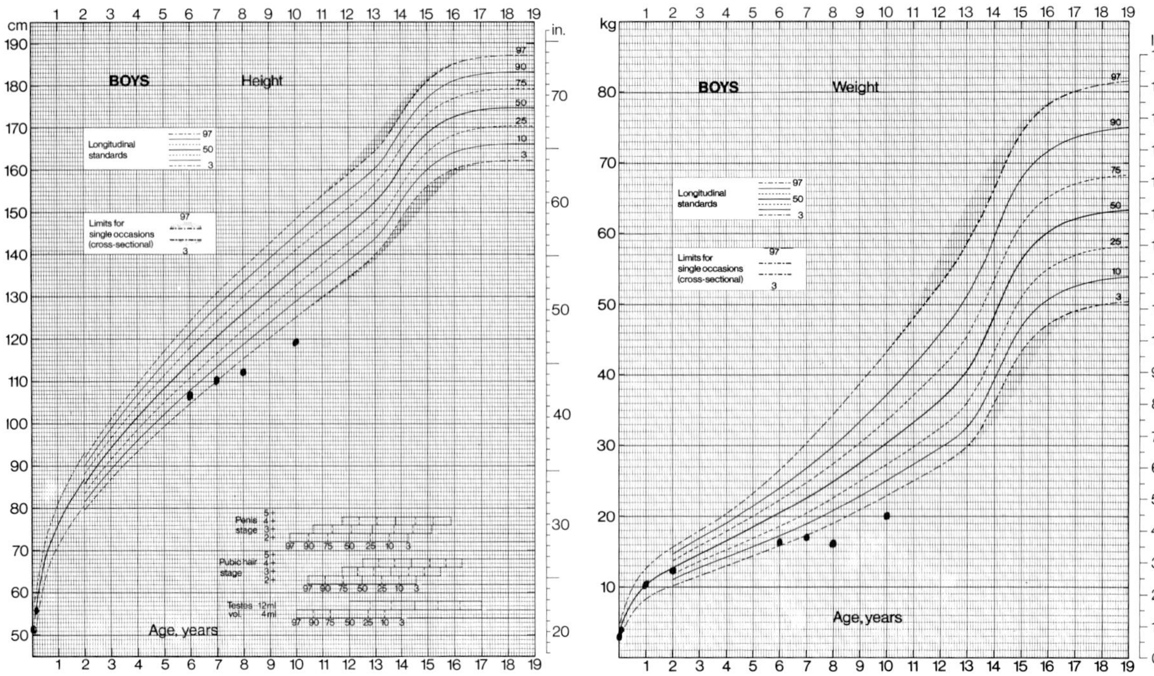

Figure 5–32. Growth curves of child (C.M.) with nephropathic cystinosis, demonstrating progressive disturbance of linear growth and weight gain.

had more energy, had grown 3 cm and radiographs showed healing of rickets (Fig. 5–34*a* and *b*). Laboratory findings are summarized in Table 5–17. Because of persistent hypophosphatemia despite correction of bicarbonate deficit, potassium phosphate was added to the regime, replacing part of the potassium citrate. In the next six months, however, serum creatinine rose and hyperphosphatemia and hypercalcemia developed. Dihydrotachysterol and supplemental phosphate were discontinued with decrease in serum calcium and phosphate concentrations. At 10 years of age kidney function had deteriorated to such a degree that he was referred to a renal dialysis and transplant unit. Renal function rapidly decompensated and after a period of hemodialysis a successful renal transplant was done. A year after transplantation he had satisfactory function.

COMMENTS: This is a fairly characteristic case of nephropathic cystinosis with the Fanconi syndrome. Growth failure, anemia and mild rickets were present at 8 years of age with reduction of GFR. Initial treatment directed at the acidosis (sodium and potassium citrate) and improvement of calcium homeostasis (dihydrotachysterol) resulted in clinical, biochemical and radiologic improvement. GFR progressively decreased and dihydrotachysterol was discontinued because of hypercalcemia. Eventually, renal function rapidly deteriorated and hemodialysis and renal transplantation were instituted. Initial response to transplantation was good with satisfactory function after one year.

Intermediate (Late Onset Juvenile or Adolescent) Cystinosis

This disorder, which probably is also transmitted as an autosomal recessive, is less severe than the infantile nephropathic form. Cultured skin fibroblasts and leukocytes of these patients also show increased concentrations of cystine, but not as great as those found in the cells of children with nephropathic cystinosis. The clinical course of these patients is milder than that of nephropathic cystinosis but with the same manifestations of renal tubular injury. Patients do not show pigmentary retinitis nor absence of

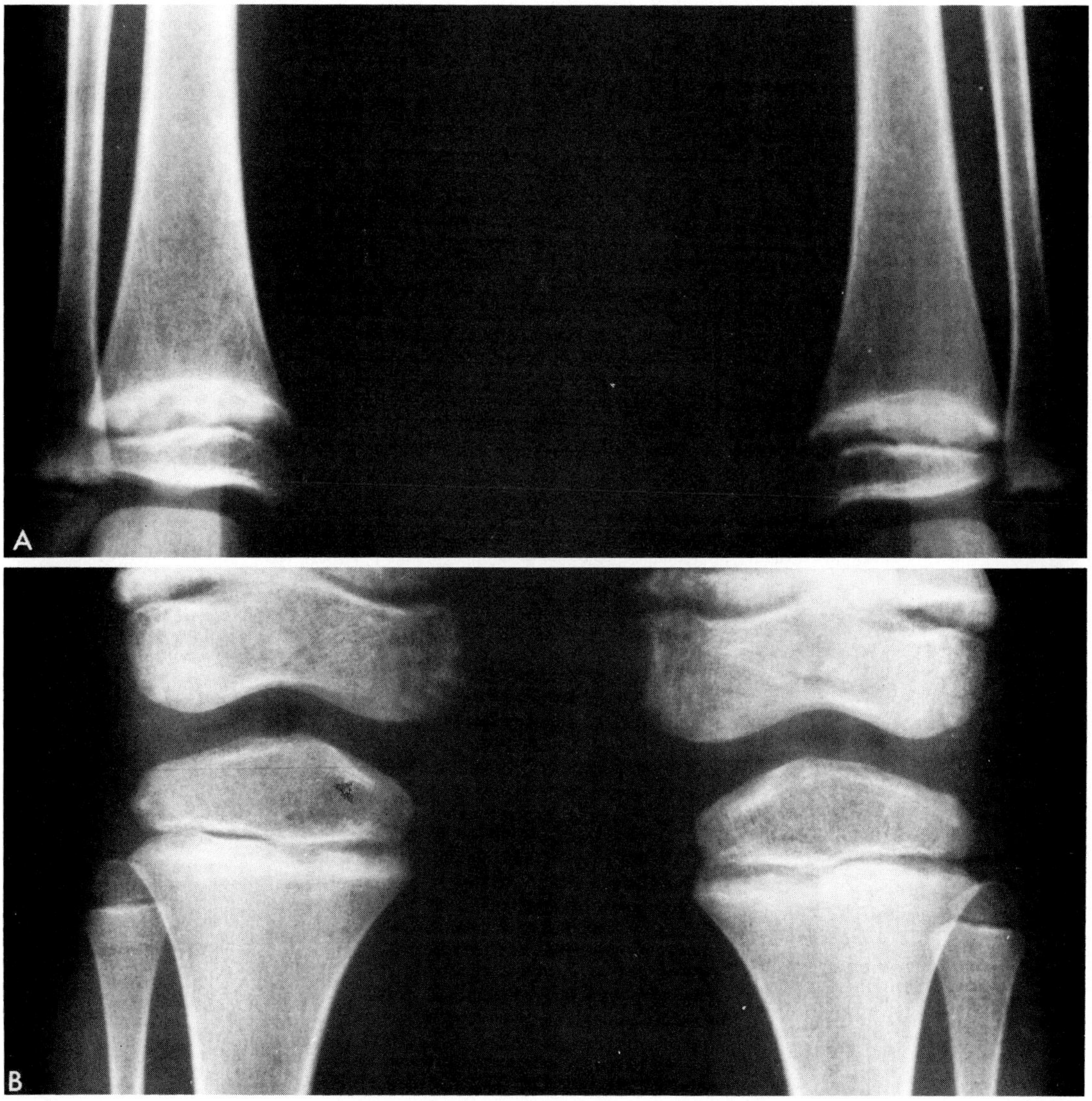

Figure 5–33. Radiographic studies of patient (C.M.) with nephropathic cystinosis, demonstrating lesions of rickets: *a*, ankle; *b*, knee.

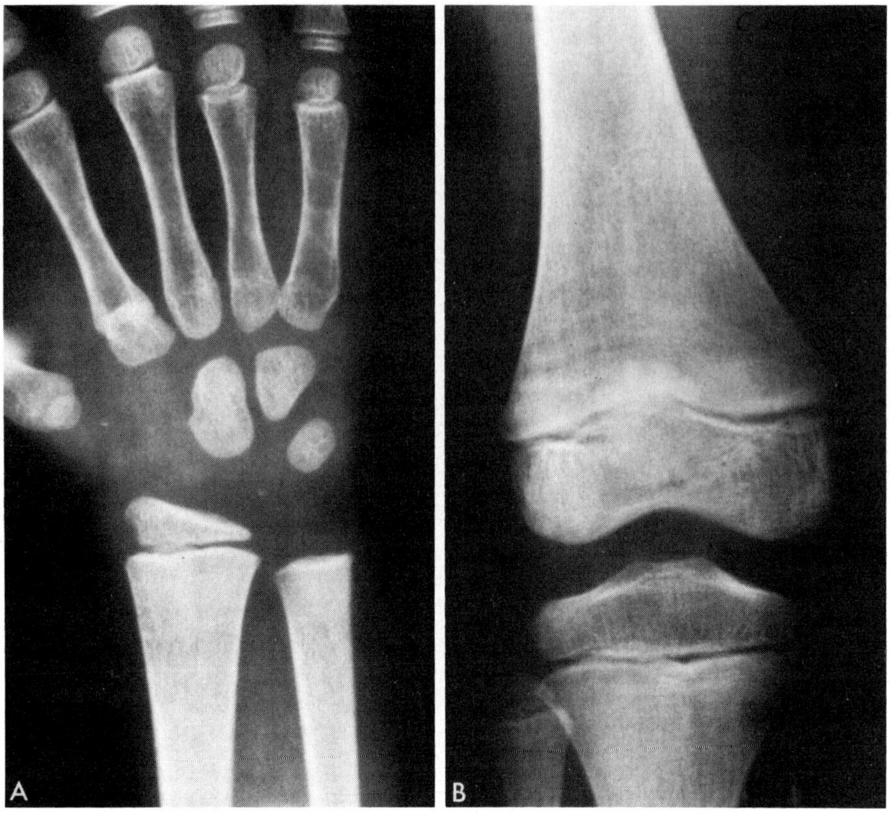

Figure 5–34. Radiographs of patient (C.M.) following symptomatic treatment, showing healing of rickets: *a*, wrist; *b*, knee.

Table 5–17. Nephropathic Cystinosis (C.M., Age 8 Years)

	SERUM							
DATE	*Na*	*Cl*	*HCO_3*	*K*	*Ca*	*P*	*Urea N*	*Creatinine*
	mEq/l				*mg/dl*			
5-5-72	137	106	17.2	2.1	8.0	4.3	56	2.5
	Dihydrotachysterol 0.25 mg/day; 30 mEq Na and 30 mEq K as citrate daily							
6-7-72	142	105	25	3.9	10.2	2.9	29	1.8
11-1-72	143	104	26	4.6	11.2	3.3	25	2.3
	Potassium phosphate (Neutra-Phos K) = 600 mg P							
1-15-73	–	–	–	–	10.4	–	37	3.4
5-9-73	142	103	21	4.6	12.5	8.5	25	3.5
	Dihydrotachysterol and potassium phosphate discontinued							
6-13-73	–	–	–	–	11.4	7.9	–	–
9-1-73	–	–	–	–	10.7	6.2	–	–
4-4-74	144	105	23	4.0	9.6	4.7	57	5.0
9-9-74	144	108	20	5.6	8.3	6.1	84	6.8

skin pigment as do these with infantile nephropathic cystinosis. Cystine crystals are found in the cornea and in bone marrow aspirates. Active rickets can be an important presenting manifestation secondary to the renal hypophosphatemia. Treatment of the renal hypophosphatemia, hypokalemia and bicarbonate deficiency is the same as in patients with nephropathic cystinosis except that the process is milder and clinical improvement is more striking. Although progressive renal insufficiency develops much later in these patients than in those with nephropathic cystinosis of the infantile type, some have shown end-stage renal failure. Whether these patients will respond more favorably to removal of cystine from tissues by large doses of reducing agents such as ascorbic acid or dithiothreitol remains to be seen.

CASE REPORT

This white female, P.M., was apparently normal until 20 months of age when her gait was noted to be peculiar. Radiographs of the hips and long bones revealed evidences of rickets. She had had an adequate intake of vitamin D since birth. Serum analyses revealed the following: Ca 10.8 mg/dl, P 3.0 mg/dl, alkaline phosphatase 62 K.A. units, Na 134 mEq/1, Cl 109.4 mEq/1, HCO_3 14.1 mEq/1 and K 4.6 mEq/1. Urine examination: protein 2^+, glucose$^+$. Urine amino acids: generalized increase. Ophthalmologic examination was reported as not showing cystine crystals in the cornea, but the bone marrow aspirate was positive for cystine crystals. The child was treated with 50,000 units of vitamin D daily, with radiologic evidence of healing of rickets, although serum phosphate concentrations remained low, 3.5 and 2.9 mg/dl. Urine calcium excretions were moderately high, 80 to 96 mg/24 hrs. Aminoaciduria continued. At 3 years of age there was minimal bowing of legs compared to findings at 20 months (Fig. 5–35).

COMMENTS: This mild case of cystinosis with rickets may be an example of the intermediate form of cystinosis or adolescent cystinosis despite evidences of rickets at 2 years of age. There was no polyuria nor potassium deficiency although serum bicarbonate concentrations were reduced. Treatment with vitamin D plus sodium and potassium citrate resulted in healing of the rickets and relatively normal growth.

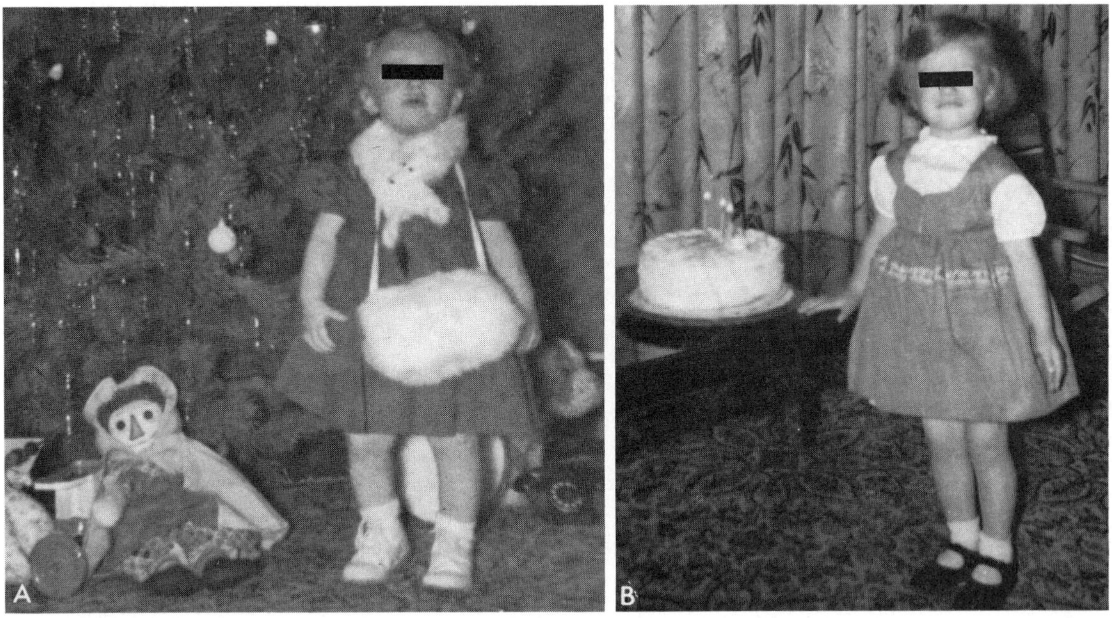

Figure 5–35. Photographs of patient with cystinosis of intermediate severity. *a*, 20 months of age – moderate bowing of legs; at this time radiographs showed evidences of rickets; *b*, 36 months of age – rickets healed following vitamin D treatment.

Adult Cystinosis

Adult subjects have been detected by slit lamp microscopy to have cystine crystals in the cornea. Cultured skin fibroblasts and peripheral leukocytes of such patients show excessive concentrations of cystine but less than one-half of that found in cells of infants and children with nephropathic cystinosis. These adults have no major manifestations of cystinosis and the discovery of their lesion is usually accidental. The occurrence of this benign form lends credence to the statement that the concentration of cystine accumulated in the kidney is the factor determining the severity and progression of renal injury. In the adult form of cystinosis the concentration is evidently below the threshold value for toxicity. The three types of cystinosis, nephropathic, intermediate and adult, indicate that several mutant alleles at one or more gene loci affect one or more gene products that control the steady state of cystine metabolism in the cell. Clinical prognosis is apparently influenced to some extent by the severity of cystine retention.

Tyrosinosis (Sakai)

This genetic disorder, also transmitted by an autosomal recessive gene, is characterized by marked increase in concentration of tyrosine in plasma with deposition of tyrosine crystals in tissues. The concentration of serum methionine is also high, possibly secondary to impaired liver metabolism due to accumulation of tyrosine. The disorder is quite rare except in a French-Canadian population in which an incidence of 0.67 per 1000 live births has been reported. Liver biopsies in affected subjects have shown a deficiency of the enzyme p-hydroxyphenylpyruvic acid oxidase, but it is doubtful that this is the primary enzyme defect in this disorder since a deficiency of this same enzyme has been reported in a benign form of defective metabolism of tyrosine, tyrosinemia (Medes variety).

In addition to renal tubular injury with the Fanconi syndrome, these patients also show severe liver injury progressing rapidly to fatal cirrhosis. By 6 months of age infants may have marked evidences of rickets, hepatosplenomegaly and severe hepatic dysfunction. Serum tyrosine concentrations are markedly increased, 3 to 12 mg/dl, and there is excessive excretion of the organic acid products of tyrosine metabolism in the urine, parahydroxyphenyllactic, parahydroxyphenylpyruvic and parahydroxyphenylacetic acid. The severity of the rickets in these patients may result from the combination of renal hypophosphatemia secondary to renal tubular dysfunction and hepatic injury with deficient 25-hydroxylation of vitamin D. At any rate treatment with vitamin D or vitamin D plus phosphate has not resulted in healing of the rickets.

The specific treatment of this disorder is special diet with a low intake of tyrosine and the tyrosine precursor, phenylalanine. If started early the low tyrosine diet has been successful in some patients in arresting the progression of the liver injury, and improvement of the Fanconi syndrome has occurred.

Lowe's Syndrome (Oculocerebrorenal Dystrophy)

The inheritance of this disorder is as a sex-linked recessive, so that only males are affected. The basic metabolic abnormality is unknown and the Fanconi syndrome is associated with severe CNS maldevelopment and congenital abnormalities of the eyes including buphthalmos, glaucoma and cataracts. The manifestations of the nervous system injury are severe muscle hypotonia, areflexia and marked mental retardation.

These patients also show the Fanconi syndrome with tubular proteinuria, aminoaciduria and renal hypophosphatemia. Glycosuria is mild, and bicarbonate wasting, hypokalemia and failure of concentration are not characteristically seen. The skeletal lesions are a combination of severe osteoporosis and rickets (Fig. 5–36*a* and *b*). The osteoporosis may result from the very poor muscle tone and the immobility of these patients because of the combination of muscle hypotonia, severe cortical dysfunction and the blindness that is a consequence of the eye abnormalities.

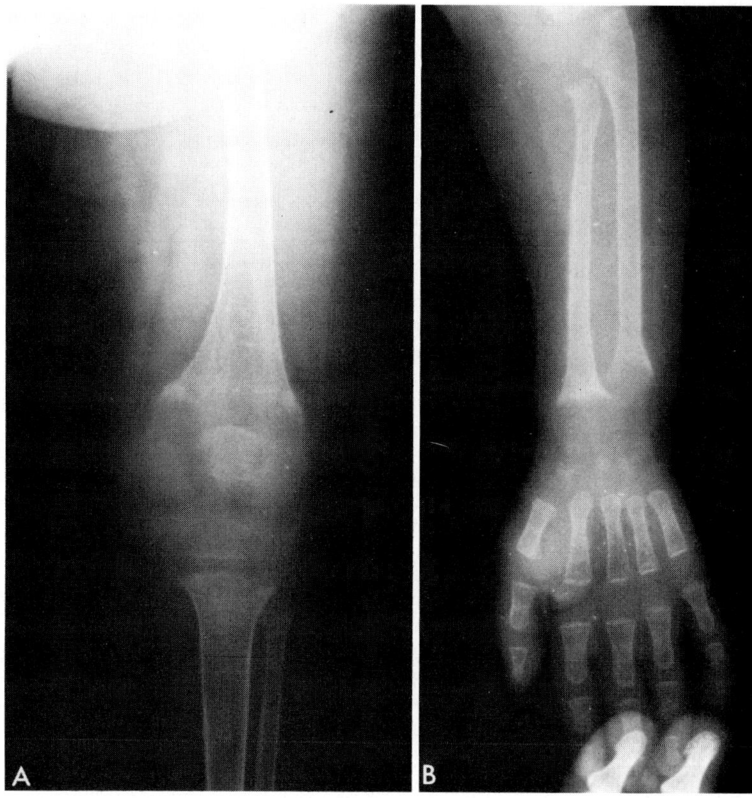

Figure 5–36. Oculocerebrorenal (Lowe's) syndrome; severe rickets and osteopenia: *a*, knee; *b*, wrist.

The disorder is usually suspected in early infancy because of the congenital anomalies of the eyes with impaired vision, and the muscle hypotonia. On urine examination proteinuria and general aminoaciduria are the early findings. Subsequently, reduced serum phosphate concentrations are found, and x-ray examination shows the bones to be very poorly mineralized with the metaphyseal irregularities of rickets.

The rickets can be treated, as in other instances of the Fanconi syndrome, with phosphate salts and moderately large doses of vitamin D which are monitored by urine calcium excretion and serum calcium concentrations. A low sodium intake with reduction of ECF volume by thiazide or other diuretic treatment has been noted to reduce renal clearance of phosphate and increase serum phosphate. This regime plus extra phosphate as the potassium salt may be effective in healing the rickets. These children usually die early because of aspiration pneumonitis or other infections.

Wilson's Disease

The major manifestations of the defect of copper metabolism in Wilson's disease usually are liver and CNS injury. Some degree of renal tubular injury also occurs as the result of the high concentrations of copper in tissue. This injury, like that of other heavy metals, invariably causes aminoaciduria and may also produce the other manifestations of the Fanconi syndrome, glycosuria and renal hypophosphatemia with rickets and osteomalacia. The treatment, of course, is the specific chelation of copper by d-penicillamine, which increases urine output of copper, plus prevention of copper absorption by potassium sulfide.

Hereditary Fructose Intolerance

The Fanconi syndrome can also be a complication of this rare inborn error of metabolism which is transmitted by an autosomal recessive gene. The enzyme deficiency is a reduced activity of hepatic fructose-1-phosphate aldolase, which results in an accumulation of fructose-1-phosphate. The manifestations of this disorder resemble to some extent those of galactosemia in which galactose-1-phosphate concentration is increased. Liver injury is predominant, but renal tubular damage with chronic acidosis and the Fanconi syndrome can occur. The role of parathyroid hormone in intensifying the nephrotoxic effect of the disturbed fructose metabolism has been learned through the coincidental occurrence of hypoparathyroidism in a patient with hereditary fructose intolerance. The renal aminoaciduria and hypophosphatemia were ameliorated as parathyroid secretion diminished and was exacerbated by PTH injection. This is compatible with the hypothesis that the action of PTH in inhibiting renal tubular reabsorption of phosphate and also of amino acids and bicarbonate intensifies the already inadequate tubular transport of these solutes due to toxic agents that injure renal tubule transport systems. This provides theoretic support to the use of vitamin D or its metabolites or a vitamin D analogue such as dihydrotachysterol in the treatment of the Fanconi syndrome, since these function to suppress PTH secretion by their action in increasing intestinal calcium absorption and extracellular calcium ion concentration.

The treatment of hereditary fructose intolerance is the careful elimination of all sources of fructose and sucrose from the diet so that patients remain symptom-free.

Multiple Myeloma (Light Chain Nephropathy)

In adult life a major etiology of the Fanconi syndrome is multiple myeloma. The renal tubular injury is caused by the kappa light chains of immunoglobulins which are present in high concentration in the plasma and excreted in the urine — the so-called Bence-Jones protein. These patients may have both hypophosphatemic osteomalacia secondary to renal tubular injury and hypercalcemia with bone demineralization as a consequence of the proliferation of plasma cells in the bone marrow. Since multiple myeloma is not seen in childhood this form of the Fanconi syndrome need not be discussed in more detail.

Idiopathic Fanconi Syndrome

A small number of patients show the renal tubular injury of the Fanconi syndrome, which cannot be explained by any of the known etiologic agents, either endogenous or exogenous. A higher proportion of adult patients with Fanconi syndrome fall into this group than do children. It is essential to evaluate the patient for cystinosis as well as other inborn errors of metabolim, and to search for exogenous toxic agents that cause renal tubular injury (discussed in the following section). Following thorough study, if none of the known causes of the Fanconi syndrome are uncovered, the diagnosis of idiopathic Fanconi syndrome remains by exclusion. The severity of the renal tubular injury in idiopathic Fanconi syndrome is quite variable, but the lesion usually does not cause renal insufficiency and the diagnosis is often made only in adult life with the presenting manifestation of bone pains caused by osteomalacia. More severe cases present in childhood with hypophosphatemic rickets, and on urine examination glycosuria and generalized aminoaciduria are found, differentiating these patients from those with primary hypophosphatemic rickets. They may also show bicarbonate wasting with chronic hyperchloremic acidosis.

Treatment is directed at compensating for the renal tubular defects since the pathogenetic factors are unknown and no specific therapy is available. The acidosis is corrected by oral feeding of sodium bicarbonate or sodium bicarbonate precursors such as sodium lactate or citrate. If potassium wasting is present a mixture of sodium and potassium citrate is used (Polycitra). The required dose depends on the degree of bicarbonate wasting and may vary from 2 to 3 mEq of cation per kg to 6 mEq/kg or more. One ml of the Polycitra solution provides 1 mEq each of sodium and potassium, so that the initial dose can be 1 to 2 ml/kg per day in divided doses and increased until serum bicarbonate concentrations are maintained in the normal range. Reduction of endogenous H^+ ion production by limitation of protein intake may be helpful in reducing the required dose of extra buffer base.

The renal hypophosphatemia is attacked by a combination of vitamin D or dihydrotachysterol and high phosphate intake in the form of a buffered solution of sodium and potassium phosphates or potassium phosphate alone. The latter may be preferable since a high sodium intake with natriuresis may increase renal phosphate clearance. Moderately large doses of vitamin D, 25,000 units per day, have been found to increase serum phosphate concentration and promote healing of the rickets. This brings up the possibility that defective vitamin D metabolism may be a factor in these patients. However, they do not show hypocalcemia as do those with vitamin D-dependent rickets who have an abnormality of renal 1-hydroxylation of 25-hydroxyvitamin D. It is possible that the vitamin D operates by suppression of normal PTH secretion, thus increasing tubular reabsorption of phosphate and amino acids. The Fanconi syndrome has been seen in infants with malabsorption and liver disease in whom secondary hyperparathyroidism is present, and treatment with vitamin D metabolites with suppression of parathyroid gland activity results in reduction of glycosuria, aminoaciduria and renal phosphate clearance.

Further study of this interesting problem is needed before we can understand the interaction of PTH and renal tubular transport systems in the Fanconi syndrome. It may well be that 25-hydroxyvitamin D or 1,25-dihydroxyvitamin D will be effective in suppressing parathyroid gland activity in these patients and 1,25-dihydroxyvitamin D is now available. The dose of vitamin D, vitamin D metabolites or dihydrotachysterol must be adjusted so that the urine calcium excretion doses not exceed 4 mg/kg/day and serum calcium concentrations remain less than 10.5 mg/dl. If suppression of parathyroid gland activity is the basis for the effectiveness of vitamin D or dihydrotachysterol, measurement of serum IPTH should be useful in adjustment of the dose of vitamin D metabolites or analogues.

Fanconi Syndrome Caused by Exogenous Toxic Agents

Heavy Metals. Renal glycosuria and aminoaciduria are early evidences of the nephrotoxicity of lead, mercury or cadmium. Chronic poisoning with lead or cadmium has also caused the Fanconi syndrome with renal hypophosphatemia, in addition to glycosuria and aminoaciduria and rickets and osteomalacia. The offending metal in infants and young children has been lead since exposure to this metal is more common in this age-group. Although most commonly the manifestations of lead poisoning in children are due to CNS injury there are reports of hypophosphatemic rickets in young children and infants. The renal tubular lesion is reversible by treatment of the lead poisoning with chelating agents, BAL and CaEDTA. Hyperuricemia is also a major manifestation of the renal tubular lesion of lead poisoning. Whether these high concentrations of serum urate play a role in the occasional progression of lead tubular injury to interstitial fibrosis and renal insufficiency with hypertension remains an unsolved question.

In adults cadmium poisoning has been a more common cause of the Fanconi syndrome and hypophosphatemic osteomalacia. Such exposure may be occupational in factories doing cadmium plating or, as in Japan, may result from contamination of water by the effluent from zinc mines, since cadmium is found along with zinc. The osteomalacia in adults manifests itself by severe bone pains and chronic disability due to poor mineralization of the skeleton, with microfractures and Looser-Milkman zones (pseudofractures) at sites of skeletal stress where unmineralized osteoid is deposited.

Prevention of further exposure to the toxic metal, together with chelation therapy, is the specific treatment for this type of Fanconi syndrome. During the healing process a high phosphate intake with moderately large doses of vitamin D can accelerate the mineralization of the undermineralized skeleton.

Organic Toxic Agents. The Fanconi syndrome has resulted from poisoning with phenols or cresol, but of more clinical interest has been the renal tubular injury resulting from ingestion of tetracyclines that have been degraded by aging. The decomposition product responsible for the renal tubular injury is thought to be anhydro-4-epitetracycline. The Fanconi syndrome in patients who have ingested outdated tetracycline is acute and short-lived once the toxic agent is discontinued. Chronic renal tubular dysfunction sufficient to produce hypophosphatemic rickets has not been reported.

Experimentally, maleic acid has been found to produce the Fanconi syndrome in animals. This organic acid however, is, not produced in the intermediary metabolism of food stuffs or other compounds in the body so that it is only of experimental interest in studies of the mechanism of the renal tubular dysfunction of the Fanconi syndrome. Most recently it has been stated that the renal injury of maleic acid reduces the 25-hydroxyvitamin D-1-hydroxylase of the renal tubule cells and thus impairs the production of the active vitamin D hormone, 1,25-dihydroxyvitamin D. If this proves to be the case it may be that defective vitamin D metabolism may be part of the disturbance in some patients with the Fanconi syndrome. Further investigation in patients will be needed to answer this question.

RENAL TUBULAR ACIDOSIS

Renal tubular acidosis (RTA) is defined as a group of specific defects of H^+ ion secretion by renal tubule cells with deficiency of associated bicarbonate reabsorption, but without the more widespread tubular transport disorders of the Fanconi syndrome and without glomerular insufficiency. Two varieties of RTA exist based on the location of the tubule in which the major defect exists: (1) proximal RTA; and (2) distal RTA. Combined lesions also occur.

It is necessary to review the role of the kidney in the maintenance of normal extracellular bicarbonate concentrations, and therefore normal body fluid H^+ ion concentration. In both proximal and distal tubule cells, filtered HCO_3^- is reabsorbed through secretion of H^+ ion into luminal fluid to convert bicarbonate to CO_2 and H_2O with diffusion of CO_2 back into the cell so that pCO_2 remains approximately the same in luminal, cellular and peritubular fluids. The intracellular CO_2 is then the source of bicarbonate which enters the peritubular fluid, as shown in Figure 5–37.

Most of the filtered bicarbonate is reabsorbed in the proximal tubule, but reabsorption of this anion can never be complete in this segment since the proximal tubule cannot develop a large H^+ ion gradient between luminal fluid and peritubular fluid, and the pH of luminal fluid at the end of the proximal tubule segment is about 7.0 at maximal rates of HCO_3^- reabsorption. This means that the luminal HCO_3^- concentration cannot be reduced to less than approximately 10 mEq/1 in this segment of the tubule. If it is assumed that 85 per cent of the glomerular filtrate volume is reabsorbed in the proximal tubule for every liter of glomerular filtrate at least $150 \times 10/1000$ or 1.5 mEq of HCO_3^- will escape reabsorption in the proximal tubule. Assuming a GFR of 144 liters per day (100 ml/min) this means over 216 mEq of HCO_3^- enters the distal tubule. In the distal tubule H^+ ion can be secreted into the lumen against an ion gradient, so that the pH of the tubular contents can be reduced to as low as 4.5 to 5.0 with complete reabsorption of remaining HCO_3^-.

The HCO_3^- reabsorption in the proximal tubule shows a threshold phenomenon, i.e., there is a maximum rate of reabsorption in terms of quantity per volume of glomerular filtrate. This threshold value in the adult is 27 mEq/1 of glomerular filtrate. At concentrations of plasma HCO_3^- above this value the excess is not reabsorbed and enters the distal tubule. Since there is no stimulus for distal tubule secretion of H^+ when plasma HCO_3^- is high, this excess of HCO_3^- is excreted in the urine and urine pH is 7.5 to 8.0. If, however, extracellular HCO_3^- is reduced appreciably below the threshold value, the HCO_3^- escaping reabsorption in the proximal tubule can be completely reabsorbed distally by the process of H^+ ion secretion and reduction of urine pH. In proximal RTA this threshold value is markedly reduced so that the rate of tubular reabsorption of bicarbonate may be only 12 to 15 mEq/1 glomerular filtrate.

Under the conditions of ordinary diet intake of protein, endogenous H+ ion is generated by metabolism of the sulfur-containing amino acids with the formation of $SO_4^=$. Some H^+ ion is also generated by the formation of $PO_4^=$ from the $OPO_3H^=$ groups of phosphorylated amino acids. During starvation or deficiency of glucose metabolism, as in diabetes mellitus, H^+ ion is also generated by the incomplete oxidation of the fatty acid portion of triglycerides, resulting in the ac-

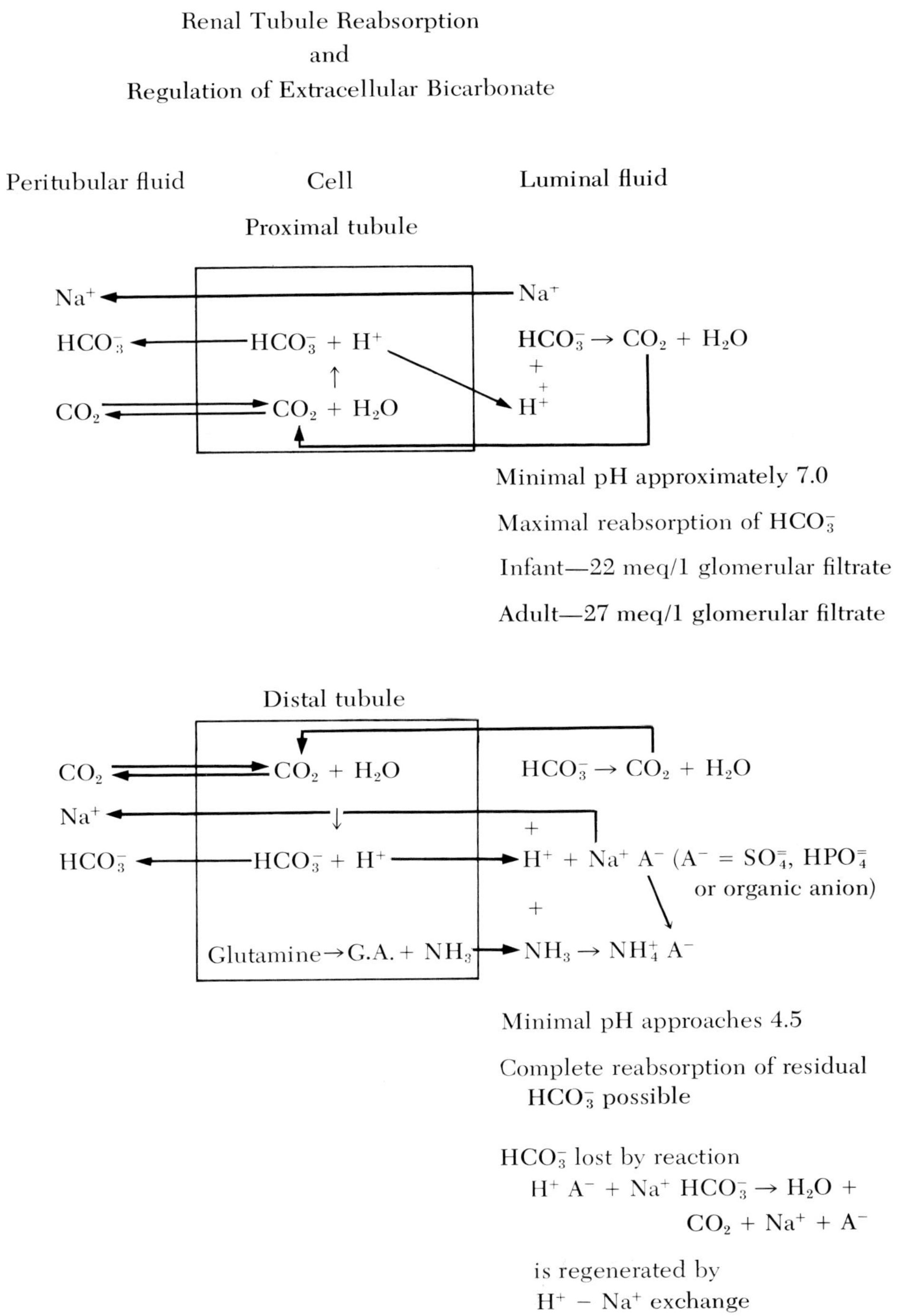

Figure 5–37. Diagram of renal tubule functions that result in reabsorption of filtered bicarbonate and in regeneration of bicarbonate lost by endogenous H^+ ion production.

cumulation of the four carbon acids, β OH butyric acid and acetoacetic acid. The H^+ ion generated endogenously reduces extracellular bicarbonate concentration through the reaction $H^+ + HCO_3^- \rightarrow H_2CO_3 \rightarrow H_2O + CO_2$, with the CO_2 formed excreted through the lungs. In order to regenerate this HCO_3^- the distal tubule function of H^+ ion secretion is required. This function of the distal tubule is not needed, however, if an excess of buffer base is ingested in an amount sufficient to compensate for the endogenous H^+ ion secretion.

If proximal tubule reabsorption of HCO_3^- is deficient the load of HCO_3^- entering the distal tubule is too large for distal tubular reabsorption, and wastage of HCO_3^- in the urine results. As a consequence of this loss serum HCO_3^- is reduced, and therefore also the filtered load of HCO_3^-. When the filtered load is reduced the proximal tubule can now reabsorb most of this reduced load, and distal tubular function can then produce an H^+ ion gradient with reduction of urine pH. The differentiation between proximal and distal tubular acidosis can be understood in terms of the two stages of bicarbonate reabsorption and the distal regeneration of bicarbonate. In pure proximal RTA the urine pH can be reduced to minimal values of 4.5 to 5.0 if serum HCO_3^- is sufficiently lowered. These patients then will excrete an acid urine if they are acidotic, either spontaneously because of endogenous H^+ ion production or made so by ammonium chloride loading. Patients with distal RTA cannot produce an H^+ ion gradient and so are unable to excrete an acid urine even at low concentrations of serum HCO_3^-. If these patients are given ammonium chloride the urine pH will not decrease below 6.5 or thereabouts despite marked reduction of serum HCO_3^- concentration.

In both forms of RTA tubular reabsorption of chloride is increased, since reabsorption of sodium requires an accompanying anion and chloride substitutes for the bicarbonate. The characteristic findings in serum concentrations are, therefore, normal serum sodium, increased serum chloride and reduced serum bicarbonate. There are also additional disorders of renal tubular function in distal RTA, which will be discussed in the description of the specific manifestations of that entity.

Proximal Renal Tubular Acidosis

Transient Renal Tubular Acidosis. One form of this psysiologic defect is transient, possibly owing to immaturity of renal tubule development with eventual maturation and normal tubule function. This transient proximal RTA can sometimes be recognized in infants who have had an episode of severe dehydration, usually due to acute diarrhea. Following rehydration and resumption of ordinary milk feedings they are found to have persistent hyperchloremic acidosis even though gastrointestinal function seems normal and there are no excessive stool losses of bicarbonate. In other instances there has been no such acute illness and the hyperchloremic acidosis is detected as part of the studies of an infant who fails to thrive. Prematurity may be one of the factors in this form of delayed maturation of renal function.

The major clinical manifestation of this form of RTA is poor growth and weight gain. Hypophosphatemia and rickets usually are not present. Urine examination is negative and generalized aminoaciduria is not found. When the serum bicarbonate is markedly reduced to 10 to 12 mEq/1, either spontaneously or after an ammonium chloride loading test, the urine pH is low, indicating that distal tubule H^+ ion gradient production is intact.

The treatment of these patients is an increased supply of bicarbonate, either as sodium bicarbonate or as sodium and potassium salts of organic acids which are rapidly metabolized to bicarbonate such as citrate, lactate or acetate. The required dose depends on the deficiency of bicarbonate reabsorption, but in the transient form this usually is not extreme and amounts of 3 to 5 mEq bicarbonate per kg per day in multiple doses suffice. The required amount is that which will maintain the serum bicarbonate concentration in a normal range — for infants 22 to 24 mEq/1. As renal tubular reabsorption of bicarbonate improves the dose is reduced, and eventually discontinued when serum bicarbonate concentrations can be maintained without extra intake of bicarbonate.

Persistent Proximal Renal Tubal Acidosis. This is presumably a congenital defect of renal tubular function but the genetics and the specific basis of the abnormality of bicarbonate transport are unknown. The major clinical manifestation is retardation of growth. In one case hypophosphatemic rickets has been reported. The mechanism of the deficiency in tubular reabsorption of phosphate and its relationship to the abnormality of tubular reabsorption of bicarbonate is unclear.

Diagnosis is made on the basis of persistent hyperchloremia and bicarbonate deficiency, with ability to reduce the pH of the urine well below 6.0 when the bicarbonate deficiency is marked <15 mEq/l. This observation can be made by watching for spontaneous reduction of serum bicarbonate on the usual high protein diet of children or by stressing the patient with ammonium chloride.

Treatment consists of the use of sodium bicarbonate or citrate in amounts that maintain serum bicarbonate concentrations in the normal range. In some patients very large amounts of bicarbonate, 5 to 10 mEq/kg/day, have been needed to maintain bicarbonate balance, indicating an extremely low renal tubule threshold for bicarbonate.

Distal Renal Tubular Acidosis

This is the more important form of RTA with respect to the occurrence of hypophosphatemia with rickets and osteomalacia, and is the classic variety of RTA first described by Albright and Butler in a child presenting with severe deformities of rickets and nephrocalcinosis. There is a definite familial aggregation of cases, indicating a genetic causative factor. These have been families in which more than one child has been affected, with both parents apparently normal, suggesting a recessive form of inheritance; and other families in which parent and child have had the disorder, indicating an autosomal dominant mode of inheritance since no definite sex predominance has been found. The specific molecular abnormality responsible for the failure of H^+ secretion is unknown. A deficiency of carbonic anhydrase has been suggested since inhibitors of carbonic anhydrase activity such as acetazolamide cause a deficiency of renal tubule H^+ ion secretion with bicarbonaturia and alkaline urine despite decrease of serum bicarbonate concentration and hyperchloremia. A lack of carbonic anhydrase or the presence of an inhibitor has not been clearly demonstrated in subjects with distal RTA, but an abnormality of the carbonic anhydrase in the erythrocytes of children with RTA has been reported. This disorder may appear in adult life without evidence of abnormality in childhood, so that an acquired lesion due to unknown injurious agents may also exist.

In addition to the inability of distal renal tubule cells to establish an H^+ ion gradient, and thus make possible the excretion of an acid urine, there are other abnormalities of renal tubule function characteristic of distal RTA. One is failure of the concentrating mechanism resulting in polyuria unresponsive to ADH. A second is excessive potassium loss with hypokalemia; this loss is secondary to the acidosis with K^+ exchange for Na^+ substituting for the H^+–Na^+ exchange which cannot be carried out by the distal tubule cells. A third is diminished renal tubular reabsorption of phosphate, with consequent renal hypophosphatemia and hypophosphatemic rickets and osteomalacia.

The defect of renal tubular absorption of phosphate, like the excessive secretion of potassium, is also secondary to the acidosis since correction of the latter by sodium bicarbonate therapy or its equivalent restores renal tubular reabsorption of phosphate to normal. If the acidosis is not corrected even large doses of vitamin D fail to increase serum phosphate concentrations and are not effective in healing the rickets. Once the acidosis is corrected healing of rickets occurs without need for extra vitamin D.

The basis for the defect of renal tubular reabsorption of phosphate in distal RTA is not completely solved. Since the major reabsorption of phosphate occurs in the proximal tubule the effect of acidosis is presumably to block phosphate reabsorption in this segment. Experimentally produced bicarbonate deficiency in dogs fed NH_4Cl has been shown to affect renal tubular reabsorption of phosphate and to lower serum phosphate concentrations. However, the hypophosphatemia in distal RTA is more severe than in other forms of chronic bicarbonate deficiency. Urine calcium excretion is increased during the acidotic state, and the original thesis of Albright was that the increased urinary loss of calcium created a deficiency of calcium that stimulated parathyroid gland activity, and the hypophosphatemia was then a manifestation of secondary hyperparathyroidism.

Other evidences of the renal action of excessive concentrations of PTH such as renal aminoaciduria are lacking, however, in this disorder. Similarly, secondary hyperpara-

Table 5–18. Tubular Reabsorption of Phosphate in Distal Renal Tubular Acidosis (S. T.—Renal Tubular Acidosis, Rickets)

			Phosphate			
Period	*Creatinine Clearance ml/min*	*Serum P. mg/dl*	*Filtered* mg/min	*Excreted* mg/min	*Reabsorbed mg/min*	*Reabsorbed mg/dl GFR*
1	16.1	2.7	0.44	0.23	0.21	1.30
2	21.8	2.7	0.59	0.35	0.24	1.10
3	20.0	2.7	0.54	0.27	0.27	1.35
			Phosphate injection started			
4	18.2	2.6	0.47	0.29	0.18	0.99
5	17.6	5.3	0.93	0.80	0.13	0.74
6	15.5	8.5	1.32	1.23	0.09	0.58
7	16.7	14.0	2.34	2.48	—	—

Filtration rate is measured by endogenous creatinine clearance; filtered phosphate is the product of serum phosphate in midperiod and filtration rate. Reabsorbed phosphate is the difference between calculated filtered phosphate and phosphate excreted during period. Serum phosphate was progressively increased after the control periods by intravenous infusion of a buffered solution of 0.1 M sodium phosphate. Tubular reabsorption of phosphate was extremely low in the control periods, averaging 1.25 mg/dl GFR in comparison with normal values of 4 to 5 mg/dl GFR. Following phosphate infusion tubular reabsorption of phosphate progressively decreased until there was a slight negative value in period 7. Whether there is true tubular secretion of phosphate, i.e., plasma to lumen phosphate transport that increases with elevation of plasma phosphate and becomes evident because of the reduced rate of tubular reabsorption of phosphate, remains to be answered definitively.

thyroidism due to a defect of renal metabolism of vitamin D does not seem likely. Serum calcium concentrations are normal or even high, whereas in known disorders of vitamin D metabolism reduction of serum calcium is the usual finding. Moreover, the degree of inhibition of renal tubular reabsorption of phosphate in some cases is greater than that due solely to increased PTH secretion (Table 5–18). Whatever the mechanism, the block in tubular reabsorption of phosphate is rapidly reversed by restoration of serum bicarbonate to normal, as indicated in Table 5–19.

A fourth renal abnormality is nephrocalcinosis, which is a secondary lesion characteristic of distal RTA (Fig. 5–38). This lesion is the result of multiple factors that cause precipitation of calcium phosphate within the tubules, in contradistinction to the nephrocalcinosis of hypercalcemic states in which calcium salt deposition occurs in the interstitial tissue as well as in the tubule. The factors responsible for the nephrocalcinosis are the concentrations of calcium and phosphate in tubular contents, the alkaline pH and the abnormally low citrate content of the tubular urine. Urine citrate concentration normally increases with alkalinization of the urine owing to decreased tubular reabsorption and utilization of citrate. The high citrate content of alkaline urine functions to prevent precipitation of calcium phosphate in the urine since citrate complexes with calcium to form an anion $(CaCit)^-$, thus removing Ca^{++} ion from the solution. In distal RTA the renal tubular reabsorption of citrate is increased and urine citrate concentration

Table 5–19. Correction of Hypophosphatemia in Distal Renal Tubular Acidosis by Administration of Bicarbonate Precursor (S. T.—Age 10 Years)

	Serum					
Date	*Ca* mg/dl	*P* mg/dl	*Alk. P'tase Bodansky Units*	CO_2 mEq/l	*Cl* mEq/l	*K* mEq/l
7-8	9.2	3.0	33.5	16.6	122	2.8
		Sodium and potassium citrate solution				
7-23	9.8	5.7	20.1	30.8	112	4.7
7-25	9.6	6.0	24.5	—	—	—
8-15	9.3	6.5	15.6	30.0	108	4.7
9-29	10.2	6.8	18.7	—	—	—

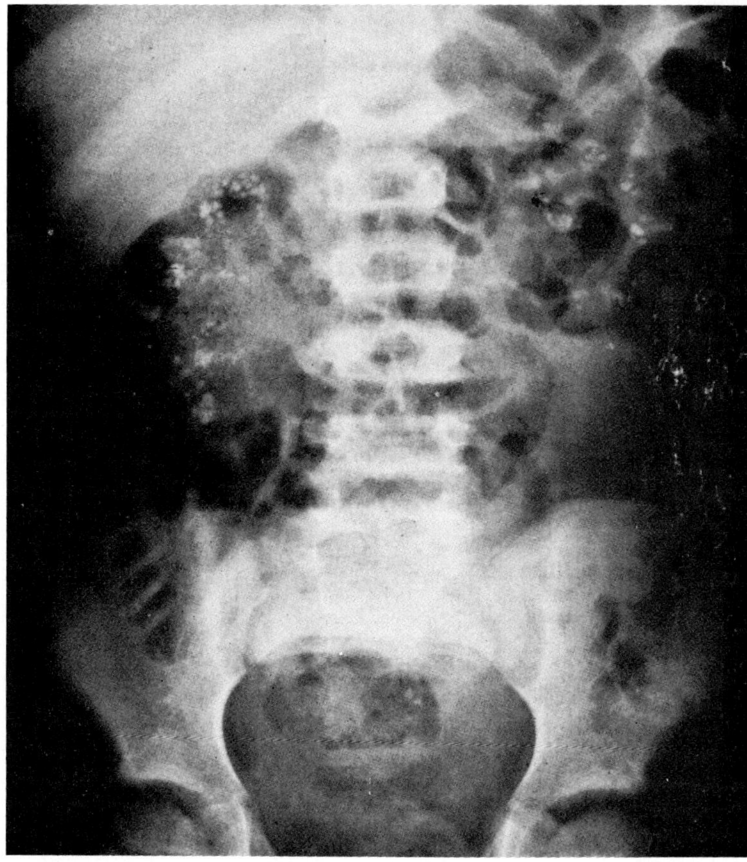

Figure 5–38. Nephrocalcinosis in patient with distal renal tubular acidosis. (From Harrison, H. E. and Harrison, H. C.: Hereditary metabolic bone diseases. Clin. Orthop. *33*:162, 1964. Reprinted with permission.)

is low despite the alkaline pH of the urine. Potassium deficiency also reduces urine citrate concentration through increased reabsorption of this ion. Both phenomena may result from intracellular acidosis of the renal tubule cells, which may be the controlling factor in citrate uptake and utilization by renal tubule cells. The combination of calciuria, phosphaturia, alkaline urine and low citrate concentrations is responsible for precipitation of calcium phosphate in the tubular lumen. Initially this is not associated with any interstitial kidney lesion, but as the calcium phosphate concretions grow the tubule is disrupted and the precipitate extended into the peritubular tissue, leading to an interstitial fibrosis.

An experimental nephrocalcinosis similar to that found in distal RTA has been produced in rats by feeding acetazolamide, a carbonic anhydrase inhibitor (Fig. 5–39). Inhibition of carbonic anhydrase activity prevents H^+ ion secretion blocking bicarbonate reabsorption and results in excretion of an alkaline urine. Citrate reabsorption is enhanced despite the alkaline pH of the tubule contents, so that this experimental RTA duplicates the important features of distal RTA and supports the mechanism of nephrocalcinosis outlined above.

CLINICAL COURSE AND DIAGNOSIS. The usual presenting manifestation of distal RTA is retardation of growth and weight gain. The infants may present early because of "failure to thrive." By the second year of life evidences of rickets may be present with characteristic deformities and x-ray findings. Polyuria and enuresis are also major complaints. Potassium deficiency with severe muscle weakness has been a serious problem in some patients, but one seen more commonly in older children and adults with untreated distal RTA.

The laboratory findings are reduced serum bicarbonate concentration, usually between 12 to 18 mEq/l, increased serum chloride

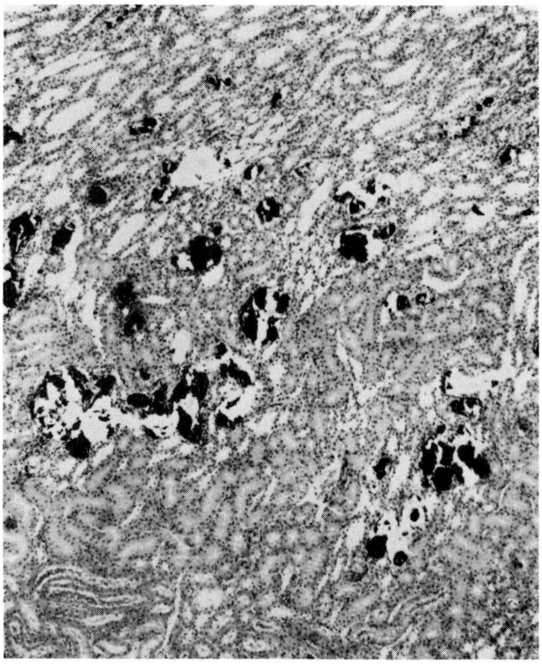

Figure 5–39. Experimental nephrocalcinosis in rats produced by feeding acetazolamide.

and usually normal serum sodium and calcium. Serum phosphate and potassium concentrations are reduced, and when rickets is present alkaline phosphatase activity is increased.

The differential diagnosis includes cystinosis, Bartter's syndrome and proximal RTA. Cystinosis is associated with the urinary findings of the Fanconi syndrome, whereas glycosuria and aminoaciduria are not present in RTA. Polyuria and potassium deficiency with growth retardation are major manifestations of Bartter's syndrome, but these patients have a hypochloremic alkalosis in contradistinction to the hyperchloremic acidosis of RTA. The differentiation between distal and proximal RTA is made by determination of urine pH during moderately severe bicarbonate deficiency, 15 mEq/l or less, either spontaneously occurring or induced by ammonium chloride in a dosage of 0.15 gm/kg. Patients with proximal RTA will excrete a urine with a pH between 5.0 and 6.0, whereas those with distal RTA will continue to excrete a neutral urine and the pH rarely drops below 6.8.

TREATMENT. The specific treatment is the administration of sufficient buffer base to cover the endogenous H^+ ion production and thus make unnecessary distal tubule H^+ ion gradient development and regeneration of bicarbonate by this means. The ordinary diet generates approximately 1 to 3 mEq/kg of H^+, chiefly from metabolism of sulfur-containing amino acids. This can be reduced by a low protein intake but severe dietary restriction is not indicated. Sodium bicarbonate, sodium lactate or sodium and potassium citrate can be given to provide the bicarbonate needed to replace that lost by endogenous H^+ ion production. We have preferred to use the last preparation in the form of a flavored solution of sodium and potassium citrate (Polycitra) providing 1 mEq/ml each of sodium and potassium. One to 2 ml/kg/day of this solution theoretically should suffice to maintain normal serum bicarbonate concentration. However, it has been found that some patients who fit the diagnostic criteria of distal RTA require much larger amounts of sodium bicarbonate or bicarbonate precursors than theoretically should be needed to replace the bicarbonate lost by endogenous H^+ ion production. This indicates that these patients also have a low renal threshold for bicarbonate, suggesting that they have both proximal and distal RTA. Doses of sodium bicarbonate or its equivalent of 6 mEq/kg/day or more may be required to maintain normal concentrations of serum bicarbonate in these patients. Although extra potassium can be given, and we prefer to use a mixture of sodium and potassium salts, the excess potassium loss is prevented if the serum bicarbonate is kept in the normal range, and patients have been adequately treated with sodium bicarbonate as the only source of base.

Treatment of the rickets also simply requires correction of the bicarbonate deficiency. The serum phosphate promptly rises when this is accomplished and normal bone mineralization follows without the need for more than physiologic quantities of vitamin D. Acceleration of growth and weight gain is associated with the return of serum bicarbonate concentrations to normal. The basis for the stunting of linear growth in chronic acidosis is not known. Whether somatomedin production by growth hormone is impaired has not been determined. Presumably many cellular enzymatic activities might be impaired by alteration of the intracellular environment in response to extracellular bicarbonate deficiency and compensating reduced pCO_2.

As the growth charts in Figure 5–40 indicate, normal growth rate is possible with adequate treatment (S.T.). However, irregular treatment with poor compliance (E.K.) resulted in major stunting of growth aside from the bowing of the legs due to rickets. Restoration of serum bicarbonate concentrations to normal also reduces urine calcium excretion and should make the development or progression of nephrocalcinosis less likely. We have seen nephrocalcinosis appear in our patients despite treatment, but these have all been referred to us for diagnosis after several years of disease and may already have had calcium phosphate precipitates in the tubules, even though not large enough to be visualized by x-ray at the initial examination. In addition, treatment may have been interrupted at times, thus permitting more deposition of calcium phosphate in the kidney. This degree of nephrocalcinosis, however, has not been of serious import and GFR has been maintained. Inadequately treated patients will show severe nephrocalcinosis and eventual renal insufficiency.

CASE REPORTS

S.T., a white male, was referred at 22 months of age for failure to thrive and awkward gait. Family history was negative for renal disorders. Polyuria and polydipsia had been present since early infancy. On exami-

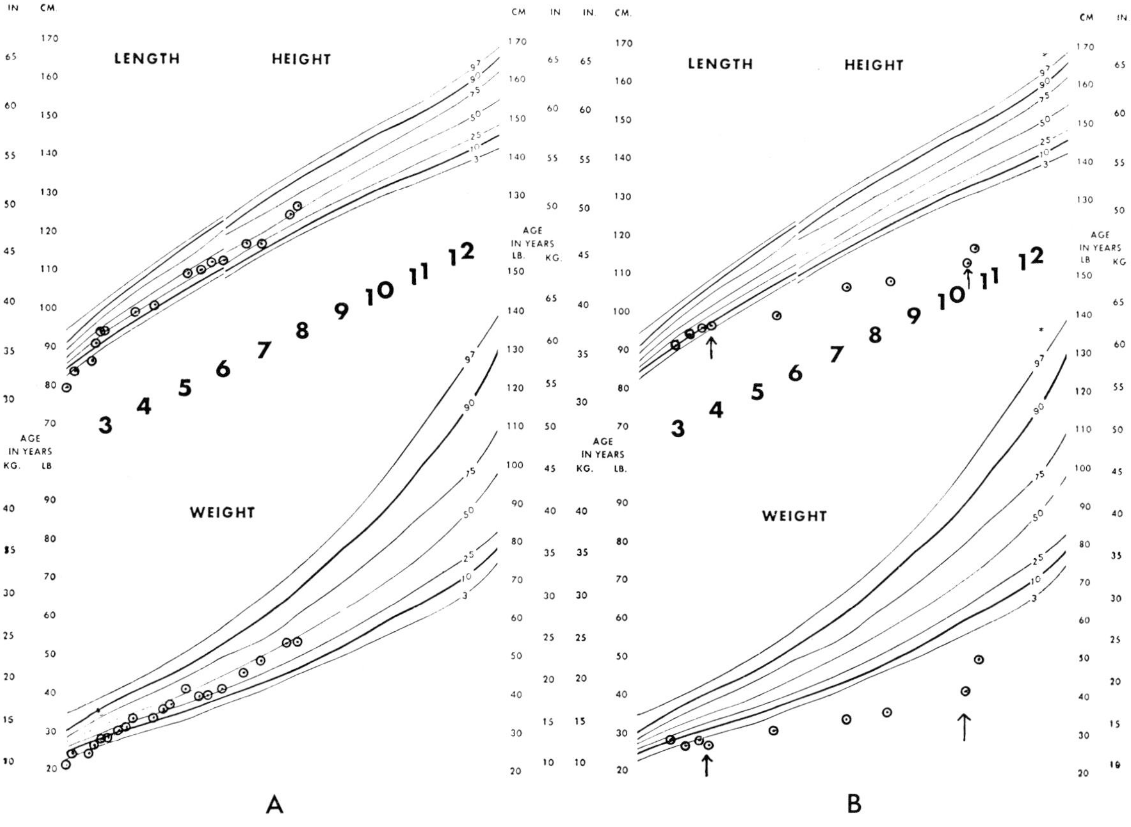

Figure 5–40. *a*, Growth curve of patient (S.T.) with renal tubular acidosis. Acceleration of growth and maintenance of normal growth pattern followed correction of acidosis by feeding sodium lactate or sodium and potassium citrates. *b*, Growth curve of patient (E.K.) with renal tubular acidosis whose treatment was inadequate because of noncompliance. Marked growth retardation is evident. (From Harrison, H. E.: The varieties of rickets and osteomalacia associated with hypophosphatemia. Clin. Orthop. 9:72, 1957. Reprinted with permission.)

Table 5–20. Distal Renal Tubular Acidosis (S.T. – Born 10/7/48)

	Serum						
	HCO_3	*Cl*	*Na*	*K*	*Ca*	*P*	*Alk. P'tase*
Date		*mEq/l*				*mg/dl*	*Bodansky Units*
5-3-50	14.1	123	–	–	9.5	3.1	26
5-29	14.9	132	151	4.1	10.6	3.6	–
5-29	Vitamin D_2 – 500,000 units p.o.						
6-5	17.4	112	144	3.5	10.2	3.6	41
7-11	–	–	–	–	9.9	2.8	–
7-11	Vitamin D_2 – 600,000 units p.o.						
7-28	–	–	–	–	10.0	3.0	–
9-15	Sodium lactate – 40 mM/day, vitamin D_2 – 50,000 units/day						
10-3	21.6	114	–	–	10.4	5.6	31
10-30	24.3	112	–	–	9.7	7.0	–
4-25-51	26.6	108	–	–	10.6	5.8	17
9-26	23.3	101	–	–	12.2	5.4	12
	Vitamin D discontinued						
6-24-54	–	–	–	–	10.6	5.2	13.5
9-14-56	–	–	–	–	10.0	5.8	–
7-8-58	16.6	122	–	2.8	9.2	3.0	33.5
	Sodium and potassium citrate						
7-23	30.8	112	–	4.7	9.8	5.7	–
8-15	30.0	108	–	4.7	9.6	6.0	15.6

nation he weighed 7.7 kg and was 74 cm in length, both far below the third percentile. The costochondral junctions were enlarged and there was obvious knobbing of wrists and ankles. The diagnosis of rickets was confirmed by x-ray examination. Urine was consistently dilute with a pH of 7.0 or greater, and there was no increase of concentration with attempted water restriction. Serum electrolytes showed a persistent hyperchloremic acidosis with normal serum calcium and reduced serum phosphate concentrations. The effect of treatment is summarized in Table 5–20.

Large doses of vitamin D failed to increase serum phosphate concentration when the bicarbonate deficiency was uncorrected. The infant was then given additional cation as sodium lactate, with restoration of normal bicarbonate concentration, along with continued high doses of vitamin D. Serum phosphate concentrations promptly rose to normal and healing of rickets ensued. Acceleration of growth was striking, as shown in Figure 5–40*a*. Because of hypercalcemia the large doses of vitamin D were discontinued and he was given only usual prophylactic amounts of vitamin D. As long as the serum bicarbonate was maintained by sodium lactate or sodium and potassium citrate (Polycitra), serum phosphate concentrations remained normal. When treatment was discontinued bicarbonate deficiency and hypophosphatemia recurred, with correction of both by Polycitra administration (Table 5–20). On continued therapy he has grown into a normal-sized adult without skeletal deformity. Polyuria with hyposthenuria persists, but there has been no deterioration of GFR despite radiologic evidence of nephrocalcinosis.

COMMENTS: The diagnosis of renal tubular acidosis of the distal type was made on the basis of urine pH consistently above 7.0 despite hyperchloremic acidosis with reduction of serum bicarbonate to 14 mEq/1. The polyuria with lack of concentrating ability and the hypophosphatemic rickets are also characteristic accompaniments of this type of renal tubular dysfunction. The lack of response of serum phosphate concentration to large vitamin D loads in the presence of bicarbonate deficiency is clearly demonstrated. When the bicarbonate deficiency was corrected by sodium lactate administration, 40 mM/day, or approximately 5 mEq/kg/day, serum phosphate promptly rose to normal concentrations and healing of the rickets resulted. This boy was continued on large doses of vitamin D, but this was unnecessary, and as shown subsequently serum phosphate concentrations decreased with recurrence of bicarbonate deficiency and increased with correction of bicarbonate deficiency without the need for additional vitamin D. In the later period hypokalemia also occurred during bicarbonate deficiency, with restoration of normal concentrations when sodium and potassium citrate was given and serum bicarbonate concentrations were increased. The negative family history is not unusual. Many cases of RTA have been sporadic although familial aggregation does occur.

K.V., a white male, was referred at 3½ years of age because of deformities suggestive of rickets. Father and mother were in good health and there were three female siblings, none of whom had chronic renal disease or rickets. He was normal at birth, weight 8 lb 6 oz. His feeding was satisfactory and he had been receiving 1000 units vitamin D daily since 6 weeks of age. At 2 years of age he weighed only 17 lb and by 3 years deformities of wrists and legs were noted. He had received 10,000 units vitamin D daily for two months prior to referral. On examination he was 11 kg in weight and 85 cm in length, well below the third percentile for a 42 month old child. The costochondral junctions, wrists and ankles were enlarged and the legs were laterally bowed. Radiographic evidences of marked rickets were present (Fig. 5–41). Serum electrolyte concentrations are summarized in Table 5–21. Urine was pH 7.0, Sp.Gr. 1.001, and protein and glucose negative. Despite the large amount of vitamin D he had received the serum phosphate concentration was low, 3.0 and 4.1 mg/dl. He was not given any more vitamin D but was started on sodium and potassium citrate solution in a quantity sufficient to correct the bicarbonate deficiency. There was a prompt rise in serum phosphate (Table 5–21) and radiologic healing of rickets followed. Response to treatment was excellent, with improvement of appetite and a growth spurt (Fig. 5–42). After one year of

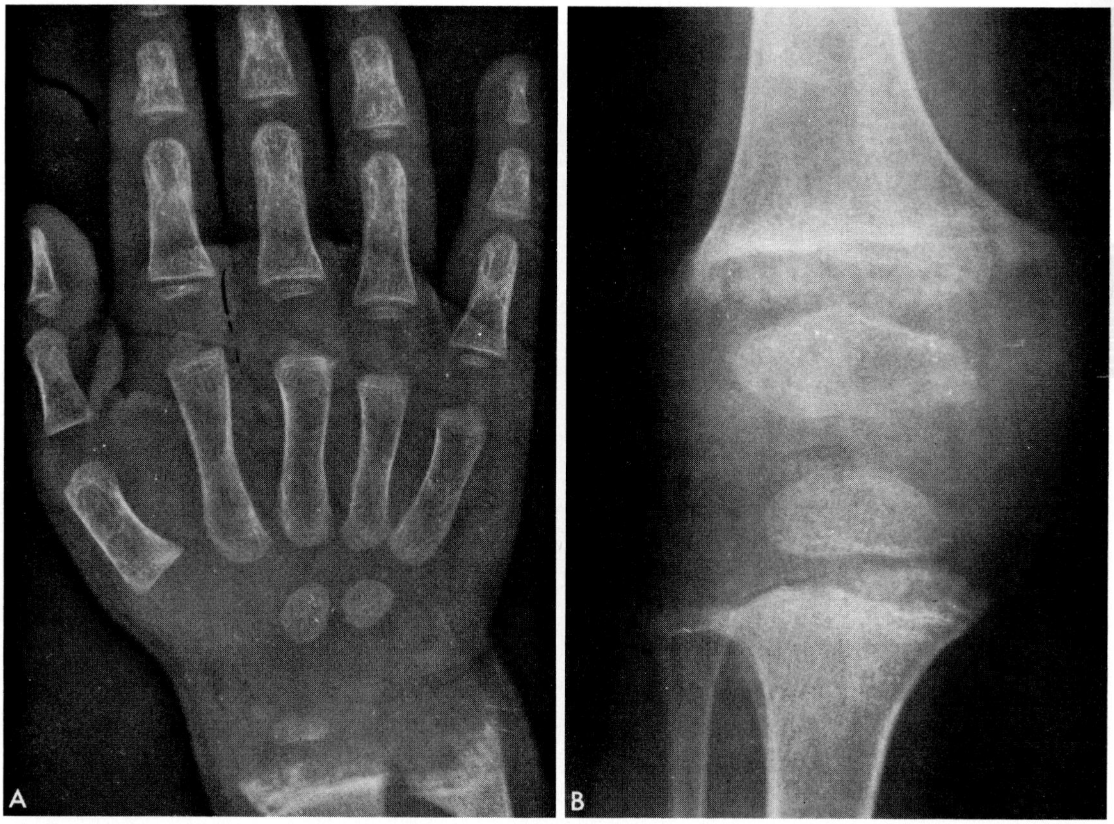

Figure 5–41. *a* and *b*, Severe rickets in patient (K.V.) with distal renal tubular acidosis.

Table 5–21. Distal Renal Tubular Acidosis Treated with Sodium and Potassium Citrate (K.V.)

	Serum						
Age Months	*HCO_3*	*Cl*	*Na*	*K*	*Ca*	*P*	*Alk. P'tase*
	mEq/l				*mg/dl*		*Bodansky Units*
42	14.9	113.0	145.0	4.9	10.8	3.0	21.4
	16.6	116.4	150.0	4.8	9.6	4.1	–
43	Sodium and potassium citrate solution – 40 ml/day (40 mEq Na and 40 mEq K/day)						
43-1/2	34.4	110.8	153.0	5.1	–	–	–
	Sodium and potassium citrate solution – 30 ml/day						
44	30.0	106.0	139.0	5.3	9.9	4.8	14.7
46	25.0	106.0	139.0	3.8	10.1	5.1	18.4
47	–	–	–	–	10.1	6.2	–
50	26.6	106.0	136.0	4.6	11.1	5.0	14.0
72	28.0	106.0	–	–	9.4	4.8	9.0

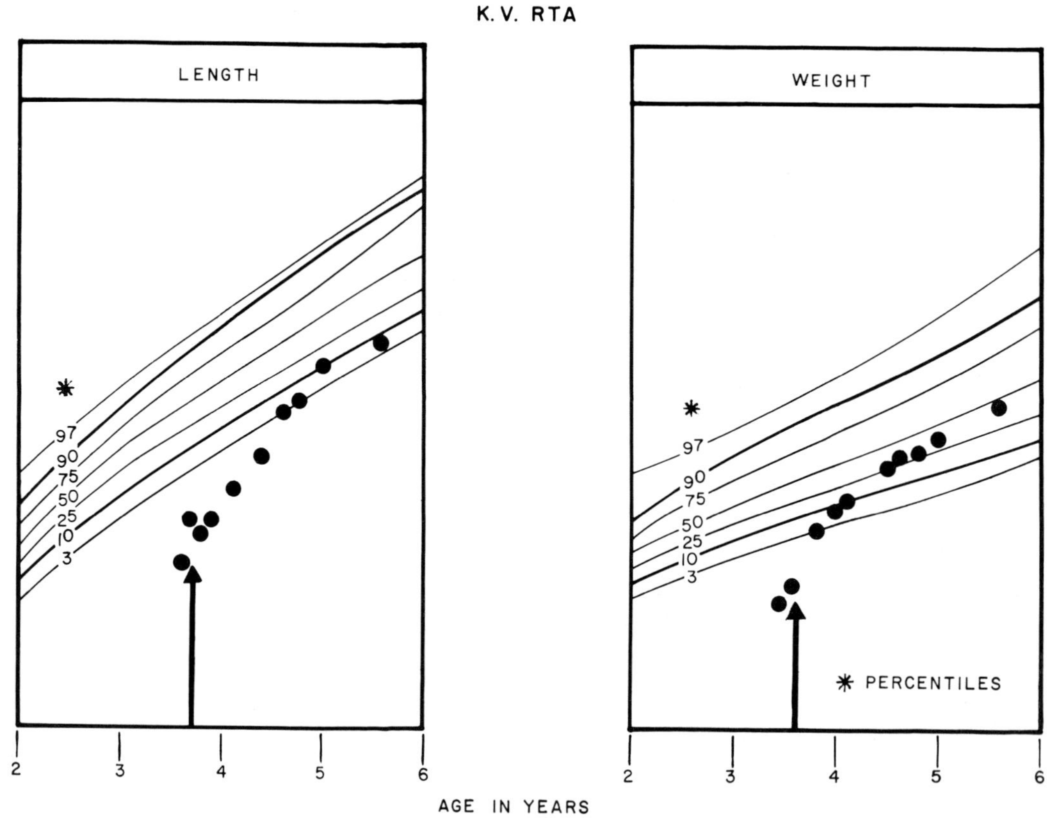

Figure 5–42. Acceleration of growth in patient with distal renal tubular acidosis following treatment of acidosis with sodium and potassium citrates. Arrows indicate beginning of treatment.

treatment the bowing of the legs, which had been the most important deformity, was considerably improved and there was little residual abnormality.

COMMENTS: This is another example of distal RTA manifested by urine pH of 7.0 concurrently with a serum HCO_3 concentration of 14.9 mEq/1. The hypophosphatemia had persisted despite large doses of vitamin D, 10,000 units/day for two months, and rickets was marked both by clinical and radiologic examination. Vitamin D was discontinued and sodium and potassium citrate treatment begun. The initial dosage, 4 mEq/kg each of Na and K as citrate, produced a metabolic alkalosis, and reduction of dosage to 3 mEq/kg of Na and K brought the serum HCO_3 concentration in the normal range. Serum phosphate promptly rose to normal values, with striking growth acceleration and complete healing of rickets.

GENETIC PRIMARY HYPOPHOSPHATEMIC RICKETS (HYPOPHOSPHATEMIC VITAMIN D-RESISTANT RICKETS)

Genetic primary hypophosphatemia is the most common form of hypophosphatemic rickets presently seen. Most of the kindreds studied are compatible with inheritance by a dominant gene on the X-chromosome. This requires that 50 per cent of the male or female offspring of an affected female are at risk, but that 100 per cent of the female and none of the male children of an affected male show the trait. It also implies that the affected females who are heterozygous for the gene have less severe disease on the average than the male who is hemizygous. This lesser severity in the female follows from the Lyons phenomenon in which one of the X-chromosomes is inactivated, so that statistically 50 per cent of the X-chromosomes with the abnormal gene will not be effective. We have encountered one family in which the genetic transmission is compatible with a dominant gene on an autosomal chromosome: affected father, two of three daughters and one son affected. There are also reports of kindreds in which the hereditary pattern is that of an autosomal recessive gene. Phenotypically there are no discernible differences among these three genotypes except that the expected lesser severity of the disorder in the female of the X-linked form would not be seen in the other two genotypes. Many of the cases are sporadic without familial occurrence, and it is possible that these are mutants who will be progenitors of a kindred with the more common X-linked dominant inheritance. On the other hand some may be examples of autosomal recessive inheritance in a single child. Only follow-up studies of the children of these patients can provide the answer.

The major physiologic disorder in this genetic abnormality is defective renal tubular reabsorption of phosphate, with consequent hypophosphatemia. There are no other associated renal tubular disorders; no glycosuria, no aminoaciduria, no bicarbonate wasting, no failure of H^+ ion secretion, no polyuria, no potassium wasting and glomerular filtration rate is normal. There is likely an associated defect of intestinal transport of phosphate so that the primary problem lies in transepithelial phosphate transport. There is no evidence of defective phosphate transport in red blood cells, so that uptake of phosphate by tissue cells probably is not abnormal except insofar as extracellular phosphate concentrations are excessively low owing to the renal tubule phosphate transport deficiency.

There is still uncertainty about the specific defect of renal tubule transport of phosphate in these patients. It has been thought that they have no abnormality of vitamin D metabolism, but there are contradictory reports about serum 1,25-dihydroxyvitamin D concentrations. There is also controversy over whether they have any degree of secondary hyperparathyroidism. There may be some degree of calcium malabsorption, thought to be secondary to failure of phosphate absorption, and this may account for the somewhat high values of serum IPTH reported by some laboratories. On the other hand other investigators have found serum IPTH concentrations in the normal range.

It is certainly clear that the defect of renal tubular reabsorption of phosphate is too great to be accounted for by secondary hyperparathyroidism. The physiologic action of parathyroid hormone, however, may play a role superimposed upon the primary abnormality of the major phosphate transport process in the proximal tubule. There is now evidence that some additional reabsorption of phosphate normally occurs in the distal tubule and that this is also inhibited by PTH. The current evidence may be summarized as indicating in primary hypophosphatemia absence of the quantitatively most important phosphate reabsorptive system in the proximal tubule, which in the normal individual is inhibited by PTH and is responsible for the largest moiety of filtered phosphate retrieved by the tubule. A system that reabsorbs a much smaller fraction of filtered phosphate remains which may be a distal tubular transport function apparently sensitive to physiologic concentrations of PTH. This may explain the increased tubular reabsorption of phosphate and consequent increased serum concentrations of phosphate observed when PTH secretion is suppressed by elevation of serum calcium. As will be pointed out in a later section, this has important implications for the treatment of these patients.

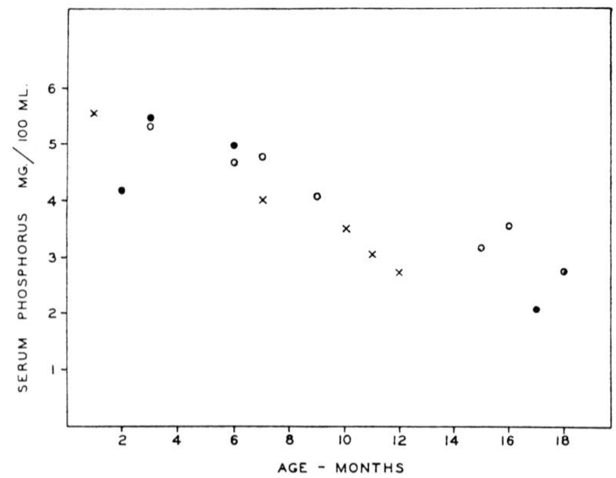

Figure 5–43. Age of onset of hypophosphatemia in infants with genetic primary hypophosphatemia. (From Harrison, H. E. et al: Growth disturbance in hereditary hypophosphatemia. Am. J. Dis. Child. *112*: 290, 1966. Reprinted with permission.)

Clinical Course and Differential Diagnosis

The infants are entirely normal at birth both in size and bone structure. Serum phosphate concentrations at this time are normal. Even though the mother is hypophosphatemic and has had rickets, placental transport of phosphate is evidently intact so that the concentration of extracellular phosphate in the fetus is normal. During the course of the next few months of life the serum phosphate decreases, but diagnostic hypophosphatemia — serum P below 4.0 mg/dl — may not be found until 6 months of age (Fig. 5–43). It therefore is not possible to predict in the first month or two of life whether an infant who has a parent with genetic hypophosphatemia will be normal or affected. The growth of the affected infant is also within the normal range for the first six to nine months of life, but as the serum phosphate decreases the growth rate diminishes and short stature is a major manifestation of this genetic disorder (Fig. 5–44). The basis for the growth retardation is presumably the low concentration of inorganic phosphate in extracellular fluid, as evidenced by the experience that slowing of growth becomes apparent at the time that serum phosphate concentrations are unequivocally low, and also by the growth response to treatment with high doses of inorganic phosphate (Fig. 5–45).

In addition to the somatic growth failure secondary to hypophosphatemia, there is further dwarfing if the lesions of rickets in the lower extremities are sufficiently severe to cause marked bowing and shortening of the legs (Fig. 5–46). The basis for the normal serum phosphate concentrations during fetal life, as mentioned above, is placental transport of phosphate whether or not maternal hypophosphatemia exists. The delayed onset of hypophosphatemia in the infant despite the genetic basis of the renal tubular defect has been assumed to be the relatively low GFR of the young infant. Since the serum phosphate concentration is a function of the balance between filtered phosphate and renal tubular reabsorption of this ion, a relatively low GFR could permit normal serum phosphate levels despite deficiency of the major phosphate reabsorptive system. Another possibility that can only be speculative is that the reduction of tubular reabsorption of phosphate is due to the accumulation of an inhibitor of this process, and several months must elapse before the inhibitor is present in sufficient concentration to reduce tubular reabsorption of phosphate to minimum values.

The progression of the rickets is slow and during the first year of life little may be noted. Unlike the infant with vitamin D deficiency rickets there is little involvement of the head or chest. Frontal and parietal bossing with head enlargement is not noticeable. Occasionally, however, premature synostosis of the sutures occurs, which may require operative intervention. The chest deformities of costochondral junction enlargement and Harrison's groove and the vertebral deformities of kyphoscoliosis seen in severe vitamin D deficiency of infancy also are not present in primary hypophosphatemic rickets. It is of importance that the pelvic deformity of vitamin D deficiency rickets, which in the infant girl results in the contracted pelvis of adult life responsible for severe dystocia, also is not a major problem in primary hypophosphatemic rickets.

The deciduous dentition is not delayed and the teeth are normally mineralized, unlike the poor enamel structure of the deciduous teeth of infants with early vitamin D deficiency rickets. The permanent teeth ordinarily are also adequately mineralized although enamel hypoplasia has been seen in some patients. This certainly is not as marked, however, as in patients with hypocalcemia due to vitamin D deficiency or dependency rickets or in children with hypoparathyroidism.

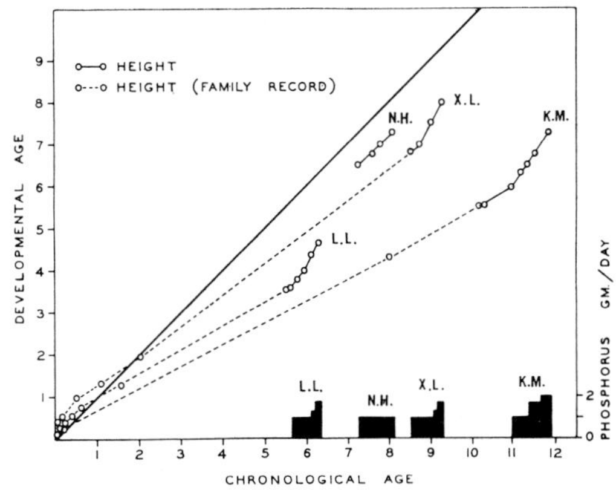

Figure 5–45. Effect of oral phosphate treatment on linear growth of patients with primary hypophosphatemia. (From Harrison, H. E. et al: Growth disturbance in hereditary hypophosphatemia. Am. J. Dis. Child. *112*:290, 1966. Reprinted with permission.)

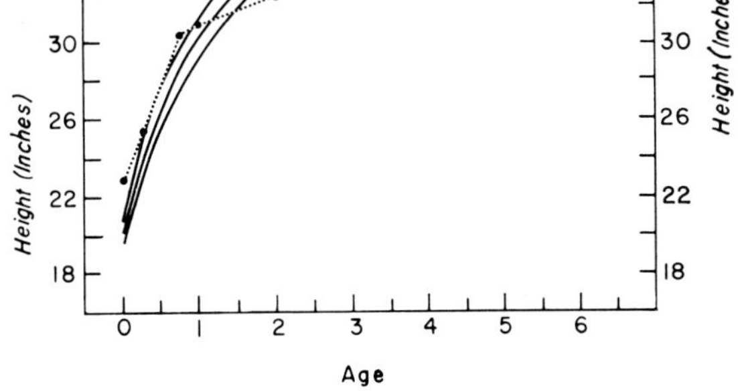

Figure 5–44. Growth chart of infant with genetic primary hypophosphatemia, showing initial normal height increase with abrupt slowing of growth at about 6 months of age. (From Tapia, J. et al: Vitamin D resistant rickets. J. Bone Joint Surg. *46-A*:935, 1964. Reprinted with permission.)

The major deformities are of the weight-bearing lower extremities, and unless there is a family history that prompts early investigation most patients come to the physician's attention after they begin to stand and walk and leg deformities become noticeable. Most commonly they exhibit lateral bowing (genu varum) but occasionally knock-knee deformity (genu valgum) is the major presenting complaint. The deformity in which one femur and tibia are laterally bowed while the other extremity is medially bowed, so that the extremities are curved in a parallel fashion, can also occur. In addition to visible bowing deformities there may be some degree of coxa vara and, probably more importantly, medial tibial torsion. The latter deformity is a significant cause of gait disability since, in order to compensate for the adduction of the feet resulting from medial torsion of the lower third of the tibia, the hips are abducted, which reduces the

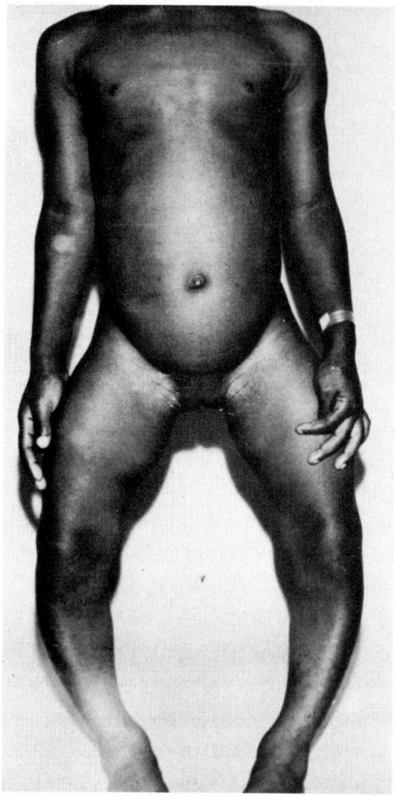

Figure 5–46. Primary hypophosphatemic rickets with severe bowing of legs and shortening of lower extremities through deformity.

stability of the hip joints. If coxa vara is also present a waddling gait results similar to that seen in congenital dislocation of both hip joints. Medial torsion of the tibias can be diagnosed by placing the child in a recumbent position on the examining table and positioning the patella so that it is in the midline. If the feet are not perpendicular to the table but deviate medially, tibial torsion is present.

The roentgenographic findings are characteristic. If the hypophosphatemia is diagnosed in the first few years of life the widening, cupping and irregular fraying of the metaphyseal ends of the long bones are seen at the distal end of the radius and ulna as well as the distal femur and proximal and distal tibia (Fig. 5–47*a* and *b*). The active lesions at the wrist do not persist, however, possibly because of slower growth of the bones of the forearm and also because weight-bearing may be an important factor in the further development of the metaphyseal lesions. In later childhood, therefore, the bony changes at the distal radius and ulna are not marked and the abnormality of metaphyseal structure in the distal femur and proximal and distal tibia is the usual finding. Radiographs of the knee and ankle joints are more important than those of the wrist in the diagnosis and follow-up of this disorder in later childhood, whereas in early vitamin D deficiency the wrist x-ray is the important diagnostic examination.

In the chronic rickets of primary hypophosphatemia the radiograph of the proximal tibia almost always shows a wedge-shaped defect of the medial portion of the metaphyseal surface (Fig. 5–48). It has been proposed that this defect results from weight stresses, since lateral bowing of the lower extremities causes increased weight thrust on the medial side of the knee joint. This is borne out by the finding of a defect in the lateral portion of the proximal tibial metaphyseal surface in a patient with genu valgum deformity, since in this deformity the lateral side of the knee joint carries the weight (Fig. 5–49).

Diagnosis

Presentation Because of Known Primary Hypophosphatemia in One of the Parents. In this situation the infant should be followed at intervals with serum phosphate and alkaline phosphatase determinations as well as careful measurements of growth. Decrease of the serum phosphate to 4.0 mg/dl or less is evidence that the infant is affected. Elevation of the serum alkaline phosphatase may occur while the serum phosphate concentration is still high enough for the diagnosis of primary hypophosphatemia to be in doubt (Table 5–22). Since the serum alkaline phosphatase activity is of bone origin, this rise indicates osteoblastic proliferation as the result of inefficient mineralization of bone and deformation of the poorly mineralized structure. That this should occur while the serum phosphate is not reduced below 4.0 mg/dl suggests that adequate mineralization of the rapidly growing bone of the young infant requires a considerably higher integrated

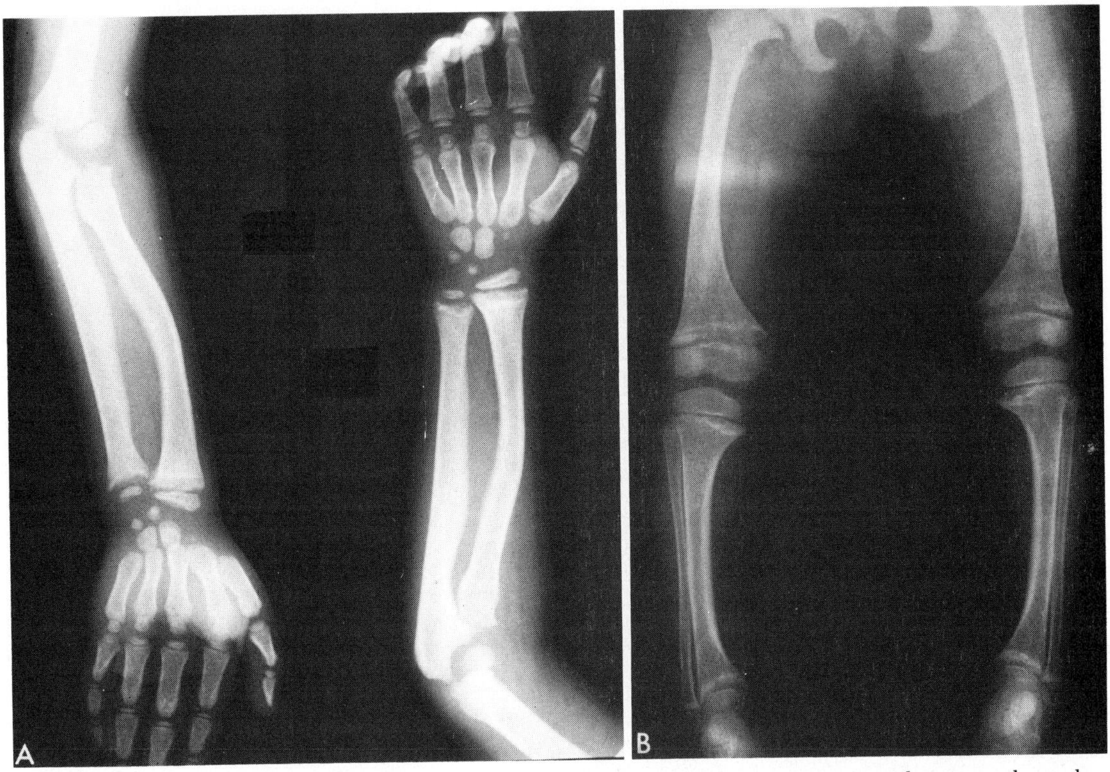

Figure 5–47. Radiographs showing lesions of rickets in a patient, age 3 years, with primary hypophosphatemia: *a*, forearm and wrist showing cupping and fraying; *b*, lower extremities showing metaphyseal irregularity of distal end of femur and of both proximal and distal ends of tibia.

extracellular phosphate concentration than 4.0 mg/dl. A single serum sample taken before a feeding does not adequately represent the integrated serum phosphate concentration during the 24 hours with increases following meals. The rise of serum alkaline phosphatase may then be a more sensitive index of reduction of the rate of bone mineralization than the postabsorptive serum phosphate concentration taken at a single time during the day. Because of the importance of a proven diagnosis of primary

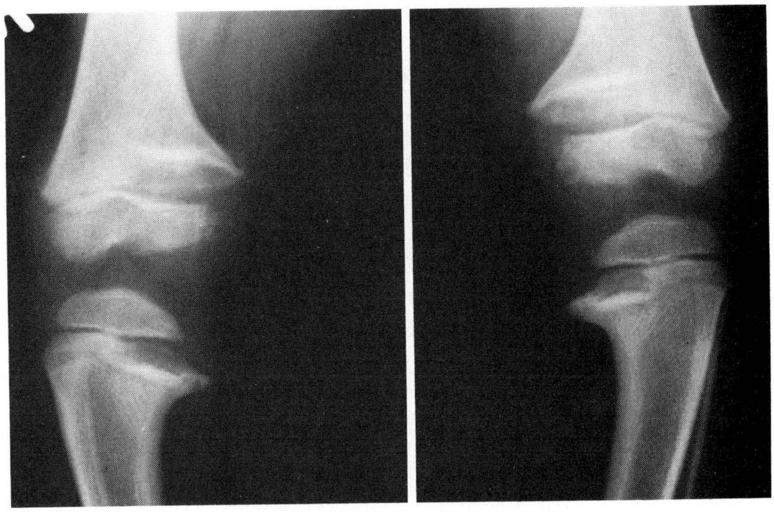

Figure 5–48. Characteristic wedge-shaped defect of medial surface of proximal tibia in patient with primary hypophosphatemic rickets and lateral bowing of legs (genu varum).

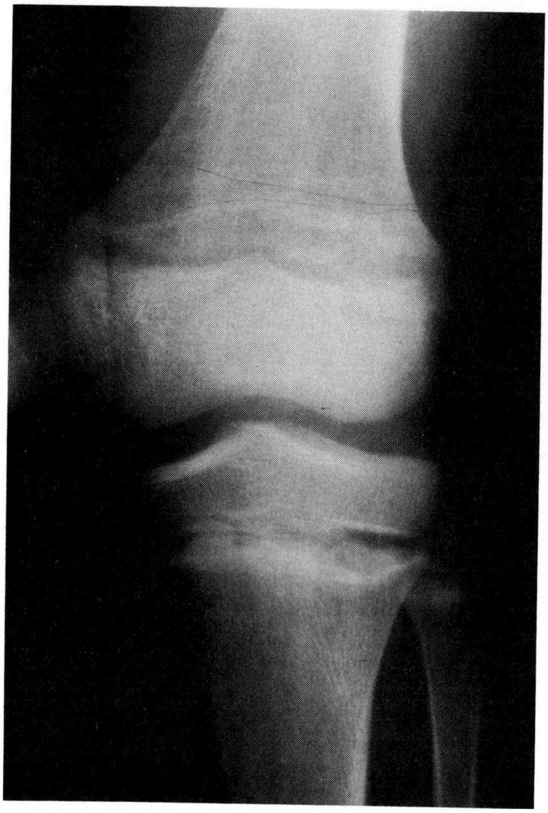

Figure 5–49. Defect of lateral surface of proximal tibia in patient with primary hypophosphatemic rickets presenting with genu valgum.

hypophosphatemia, we have not started treatment until the serum alkaline phosphatase is elevated on two successive monthly determinations and the serum phosphate is 4.0 mg/dl or less. Radiographs at this time are usually negative or may show equivocal changes of rickets.

Presentation Without Known Family History of Primary Hypophosphatemia. These children generally are seen after the age of 1 year because of leg deformities — usually lateral bowing and retardation of growth. On occasion the latter may be sufficiently marked for the patient to be referred because of growth failure, and deformities of the lower extremities may be minimal. The differential diagnosis includes physiologic bowing of the legs commonly seen in the second year of life, Blount's disease, metaphyseal dysostosis and various other forms of rickets. In physiologic bowing the radiograph shows normal metaphyseal structure, and serum calcium, phosphate and alkaline phosphatase are normal. In Blount's disease there is beaking of the medial portion of the upper tibial metaphyseal plate, but no irregularity of the metaphysis and no increased distance between the provisional zone of calcification at the metaphysis and the epiphysis. Serum calcium, phosphate and alkaline phosphatase concentrations are normal. Metaphyseal dysostosis may present with lower extremity deformities similar to those of late rickets, and the irregular calcification at the metaphyseal end of the long bones suggests rickets. Again, however, serum calcium, phosphate and alkaline phosphatase are normal.

The various forms of rickets that may present at this age are vitamin D deficiency rickets of the chronic type with periods of partial healing and activity, vitamin D-dependent rickets, rickets associated with the Fanconi syndrome or with renal tubular acidosis, as well as primary hypophosphatemic rickets. Serum and urine examinations as well as history, physical examination and x-rays are needed for differential diagnosis. Serum calcium concentrations are normal in primary hypophosphatemic rickets and are often reduced in vitamin D deficiency and vitamin D-dependent rickets. However, as mentioned above, in chronic vitamin D deficiency rickets, and possibly also in some cases of vitamin D-dependent rickets, secondary hyperparathyroidism is marked and serum phosphate concentrations are reduced so that serum calcium concentrations are returned to the lower limits of normal despite deficiency of the vitamin D metabolites.

Urine examination in these patients reveals marked generalized aminoaciduria and increased cyclic AMP excretion. Aminoaciduria is not found in primary hypophosphatemic rickets. Studies of cyclic AMP excretion are too few to permit of generalization but it is not consistently increased in our own determinations in patients with primary hypophosphatemia. In the Fanconi syndrome urine glucose and amino acids are increased, and in addition the multiple defects of renal tubule function

Table 5–22. Age of Appearance of Hypophosphatemia in Infant with Primary Hypophosphatemia (R. C.). Correlation with Alkaline Phosphatase in Serum

AGE DAYS	SERUM *Ca* (mg/dl)	SERUM *P* (mg/dl)	*Alk. Phosphatase Bodansky Units*
0	10.1	7.3	–
15	10.1	6.3	–
45	10.5	5.2	–
75	10.1	6.0	36.9
105	9.7	4.2	57.4
135	10.5	3.0	64.0
	Dihydrotachysterol = 0.1 mg/day; Neutra-Phos K = 1 gm P/day		
150	10.8	3.4	52.6
165	9.7	3.8	48.2
195	9.9	4.6	42.6
255	11.8	6.3	28.9
	Dihydrotachysterol reduced to 0.05 mg/day		

may result in reduced concentrations of serum bicarbonate and potassium. The rickets of renal tubular acidosis is associated not only with hypophosphatemia but with hyperchloremic acidosis, which is not present in primary hypophosphatemia. Another entity in the differential diagnosis is tumor-associated acquired hypophosphatemia (oncogenous hypophosphatemic rickets). This disorder will be discussed in detail in a subsequent section (p. 246).

A thorough family study is of the utmost importance in the total diagnostic procedure when a tentative diagnosis of primary genetic hypophosphatemic rickets is made. Although the transmission is usually as a dominant characteristic, there may be incomplete penetrance and the parent responsible for the gene may not have an obvious picture of chronic rickets and osteomalacia. The only manifestation in the affected parent may be short stature and hypophosphatemia (Fig. 5–50), with such slight evidences of rachitic deformities or osteomalacic change that ordinary examination may be negative. Bone x-rays may reveal the scars of old inactive osteomalacia (Fig. 5–51*a* and *b*). In other instances there may be neither hypophosphatemia nor unusual short stature in either parent, but a grandparent or a sibling of one of the parents may have the clinical picture of primary hypophosphatemic rickets. In such instances it must be assumed that the gene has failed of expression in the parent. If the parent in question is the mother this may be a consequence of the Lyons phenomenon with, by chance, inactivation in most cells of the X-chromosome with the responsible gene. If there is no evidence of familial incidence either of hypophosphatemia, short stature or the deformities of rickets, the possibilities of spontaneous mutation or of recessive inheritance remain.

Treatment

The original treatment employed when this entity was simply termed "vitamin D-resistant rickets" was vitamin D in pharmacologic dosage, 50,000 to 100,000 units or more per day. Some degree of healing resulted from this treatment and in the milder cases fairly good results were obtained. The presumption is that these large doses of vitamin D resulted in manifold increases in the concentration of 25-hydroxyvitamin D in plasma, which elevated serum phosphate concentrations by increased intestinal absorption of calcium and phosphate from the intestinal tract and suppression of parathyroid gland activity by the elevated serum calcium levels, and possibly also by increased concentration in plasma of vitamin D metabolites that may directly inhibit parathyroid gland secretion. In addition to the actions of large doses of vitamin D in increasing availability of dietary phosphate

Kindred—Primary Hypophosphatemia: R Family

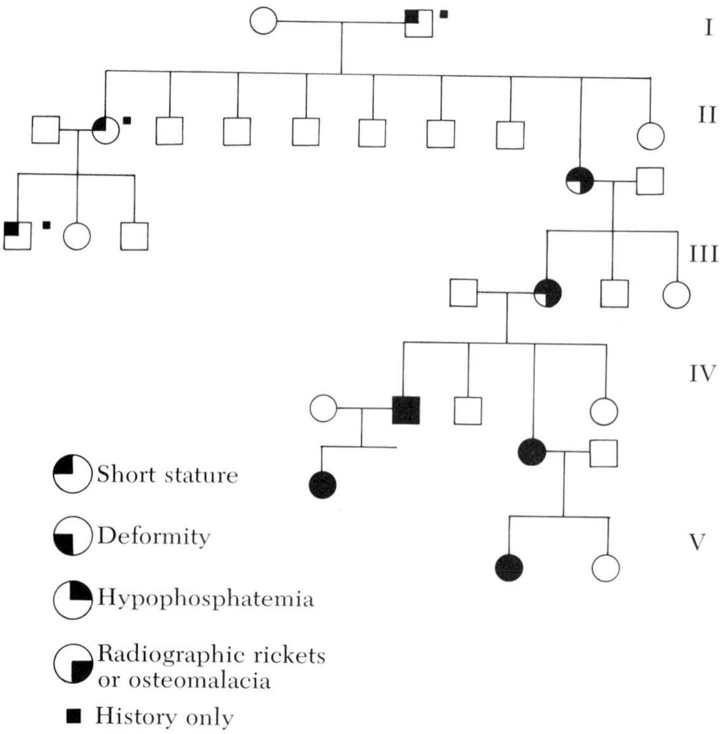

Figure 5–50. Kindred of genetic hypophosphatemia in which affected members of generations I, II and III never presented for medical care. They were short in stature but had only minimal deformity. Hypophosphatemia was discovered only when two siblings of generation IV were diagnosed as having primary hypophosphatemic rickets, and the mother and grandmother were then studied.

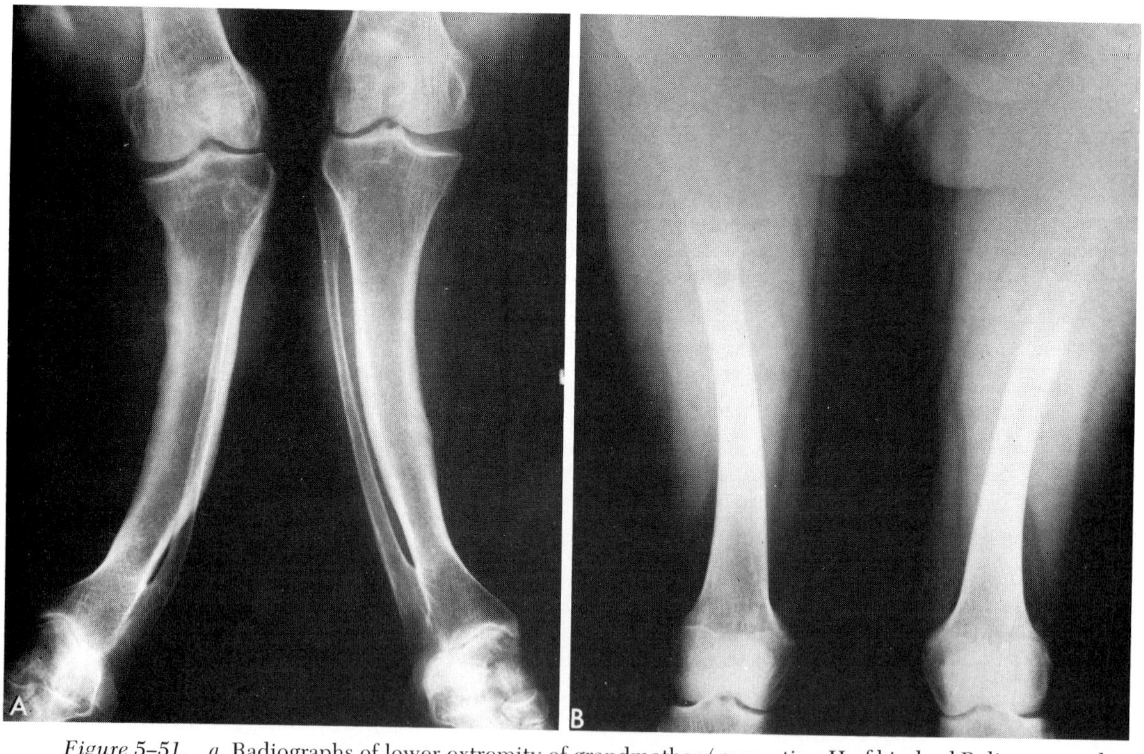

Figure 5–51. *a*, Radiographs of lower extremity of grandmother (generation II of kindred R diagrammed in Fig. 5–50). There are clear evidences of old osteomalacia and deformity despite lack of symptoms.

b, Radiographs of lower extremity of mother (generation III of kindred R). These films also show minimal evidences of osteomalacic deformity in an apparently well woman whose only manifestation was short stature.

and removing any inhibitory effect of PTH on tubular reabsorption of phosphate, a further effect could be responsible for raising serum phosphate concentrations and thus promoting healing of rickets, namely, a reduction of glomerular filtration rate. This is a toxic action of vitamin D caused by the arteriolar constriction resulting from hypercalcemia. The vitamin D treatment of primary hypophosphatemia exposed the patient to the major toxic effect of vitamin D, i.e., increased concentrations of extracellular calcium with hypercalciuria and renal calcification eventually resulting in progressive renal insufficiency. Sufficient examples of renal calcification and renal injury subsequent to such treatment stimulated a search for alternative methods of treatment.

One alternative we examined was the use of dihydrotachysterol. As discussed in the section on renal osteodystrophy, this sterol is obtained by the reduction of tachysterol which, like vitamin D, is derived from the ultraviolet irradiation of provitamin D. The dihydrotachysterol used pharmaceutically is dihydrotachysterol$_2$ produced by irradiation of ergosterol. Dihydrotachysterol resembles vitamin D in its pharmacologic effects, i.e., producing hyperabsorption of calcium and phosphate from the gut, and increasing the concentration of calcium in serum, thus inhibiting parathyroid gland activity. Dihydrotachysterol is metabolized rapidly in the liver to 25-hydroxydihydrotachysterol, which is presumably the active metabolite inasmuch as it is not acted upon by the kidney and is active in the absence of the kidney. The 25-hydroxydihydrotachysterol, therefore, may substitute for the vitamin D hormone, 1,25-dihydroxyvitamin D.

The property of dihydrotachysterol that makes it an attractive substitute for pharmacologic doses of vitamin D is that it is more rapidly metabolized and inactivated in the body than is vitamin D. The metabolism of vitamin D to 25-hydroxyvitamin D is limited, and when large doses are given, although the concentration of 25-hydroxyvitamin D in plasma can be increased ten- to twenty-fold, a considerable fraction of the vitamin D is stored as such in adipose tissue. The consequence of this is that cessation of vitamin D treatment because of early toxic effects, hypercalciuria and hypercalcemia, does not stop the continued conversion to active metabolites of the large amounts of vitamin D stored in the body, so that there may be continued toxicity for months. Because dihydrotachysterol is metabolized and inactivated much more rapidly this cumulative toxicity is less important, and manifestations of excessive dosage disappear within ten days to two weeks after the sterol is discontinued. Manipulation of the dose to achieve therapeutic effects without dangerous hypercalciuria and hypercalcemia is thus simplified.

When crystalline dihydrotachysterol became available so that a pure preparation could be used, we examined its action in patients with primary hypophosphatemia. We found that it could indeed produce healing of the rickets in such patients, and that on a weight basis it was several times more potent than vitamin D, so that doses of 0.5 to 1 mg were approximately equivalent to 1.5 to 3 mg of vitamin D (60,000 to 120,000 units). Treatment with crystalline dihydrotachysterol alone has produced healing, as indicated in the case reports, G.M. and G.H. This treatment requires careful continued supervision of the patient, with monitoring of urine calcium excretion and serum calcium concentrations, to achieve a dose that will raise serum phosphate sufficiently to maintain healing but which will not cause hypercalciuria and renal calcification.

As evidence mounted that the functional lesion of primary hypophosphatemia was a defective tubular phosphate transport system, attempts were made to treat this entity by large phosphate loads. Healing could be produced by continued intravenous infusion of phosphate sufficient to maintain normal concentrations of extracellular phosphate, but this of course was not a solution for the treatment of this chronic disorder. In adults with acquired hypophosphatemic osteomalacia large doses of oral phosphate could raise extracellular phosphate sufficiently to improve bone mineralization. In children this treatment by itself requires such large doses of phosphate salts that it is not practicable. In addition these large phosphate loads interfere with calcium absorption and result in secondary hyperparathyroidism on this basis, thus antagonizing the healing action of the phosphate treatment. For this reason a combined treatment protocol combining oral phosphate loading plus dihydrotachysterol or vitamin D has become the more common method of therapy. In our clinic we have preferred dihydrotachysterol to vitamin D, but others have continued to use vitamin D in moderately large dose. In the future 1,25-dihydroxy vitamin D may be an effective steroid to use along with oral phosphate treatment, but there is as yet inadequate long-term experience to make a recommendation.

When the diagnosis has been confirmed, treatment is begun with gradually increasing amounts of phosphate given either as a buffered solution of sodium and potassium phosphate or potassium phosphate alone. Since large sodium loads increase ECF volume and GFR and reduce reabsorption of sodium and water in the proximal tubule, potassium phosphate might be theoretically preferable, and for this reason we treat our patients with such a preparation (Neutra-Phos K). This is given as a solution (300 ml= 1 gm P) in four or five divided doses during the day, the first on arising and the last before bedtime. The quantity of phosphate is gradually increased starting with 0.5 gm P per day up to 1.0 to 1.5 gm P per day under 1 year of age, and 1.5 to 2.5 gm P per day in older children and adolescents. The major side-effects of phosphate salt treatment are hyperperistalsis and diarrhea. By gradually increasing the dose over a period of several weeks intestinal tolerance can usually be achieved, although many patients have occasional colicky pains and watery stools. One of our patients, however, has developed profuse watery diarrhea whenever his daily dose has been raised above 1 gm P per day, so that he has never been able to take an adequate amount. Simultaneously with the phosphate treatment either dihydrotachysterol, 0.02 mg/kg/day, or vitamin D, 0.05 mg (2000 units)/kg/day, is started.

Monitoring of these patients requires measurement of urine excretion of calcium and phosphate as well as serum calcium, phosphate and alkaline phosphatase concentrations, plus x-ray examinations, as a guide to the effectiveness of treatment in the mineralization of the growing bone. The 24-hour urine excretion of phosphate serves as an indication of treatment compliance since most of the ingested phosphate will appear in the urine. The 24-hour urine excretion of calcium is the most sensitive guide to the appropriate dose of dihydrotachysterol or vitamin D. The urine calcium excretion in the untreated patient or the patient given phosphate alone is low, usually less than 10 to 15 mg per day. As the treatment with dihydrotachysterol or vitamin D continues urine calcium excretion rises, and a safe amount is in the range of 50 to 100 mg/day. If the excretion of calcium exceeds 4 mg/kg/day or 0.25 mg/mg urine creatinine per day, hypercalciuria is present and the dose of dihydrotachysterol should be reduced. If vitamin D is being given it should be discontinued, and when urine calcium excretion has decreased to the safe range it is restarted at a lower dose.

In this way each patient's dihydrotachysterol or vitamin D requirements can be determined and toxic effects avoided. Serum calcium concentrations should remain normal, not exceeding 10.8 mg/dl if urine calcium excretion is monitored. If higher concentrations are found the activated sterol treatment, whether dihydrotachysterol or vitamin D, should be discontinued until serum calcium is in the normal range. Serum phosphate in the fasting state may remain well below 4 mg/dl despite radiographic evidences of healing. If, however, determinations of serum phosphate are made one or two hours following a standard dose of phosphate, it will be found that the concentration will be increased about 1 mg/dl above the pre-dose level and remain elevated for several hours (Table 5–23). It should be possible, therefore, with the large loads of phosphate recommended, to maintain extracellular phosphate concentrations between early morning and late evening above the minimum value needed to initiate and maintain deposition of bone mineral. The serum alkaline phosphatase frequently remains somewhat elevated, indicating that complete healing has not

Table 5–23. Response of Serum Phosphate Concentration to Oral Phosphate – Primary Hypophospatemia

TIME HRS	SERUM Ca mg/dl	SERUM P mg/dl
0	10.1	2.5
	Neutra-Phos K = 400 mg P	
1	9.6	3.0
2	9.6	3.0
3	9.7	3.0
	Neutra-Phos K = 400 mg P	
4	9.7	4.2

been accomplished. When growth ceases the serum alkaline phosphatase can return to normal.

The combined treatment outlined above requires great motivation on the part of the patient and family if it is to succeed. The ingestion four to five times a day of a solution which, although flavored, cannot completely disguise the somewhat unpleasant taste of the phosphate salts, plus the occasional uncomfortable gastrointestinal effects of these salts, is not an easy task. Moreover, the necessity for close supervision by the physician with collection of 24-hour urine samples and determinations of serum constituents is also a trial to patient and family. These studies of urine and blood serum are initially done at monthly intervals, but this can be increased to three to four months after the patient has achieved stable values. Radiographs should not be repeated too often to avoid unnecessary exposure, and after the initial studies examination only of the knee joints is usually needed to determine the degree of healing. A repeat examination six months after beginning of treatment, and thereafter at yearly intervals, is adequate.

If the side-effects of phosphate salt loading are too great, treatment with dihydrotachysterol alone can be considered. In our initial studies healing was obtained with sterol treatment without added phosphate, but partly for theoretic reasons we chose to continue with the combined phosphate salt and dihydrotachysterol regime. Our associate in these early studies, Dr. Laurence Finberg, has maintained a group of patients at the Montefiore Hospital on dihydrotachysterol without phosphate salt treatment, and by close supervision of urine calcium excretion has been able to give doses of dihydrotachysterol that maintained continued healing and adequate growth without the addition of phosphate salts. This treatment requires further study, but may be the method of choice in patients who cannot tolerate phosphate salts without severe gastrointestinal disturbance.

Treatment must be continued until growth has ceased, in the female usually by 16 to 17 years of age and in the male 18 to 20 years. After this time treatment can be stopped, but the patient should be warned that osteomalacia may recur in later years, with back and hip pains. Continued supervision by a physician who knows the problem is required, and treatment may be reinstituted in some adults because of symptomatic osteomalacia.

Course and Prognosis

If the diagnosis is made early because of the familial story before radiographic evidences of rickets are present, and treatment is systematic with good compliance, there may be minimal deformity. The patients are usually in the lower percentiles for height but not so short as to be seriously handicapped. This is the ideal result but is not achieved in most patients, either because the diagnosis is not made until major deformity is already present or because the rigorous treatment and supervisory requirement are too much for the patient and lapses of treatment occur. With such lapses recurrence of active rickets, retardation of growth and progressive deformity of the lower extremities result. Such patients may be seriously handicapped because of short stature and leg deformities, which can not only be cosmetically unpleasant but also cause difficulty in walking and running.

If the leg deformities are severe corrective osteotomies are necessary. The preferred time for surgical correction depends on the severity of the deformity as well as the adequacy of medical treatment. Osteotomy in early childhood without good medical treatment may have disastrous results, with poor healing and recurrence of deformity. It is our practice to defer osteotomy if feasible until growth has ceased, meanwhile maintaining as good medical treatment as possible. In some instances earlier osteotomy has been necessary because the leg deformities have been so marked as to interfere with normal childhood activities. As good healing as possible should be achieved by intensive treatment prior to osteotomy. Vitamin D or dihydrotachysterol should be discontinued four to six weeks before operation to avoid immobilization hypercalcemia following surgery. The oral phosphate treatment should be continued, however, before and after osteotomy. The sterol treatment is resumed when the patient is up on crutches and no longer immobilized.

Patients with severe deformity of the lower extremities and consequent abnormal weight-bearing stresses have a high incidence of chronic degenerative joint disease in later years, particularly severe hip joint degeneration. Whether joint replacement can be possible in such patients remains to be seen. Active osteomalacia requiring continued treatment may also be seen during adult life. Phosphate treatment alone may be adequate to improve bone mineralization, but with proper monitoring there is no reason why hypophosphatemic osteomalacia should not be treated with dihydrotachysterol or vitamin D as well as oral phosphate loading.

CASE REPORTS

D.M., a white female, was first seen in consultation at 18 months of age because of severe rachitic deformities despite vitamin D supplementation. The family history was important since the father, now age 30, had had severe rickets in childhood, with osteotomies at age 11 years. His legs were very short and he walked with difficulty because of hip joint pain. The mother was of average height and had no deformities. D.M. was the first-born. She was a full term infant and had apparently done satisfactorily during the first year of life. Subsequently, marked bowing of the legs was noted that progressed despite conventional doses of vitamin D. On examination there was marked lateral bowing of the legs with medial tibial torsion, so that the child's gait was unstable. The pertinent laboratory findings in both child and father were normal serum calcium concentrations with low serum phosphate: D.M.'s serum P ranged between 2.5 and 3.5 mg/dl and the father's between 1.5 and 2 mg/dl. X-ray studies in D.M. revealed advanced rickets, with metaphyseal changes most marked at distal end of femur and at proximal and distal metaphyses of tibia. In the father bone x-rays revealed the coarse trabeculation and poor mineral density of osteomalacia.

The child was placed on vitamin D as the only form of treatment, and since healing of rickets was slow the dose was gradually increased until at age 3½ years she was receiving 150,000 units per day or approximately 10,000 units/kg/day. Healing of rickets did occur, as indicated by the radiograph, but after several months of this high dose vitamin D therapy serum calcium was found to be in the 11 to 12 mg/dl range and serum phosphate was also elevated above 4 mg/dl. The urine showed proteinuria and increased numbers of WBC but was sterile on culture. Vitamin D was discontinued and serum calcium concentrations slowly fell, but proteinuria and increased numbers of WBC persisted in the urine, and serum urea nitrogen and creatinine rose. Creatinine clearance was reduced to 40 to 50 ml/min/1.73M^2. The rickets was completed healed by x-ray, presumably because of the elevation of serum phosphate concentration secondary to reduced GFR. The diagnosis of hypervitaminosis D with renal injury was made despite the fact that hypercalcemia had never been severe.

Vitamin D treatment was never resumed in this patient. The serum phosphate dropped below 4 mg/dl and radiographic evidence of reactivation of rickets was seen. She was given phosphate salts to maintain normophosphatemia and healing of rickets resulted. On this regime it was noted that the growth rate, which was low since 1 year of age and had become even more reduced during the period of hypercalcemia, now increased so that the child's height reached the 50th percentile (Fig. 5–52). At age 6 years bilateral tibial osteotomies were done because the lateral bowing and medial torsion of the tibias were so severe that the child had an extremely awkward gait and could not run at all. The results of the operation were excellent and there was no residual deformity of consequence (Fig. 5–53). Kidney function remained stable through adolescence although creatinine clearance was consistently less than 50 per cent of normal. She had many problems during adolescence and early adult life, both emotional and physical. A major complaint was peptic ulcer with epigastric pain and bleeding, for which she was treated medically. There was no bone pain or osteomalacic change and serum phosphate concentration remained in high normal range.

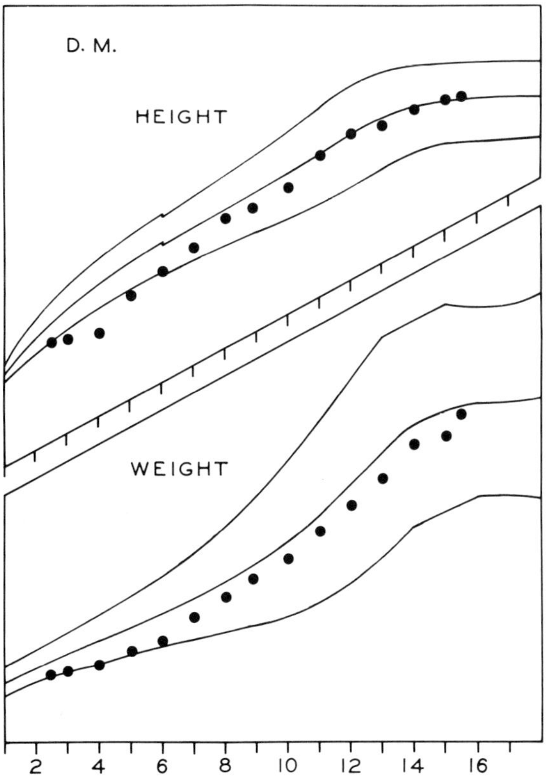

Figure 5–52. Growth curve of patient with hypophosphatemic rickets (D.M). Following recovery from hypercalcemia, growth rate accelerated and normal adult height was achieved with oral phosphate as the only treatment. (From Harrison, H. E. et al: Growth disturbance in hereditary hypophosphatemia. Am. J. Dis. Child. *112*:290, 1966. Reprinted with permission.)

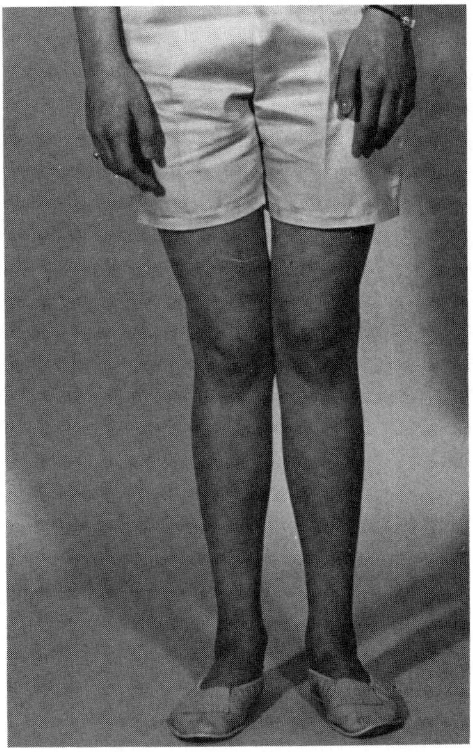

Figure 5–53. Patient (D.M.) as young adult. There are no apparent deformities of lower extremities.

At age 24 D.M. married. She and her husband were given genetic counseling concerning the chances of her bearing a child with genetic hypophosphatemic rickets. D.M. did become pregnant and her pregnancy was complicated by diminishing renal function. She delivered a full term, but small, male infant of 2500 gm who seemed normal at birth and who had normal serum phosphate concentrations (Table 5–22, R.C.) for the first three to four months of life. However, serum alkaline phosphatase concentrations were elevated at 3 months of age, and subsequently the infant showed hypophosphatemia and was started on phosphate salt and dihydrotachysterol treatment. When this child was about 1 year of age the family moved away from Baltimore and we have not been able to follow him. Treatment apparently has not been consistent and this boy has active rickets.

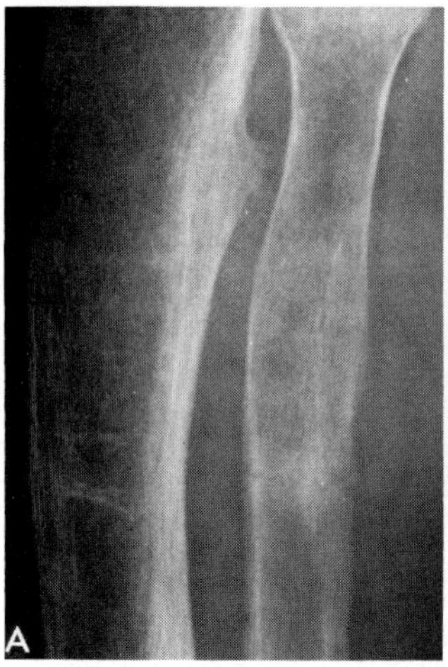

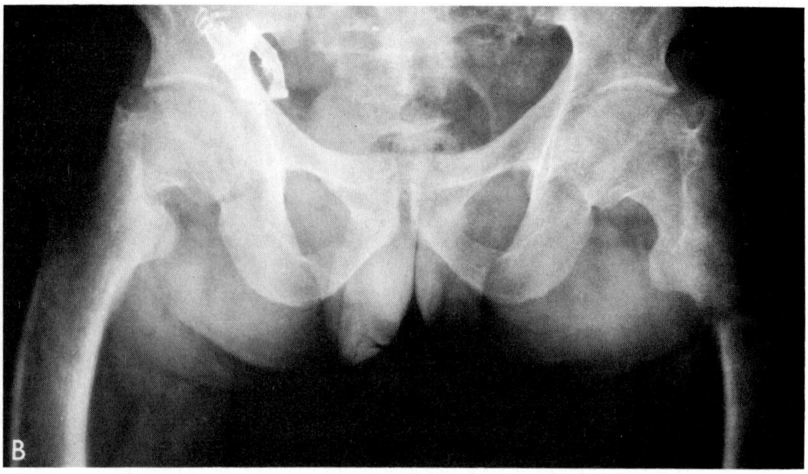

Figure 5–54. *a* and *b*, Osteomalacia in adult with primary hypophosphatemia. (Generation I of kindred M diagrammed in Fig. 5–55). The pseudofracture or Looser-Milkman zone of unmineralized matrix at site of muscle pull on bone is clearly demonstrated.

The mother's renal function deteriorated progressively following delivery and she is now in a renal dialysis program.

Family Study

As noted, the father of D.M. had hypophosphatemic rickets and osteomalacia for which he had had osteotomies as a child. At age 30 he had manifestations of active osteomalacia with severe pains in hips and legs on weight-bearing and neck pain. He had several automobile accidents and other trauma resulting in injury to cervical spine and fractures of extremities. Radiographs showed marked osteomalacia with Looser-Milkman zones (Fig. 5–54). He was treated with phosphate salts and dihydrotachysterol; the latter had to be carefully monitored since he had hypercalciuria and renal stones on doses of 1 mg/day. At present, at 50 years of age, he is relatively stable, walks with the aid of canes, has limited motion at both hip joints and fusion of articular facets of cervical spine.

The incidence of genetic primary hypophosphatemic rickets and osteomalacia in this kindred is shown in the diagram (Fig. 5–55). Of three daughters, two are affected and one is normal, both with respect to growth and serum phosphate concentrations. The one son is affected. This family, then, represents the less usual form of inheritance by a dominant gene on an autosomal chromosome. Phenotypically there is no obvious difference between this kindred and those with the more common mode of inheritance as an X-linked dominant characteristic, except possibly for the more severe osteomalacia in the father continuing during adult life.

The other affected daughter, Da.M., has had no unusual complications. Her treatment consistently has been oral phosphate salts and dihydrotachysterol. She has never had hypercalcemia and renal function is normal. At age 18 years she is in the 25th percentile for height and has minimal evidences of rachitic deformity, slight lateral bowing of the legs. She has been continued on treatment with both phosphate and dihydrotachysterol because of the father's history. Representative x-ray studies of the bones of this patient are shown in Figure 5–56.

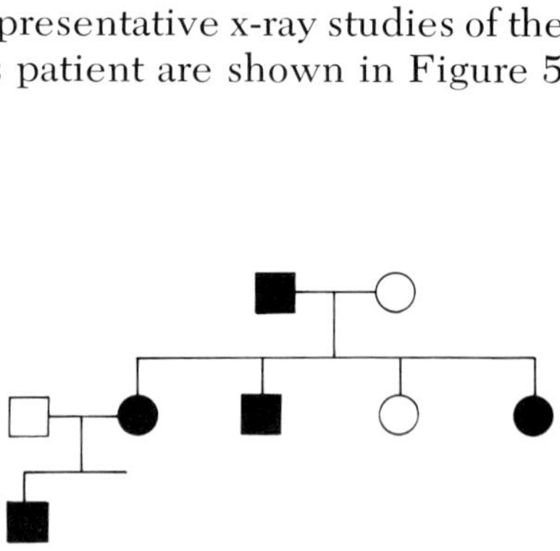

Figure 5–55. Diagram of kindred M showing autosomal dominant transmission of primary hypophosphatemia.

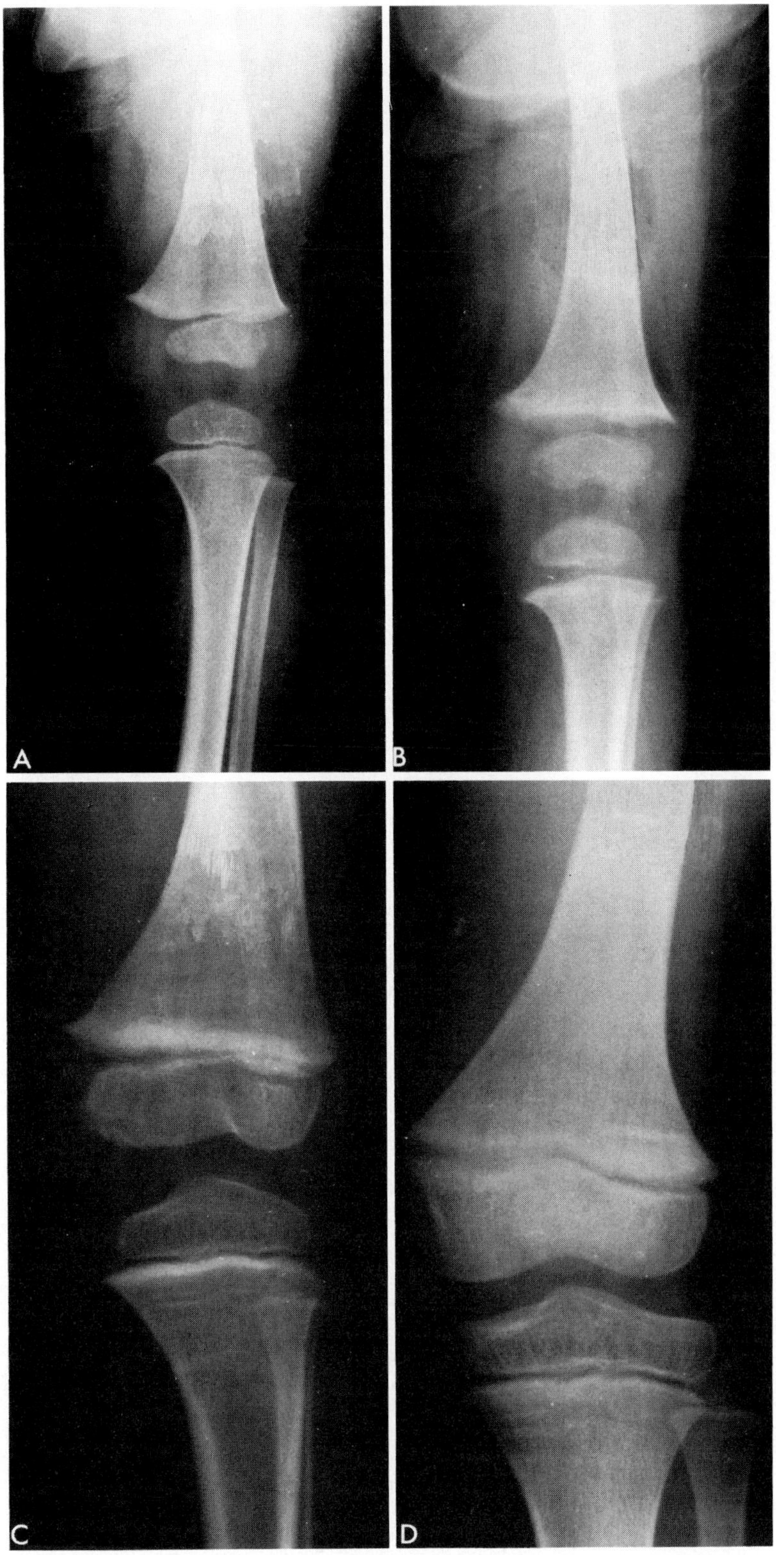

Figure 5–56. *a* to *d*, Serial radiographs of patient (Da. M.) of kindred M (Fig. 5–55). This girl had relatively mild hypophosphatemic rickets and achieved normal adult height. Lyonization could not account for her relatively mild course since the genetic transmission in this family was not X-linked.

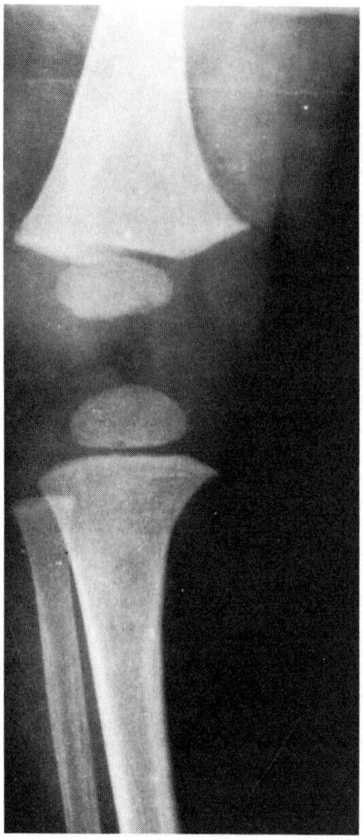

Figure 5–57. Radiographs of legs at 6 months of age (G.M. III). No evidence of rickets at this time.

The son, G.M. III, was followed from early infancy and showed no evidence of hypophosphatemia during the first months of life. At 6 months of age the serum phosphate concentration was slightly low but a definitive diagnosis was not made. Radiographs at this time did not show distinct lesions (Fig. 5–57). He was not reexamined until 18 months of age when he was clearly hypophosphatemic and showed radiographic evidence of active rickets (Fig. 5–58). He was treated with dihydrotachysterol alone without phosphate salt loading; the effect of this treatment on serum calcium, phosphate and alkaline phosphatase concentrations is shown in Table 5–24 and the healing of rickets by x-ray is documented in Figure 5–59. Apparent complete healing of rickets resulted, with dihydrotachysterol alone as the modality of treatment. Serum phosphate concentrations were increased without evidences of toxicity, i.e., no hypercalciuria, hypercalcemia or impairment of renal function. However, it was elected to continue treatment with the combination of phosphate salts and dihydrotachysterol. On

Text continued on page 238

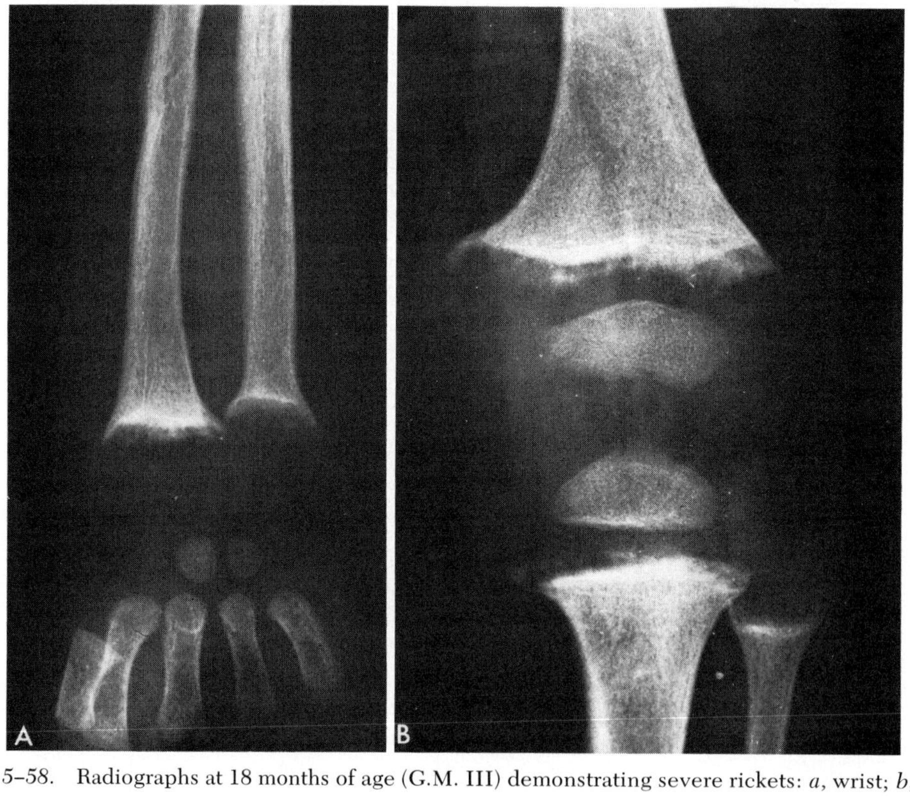

Figure 5–58. Radiographs at 18 months of age (G.M. III) demonstrating severe rickets: *a*, wrist; *b*, knee.

Table 5–24. Effect of Dihydrotachysterol Administration on Serum Calcium, Phosphate and Alkaline Phosphatase Concentrations of Patient with Genetic Primary Hypophosphatemia. No Added Phosphate

AGE MONTHS	SERUM *Ca* *mg/dl*	SERUM *P* *mg/dl*	*Alk. Phosphatase Bodansky Units*
	Patient G. M. III		
18	10.2	2.1	34.9
	DHT – 0.5 mg/day		
20	10.1	3.2	32.3
23	10.8	4.3	22.2
	DHT – 0.25 mg/day for one month followed by DHT 0.125 mg/day		
28	9.4	3.7	13.3
32	9.9	3.7	8.8
	Patient N. S.		
11	–	2.9	34.9
	Vitamin D_2 – 0.25 mg/day		
13	9.1	2.5	36.5
	DHT – 0.25 mg/day		
13-1/2	10.5	3.4	39.9
14-1/2	10.0	3.7	33.6
15	10.0	4.3	21.7

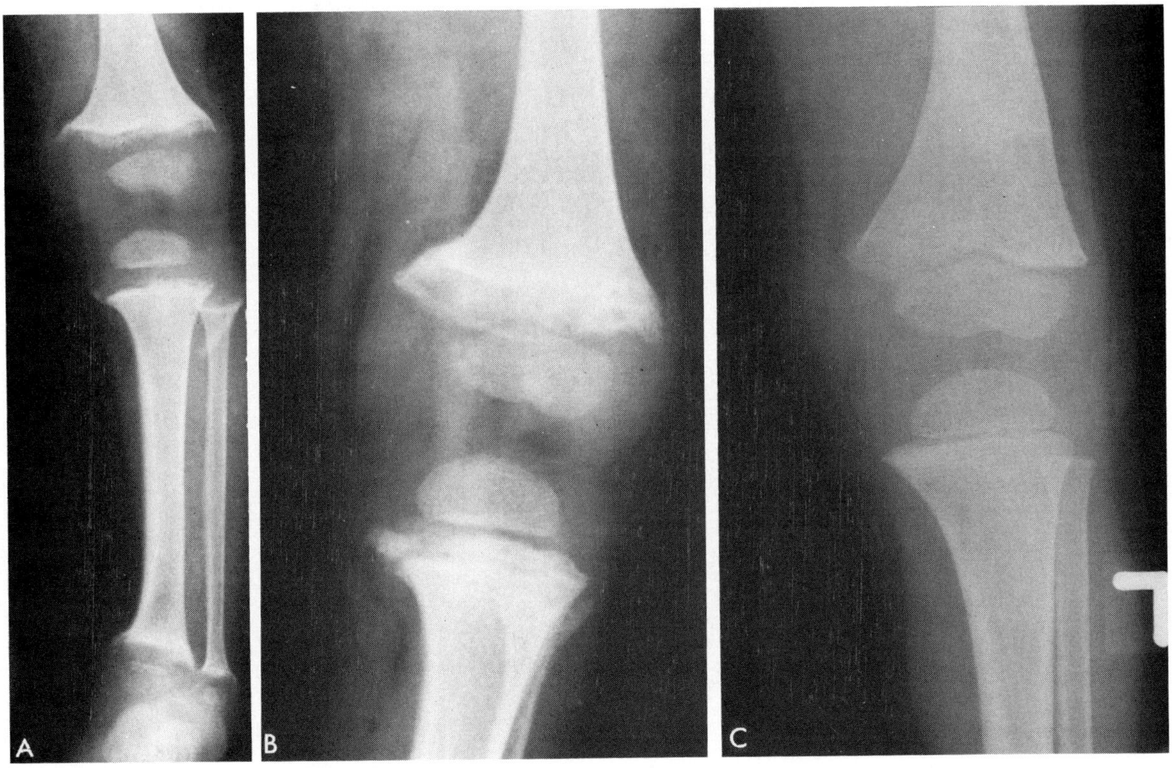

Figure 5–59. Healing of primary hypophosphatemic rickets by treatment with dihydrotachysterol alone (G.M. III), as indicated in Table 5–24: *a*, deposition of mineral in chondro-osteoid after two months of treatment; *b*, continued healing of rickets; *c*, complete healing of rickets by x-ray following 12 months of treatment. Patient's age 30 months.

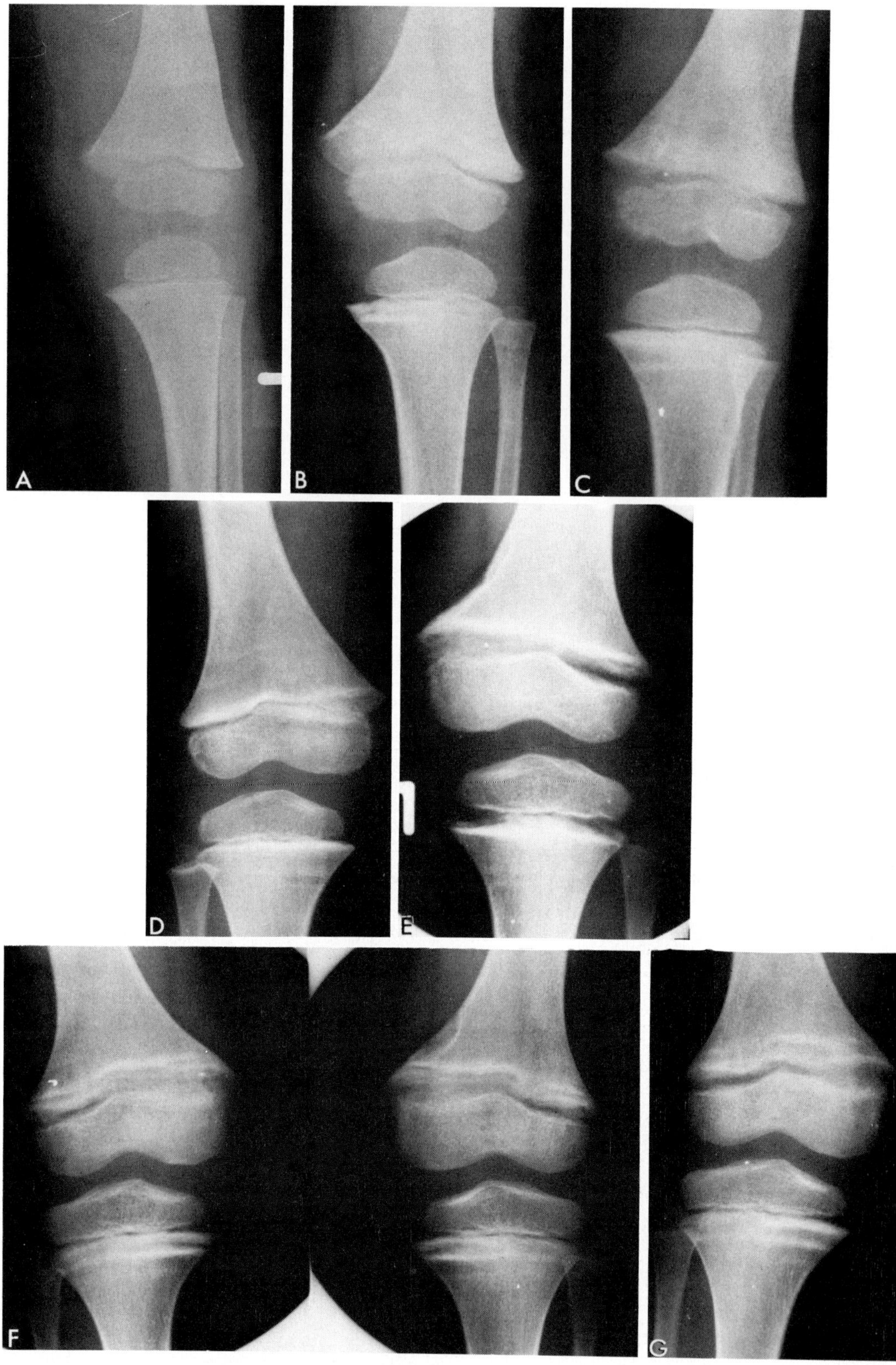

Figure 5–60. *a* to *p*, Sequence of radiographs of legs of patient (G.M. III) showing repeated relapses and healing of primary hypophosphatemic rickets, with development of radiolucent lesions diagnosed by biopsy as nonossifying fibromas. These cystic lesions spontaneously mineralized without surgical treatment.

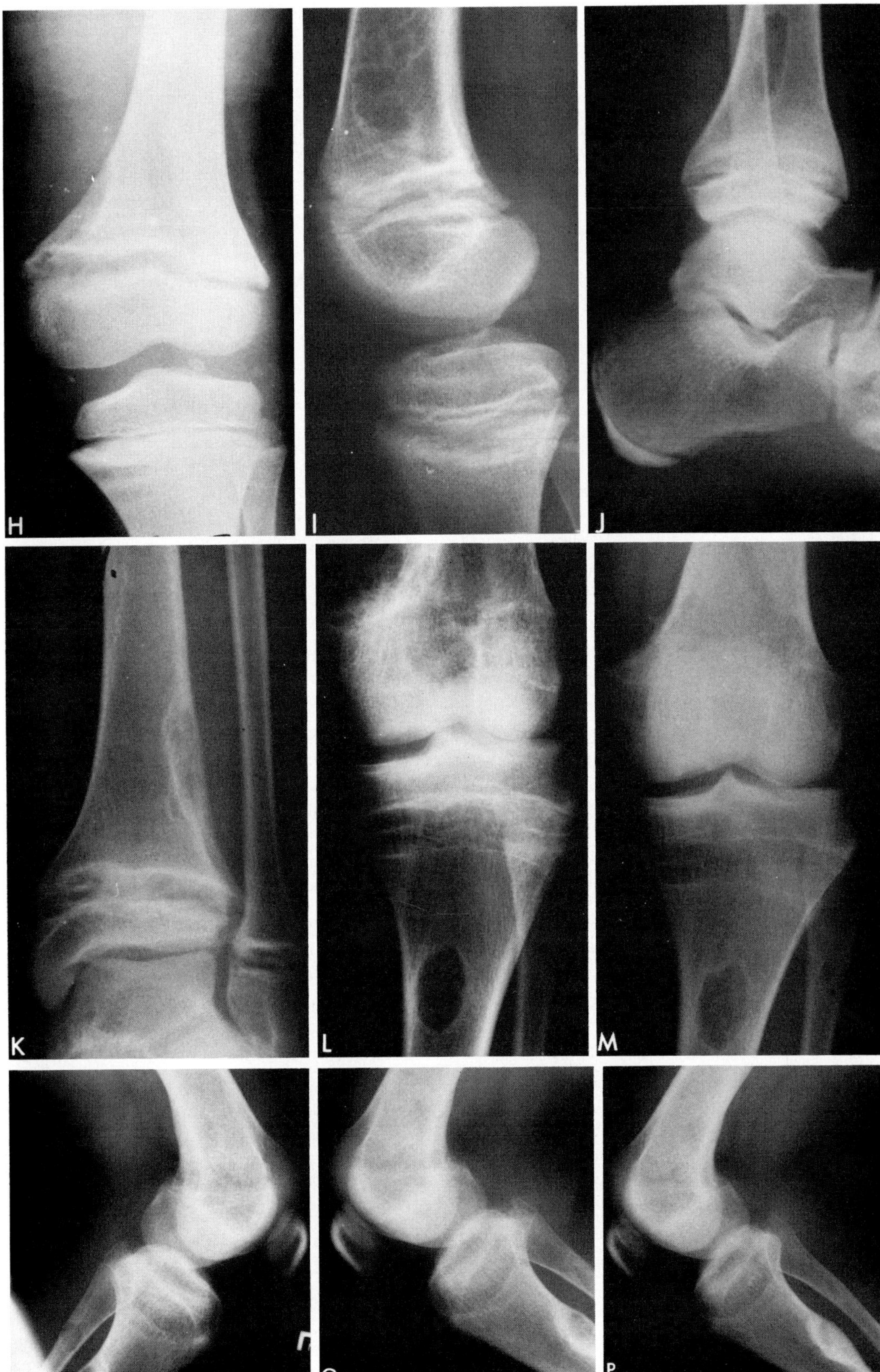

Figure 5–60. Continued.

this regimen he has had relapses and rehealing of rickets as documented in the radiographs shown in Figure 5–60. He has also shown unusual lesions, namely, radiolucencies in femurs and tibias. One of these was biopsied and was histologically a nonossifying fibroma without evidence of osteoclastosis, so that it did not suggest hyperparathyroidism. These lesions slowly remineralized without surgical treatment. Currently at age 20 he is a short adult, with slight bowing of the legs. There are no bone pains and no gait difficulties. He remains on treatment, also because of his father's history.

COMMENTS: This family with genetic primary hypophosphatemic rickets and osteomalacia presents a number of unusual features: (1) inheritance as an autosomal dominant rather than an X-linked dominant trait; (2) the severity of the osteomalacia in the father, which may be partly due to lack of any specific treatment until the age of 30; (3) the occurrence of hypervitaminosis D in the proband with renal injury that remained stable during childhood and adolescence, so that she grew and developed normally, but which progressed during pregnancy with resultant renal insufficiency; and (4) the unusual occurrence of multiple nonossifying fibromas in G.M. III which mineralized with continued treatment.

The good response of G.M. III to treatment with dihydrotachysterol is also of significance. It will be noted that serum phosphate concentrations increased moderately on this therapy without hypercalcemia or impairment of kidney function. The possibility needs to be considered that this effect is exerted through suppression of physiologic parathyroid function or through a direct action of calcium on one component of tubular transport of phosphate. It is possible that one of the vitamin D metabolites, which like dihydrotachysterol has a shorter survival time in the body than does vitamin D itself, may be as good or better than dihydrotachysterol in the treatment of primary hypophosphatemia. A form of treatment that would not require the intake of large amounts of phosphate salts would be highly desirable. However, it must be emphasized that, whether dihydrotachysterol or a vitamin D metabolite is employed, close monitoring of calcium metabolism is required to avoid hypercalciuria, hypercalcemia and renal damage.

N.S., a female infant, was born of a mother with known hypophosphatemic vitamin D rickets. The mother had been followed by Dr. Harriet Guild at the Johns Hopkins Hospital and had moderately severe deformities of the lower extremities. N.S. was born at term and showed no abnormalities. Serum calcium and phosphate concentrations during the first year of life are shown in Table 5–25. By 1 year of age it was evident that this child had genetic hypophosphatemia and mild rickets despite receiving vitamin D, 10,000 units daily. She was then placed on dihydrotachysterol treatment (Table 5–24) which was continued without complication until she ceased growing at 15 years of age. Her growth curve followed the third percentile (Fig. 5–61) and her adult height was 151 cm. The only deformity was minimal lateral bowing of the legs. She married at 21 years of age and was given genetic counseling.

Following three pregnancies which ended in early spontaneous abortions she carried a male infant to term, S.W. He was entirely normal at birth, weighing 4000 gm and 50 cm in length. Serum calcium, phosphate and alkaline phosphatase concentrations are given in Table 5–26. Genetic hypophosphatemia was evident by 4 months of age and skeletal x-rays at this time showed no obvious evidences of rickets (Fig. 5–62). The growth curve, shown in Figure 5–63, indicated continued growth at the 50th percentile. At 4 months of age he was begun on potassium phosphate (Neutra-Phos K) in dosage increasing gradually to 2 gm per day, and dihydrotachysterol starting at 0.1 mg and increasing to 0.25 mg per day. For 12 months he did very well on this regimen, and then began to have profuse watery diarrhea with abdominal cramping pains which necessitated discontinuance of the potassium phosphate. Several attempts were made to reinstitute phosphate treatment, and finally by gradual increase of dosage he was able to tolerate potassium phosphate in an amount

Table 5–25. Genetic Hypophosphatemia.
Age at which Hypophosphatemia is Manifest (N.S.)

AGE MONTHS	*Ca*	SERUM *P* mg/dl	*Alk. P'tase Bodansky Units*	*Treatment*
1	11.0	5.6	19.8	Vitamin D_2 – 0.1 mg/day
7	10.9	4.0	44.0	
10	–	3.6	37.4	
11	–	2.9	34.9	Vitamin D_2 – 0.25 mg/day
Mother	11.1	2.0	5.2	

equivalent to 1 gm P per day. This phosphate intake, however, was less than that usually required to maintain bone mineralization in such patients, so that dihydrotachysterol dosage was increased to 1 mg/day since urine calcium excretion remained low. The

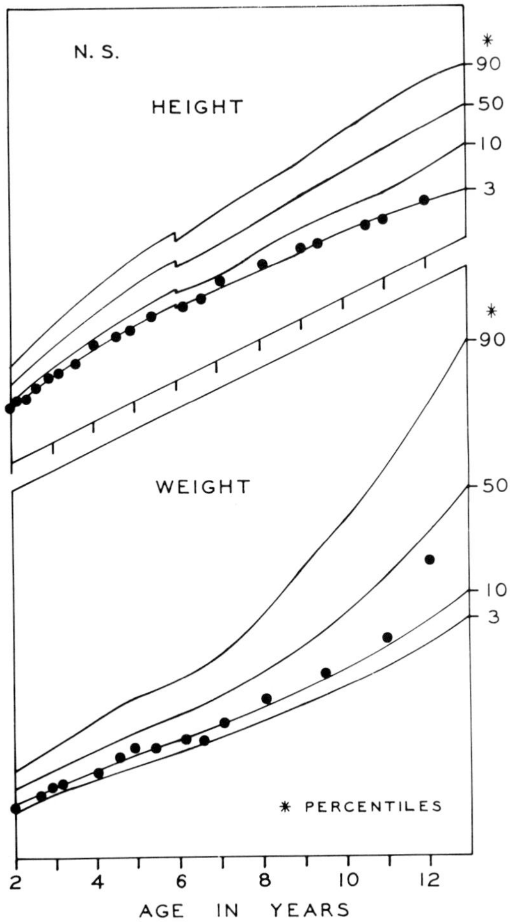

Figure 5–61. Primary hypophosphatemic rickets (N.S.). Growth curve.

further changes in serum calcium, phosphate and alkaline phosphatase concentrations are given in Table 5–26. Radiographic evidences of active rickets appeared (Fig. 5–64), but as dihydrotachysterol dosage was increased healing of the rickets was seen (Fig. 5–65). This child at 3 years of age shows mild lateral bowing of the extremities without other deformity.

COMMENTS: Both mother and son were studied from early infancy, and in both instances the reduced serum phosphate concentrations diagnostic of genetic hypophosphatemia were not evident until 11 and 4 months of age respectively. The mother had an uneventful response to treatment with a good result, namely, minimal deformities and growth at the third percentile. The son was one of the few patients unable to tolerate phosphate salt treatment. This was unfortunate since his early course was excellent with continued good growth and normal bone mineralization as seen radiographically. Because of the phosphate intolerance dihydrotachysterol dosage was increased, with healing of the roentgenographic rickets that had developed during the period of adjustment of phosphate and dihydrotachysterol treatment.

J.W., a white female, was referred at the age of 11 years because of progressively worsening genu valgum. Knock-knee deformity had been noted at the age of 4 or 5 years and had been diagnosed as a congenital deformity. X-ray studies had revealed some irregularity of the metaphyses of the proximal tibia, thought to represent a metaphyseal dysplasia. Serum calcium concentrations were consistently normal and

Table 5–26. Primary Hypophosphatemic Rickets (S.W.)

AGE MONTHS	*Ca*	SERUM *P* mg/dl	*Alk. P'tase* *Bodansky Units*	*Remarks*
2-1/2	10.0	4.2	28.8	
4	10.9	3.4	32.5	Treatment begun: Neutra-Phos K = 0.5 gm P, dihydrotachy-sterol 0.25 mg/day
5	10.5	3.7	25.2	
7	10.3	4.1	30.4	Neutra-Phos K increased = 1.0 gm P
8	10.4	4.5	24.0	
10	10.0	6.0*	20.9	Neutra-Phos K = 1.2 gm P
11	10.1	5.4*	23.6	
13	10.5	3.0	18.8	Moderate diarrhea
17[1]	9.7	2.4	19.1[1]	Severe diarrhea—Neutra-Phos K discontinued for 1 month then resumed at lower dose
19	10.2	3.3	19.6	Neutra-Phos K = 0.6 gm P/day
22[2]	9.6	2.5	24.8	Dihyrotachysterol increased to 0.4 mg/day
24	9.8	2.6	22.4	
26	9.5	3.7	32.5	Neutra-Phos = 1.0 gm P/day Dihydrotachysterol 0.5 mg/day
28	9.9	2.9	20.5	Dihydrotachysterol 0.625 mg/day
30	9.4	1.8	21.4	Dihydrotachysterol 0.75 mg/day
33	10.0	3.4	10.7	Dihydrotachysterol 0.875 mg/day
36[3]	10.9	3.4	—	

*Serum P determined following a.m. dose of Neutra-Phos K.
[1]X-ray = no definite evidence of rickets.
[2]X-ray = active rickets present.
[3]X-ray = healed rickets.

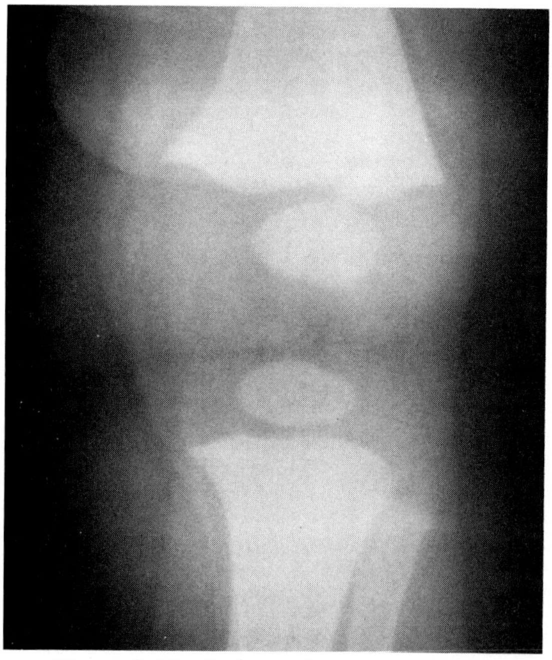

Figure 5–62. Radiographs at 7 months of age (S.W.). No lesions of rickets evident.

serum phosphate was in the normal adult range, 3.5 to 4.0 mg/dl. The genu valgum deformity became sufficiently marked for osteotomy to be recommended. The parents asked for a further opinion and were referred to the Orthopedic Service of the Johns Hopkins Hospital, where the x-ray findings were considered indicative of hypophosphatemic rickets (Fig. 5–66). On review of the data it was apparent that the serum phosphate was indeed low for a growing child, in the range of 3.0 mg/dl. Serum calcium and bicarbonate concentrations were normal. Urine examination was negative without glucose or unusual concentration of amino acids. The family history was of interest. Both parents were normal in height, and serum calcium and phosphate were normal. However, a maternal great-aunt had deformities highly suggestive of severe rickets; she was not available for further investigation.

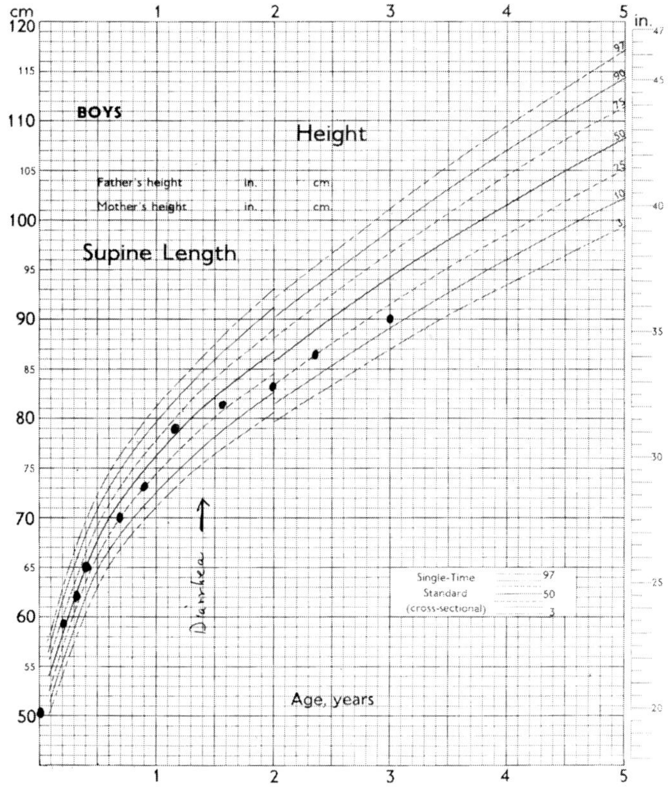

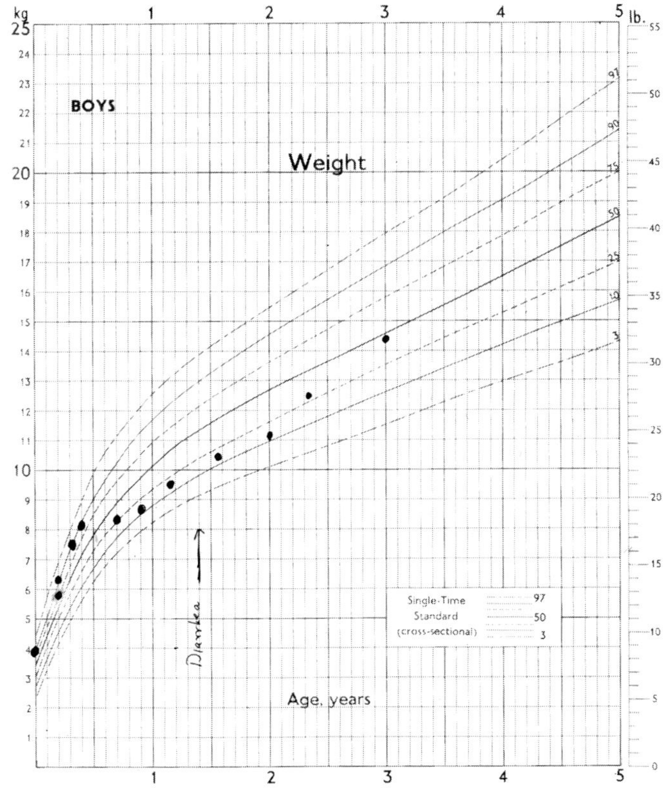

Figure 5–63. Primary hypophosphatemic rickets (S.W.). Growth curve.

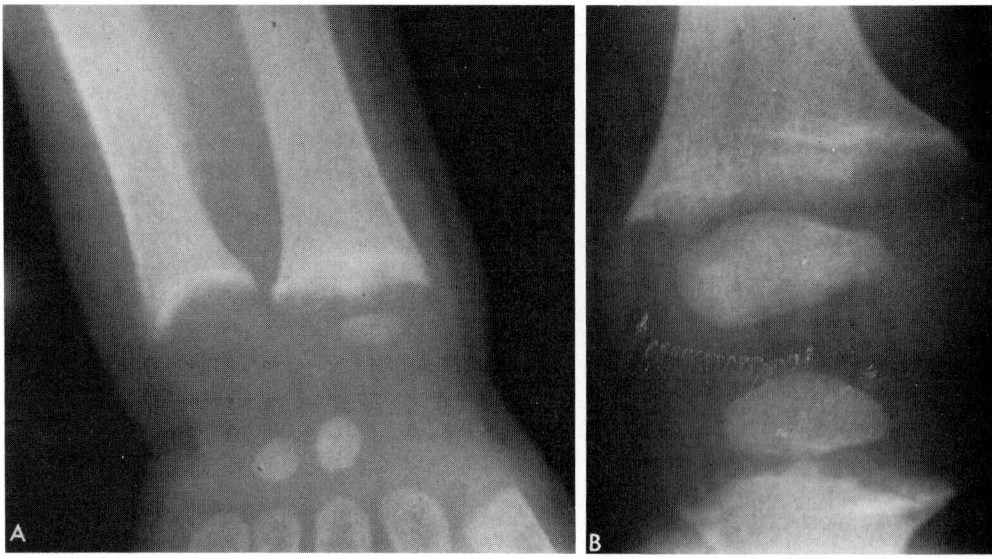

Figure 5–64. *a* and *b*, Patient (S.W.) age 24 months. Signs of active rickets present on radiograph.

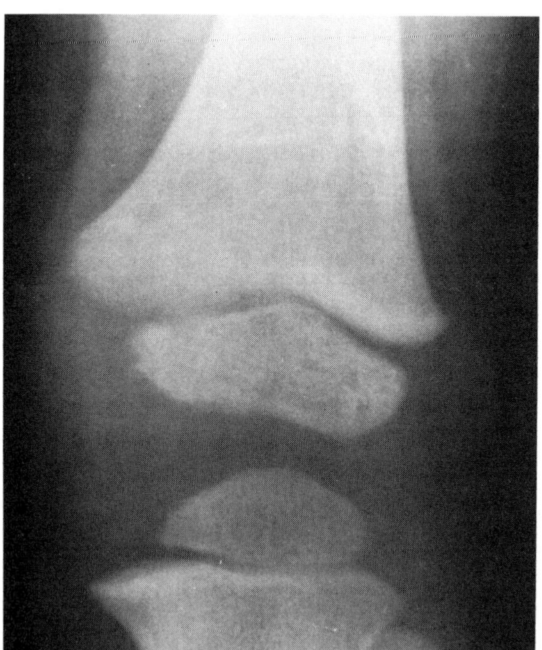

Figure 5–65. Patient (S.W.) age 28 months. Healing of rickets with dihydrotachysterol treatment and relatively small amounts of supplemental phosphate, 1 gm P per day.

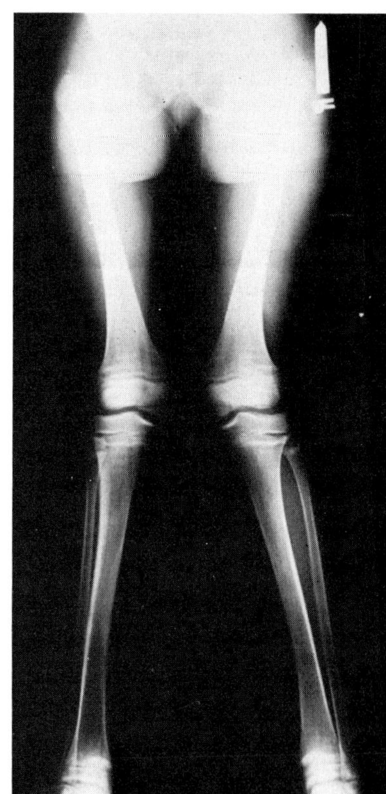

Figure 5–66. Radiographs of lower extremities showing genu valgum and abnormalities of metaphyseal mineralization compatible with hypophosphatemic rickets (J.W.).

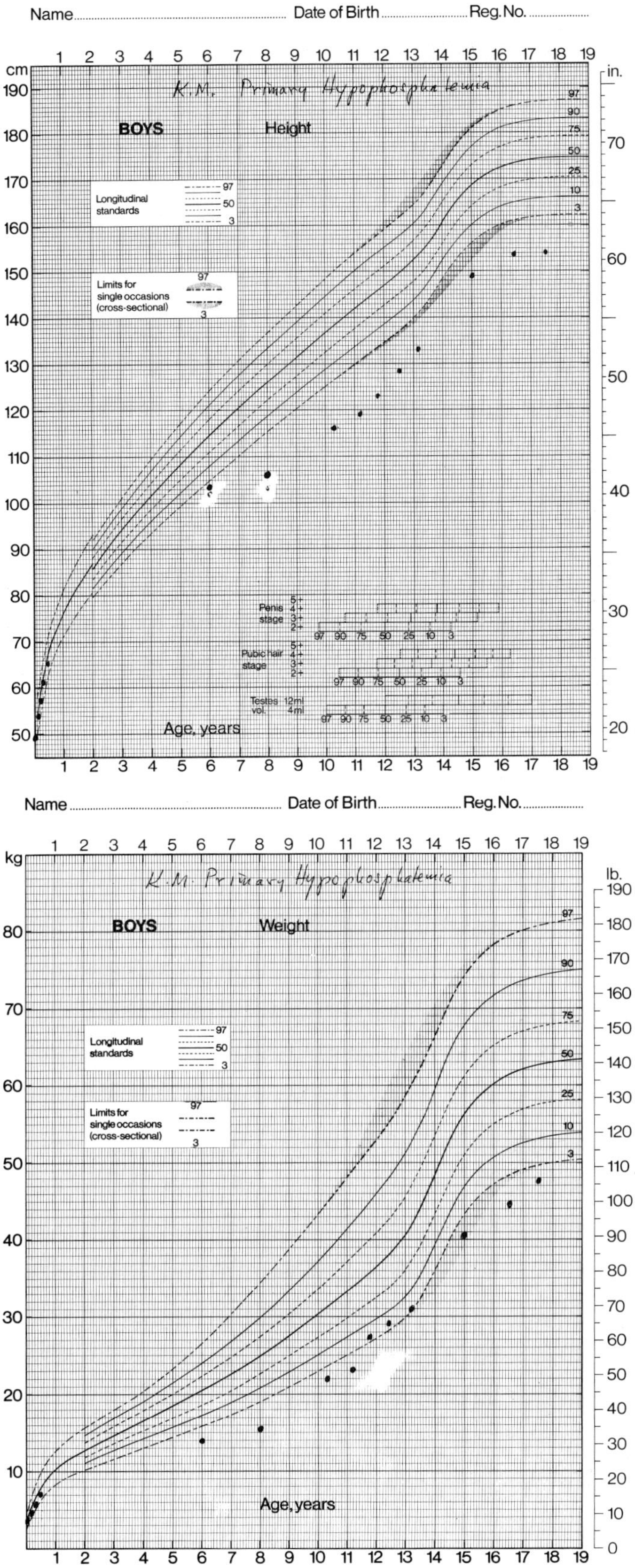

Figure 5–67. Genetic primary hypophosphatemia with growth retardation (K.M.). Growth curve.

The patient was started on oral potassium phosphate increasing to 2 gm of phosphorus per day, and dihydrotachysterol increasing to 0.75 mg/kg. On this regime healing of the rickets by x-ray occurred. There has been no improvement of the genu valgum deformity and osteotomies are scheduled.

COMMENTS: This is an example of mild primary hypophosphatemia with slowly progressive, somewhat atypical deformity of the lower extremities. The family history is of interest since, if the maternal great-aunt did have primary hypophosphatemia, this would suggest lack of penetrance of the gene in the mother of the patient but with expression in the patient. Failure of early diagnosis was apparently due to lack of realization that serum phosphate concentrations between 3 and 4 mg/dl in a young child are abnormally low. The somewhat unusual genu valgum deformity also may have been a source of confusion.

Although this child's disease is mild, and healing should be readily maintained, the genu valgum deformity is sufficiently pronounced for its correction by normal growth and remodeling of bone to be unlikely. Osteotomy will be necessary to achieve normal function.

K.M., a white male age 10 years, was referred because of short stature. He was born at term weighing 3400 gm and 49 cm in length. During the first six months of life he grew normally but subsequently growth rate markedly diminished (Fig. 5–67). By 6 years of age he was "tiny" but, because he was not sickly, medical attention was not sought. At 10 years of age his retardation of growth caused the school nurse to urge an evaluation. Except for short stature, physical examination was negative.

Laboratory data: serum electrolyte concentrations normal except for low concentrations of serum phosphate, 2.4 and 2.6 mg/dl; serum calcium 10.6 mg/dl, alkaline phosphatase 12.6 Bodansky Units; urine analysis negative including normal

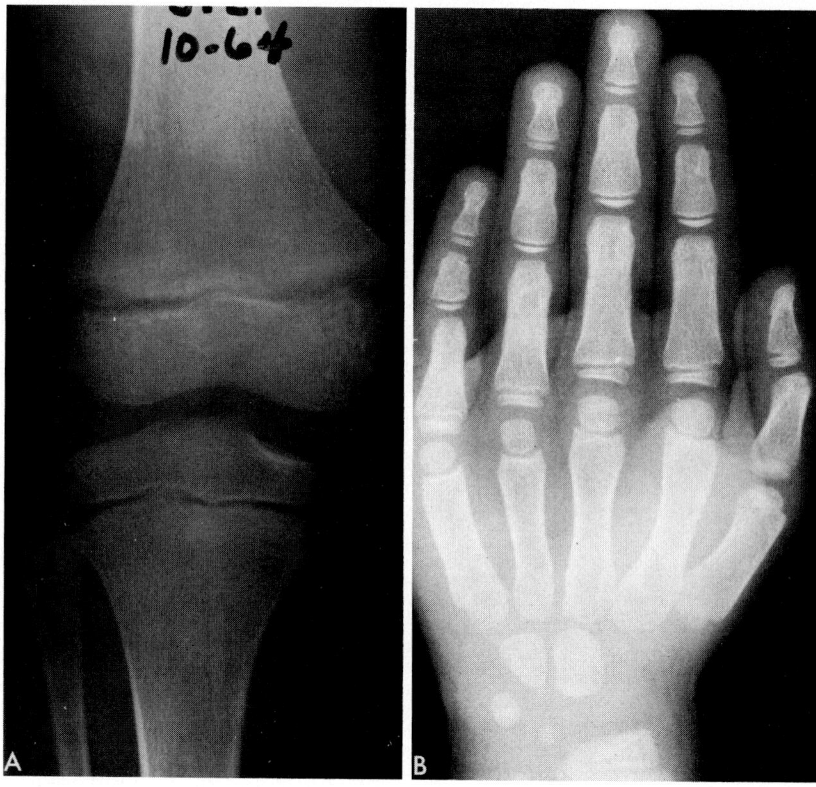

Figure 5–68. *a* and *b*, Radiographs of knee and wrist showing only questionable evidences of rickets (K.M.).

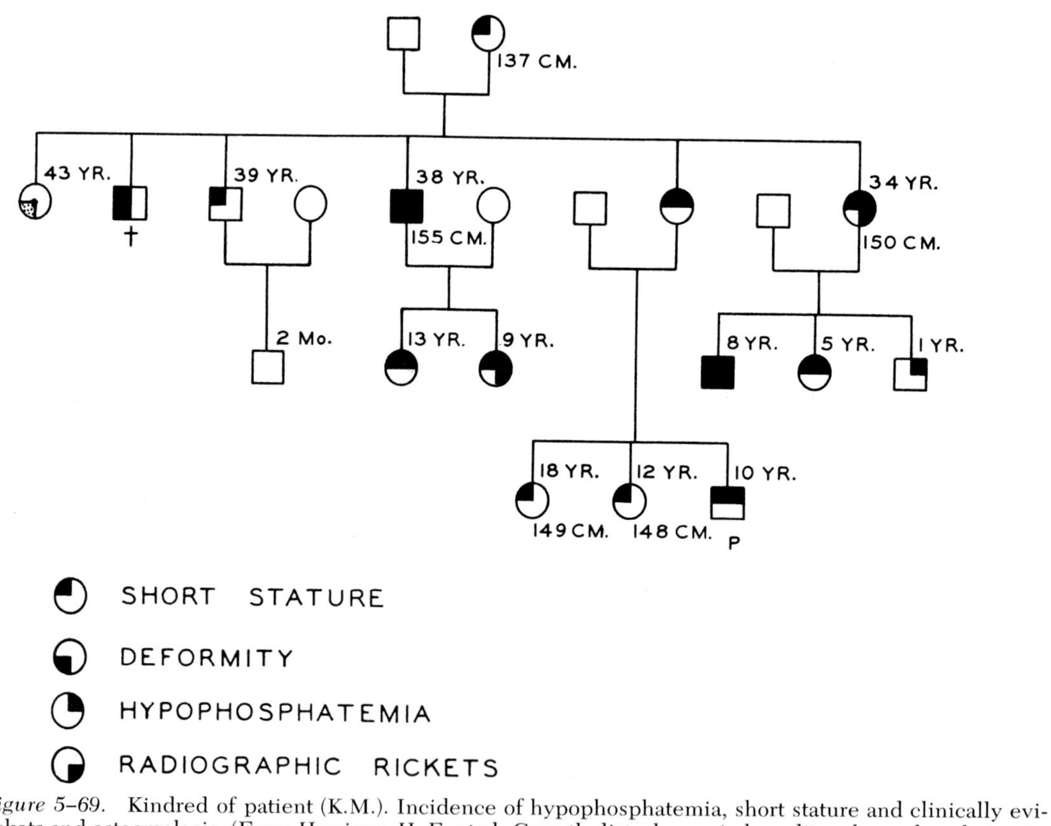

Figure 5–69. Kindred of patient (K.M.). Incidence of hypophosphatemia, short stature and clinically evident rickets and osteomalacia. (From Harrison, H. E. et al: Growth disturbance in hereditary hypophosphatemia. Am. J. Dis. Child. *112*:290, 1966. Reprinted with permission.)

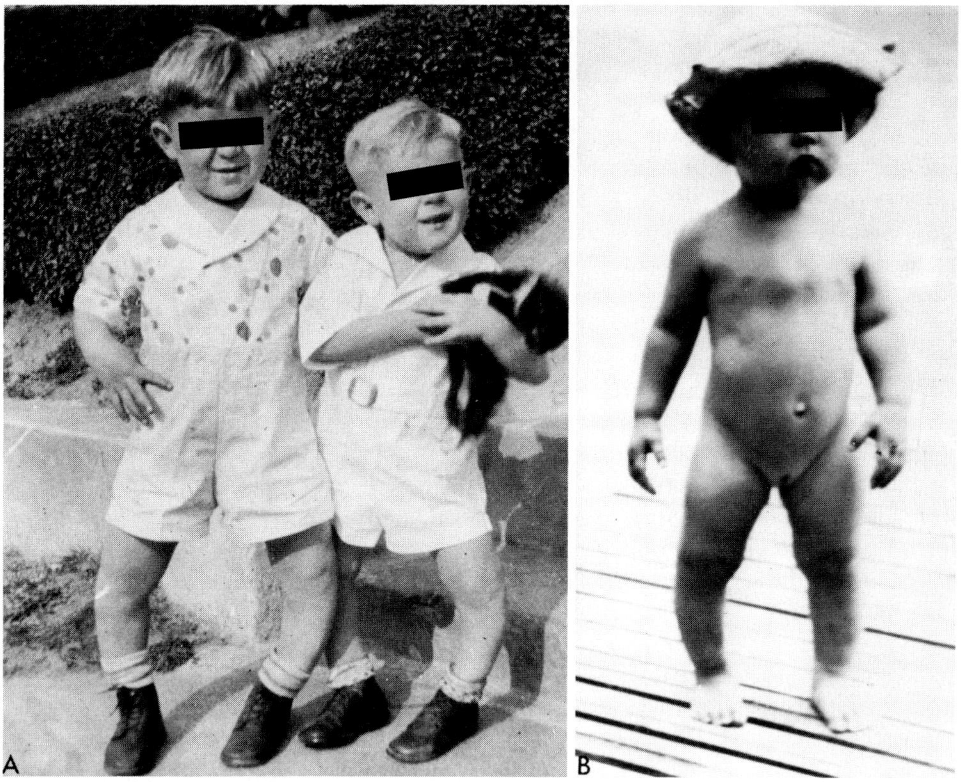

Figure 5–70. *a* and *b*, Photographs of children in kindred of patient (K.M.) with bowing of legs characteristic of rickets.

amino acid excretion. X-ray examination showed delayed bone age at the 5-year level but minimal, if any, evidence of rickets or osteomalacia (Fig. 5–68). Studies of pituitary function ruled out panhypopituitarism, although growth hormone concentrations following insulin hypoglycemia showed only borderline increase.

The family investigation proved to be of extreme importance. Although none of the family had been evaluated for rickets or osteomalacia there was a high incidence of short stature, hypophosphatemia and mild rachitic deformities in the three generations of this kindred diagrammed in Figure 5–69. Nine individuals of 13 tested had significant hypophosphatemia, including two adults with clinical and radiologic evidences of osteomalacia that had not been previously suspected and three children with radiographic rickets and mild deformities (Fig. 5–70). The family study supported the diagnosis of genetic hypophosphatemia expressing itself in the proband K.M. primarily as growth retardation. It is possible that there was genetic short stature independent of hypophosphatemia, and that K.M. presented without rickets because of extremely slow growth rate.

K.M. was given a course of oral phosphate supplying 1.8 gm phosphorus per day with apparent acceleration of growth, as indicated in Figures 5–67 and 5–45. The growth curves of three other patients with primary hypophosphatemia are also diagrammed in Figure 5–45, and two of the three had increased growth rate following phosphate treatment. The one patient who showed no response had received phosphate equivalent to only 1 gm phosphorus per day, which is a low dose, and would possibly have responded to a more adequate dose. Glorieux and Scriver have given patients with genetic hypophosphatemia larger amounts of phosphate, providing 2 to 4 gm phosphorus per day, with significant growth acceleration. K.M. was given continued treatment with phosphate and dihydrotachysterol, 0.5

mg/day. He reached an adult height of only 154 cm but was otherwise in good health. There were never any skeletal manifestations.

COMMENTS: This patient and family represent an unusual variant of primary hypophosphatemia. None of the members of the family had been previously diagnosed despite evidences of mild osteomalacia and rickets in a number of them. Short stature and hypophosphatemia were the dominant features. The proband had such marked growth retardation that x-ray evidences of rickets were not present despite serum phosphate concentrations in the range of 2.5 mg/dl. It has been emphasized by Winters et al that short stature and hypophosphatemia may be the major findings in many members of kindreds with X-linked genetic hypophosphatemia.

ACQUIRED PRIMARY HYPOPHOSPHATEMIC RICKETS AND OSTEOMALACIA

A form of primary hypophosphatemic rickets and osteomalacia occurs that is indistinguishable from genetic hypophosphatemia except that its onset is in later childhood or adult life. There is no familial history. These patients grow normally and have no rachitic deformities in early childhood. During later childhood or adult life complaints of bone pain, muscle weakness and eventual deformities develop. The muscle weakness is more marked than in genetic hypophosphatemia and may be so dominant a complaint as to suggest some form of muscle dystrophy. X-ray examination reveals the findings of rickets if the bones are still growing, and of osteomalacia if growth has ceased. Serum analyses show normal concentrations of serum calcium with very low serum phosphate, often less than 2 mg/dl. The muscle weakness has been thought to be a consequence of this extreme hypophosphatemia, which is renal in origin. Renal phosphate clearances are high and calculated tubular reabsorption of phosphate greatly reduced. Other renal function studies are normal except that a mild degree of renal glycosuria may occur. Serum bicarbonate concentrations are normal. There is no generalized aminoaciduria. The excretion of increased amounts of a single amino acid, glycine, has been reported, so that hyperglycinuria has been considered part of this syndrome.

The differential diagnosis includes renal tubular acidosis and the Fanconi syndrome secondary to a nephrotoxic agent. The former is ruled out by the normal serum bicarbonate concentrations and the latter by the absence of renal aminoaciduria. The renal tubule transport systems for amino acids in the proximal tubule are extremely sensitive to nephrotoxic agents. Generalized aminoaciduria is an early and consistent manifestation of the renal tubule injury of heavy metal poisoning, multiple myeloma or the various organic compounds that can produce the Fanconi syndrome. The hyperglycinuria of acquired primary hypophosphatemia probably is not renal in origin since increased concentrations of serum glycine have been reported.

Many of the cases of acquired primary hypophosphatemia have been found to be associated with connective tissue tumors, both within bone and extraskeletal, and it may be that tumor is the pathogenetic factor in all the cases, but because of the benign nature and small size of the tumor it has not been discovered. Further discussion of acquired primary hypophosphatemia will therefore be under the heading of tumor-associated hypophosphatemic rickets and osteomalacia.

TUMOR-ASSOCIATED (ONCOGENOUS) PRIMARY HYPOPHOSPHATEMIC RICKETS AND OSTEOMALACIA

Prader and his colleagues originally described a case of hypophosphatemic rickets associated with a bone tumor visualized by x-ray. When the tumor was excised the serum phosphate concentration returned to normal and the rickets healed. Since then there have been several reports of acquired primary hypophosphatemic rickets and osteomalacia in which tumors have been found within bone, and cure followed removal of the tumor. More recently, extraskeletal tumors have been found in such patients, again with restoration of normophosphatemia and remineralization of bone following removal. We suggested that such tumors might be the cause of most cases of acquired primary hypophosphatemia, and in a personal communication from the late Dr. Charles Dent we learned that, of five cases of acquired primary hypophosphatemic osteomalacia originally reported by him, two subsequently were found to have tumors in the nasal antrum, and an additional patient with the disease had a tumor of the rectus abdominis sheath. In all instances removal of the tumor resulted in cure of the disease.

These tumors, whether within the bone or in extraskeletal connective tissue, are all of mesenchymal origin. The histologic picture varies, so that they have been called nonossifying fibromas or giant cell tumors when within bone, or ossifying mesenchymal tumors, fibroangiomas, histiocytomas, etc. if extraskeletal. All tumors have in common a connective tissue origin and can be classified as pericytomas. It is assumed that these tumors secrete a humoral agent that inhibits tubular reabsorption of phosphate, since within 24 hours following removal of such a tumor tubular reabsorption of phosphate increases, serum phosphate returns to normal, and progressive healing of the rickets and osteomalacia results without further treatment. The nature of this inhibitory agent is unknown but is not parathyroid hormone since serum IPTH concentrations are not elevated. Recently Auschinberg et al have reported a case of rickets in a 12 year old boy with epidermal nevus syndrome with several fibroangiomas of face and one leg. When these lesions were removed the serum phosphate returned to normal and the severe rickets improved. The lesions histologically showed increased vascularity and fibrosis of the dermis, and were classified as fibroangiomas. Extracts of these tumors were prepared and when injected into puppies were found to have a phosphaturic effect. No further information on the nature of this phosphaturic substance is available. The basis for the increased plasma glycine concentrations and hyperglycinuria in some of these patients is also unknown.

In all cases of acquired hypophosphatemic rickets and osteomalacia it is necessary to make a careful search for a tumor of mesenchymal origin that might be the cause of the disturbed renal tubular reabsorption of phosphate. X-ray studies of the entire skeleton are required since tumor has been found in the ribs as well as in various long bones. The extraskeletal tumors may be extremely hard to find since they can be quite small and have been present within the skin, in muscle sheaths, in the nasal antrum and in the pharyngeal wall, among other locations. If a mesenchymal tumor is found it should of course be removed in toto, and the serum phosphate concentration of the patient determined at intervals postoperatively. If the tumor is the basis of the hypophosphatemia a rising serum phosphate may be evident by 24 hours after its removal. In such patients no further specific treatment is needed.

There remain, however, patients with the same syndrome of acquired primary hypophosphatemic rickets and osteomalacia in whom no tumor can be found after careful search. It may be that such a tumor exists but is in a location where it cannot be palpated or visualized. Since these tumors are slow-growing, benign and never reach large size it is possible that they may never be found.

There may also be other mechanisms to account for acquired hypophosphatemia besides a "nonendocrine" tumor secreting a phosphaturic agent. Whatever the reason, if no tumor is located the treatment of the patient must be the same as that of those with genetic primary hypophosphatemic rickets and osteomalacia. Treatment with phosphate salts and dihydrotachysterol or vitamin D has been successful in producing improved mineralization of bone and restoration of muscle strength and well-being in these subjects.

CASE REPORT

J.M. was a 30 year old white male who had been in good health until two or three years before referral. He had been a good athlete and an energetic individual. His symptoms were insidious in onset with excess fatigue, weakness and vague skeletal pains in back, hips and legs. Repeated examinations were apparently negative, so that an emotional basis for his symptoms was considered and referral for psychiatric evaluation was made. His symptoms progressed and, because weakness seemed to predominate over bone pain, he was referred to the neurologic clinic for possible muscle dystrophy. No evidence of a neurologic or primary muscle disorder was found and determinations of serum electrolytes showed a serum phosphate concentration hovering about 2 mg/dl. Serum calcium was normal as were serum bicarbonate, chloride, sodium and potassium concentrations. Skeletal x-rays at this time showed poor mineral density, coarse trabeculation and Looser-Milkman zones consistent with osteomalacia.

Urine examination showed mild glycosuria at normal concentrations of serum glucose. There was no generalized aminoaciduria. The phosphate to creatinine clearance ratio was high in relation to a serum phosphate concentration of 2 mg/dl, confirming the diagnosis of renal hypophosphatemia. The creatinine clearance was normal. Renal tubular acidosis had been ruled out by the normal serum bicarbonate concentration. Although unlikely in view of absence of aminoaciduria, the various causes of Fanconi syndrome such as heavy metal poisoning and multiple myeloma were searched for, and no evidence found for lead, cadmium or mercury poisoning or clonal gammopathy. A complete skeletal survey was made in search of a radiolucency caused by a mesenchymal tumor, but none was found. No extraskeletal masses could be visualized or palpated. A diagnosis of acquired primary hypophosphatemic osteomalacia was made. Since the patient had grown normally in childhood and adolescence and had no deformities characteristic of rickets, genetic hypophosphatemia was eliminated. There was also no family history of growth retardation or skeletal abnormalities, and the patient's two children had normal serum phosphate concentrations as well as normal physical examinations.

Since no tumor of connective tissue origin could be found the patient was treated with phosphate salts and dihydrotachysterol. Potassium phosphate solution providing 2.5 gm phosphorus per day was taken daily, divided into five doses, and dihydrotachysterol in doses ranging from 1.0 to 2.0 mg per day. A dose of 1.25 mg/day was decided upon on the basis of urine calcium excretion, since 1.5 to 2 mg/day resulted in a urine calcium output of 420 mg/day or 0.3 mg calcium per mg creatinine. Studies of serum phosphate concentration in the morning before any phosphate solution was taken, and one hour following his standard dose of phosphate equivalent to 0.5 gm P, were made at each visit and are shown in Figure 5–71. There is the suggestion that at the highest doses of dihydrotachysterol which could be safely given both values were higher than at lower doses of sterol. A more detailed study of response of serum phosphate to phosphate loading was made in which serum phosphate was determined prior to the standard dose of phosphate solution, one and two hours following phosphate intake, and then one hour after a second dose of phosphate and three hours after the initial dose. The data, given in Table 5–23, indicate that the serum phosphate is increased about 1 mg/dl by a single dose of phosphate (0.4 gm P) and remains at this increased concentration for several hours, so that following a second dose of phosphate the serum phosphate is above 4.0 mg/dl, a concentration adequate to induce mineralization of bone matrix. On the treatment outlined J.M. has made a complete recovery and is pursuing an entirely normal life including sports. Radiographs of the bones show improved mineralization.

COMMENTS: This example of acquired primary hypophosphatemic osteomalacia is characteristic in that the symptoms of muscle weakness predominate over those of bone pain. The weakness may be the result of severe hypophosphatemia, but also in part inhibition of muscle contraction by pain. The hypophosphatemia had not been properly evaluated until his referral to the neurologist.

Although no tumor was found in this patient the findings are similar to those seen in tumor-associated osteomalacia, and it is still possible that a small tumor is present in a position inaccessible to examination. The failure to find a tumor that could be removed with cure of the hypophosphatemic osteomalacia was disappointing, but the treatment of this patient with oral phosphate

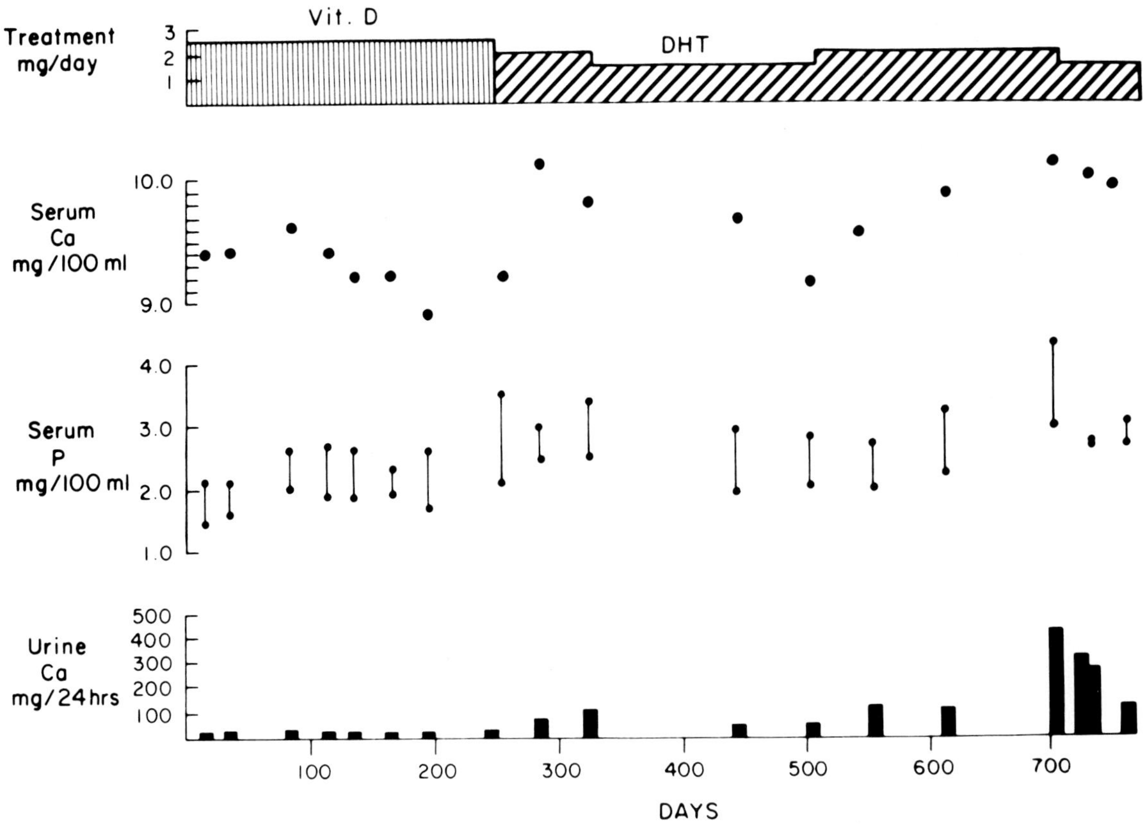

Figure 5–71. Acquired hypophosphatemic osteomalacia. Diagram of effect of dihydrotachysterol dosage on urine calcium excretion and serum phosphate concentrations.

loading and dihydrotachysterol was successful. The methods used to find the largest safe dose of dihydrotachysterol have been well illustrated by the course of this patient, and the effect of dihydrotachysterol in increasing the serum phosphate concentration achieved following phosphate ingestion is also demonstrated. This patient must still remain under close observation and is monitored every three to four months by determination of urine calcium excretion in a 24-hour collection, as well as of serum calcium and phosphate concentrations.

HYPOPHOSPHATEMIC RICKETS SECONDARY TO DEFICIENT PHOSPHATE INTAKE

Nutritional phosphate deficiency is almost an impossibility in children on mixed diets because of the high content of phosphate in all foods containing animal or vegetable cells or derived from animal sources such as milk and eggs. However, occasional instances of phosphate deficiency are seen in prematurely born infants treated with prolonged intravenous hyperalimentation or fed human milk at minimal caloric requirements, or a combination of the two. Human milk is unusual in its low calcium and phosphate content, 33 and 15 mg/dl respectively, or 50 and 23 mg/100 cal. This compares with the content of these elements in cow's milk of 125 and 100 mg/dl, or 187 and 150 mg/100 cal. The phosphate content of human milk is obviously adequate for the nutrition of the full term human infant, but is barely sufficient for the prematurely born infant who is receiving only the minimal amount of calories required for growth. It has been suggested that the phosphate concentration of human milk is a limiting factor in the mineralization of the skeleton of the prematurely born because the phosphate requirements of the rapidly growing soft tissues are so great.

In recent years prematurely born infants with severe respiratory distress and with necrotizing enterocolitis have survived after prolonged periods of intravenous alimentation. The various intravenous hyperalimentation solutions used supply only a minimal amount of phosphate because of the relative insolubility of calcium phosphate in such solutions. Phosphate deficiency has been seen in such infants with hypophosphatemic rickets. Specially prepared hyperalimentation solutions for prematurely born infants can be formulated, providing 1 mM phosphate per kg when the usual volume is infused. The following case report is an example of phosphate deficiency rickets in a prematurely born infant.

CASE REPORT

J.Z. was a prematurely born infant, birth weight 1235 gm, who was found at 2 months of age to have poorly mineralized bone with metaphyseal changes suggestive of rickets (Fig. 5–72). At this time serum phosphate was determined to be 3.2 mg/dl. The infant had had respiratory distress followed by diarrhea, and a diagnosis of necrotizing enterocolitis had been made. Oral feedings had been stopped and the patient had been maintained on intravenous fluids for six days. Following this, feedings with human milk were begun with calorie intake restricted to minimal requirements. Vitamin D supplement was given. At the time that the skeletal changes of rickets and hypophosphatemia were detected, serum calcium concentration was 10.4 mg/dl and

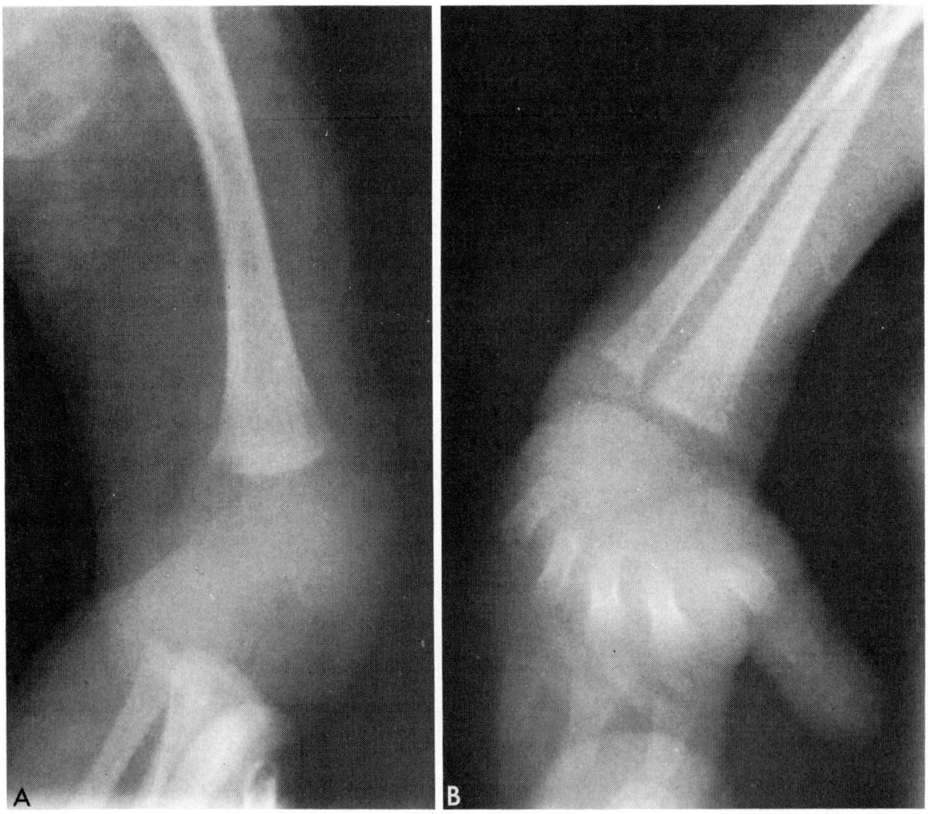

Figure 5–72. *a* and *b*, Prematurely born infant (J.Z.) with phosphate deficiency. Radiographs show poorly mineralized bone.

alkaline phosphatase 33.2 Bessey-Lowry Units. The 25-hydroxyvitamin D concentration in serum was 31 ng/ml. The serum calcium in the mother was 10.2 mg/dl and serum phosphate 3.7 mg/dl, which indicated that the mother's vitamin D status was normal. It was thought that the infant had true nutritional phosphate deficiency resulting from low phosphate intake in the intravenous fluids followed by a low phosphate feed of human milk, which provided insufficient phosphate to make up for the deficit during the period of food withdrawal. Tubular reabsorption of phosphate was almost 100 per cent, with a 24-hour phosphate excretion of only 0.19 mg. This showed that the hypophosphatemia was not due to inadequate renal tubule function.

The diagnosis was phosphate deficiency without vitamin D deficiency. The infant was treated by phosphate supplementation of the feeds without any change in vitamin D intake. Sufficient phosphate was added to the breast-milk feeding to provide a total of 107 mg P per day (60 mg/kg/body weight). On this diet serum phosphate increased to 6.7 mg/dl, and subsequent skeletal x-rays showed increased density of mineral at the metaphyseal ends consistent with healing rickets (Fig. 5–73). The further course is summarized in Table 5–27.

COMMENTS: True phosphate deficiency is rare, but is seen particularly in prematurely born infants maintained on intravenous alimentation because of intestinal problems, either with or without surgery. This infant had been given phosphate-free intravenous fluids to maintain hydration during a period of diarrhea, followed by a minimal intake of a low phosphate food, human milk. Vitamin D deficiency was ruled out, not only by a history of vitamin D supplementation but by normal concentrations of calcium and 25-hydroxyvitamin D in serum. Increased vitamin D requirement (vitamin

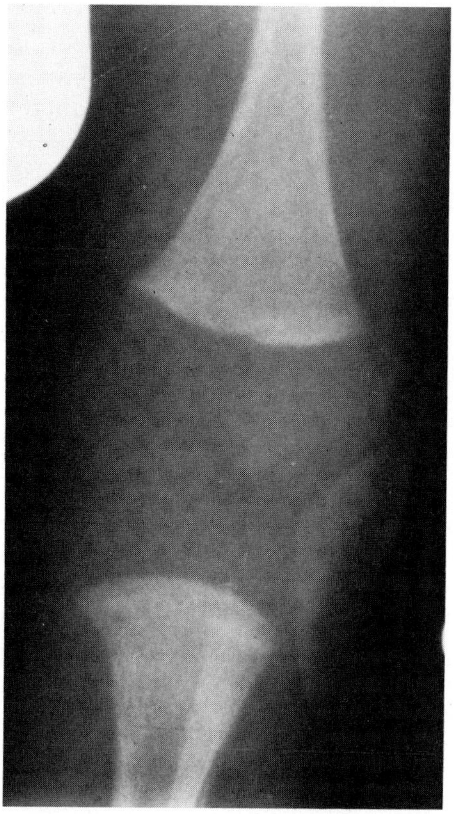

Figure 5–73. Radiographs showing improved mineralization of bone following phosphate addition to diet (J.Z.) (compare with Fig. 5–72).

Table 5–27. Serum Calcium and Phosphate Concentrations in Prematurely Born Infant with Phosphate Deficiency (J. Z.)

AGE DAYS	*Wt.* *gm*	SERUM *Ca* *mg/dl*	*P*	*Treatment*
49	1350	9.3	2.9	Human milk 260 ml/day (41 mg P)
53	X-ray—osteopenia, ? rickets			
64	1700	10.4	3.2	
70				Human milk 336 ml/day (54 mg P)
71	—	8.4	5.0	Neutra-Phos K = 53 mg P
77	2100	9.9	6.7	
84	X-ray—healing rickets			
91	2380	9.5	6.1	
95	—	9.3	5.6	Phosphate supplement discontinued
128	—	10.5	4.5	

D dependency) was unlikely in view of the normal serum calcium concentration. Addition of phosphate to the diet restored serum phosphate to normal without reduction of serum calcium, which was independent evidence of normal vitamin D status. Improved bone mineralization followed this increase of phosphate intake.

Solutions for intravenous alimentation designed for adults or older children provide too little phosphate for the growth requirements of the prematurely born infant, and hypophosphatemia with impaired bone mineralization has been seen in such infants. A phosphate intake of 1 mM/kg (30 to 35 mg/kg) is required by these infants; this is the amount of phosphate contained in 200 ml of human milk/kg providing 136 cal/kg. Intestinal absorption of phosphate by the young infant fed human milk is quite efficient, so that the amount given intravenously cannot be reduced much below the required oral intake since there is little correction for unabsorbed phosphate.

Hypophosphatemic Rickets and Osteomalacia due to Malabsorption of Phosphate

Phosphate absorption is reduced in vitamin D deficiency, of course, and may be reduced in genetic primary hypophosphatemia, but malabsorption of phosphate sufficient to produce hypophosphatemia of itself without renal loss of phosphate occurs only as the result of ingestion of a phosphate-precipitating cation. The usual circumstance in which this occurs is the intake of large amounts of aluminum hydroxide as an antacid. This is rarely a problem in childhood but has been seen in adults, and osteomalacia secondary to such hypophosphatemia has occurred.

REFERENCES

VITAMIN D DEFICIENCY RICKETS

Arnaud, S. B., Stickler, G. B. and Haworth, J. C.: Serum 25-hydroxyvitamin D in infantile rickets. Pediatrics 57:221–225, 1976.

Bordier, P. Pechet, M. M., Hesse, R. et al: Response of adult patients with osteomalacia to treatment with crystalline 1α-hydroxy vitamin D_3. N. Engl. J. Med. *291*:866–871, 1974.

Dent, C. E. and Smith, R.: Nutritional osteomalacia. Q. J. Med. New Series *38*:195–209, 1968.

Doxiadis, S. Angelis, C., Karatzas, P. et al: Genetic aspects of nutritional rickets. Arch. Dis. Child. *51*:83–90, 1976.

Glasgow, J. F. T. and Thomas, P. S.: Rachitic respiratory distress in small preterm infants. Arch. Dis. Child. *52*:268–273, 1977.

Harrison, H. E.: The disappearance of rickets. Am. J. Public Health *56*:734–737, 1966.

Hassanein, E. A. and Patel, H.: The parathyroid hormone and aminoaciduria in vitamin D deficiency rickets. Acta Paediatr. Scand. *56*:445–449, 1967.

Holmes, A. M., Enoch, B. A., Taylor, J. L. et al: Occult rickets and osteomalacia amongst the Asian immigrant population. Q. J. Med. New Series *165*:125–149, 1973.

Illig, R., Uehlinger, E. and Prader, A.: Sekundärer Hyperparathyroidismus bei Vitamin D Mangel Rachitis. Helv. Paediatr. Acta *14*:566–579, 1959.

Jeans, P. C.: Vitamin D milk. J.A.M.A. *106*:2066–2069, 2149–2159, 1936.

Lester, E., Skinner, R. K. and Wills, M. R.: Seasonal variation in serum 25-hydroxyvitamin D in the elderly in Britain. Lancet *1*:979–980, 1977.

Miller, C. G. and Chutkan, W.: Vitamin D deficiency rickets in Jamaican children. Arch. Dis. Child. *51*:214–218, 1976.

Shany, S., Hirsh, J. and Berlyne, G. M.: 25-Hydroxycholecalciferol levels in Bedouins in the Negev. Am. J. Clin. Nutr. *29*:1104–1107, 1976.

Stamp, T. C. B., Haddad, J. G. and Twigg, C. A.: Comparison of oral 25-hydroxycholecalciferol, vitamin D, and ultraviolet light as determinants of circulating 25-hydroxyvitamin D. Lancet *1*:1341–1343, 1977.

Steendijk, R.: The effect of parathyroid extract on the serum concentrations of calcium and inorganic phosphate in active and healing rickets. Acta Paediatr. *53*:105–108, 1964.

Vainsel, M., Manderlier, Th., Corvelain, J. et al: Study of the secondary hyperparathyroidism in vitamin D deficiency rickets. Biomedicine *21*:368–371, 1974.

Vainsel, M., Manderlier, Th. and Otten, J.: Urinary excretion of adenosine 3′5′ monophosphate in vitamin D deficiency. Eur. J. Clin. Invest. *6*:127–130, 1976.

HEPATIC RICKETS

Glasgow, J. F. T. and Thomas, P. S.: The osteodystrophy of prolonged obstructive liver disease in childhood. Acta Paediatr. Scand. *65*:57–64, 1976.

Kopelman, A. E., Halsted, C. C. and Minnefor, A. B.: Osteomalacia and spontaneous fractures in twins with congenital cytomegalic inclusion disease. J. Pediatr. *81*:101–105, 1972.

Skinner, R. K., Sherlock, S., Long, R. G. et al: 25-hydroxylation of vitamin D in primary biliary cirrhosis. Lancet *1*:720–724, 1977.

Spirer, A. Heiman, I., Shorr, S. et al: Rickets and protracted neonatal obstructive jaundice. Helv. Paediatr. Acta *28*:437–442, 1973.

ANTICONVULSANTS

Borgstedt, A. D., Bryson, M. F., Young, L. W. et al: Long-term administration of antiepileptic drugs and the development of rickets. J. Pediatr. *81*:9–15, 1972.

Bouillon, R. Reynaert, J., Claes, J. H. et al: The effect of anticonvulsant therapy on serum levels of 25-hydroxyvitamin D, calcium and parathyroid hormone. J. Clin. Endocrinol. Metab. *41*:1130–1135, 1975.

Caspary, W. F.: Inhibition of intestinal calcium transport by diphenylhydantoin in rat duodenum. Naunyn-Schmiedebergs Arch. Pharmakol. *274*: 146–153, 1972.

Christiansen, C., Rødbro, P. and Lund, M.: Incidence of anticonvulsant osteomalacia and effect of vitamin D: controlled therapeutic trial. Br. Med. J. *4*:695–701, 1973.

Corradino, R. A. Diphenylhydantoin: direct inhibition of the vitamin D_3-mediated calcium absorptive mechanism in organ-cultured duodenum. Biochem. Pharmacol. *25*:863–864, 1975.

Dent, C. E., Richens, A., Rowe, D. J. F. et al: Osteomalacia with long-term anticonvulsant therapy in epilepsy. Br. Med. J. *4*:69–72, 1970.

Friis, B. and Sardemann, H.: Neonatal hypocalcemia after intrauterine exposure to anticonvulsant drugs. Arch. Dis. Child. *52*:239–247, 1977.

Hahn, T. J., Birge, S. J., Scharp, C. R. et al: Phenobarbital-induced alterations in vitamin D metabolism. J. Clin. Invest. *51*:741–748, 1972.

Hahn, T. J., Hendin, B. A., Scharp, C. R. et al. Effect of chronic anticonvulsant therapy on serum 25-hydroxycalciferol levels in adults. N. Engl. J. Med. *287*:900–904, 1972.

Hahn, T. J. Hendin, B. A., Scharp, C. R. et al: Serum 25-hydroxycalciferol levels and bone mass in children on chronic anticonvulsant therapy. N. Engl. J. Med. *292*:550–554, 1975.

Harrison, H. C. and Harrison, H. E. Inhibition of vitamin D-stimulated active transport of calcium of rat intestine by diphenylhydantoin-phenobarbital treatment. Proc. Soc. Exp. Biol. Med. *153*:220–224, 1976.

Hunter, J. Maxwell, J. D., Stewart, D. A. et al: Altered calcium metabolism in epileptic children on anticonvulsants. Br. Med. J. *4*:202–204, 1971.

Jubiz, W., Haussler, M. R., McCain, T. A. et al: Plasma 1,25-dihydroxyvitamin D levels in patients receiving anticonvulsant drugs. J. Clin. Endocrinol. Metab. *44*:617–621, 1977.

Koch, H-U., Kraft, D., Von Herrath, D. et al: Influence of diphenylhydantoin and phenobarbital on intestinal calcium transport in the rat. Epilepsia *13*:829–834, 1972.

Kruse, R.: Osteopathien Kalzium-und Vitamin D Stoffwechelstorungen unter antiepileptischer Langzeittherapie. Bibl. Psychiatr. *151*:114–143, 1975.

Lifshitz, F. and Maclaren, N. F.: Vitamin D-dependent rickets in institutionalized, mentally retarded children receiving long-term anticonvulsant therapy. I. A survey of 288 patients. J. Pediatr. *83*:612–620, 1973.

Medlinsky, H. L.: Rickets associated with anticonvulsant medication. Pediatrics *53*:91–95, 1974.

Richens, A. and Rowe, D. J. F.: Disturbance of calcium metabolism by anticonvulsant drugs. Br. Med. J. *4*:73–76, 1970.

Schaeffer, K. Flury, W. H., von Herrath, D. et al: Vitamin D-Stoffwechsel und Antiepileptika. Schweiz. Med. Wochenschr. *102*:785–791, 1972.

Schmid, F.: Osteopathien bei antiepileptischer Dauerbehandlung. Fortschr. Med. *9*:381, 1967.

Shafer, R. B. and Nuttall, F. Q.: Calcium and folic acid absorption in patients taking anticonvulsant drugs. J. Clin. Endocrinol. Metab. *41*:1125–1129, 1975.

Silver, J. Davies, T. J., Kupersmitt, E. et al: Prevalence and treatment of vitamin D deficiency in children on anticonvulsant drugs. Arch. Dis. Child. *49*:344–350, 1974.

Sohn, R. S. and Ferrendelli, J. A.: Inhibition of Ca^{++} transport into rat brain synaptasomes by diphenylhydantoin (DPH). J. Pharmacol. Exp. Ther. *185*:272–275, 1973.

Sotaniemi, E. A., Hakkarainen, H. K., Puranen, J. A. et al: Radiologic bone changes and hypocalcemia with anticonvulsant therapy in epilepsy. Ann. Intern. Med. *77*:389–394, 1972.

Stamp, T. C. B., Round, J. M., Rowe, D. J. F. et al: Plasma levels and therapeutic effect of 25-hydroxycholecalciferol in epileptic patients taking anticonvulsant drugs. Br. Med. J. *4*:9–12, 1972.

Villareale, M. Gould, L. V., Wasserman, R. H. et al.: Diphenylhydantoin: effects on calcium metabolism in the chick. Science *183*:671–673, 1974.

VITAMIN D–DEPENDENT RICKETS

Arnaud, C. Maijer, R., Reade, T. et al: Vitamin D dependency: an inherited postnatal syndrome with secondary hyperparathyroidism. Pediatrics *47*: 871–880, 1970.

Brooks, M. H., Bell, N. H., Love, L. et al.: Vitamin D-dependent rickets type II. Resistance of target organs to 1-25-dihydroxy vitamin D. N. Engl. J. Med. *298*:996–999, 1978.

Fanconi, A. and Prader, A.: Die Hereditäre Pseudomangelrachitis. Helv. Paediatr. Acta *24*:423–447, 1969.

Fraser, D. Kooh, S. W., Kind, H. P. et al: Pathogenesis of hereditary vitamin D-dependent rickets. N. Engl. J. Med. *289*:817–822, 1973.

Scriver, C. R.: Commentary: vitamin D dependency. Pediatrics *45*:361–363, 1970.

Stoop, J. W., Schraagen, M. J. C. and Tiddens, H. A. W. M.: Pseudo vitamin D deficiency rickets. Acta Paediatr. Scand. *56*:607–616, 1967.

RENAL INSUFFICIENCY

Castleman, B. and Mallory, T. B.: Parathyroid hyperplasia in chronic renal insufficiency. Am. J. Pathol. *13*:553–574, 1937.

Chan, J. C. M., Oldham, S. B. and DeLuca, H.: Effectiveness of 1α-hydroxyvitamin D_3 in children with renal osteodystrophy associated with hemodialysis. J. Pediatr. *90*:820–824, 1977.

Coburn, J. W., Koppel, M. H., Brickman, A. S. et al: Study of intestinal absorption of calcium in patients with renal failure. Kidney Int. *3*:264–272, 1973.

Davie, M. W. S., Chalmers, T. M., Hunter, J. O. et al: 1-Alphahydroxycholecalciferol in chronic renal failure. Studies of the effect of oral doses. Arch. Intern. Med. *84*:281–285, 1976.

Drüeke, T., Bordier, P. J., Man, N. K., et al: Effects of high dialysate calcium concentration on bone remodeling, serum biochemistry, and parathyroid hormone in patients with renal osteodystrophy. Kidney Int. *11*:267–274, 1977.

Follis, R. H., Jr.: Renal rickets and osteitis fibrosa in children and adolescents. Bull. Johns Hopkins Hosp. *87*:593–615, 1950.

Gruskin, A. B., Root, A. W., Duckett, G. E. et al: Parathyroid function in uremic children during periods of renal insufficiency, hemodialysis, and transplantation. J. Pediatr. *89*:755–762, 1976.

Habener, J. F. and Schiller, A. L.: An editorial—pathogenesis of renal osteodystrophy—a role for calcitonin? N. Engl. J. Med. *296*:1112–1113, 1977.

Henderson, R. G., Russell, R. G. G., Ledingham, J. G. G. et al: Effects of 1,25-dihydroxycholecalciferol on calcium absorption, muscle weakness and bone disease in chronic renal failure. Lancet *1*:379–384, 1974.

Kaye, M., Chatterjee, G., Cohen, G. F. et al: Arrest of hyperparathyroid bone disease with dihydrotachysterol in patients undergoing chronic hemodialysis. Ann Intern. Med. *73*:225–233, 1970.

Llach, F. Massry, S. G., Singer, F. R. et al: Skeletal resistance to endogenous parathyroid hormone in patients with early renal failure. A possible cause for secondary hyperparathyroidism. J. Clin. Endocrinol. Metab. *41*:339–345, 1975.

Lumb, G. A., Mawer, E. B. and Stanbury, S. W.: The apparent vitamin D resistance of chronic renal failure. Am J. Med. *50*:421–441, 1971.

Mankin, H. J.: Rickets, osteomalacia and renal osteodystrophy—part I, review article. J. Bone Joint Surg. *56A*:101–128, 1974.

Mankin, H. J.: Rickets, osteomalacia and renal osteodystrophy—part II, review article. J. Bone Joint Surg. *56-A*:352–386, 1974.

Meyrier, A., Marsac, J. and Richet, G.: The influence of a high calcium carbonate intake on bone disease in patients undergoing hemodialysis. Kidney Int. *4*:146–153, 1973.

Slatopolsky, E. and Bricker, N. S.: The role of phosphorus restriction in the prevention of secondary hyperparathyroidism in chronic renal disease. Kidney Int. *4*:141–145, 1973.

Slatopolsky, E., Hruska, K. and Rutherford, W. E.: Current concepts of parathyroid hormone and vitamin D metabolism: perturbations in chronic renal disease. Kidney Int. (Suppl.)2:90–96, 1975.

CALCIUM DEFICIENCY RICKETS

Berlyne, G. M., Ari, J. B., Nord, E. et al: Bedouin osteomalacia due to calcium deprivation caused by high phytic acid content of unleavened bread. Am. J. Clin. Nutr. *26*:910–911, 1973.

Kooh, S. W., Fraser D., Reilly, B. J. et al: Rickets due to calcium deficiency. N. Engl. J. Med. *297*:1264–1266, 1977.

Maltz, H. E., Fish, M. B. and Holliday, M. A.: Calcium deficiency rickets and the renal response to calcium infusion. Pediatrics *46*:865–870, 1970.

PHOSPHATE DEFICIENCY RICKETS

Leape, L. L. and Valaes, T.: Rickets in low birth weight infants receiving total parenteral nutrition. J. Pediatr. Surg. *11*:665–674, 1976.

Rowe, J. C., Wood, D. H. and Rowe, D. W. et al: Hypophosphatemic rickets in breast milk fed prematures (Abstract). Pediatr. Res. *12*:441, 1978.

Sydow, G.: The development of rickets in premature infants. Acta Paediatr. Scand. *35*(Suppl. 1):169–176, 1948.

RENAL TUBULAR HYPOPHOSPHATEMIA

Harrison, H. E.: The varieties of rickets and osteomalacia associated with hypophosphatemia. Clin. Orthop. *9*:61–74:1957.

Scriver, C. R.: Rickets and the pathogenesis of impaired tubular transport of phosphate and other solutes. Am. J. Med. *57*:43–49, 1974.

PRIMARY HYPOPHOSPHATEMIA

Arnaud, C., Glorieux, F. and Scriver, C.: Serum parathyroid hormone in X-linked hypophosphatemia. Science *173*:845–847, 1971.

Blackard, W. G., Robinson, R. R. and White, J. E.: Familial hypophosphatemia. New Engl. J. Med. *266*:899–905, 1962.

Condon, J. R., Nassim, J. R. and Rutter, A.: Defective intestinal phosphate absorption in familial and non-familial hypophosphatemia. Br. Med. J. *3*:138–141, 1970.

Fanconi, A., Fischer, J. A. and Prader, A.: Serum parathyroid hormone concentrations in hypophosphatemic vitamin D resistant rickets. Helv. Paediatr. Acta *29*:187–194, 1974.

Frame, B. and Smith, R. W., Jr.: Phosphate diabetes—a case study of osteomalacia. Am. J. Med. *25*:771–779, 1958.

Freeman, S. and Dunsky, I.: Resistant rickets. Am. J. Dis. Child. *79*:409–427, 1950.

Glorieux, F. H., Holick, M. F., Scriver, C. R. et al: X-linked hypophosphatemic rickets: inadequate therapeutic response to 1,25-dihydroxycholecalciferol. Lancet *II*:287–289, 1973.

Glorieux, F. H., Morin, C. L., Travers, R. et al: Intestinal phosphate transport in familial hypophosphatemic rickets. Pediatr. Res. *10*:691–696, 1976.

Glorieux, F. H. and Scriver, C. R.: Loss of a parathyroid hormone-sensitive component of phosphate transport in X-linked hypophosphatemia. Science *175*:997–999, 1972.

Glorieux, F. H., Scriver, C. R., Reade, T. M. et al: Use of phosphate and vitamin D to prevent dwarfism and rickets in X-linked hypophosphatemia. New Engl. J. Med. *287*:481–487, 1972.

Hahn, T. J., Scharp, C. R., Halsted, L. R. et al: Parathyroid hormone status and renal responsiveness in familial hypophosphatemic rickets. J. Clin. Endocrinol. Metab. *41*:926–937, 1975.

Harrison, H. E., Harrison, H. C., Lifshitz, F. et al: Growth disturbance in hereditary hypophosphatemia. Am. J. Dis. Child. *112*:290–297, 1966.

Lewy, J. E., Cabana, E. C., Repetto, H. A. et al: Serum parathyroid hormone in hypophosphatemic vitamin D resistant rickets. J. Pediatr. *81*:294–300, 1972.

Morgan, J. M., Hawley, W. L., Chenoweth, A. I. et al: Renal transplantation in hypophosphatemia with vitamin D-resistant rickets. Arch. Intern. Med. *134*:549–552, 1974.

Parfitt, A. M.: Hypophosphatemic vitamin D refractory rickets and osteomalacia. Orthop. Clin. North Am. *3*:653–679, 1972.

Reitz, R. E. and Weinstein, R. L.: Parathyroid hormone secretion in familial vitamin-D-resistant rickets. N. Engl. J. Med. *289*:941–945, 1973.

Riggs, B. L., Sprague, R. G., Jowsey, J. et al: Adult-onset vitamin-D resistant hypophosphatemic osteomalacia. Effect of total parathyroidectomy. N. Engl. J. Med. *281*:762–766, 1969.

Rose, G. A.: Role of phosphate in treatment of renal tubular (hypophosphatemic) rickets and osteomalacias. Br. Med. J. *2*:857–861, 1964.

Russell, R. G. G., Smith, R., Preston, C. et al: The effect of 1,25-dihydroxycholecalciferol on renal tubular reabsorption of phosphate, intestinal absorption of calcium and bone histology in hypophosphatemic renal tubular rickets. Clin. Sci. Mol. Med. *48*:177–186, 1975.

Sheridan, R. M., Chiroff, R. T. and Friedman, E. M.: Operative and non-operative treatment of rachitic lower extremity deformities. Clin. Orthop. *116*:66–69, 1976.

Short, E., Morris, R. C., Sebastian, A. et al: Exaggerated phosphaturic response to circulating parathyroid hormone in patients with familial X-linked hypophosphatemic rickets. J. Clin. Invest. *58*:152–163, 1976.

Short, E. M., Binder, H. J. and Rosenberg, L. E.: Familial hypophosphatemic rickets: defective transport of inorganic phosphate by intestinal mucosa. Science *179*:700–702, 1973.

Stamp, T. C. B. and Baker, L. R. I.: Recessive hypophosphatemic rickets and possible aetiology of the vitamin D-resistant syndrome. Arch. Dis. Child. *51*:360–365, 1976.

Tapia, J., Stearns, G. and Ponseti, I. V.: Vitamin D resistant rickets. J. Bone Joint Surg. *46-A*:935–958, 1964.

Teitelbaum, S. L., Rosenberg, E. M., Bates, M. et al: The effects of phosphate and vitamin D therapy on osteopenic hypophosphatemic osteomalacia of childhood. Clin. Orthop. *116*:38–47, 1976.

FANCONI SYNDROME

Bickel, H. and Smellie, J. M.: Cystine storage disease with aminoaciduria. Lancet *1*:1093–1095, 1952.

Bickel, H.: Die Entwicklung der biochemischen Läsion bei der Lignac-Fanconischen Krankheit. Helv. Paediatr. Acta *10*:259–268, 1955.

Chesney, R. W. and Harrison, H. E.: Fanconi syndrome following bowel surgery and hepatitis reversed by 25-hydroxycholecalciferol. J. Pediatr. *86*:857–861, 1975.

Chisolm, J. J., Jr., Harrison, H. C., Eberlein, W. R. et al: Amino-aciduria hypophosphatemia, and rickets in lead poisoning. Am. J. Dis. Child. *89*: 159–168, 1955.

Engle, R. L. and Wallis, L. A.: Multiple myeloma and the adult Fanconi syndrome. Am. J. Med. *22*: 5–12, 1957.

Etches, P., Pickering, D. and Smith, R.: Cystinotic rickets treated with vitamin D metabolites. Arch. Dis. Child. *52*:661–664, 1977.

Fanconi, G.: Der frühinfantile nephrotischglykosurische Zwergwuchs mit hypophosphatämischer Rachitis. Jahrb. f. Kinderheilkd. *147*:299–338, 1936.

Garty, R., Cooper, M. and Tabachnik, E.: The Fanconi syndrome associated with hepatic glycogenosis and abnormal metabolism of galactose. J. Pediatr. *85*:821–823, 1974.

Goldman, H., Scriver, C. R., Aaron, K. et al: Adolescent cystinosis: comparisons with infantile and adult forms. Pediatrics *47*:979–988, 1971.

Harrison, H. E.: The Fanconi syndrome, J. Chronic Dis. *7*:346–355, 1958.

Harrison, H. E. and Harrison, H. C.: Experimental production of renal glycosuria, phosphaturia, and aminoaciduria by injection of maleic acid. Science *120*:606–608, 1954.

Harrison, H. E. and Harrison, H. C.: Hereditary metabolic bone diseases. Clin. Orthop. *33*:147–163, 1964.

Hunt, D. D., Stearns, G., McKinley, J. B. et al.: Long-term study of family with Fanconi syndrome without cystinosis (DeToni-Debré-Fanconi syndrome). Am. J. Med. *40*:492–510, 1966.

Lee, D. B. N., Drinkard, J. P., Rosen, V. J. et al: The adult Fanconi syndrome. Medicine *51*:107–138, 1972.

Lowe, C. U., Terrey, M. and MacLachlan, E. A.: Organic aciduria, decreased renal ammonia production, hydrophthalmos and mental retardation. Am. J. Dis. Child. *83*:164–184, 1952.

Maldonado, J. E., Velosa, J. A., Kyle, R. A. et al: Fanconi syndrome in adults. A manifestation of a latent form of myeloma. Am. J. Med. *58*:354–364, 1975.

Malekzadeh, M. H., Neustein, H. B., Schneider, J. A. et al: Cadaver renal transplantation in children with cystinosis. Am. J. Med. *63*:525–533, 1977.

Morris, R. C., Jr.: An experimental renal acidification defect in patients with hereditary fructose intolerance. J. Clin. Invest. *47*:1648–1662, 1968.

Morris, R. C., Jr., McSherry, E. and Sebastian, A.: Modulation of experimental renal dysfunction of hereditary fructose intolerance by circulating parathyroid hormone. Proc. Natl. Acad. Sci. U.S.A. *68*:132–135, 1971.

Salassa, R. M., Power, M. H., Ulrich, J. A. et al: Observations on the metabolic effects of vitamin D in Fanconi's syndrome. Proc. Staff Meetings Mayo Clinic *29*:214–224, 1954.

Schneider, J. A., Verroust, F. M., Kroll, W. A. et al.: Prenatal diagnosis of cystinosis. N. Engl. J. Med. *290*:878–882, 1974.

Scriver, C. R., Chesney, R. W. and McInnes, R. R.: Genetic aspects of renal tubular transport: diversity and topology of carriers. Kidney Int. *9*:149–171, 1976.

Soriano, J. R., Houston, I. B., Boichis, H. et al: Calcium and phosphorus metabolism in the Fanconi syndrome. J. Clin. Endocrinol. Metab. *28*:1555–1563, 1968.

VanBiervliet, J. P. G. M., Donckerwolcke, R. A. M. G., Van Stekelenburg, G. J. et al: Sodium chloride restriction and extracellular fluid volume contraction in hyperphosphaturic vitamin D resistant rickets in the Lowe syndrome. Helv. Paediatr. Acta *30*:365–375, 1975.

Wallis, L. A. and Engle, R. L.: The adult Fanconi syndrome. Am. J. Med. *22*:13–18, 1957.

RENAL TUBULAR ACIDOSIS

Albright, F., Burnett, C. H., Parson, W. et al: Osteomalacia and rickets: the various etiologies met in United States with emphasis on that resulting from a special form of acidosis: the therapeutic implications for each etiological sub-group and the relationship between osteomalacia and Milkman's syndrome. Medicine *25*:399–479, 1946.

Albright, F., Consolazio, W. V., Coombs, F. S. et al: Metabolic studies and therapy in a case of nephrocalcinosis with rickets and dwarfism. Bull. Johns Hopkins Hosp. *66*:7–33, 1940.

Graziani, G., deVecchi, A. and Rosti, D.: Primary proximal renal tubular acidosis. Helv. Paediatr. Acta *31*:427–434, 1976.

Harrison, H. E.: The varieties of rickets and osteomalacia associated with hypophosphatemia. Clin. Orthop. *9*:61–74, 1957.

Harrison, H. E. and Harrison, H. C.: Inhibition of urine citrate excretion and the production of renal calcinosis in the rat by acetazolamide (diamox) administration. J. Clin. Invest. *34*:1662–1670, 1955.

Huguenin, M., Schacht, R. and David, R.: Infantile rickets with severe proximal renal tubular acidosis, responsive to vitamin D. Arch. Dis. Child. *49*:955–959, 1974.

Lightwood, R., Payne, W. W. and Black, J. A.: Infantile renal acidosis. Pediatrics *12*:628–644, 1953.

McSherry, E. and Morris, R. C., Jr.: Attainment and maintenance of normal stature with alkali therapy in infants and children with classic renal tubular acidosis. J. Clin. Invest. *61*:509–527, 1978.

McSherry, E., Sebastian, A. and Morris, R. C., Jr.: Renal tubular acidosis in infants — several kinds, including bicarbonate wasting, classic renal tubular acidosis. J. Clin. Invest. *51*:499–514, 1972.

Morris, R. C., Jr.: An experimental renal acidification defect in patients with hereditary fructose intolerance. I. Its resemblance to renal tubular acidosis. J. Clin. Invest. *47*:1389–1397, 1968.

Morris, R. C., Jr.: Renal tubular acidosis. N. Engl. J. Med. *281*:1405–1413, 1969.

Rodriguez-Soriano, J. and Edelmann, C. M., Jr.: Renal tubular acidosis. Annu. Rev. Med. *20*:363–382, 1969.

TUMOR-ASSOCIATED PRIMARY HYPOPHOSPHATEMIC RICKETS AND OSTEOMALACIA

Aschinberg, L. C. Solomon, L. M., Zeis, P. M. et al: Vitamin D-resistant rickets associated with epidermal nevus syndrome: demonstration of a phosphaturic substance in the dermal lesions. J. Pediatr. *91*:56–60, 1977.

Dent, C. E. and Stamp, T. C. B.: Hypophosphatemic osteomalacia presenting in adults. Q. J. Med. *40*:303–329, 1971.

Evans, D. J. and Azzopardi, J. G.: Distinctive tumours of bone and soft tissue causing acquired vitamin D-resistant osteomalacia. Lancet *1*:353–354, 1972.

Frymoyer, J. W. and Hodgkin, W.: Adult-onset vitamin-D resistant hypophosphatemic osteomalacia. J. Bone Joint Surg. *59-A*:101–106, 1977.

Harrison, H. E.: Oncogenous rickets. Possible elaboration by a tumor of a humoral substance inhibiting tubular reabsorption of phosphate. Pediatrics *52*:432–433, 1973.

Olefsky, J., Kempson, R., Jones, H. et al: Tertiary hyperparathyroidism and apparent "cure" of vitamin-D-resistant rickets after removal of an ossifying mesenchymal tumor of the pharynx. N. Engl. J. Med. *286*:740–745, 1972.

Pollack, J. A., Schiller, A. L. and Crawford, J. D.: Rickets and myopathy cured by removal of a non-ossifying fibroma of bone. Pediatrics *52*:364–371, 1973.

Prader, A., Illig, R., Uehlinger, E. et al: Rachitis infolge Knochentumors. Helv. Paediatr. Acta *14*: 554–565, 1959.

Salassa, R. M., Jowsey, J. and Arnaud, C. D.: Hypophosphatemic osteomalacia associated with "non-endocrine" tumors. N. Engl. J. Med. *283*:65–70, 1970.

Chapter Six

OSTEOPENIC AND OSTEOSCLEROTIC DISORDERS OF BONE

OSTEOPENIA

Osteopenia means a deficiency of bone including both mineral and matrix, and therefore is distinguished from osteomalacia in which mineral is deficient but there is an abundance of unmineralized matrix. Histologic examination of osteomalacic bone shows this excess of unmineralized matrix osteoid covering the mineralized matrix, whereas osteopenic bone shows no accumulation of osteoid, and aside from paucity of bone per unit volume is not different from normal bone on ordinary histologic examination. On chemical examination osteomalacic bone is distinguishable from normal bone by the proportion of bone ash to total bone solids; osteopenic bone cannot be differentiated from normal bone in this way. The concentration of bone ash or bone calcium expressed in terms of fat-free dried weight is low in osteomalacic bone whereas this value is normal in osteopenic bone. The density of osteopenic bone, however (i.e., the weight of bone per volume of anatomic bone structure), is reduced, indicating a paucity of total bone.

The diagnosis of osteopenia can be made radiographically if the reduction of bone density is sufficiently great. However, this is a rather insensitive method of determining bone density and a decrease of 30 per cent or more is said to be needed before the changes can be detected by ordinary x-ray examination. The characteristic features are reduced opacity of bone structure with thin cortices and a ground-glass trabecular structure, rather than the coarse cystic structure of osteomalacia. A method of quantitatively measuring the width of the cortex of a metacarpal bone related to the total width of the bone has been developed as an index of osteopenia.

Other methods of determining bone density without biopsy of bone are the photon absorption technique and neutron activation analysis. In the former the forearm is clamped in a fixed position over a source of gamma radiation such as I^{131} and the radiation energy penetrating the tissue is measured by a scintillation counter moving transversely over the forearm. Since bone mineral absorbs this radiation to a much greater extent than does soft tissue, the density of the bone mineral can be estimated if the cross-sectional area of the bone is also determined. This method is highly useful in measurement of change of bone density as an index of effect of treatment. It cannot be applied, however, to the axial skeleton which may be primarily affected in osteoporosis. Neutron activation analysis can be used for determination of total skeletal calcium. Exposure of the body to neutron flux causes activation of calcium with emission of electrons of specific energy that can be measured, and is an index of total body calcium. Although a valuable research tool, this method requires costly apparatus and an expertise that makes it unavailable for clinical purposes.

The osteopenias can be considered in two major groups: (1) those in which there is a congenital defect of the bone-forming cells so that normal matrix is not produced and there is an inadequate quantity of matrix in which mineral can be deposited (the forms of osteogenesis imperfecta); and (2) the acquired abnormalities of bone formation and resorption in which an imbalance develops between bone matrix formation and mineralization, and bone resorption, so that the resorptive process dominates and the bone becomes less dense (the osteoporoses).

OSTEOGENESIS IMPERFECTA

In all forms of osteogenesis imperfecta the basic abnormality probably is inadequate production of matrix by bone cells. The possibility has also been proposed that the matrix is abnormal in its composition owing to lack of maturation of cross-linking of collagen, but much of the evidence suggests a quantitative matrix deficiency. A reduction of type I collagen, the major collagen type in bone, has been reported in osteogenesis imperfecta.

On histologic examination the bone is more primitive than normal with an increased number of osteocytes per unit bone volume that would be compatible with a reduced volume of matrix. The congenital, extremely severe form of osteogenesis imperfecta (Vrolik type) results in such fragile bone that intrauterine fractures and attempted repair occur, and the infant is born with multiple fractures in various stages of healing with severe deformity. The skull is almost completely unmineralized. These infants frequently die in utero or soon after birth. If they live they are subject to fractures even with the most careful management. This disorder is probably inherited as an autosomal recessive characteristic. There is also a milder variant of recessively inherited osteogenesis with fractures at birth, but the infant survives and develops progressive deformities in childhood.

A less severe form of osteogenesis imperfecta is transmitted by an autosomal dominant gene (Lobstein type). Occasional intrauterine fractures may occur but most of the fractures result from mild trauma in infancy and childhood. The bones on x-ray examination are delicate with extremely thin cortices (Fig. 6–1). The skull shows a characteristic pattern of multiple wormian bones with so-called lacunar skull. Hydrocephalus can be a complication, possibly because of basilar obstruction due to flattening of the posterior fossa, platybasia. Scoliosis occurs in a significant proportion of these patients, starting in later childhood. There is a general abnormality of collagen structure, one of the manifestations of which is the thin scleral coat of the eyeball, resulting in the blue sclerae caused by the dark choroid pigment visible through the relatively transparent delicate scleral layer. Another associated anomaly is otosclerotic deafness occurring in later life. There may be formes frustes with otosclerosis and blue sclerae, but without extremely fragile bones, so that a family history of deafness may be a clue in the diagnosis. Poor mineralization of the teeth, dentinogenesis imperfecta, may also be present with early caries and loss of teeth; it has been suggested that this condition represents a separate subtype of dominantly inherited osteogenesis imperfecta.

These patients suffer multiple fractures that heal with adequate callus formation but result in marked deformity and shortening of the limbs. They show spontaneous improvement following puberty and cessation of growth. This may represent the lessened stress on the bone formation process when the bone is no longer increasing in volume. Serum calcium and phosphate concentra-

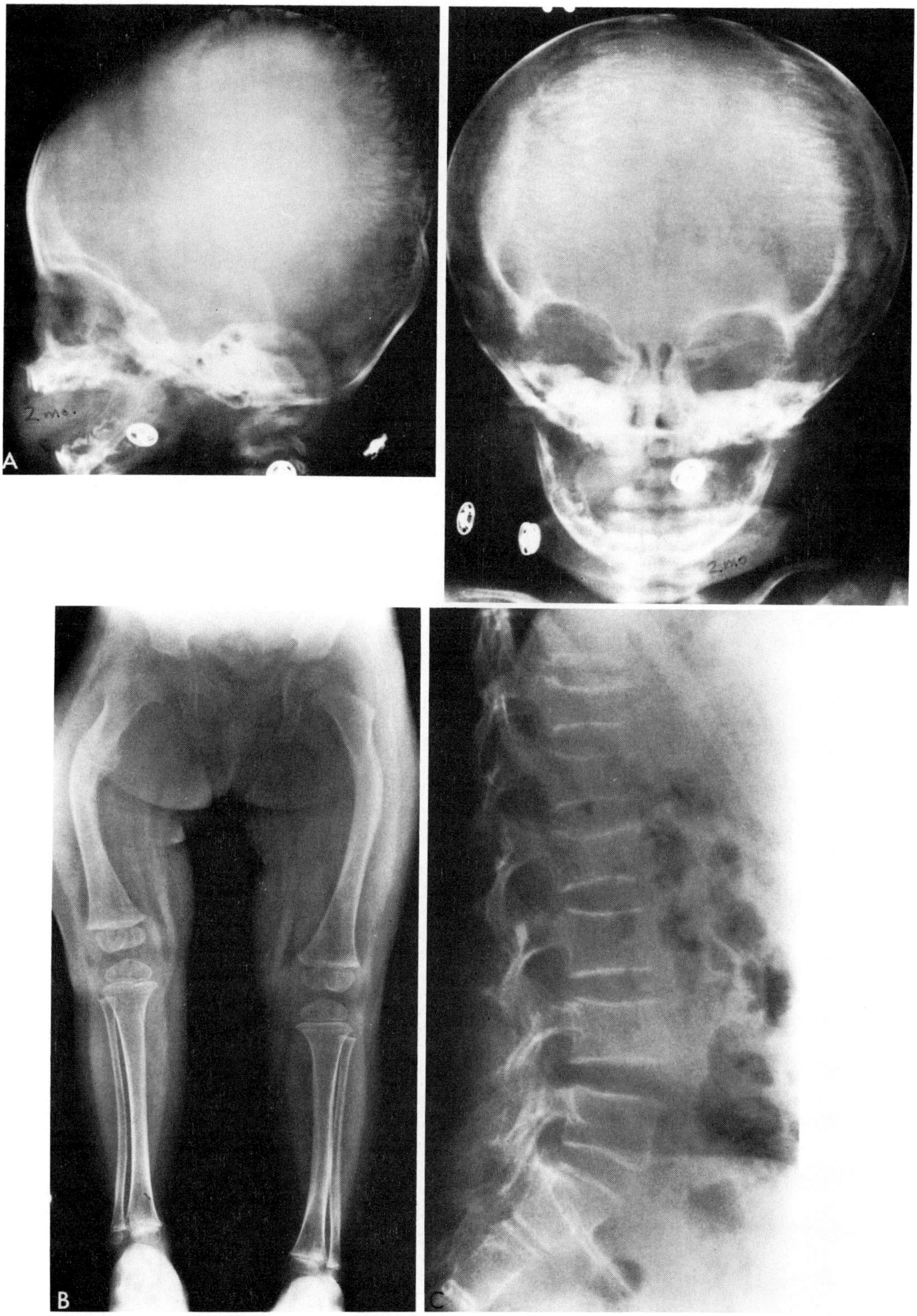

Figure 6–1. Osteogenesis imperfecta: *a,* P.A. and lateral of skull of an infant, age 2 months, with characteristic mosaic pattern of multiple wormian bones; *b,* child, age 16 months—long bones demonstrating thin cortices and low mineral density—fracture of right femur with callus formation; *c,* vertebrae—low mineral density and collapse of bodies.

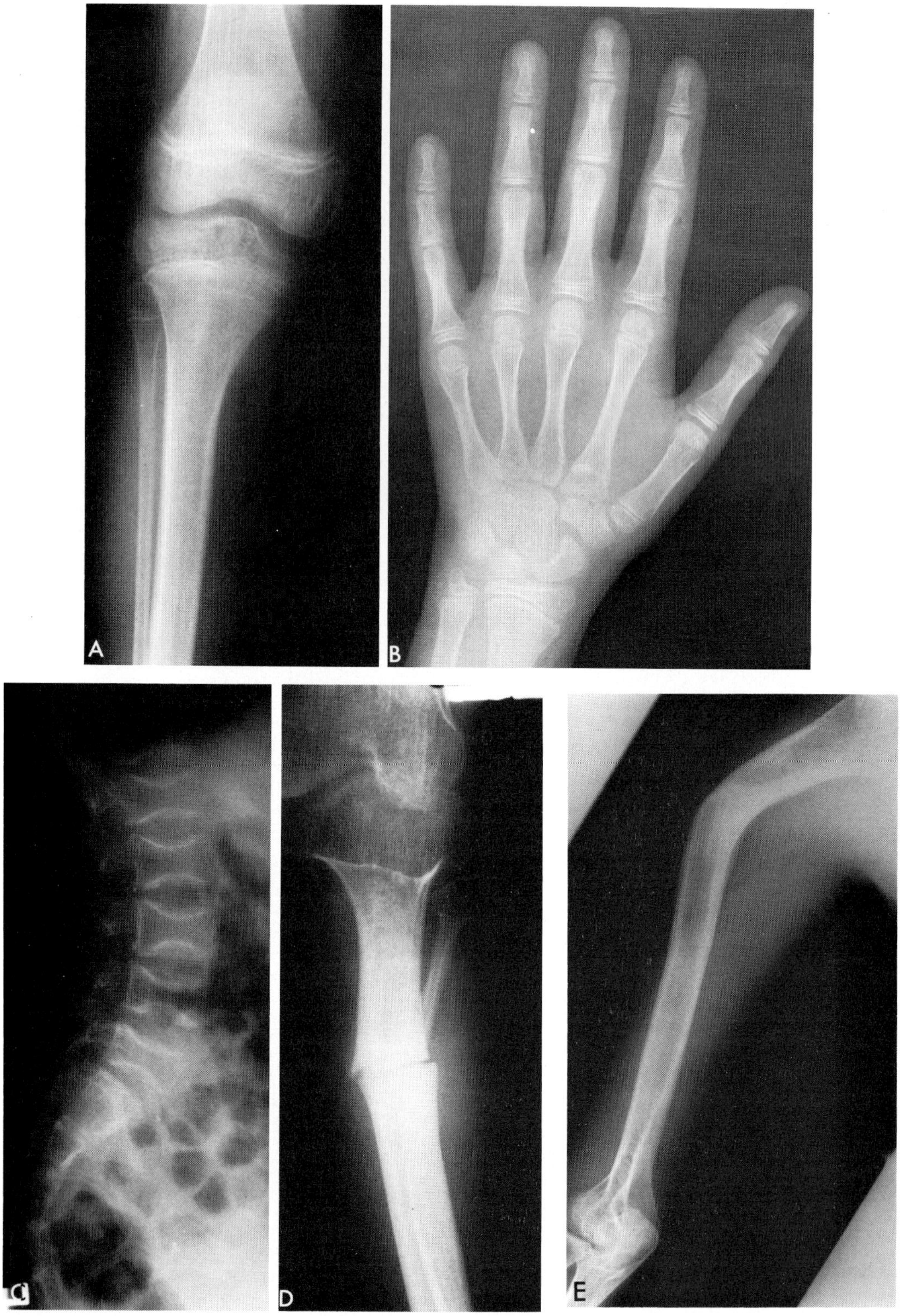

Figure 6–2. Osteogenesis imperfecta in child with pseudoglioma: *a*, knee, age 11 years—reduced bone mineral density; *b*, hand—thin cortices of metacarpals and phalanges; *c*, vertebrae, age 13 years—demineralization and invagination of vertebral bodies by nucleus pulposus; *d*, tibial fracture, age 18 years; *e*, humerus with severe deformity following fracture.

tions are normal. Alkaline phosphatase concentrations may be increased in association with multiple fractures.

A variety of treatments of osteogenesis imperfecta have been proposed. There is no evidence at this time that vitamin D or any of its metabolites can improve the density of bone lacking in matrix, and excessive vitamin D intake may result in hypercalcemia and hypercalciuria. Sodium fluoride has been tried because of the evidence that fluoride toxicity is associated with excessively dense bone with inhibition of bone resorption. This treatment has not produced any significant change in the course of the disorder. Magnesium salt treatment has also been recommended but clear evidence of its value is still lacking. A prolonged course of calcitonin injections has also been tried with conflicting results. Hormonal treatment to produce precocious puberty and cessation of growth has been suggested in very severe cases but is a drastic remedy that does not seem warranted. Careful protection of the child and prompt treatment of fractures as they occur appear to be the only measures available at present.

There may be quite mild variants of osteogenesis imperfecta of the dominant variety with fractures occurring after trauma as the only manifestation, and the only feature suggesting an abnormality is the frequency of fractures during childhood and the x-ray findings of thin bones, particularly reduced cortical width.

Osteogenesis Imperfecta Associated with Congenita Pseudoglioma

We have had one patient with this variation and have knowledge of several others. Retinal separation is seen during infancy owing to masses behind the retina. The appearance is that of retinoblastoma, and enucleation of the eyes may be done because of the difficulty in differential diagnosis. On histologic examination the lesion is a benign proliferation of glial tissue.

During childhood multiple fractures occur following slight trauma and severe deformity may result (Fig. 6–2). Because of the blindness it is difficult to protect these patients from falls that predispose to a high incidence of fractures. The basis of the abnormality of bone structure is unknown. Serum calcium, phosphate and alkaline phosphatase concentrations are within normal limits. Bone biopsy in one case showed no characteristic lesions differing from other types of osteogenesis imperfecta. There is no known treatment of this disorder. It is possible that improvement may occur spontaneously after growth of bone volume ceases.

OSTEOPOROSIS

By definition, osteoporosis is an acquired form of osteopenia in which there is progressive loss of bone density resulting from an imbalance between bone matrix formation and mineralization of this matrix and resorption of mineral and matrix. Whether the defect is primarily decreased matrix formation or increased mineral and matrix resorption may vary with the pathogenesis of the osteoporosis. The various forms of osteoporosis encountered in childhood and adolescence are listed in Table 6–1.

Steroid Osteoporosis

This is probably the most common of the osteopenic problems of childhood. Cushing's disease, whether due to adrenal carcinoma or hypersecretion of adrenal cortex secondary to excess ACTH production, is relatively rare, but the use of suppressive doses of corticosteroids in a variety of diseases is of course more frequent. Since the

Table 6–1. Causes of Osteoporosis in Childhood and Adolescence

Corticosteroid Excess
Cushing's disease
Steroid treatment
Immobilization
Nutritional Deficiency
Protein-calorie malnutrition
Vitamin C deficiency
Copper deficiency
Homocystinuria
Idiopathic Juvenile Osteoporosis

bone lesion is the same in both spontaneous and iatrogenic corticosteroid excess the following discussion will not separate these two entities.

The primary mechanism of this form of osteopenia is failure of bone cell development and matrix formation to keep pace with the normal rate of bone resorption occurring in growing and remodeling bone. This deficiency of osteoblast development from progenitor cells and lack of osteoblast production of matrix is presumably another manifestation of the growth-suppressing action of corticosteroid excess. Whether this effect is mediated through inhibition of somatomedin production or other mechanisms is still under investigation.

An additional factor in corticosteroid osteopenia is reduction of intestinal calcium absorption. High doses of corticosteroids inhibit vitamin D-stimulated intestinal calcium transport. This has been attributed both to a primary effect on the calcium transport system and to an abnormality in the sequence of vitamin D metabolism and action of the active vitamin D metabolite on the calcium absorptive mechanism. One suggestion has been that corticosteroids accelerate the metabolism of vitamin D to inactive products, thus increasing the requirement for vitamin D to maintain a normal concentration of 1,25-dihydroxyvitamin D. There is recent evidence indicating that the serum concentrations of 1,25-dihydroxyvitamin D are reduced in patients receiving suppressive corticosteroid treatment for nephrosis or other forms of renal disease. It has also been shown that the reduced calcium absorption of steroid-treated patients can be increased by high doses of vitamin D or its metabolites, 25-hydroxyvitamin D or 1,25-dihydroxyvitamin D. The elevated concentrations of serum IPTH following corticosteroid dosage are in accordance with the evidence of impaired calcium absorption resulting from high concentrations of corticosteroids. This represents the physiologic response to reduced entrance of calcium into extracellular fluid, and as a result of the augmented secretion of parathyroid hormone increased bone resorption occurs as well as reduced matrix formation.

The severity of corticosteroid osteopenia differs to some extent according to the underlying disease for which steroid treatment is given. This may depend partly on the metabolic consequences of systemic disease but is also related to the dosage and duration of corticosteroid treatment. Very severe osteopenia has been seen in patients with disseminated lupus who require large doses of corticosteroids on a daily basis to control the disease, whereas this is a less frequent complication in children with minimal change nephrosis who can be treated with corticosteroids on an alternate day basis or other intermittent treatment schedule, particularly if remission is achieved by short courses of treatment.

The bone structure at great risk in steroid osteoporosis is the vertebral body. The vertebrae depend on trabecular bone for their resistance to compression, and trabecular bone disappears most rapidly under conditions of imbalance between bone formation and resorption. Severe back pain, compression fractures of the vertebrae and kyphosis are major manifestations of steroid osteopenia. The vertebral lesions are evident on x-ray examination, namely, the "codfish" vertebral bodies representing the compression of the vertebrae by intervertebral discs and the crushing of the vertebrae following fracture. The appendicular skeleton also shows the characteristic loss of bone density.

Treatment of steroid osteopenia is difficult unless the disease process for which corticosteroids are being used can be kept under control by other treatment. Immobilization and traction treatment of the vertebral compression fractures may be necessary, but introduces another cause of osteopenia—immobilization. High calcium, high vitamin D intake is not effective in the treatment of far-advanced osteopenia, but should be employed in the prevention of corticosteroid bone disease on the basis of the evidence that the impaired calcium absorption resulting from corticosteroid excess is improved by pharmacologic doses of vitamin D. There is inadequate experience to indicate whether salmon calcitonin injection can produce improvement in the bone structure of patients on steroid treatment.

Intermittent steroid therapy rather than continuous treatment should be used if the disease process can be controlled by such a schedule.

Immobilization Osteopenia

This form of loss of bone density has been carefully studied both clinically and experimentally. As discussed in Chapter 1 the development and activity of osteoblasts are sensitive to the forces of compression produced by weight-bearing and of bending caused by muscle pull. Thickening of the bone at the sites of muscle insertion is well-known. Abrupt cessation of compressive and bending stresses by immobilization results in reduction of bone cell formation of matrix without comparable reduction, or even with increase, in the continued bone resorption that is part of the normal physiologic activity of osteocytes and osteoclasts. In subjects who previously have been engaging in intensive muscular exercise and who are also rapidly growing, such sudden imbalance between bone formation and resorption may lead to hypercalciuria and hypercalcemia (see Chapter 4). Even in the absence of overt hypercalcemia, increased urine calcium excretion, sometimes associated with stone formation, will be seen as a consequence of the reduced uptake of calcium by bone during immobilization. This can be an important complication of extensive paralysis as in patients with poliomyelitis, spinal cord injury or Guillain-Barré syndrome, or in children with severe muscular dystrophy.

Immobilization of a single extremity as a result of trauma can lead to sufficient loss of bone density for the affected bone to be susceptible to recurrent fracture when activity is resumed, particularly if a fall should occur. Attempts have been made to accelerate fracture healing so as to reduce the period of immobilization, and one of the most interesting approaches has been the use of applied electrical fields on the basis of the theory that the bone mineral crystals act as a mechanical-electrical transducer when subjected to stress with resulting deformation. External fluctuating electrical fields may then substitute for the electrical potentials produced by the normal compressive and bending stresses of weight-bearing and muscle pull.

When immobilization osteopenia is associated with hypercalciuria and hypercalcemia, salmon calcitonin by intramuscular or subcutaneous injection, 4 to 6 MRC units/kg given at eight- to 12-hour intervals, has been found to be effective in reducing the excessive calcium concentrations in body fluids. It is possible that prolonged daily IM injection of salmon calcitonin might be beneficial in preventing severe immobilization osteopenia in patients with paralysis.

Osteopenia Secondary to Hyperplasia of Bone Marrow

In subjects with chronic hemolytic anemia, stimulation of erythropoiesis can result in such extensive marrow hyperplasia that increased resorption of bone occurs, with marked thinning of cortices and extreme fragility of long bones. This is seen most strikingly in patients with thalassemia major, but can also be found in children with sickle cell disease and familial spherocytosis.

Nutritional Osteopenia

Protein-Calorie Malnutrition. Severely malnourished children, particularly those with evidences of protein deficiency, exhibit loss of bone density. Although these children may also be on diets low in calcium it is likely that the osteopenia is due more to lack of formation of collagen matrix than to deficiency of calcium. However, the two factors may potentiate each other since severe calcium deficiency does cause marked osteopenia owing to increased resorption of bone along with failure of bone formation. Patients with severe protein malnutrition do exhibit hypocalcemia, but the decrease of concentration of total serum calcium is chiefly the result of reduction of serum albumin concentration and therefore of protein-bound calcium, rather than of ionized calcium concentration. The prevention and treatment of this form of osteopenia is obviously directed at correcting the primary nutritional deficiency.

Vitamin C Deficiency. Osteopenia is a characteristic feature of the bone lesions of scurvy. One of the biochemical functions of ascorbic acid is the facilitation of the hydroxylation of proline to hydroxyproline. This reaction is essential in the synthesis of collagen so that disorders of connective tissue structure are a prominent feature of deficiency of this essential factor. The roentgenographic features of scurvy are those of osteopenia: ground-glass appearance of the shafts of the long bones owing to lack of trabecular structure and thinning of the cortex. In addition there are other features that distinguish the bone lesions of scurvy from other forms of osteopenia: the brittle calcified cartilage plate at the metaphyseal ends of the bones with fractures; the submetaphyseal zone of rarefaction due to disorganization of the bone structure; and the elevation of the periosteum by hemorrhage with calcification of the subperiosteal hematomas as healing occurs. The roentgenographic features are so distinctive that the diagnosis can be made in the young child by x-ray examination of the bones. Treatment is of course the feeding of ascorbic acid, and repair of the osseous lesions is rapid.

Copper Deficiency. Early studies of copper deficiency states were directed at understanding the anemia of this form of malnutrition. As experimental studies of copper deficiency in larger animals, dogs and pigs, were pursued, it became evident that skeletal deformities were characteristic manifestations of this disorder (Fig. 6–3). The gross deformities of bending of the extremities were suggestive of rickets, but x-ray studies of the bones did not show the specific metaphyseal lesions of rickets nor the gross trabeculation and pseudocystic appearance of the shafts diagnostic of osteomalacia. Histologic and chemical analysis of the bones confirmed that the lesion was that of severe osteoporosis, i.e., deficiency of matrix as well as mineral, and that the ratio of bone mineral to matrix as measured by the concentration of calcium in fat-free dried bone was normal.

The osteoporosis of experimental copper deficiency had not been thought to arise in man until studies of infants and young children with copper deficiency were made by Graham and his associates. Copper deficiency had occurred in these patients as the result of chronic intestinal malabsorption and feeding of restricted diets in the attempt to control steatorrhea or chronic watery diarrhea. Copper deficiency was suspected because of the severe anemia and neutropenia that these children showed, and x-ray studies of the bones revealed severe osteopenia (Fig. 6–4), which in all respects resembled the osteopenia seen in experimental copper deficiency in animals. Determination of serum copper concentrations was confirmatory: total serum copper and ceruloplasmin concentrations were reduced. Feeding of copper corrected the neutropenia and anemia and improved bone mineralization.

The experimental and clinical studies of copper deficiency indicate that the osteopenia is a specific abnormality and not a nonspecific disorder secondary to severe malnutrition. The specific basis for the matrix deficiency has been thought to be the requirement for copper by the enzymes that hydroxylate proline and lysine to form hydroxyproline and hydroxylysine, which are essential components of the mature collagen polypeptide chain. Ascorbic acid, as indicated above, is also an essential cofactor in this hydroxylation, and the osteopenia of severe copper deficiency does resemble the bone matrix deficiency found in scurvy. The hemorrhagic lesions of scurvy however, are not part of the syndrome of copper deficiency.

Osteopenia is also seen in an inborn X-linked recessive error of copper metabolism, Menkes' kinky hair syndrome. In addition to the abnormality of the hair structure and osteopenia these patients also have severe neurologic abnormalities. Serum copper and ceruloplasmin concentrations are reduced in this disorder and not raised by oral administration of copper salts. Red blood cell copper concentration is normal, however, and these patients do not show the anemia and neutropenia of nutritional copper deficiency. Balance studies have shown that intestinal absorption of copper is blocked, and fecal excretion of copper may actually exceed copper intake when parenteral copper is given. It has been suggested

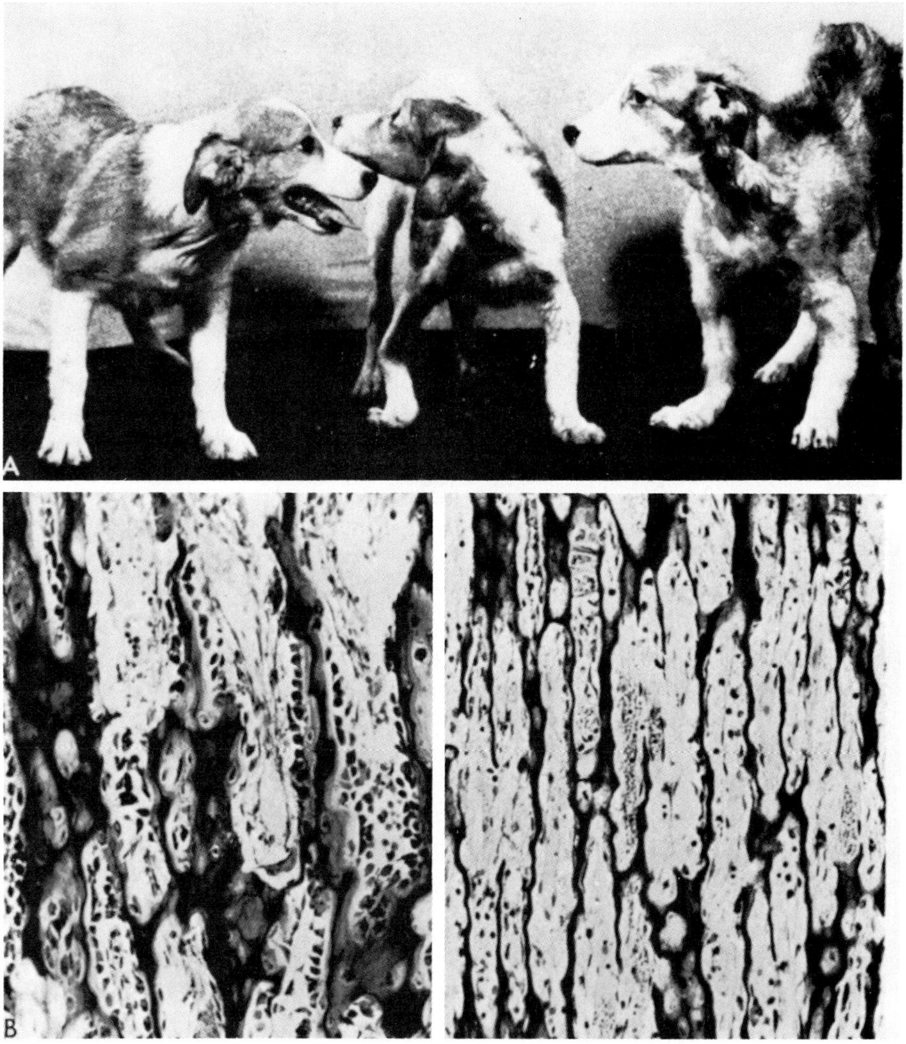

Figure 6–3. Osteopenia in experimental copper deficiency: *a*, deformities of legs in copper-deficient dogs resembling rickets; *b*, bone structure in copper-deficient pigs. On *left* is section of normal bone for comparison. The bone section on *right* from a copper-deficient animal shows the thin trabeculae and low density of bone per area of section.
(From Baxter, H. J. and Van Wyk, J. J.: Bull. Johns Hopkins Hosp. 93:1, 1953.)

that an abnormal copper-binding protein is present in intestinal mucosa and other tissues that complexes copper and makes it unavailable for copper-requiring enzymatic processes. Improvement in some patients has been achieved by intravenous or subcutaneous injections of copper in the form of cupric chloride stabilized with 1-histidine.

Homocystinuria

Osteoporosis with compression deformity of the vertebrae is one of the major manifestations of this inborn error of amino acid metabolism. This is transmitted as an autosomal recessive character, and although there is genetic heterogeneity the most common enzymatic deficiency is that of cystathionine synthetase, which condenses homocysteine with serine to form cystathionine. A deficiency of cystathionine therefore results, and both homocysteine and its precursor methionine accumulate in body fluids. Mental deficiency may be an associated manifestation but is not invariable. In addition to osteoporosis, ectopia lentis and thromboembolic episodes are important features. The basis for the osteoporosis is not understood. In some patients with cystathionine synthetase deficiency large doses

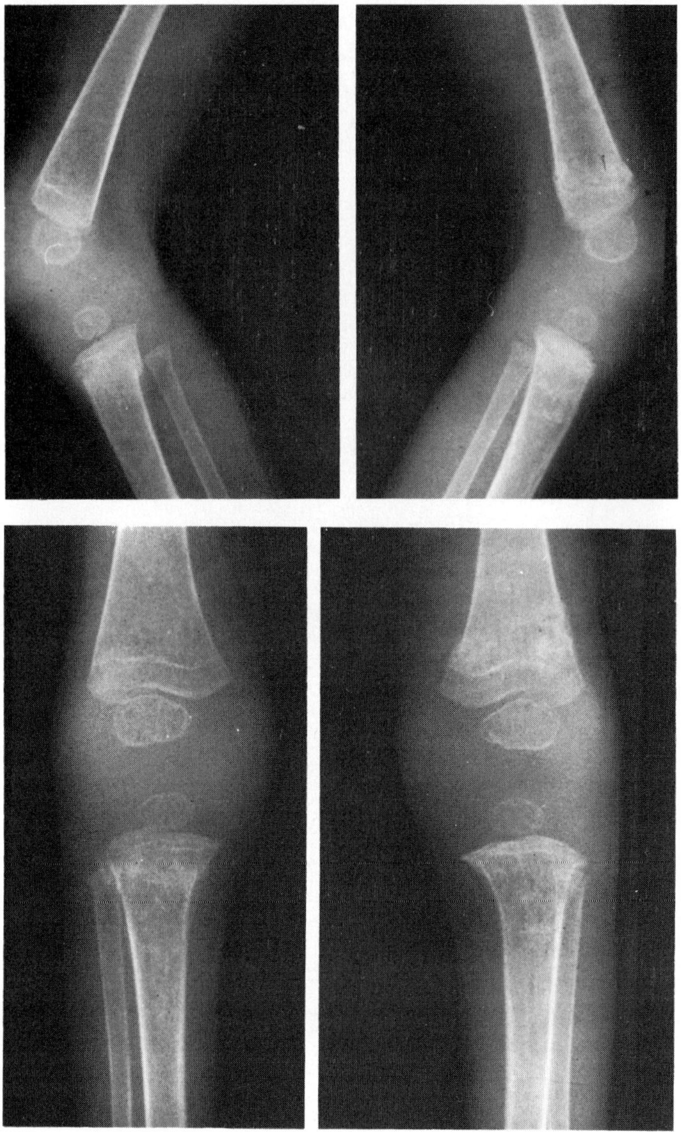

Figure 6–4. Bone radiographs of child with copper deficiency secondary to intestinal malabsorption and deficient diet. There is marked osteopenia with fracture of lower end of femur. (From Cordano, A. and Graham, G. G.: Pediatrics 38:596–604, 1966.)

of pyridoxine (vitamin B_6) reduce the homocystinuria and result in clinical improvement. There is no specific therapy for the osteoporosis.

Juvenile Osteoporosis

The differentiation between mild forms of osteogenesis imperfecta, which become evident only when children are physically very active and subject to trauma, and acquired osteoporosis is always difficult. The term *juvenile osteoporosis* has been applied specifically to a disorder of late childhood or early adolescence characterized by bone pains, fractures and x-ray evidence of osteopenia. Serum calcium, phosphate and alkaline phosphatase concentrations are normal, and there is no evidence of nutritional inadequacy. In the original report by Dent et al it was stressed that spontaneous improvement occurred after puberty with cessation of growth. This is similar to the course of the skeletal abnormalities in the less severe forms of osteogenesis imperfecta. At the moment the diagnosis of juvenile osteoporosis is made on clinical and x-ray evidence of osteopenia in children and adolescents who have not had previous manifestations of skeletal disorder and have no family history of bone fragility.

There is no specific treatment except protection against trauma in the expectation that improvement will occur when growth has stopped.

CASE REPORT

V.G., a 10 year old girl, came with a complaint of "growing pains" and limitation of joint motion. One year prior to admission it was noted that she could no longer jump rope because of limitation of motion of shoulder joints. No swelling or tenderness of the joints was noted. Later she complained of aching pain in feet after walking and stiffness of back. The father and mother and four siblings were well without history of bone or joint pains. The child's nutritional history was of interest in that she had a distaste for milk and had drunk none since about 14 months of age. On examination she was adequately nourished and developed. The major findings were stiffness and partial limitation of motion of shoulders, knees and ankles without swelling, redness or tenderness of the joints. Laboratory data: serum electrolytes and urea nitrogen normal, cortisol, 13.6 μg/dl; serum calcium 9.4 and phosphate 4.4 mg/dl; alkaline phosphatase 4.0 Bodansky Units; urine amino acids — normal pattern. On a low calcium diet that provided about 200 mg calcium per day she was in constant negative calcium balance with a stool calcium averaging 415 mg per day and urine calcium excretion of 88 mg per day. Bone x-rays showed diffuse demineralization of vertebrae and long bones (Fig. 6–5). She was treated with calcium supplement providing 1000 mg calcium

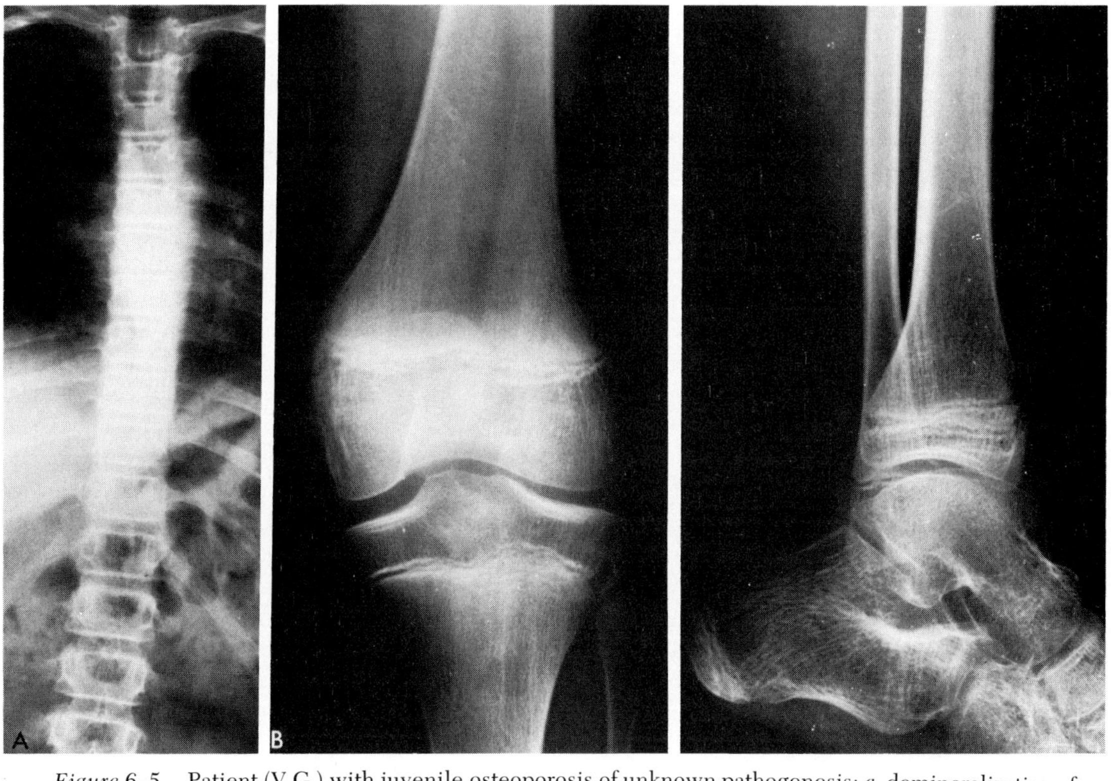

Figure 6–5. Patient (V.G.) with juvenile osteoporosis of unknown pathogenesis: *a*, demineralization of vertebral bodies; *b*, osteoporosis of long bones with vertical striations at growing ends of weight-bearing bones.

per day and vitamin D 1000 units per day. Over the next three years there was no progression of the symptoms but no striking remineralization of the skeleton.

COMMENTS: This patient with juvenile osteoporosis resembles others with this disorder in that the onset of manifestation of pain in extremities was in the latter part of the first decade of life. The limitation of joint motion was somewhat atypical but may have represented a response to bone pain. The x-ray findings were characteristic of marked osteopenia without evidence of rickets or osteomalacia. An unusual feature was the low calcium diet and, as the balance studies indicated, an apparent inability to adapt to such a diet by increased efficiency of calcium absorption or by reduction of urine calcium loss. Such failure of adaptation to low calcium intake has been suggested as a possible mechanism for osteoporosis in the adult. There was no dramatic improvement, however, following addition of calcium to the diet.

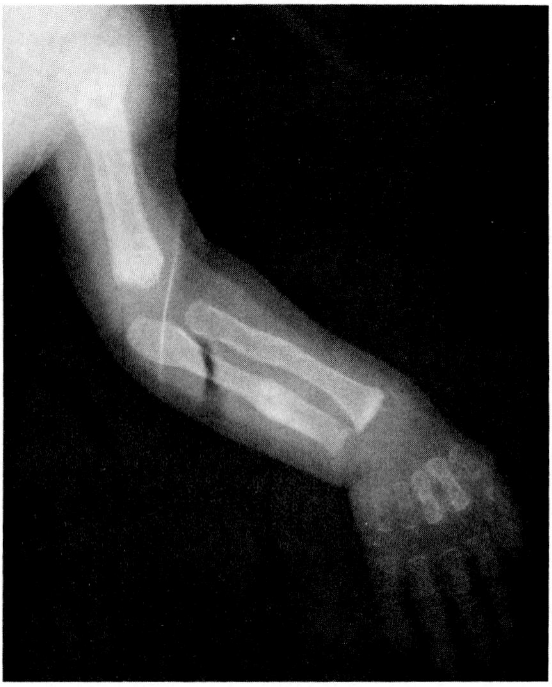

Figure 6–6. Long bones of forearm of infant with primary hyperphosphatasia. Proliferation of poorly mineralized subperiosteal osteoid with separation of periosteum from cortex. Fracture of ulna.

PRIMARY HYPERPHOSPHATASIA

Elevated serum alkaline phosphatase concentration is a secondary phenomenon in vitamin D deficiency and hyperparathyroidism, and is correlated with osteoblastic proliferation and repair of weakened bone. A mild degree of increase of serum alkaline phosphatase is seen after fractures and can be quantitatively significant following multiple fractures. Primary hyperphosphatasia is a congenital abnormality of bone structure associated with marked elevation of serum alkaline phosphatase of bone origin. The long bones are osteopenic and traumatic fractures may occur, but the striking feature is the marked subperiosteal proliferation of osteoid with characteristic separation of the periosteum from the mineralized cortex, which is apparent because of mineralization of the layer of osteoid directly below the periosteum (Fig. 6–6).

In infants the differential diagnosis includes Caffey's disease (infantile cortical hyperostosis), severe rickets (particularly if the patient has been prematurely born), and child abuse with traumatic periosteal separation. Caffey's disease is excluded by the late onset and progressive character of the disorder after 1 year of age and by the lack of involvement of the mandible, which is almost always a prominent part of the syndrome of infantile cortical hyperostosis. In prematurely born infants rapid bone growth, once adequate calorie intake has been established, can result in laying down of subperiosteal matrix more rapidly than it can be mineralized, resulting in the picture of apparent periosteal elevation. If vitamin D intake has been inadequate during the early postpartum period the combination of partial vitamin D deficiency and rapid growth may result in the x-ray appearance of periosteal separation and the biochemical finding of elevated serum alkaline phosphatase. Reduced concentrations of serum calcium or phosphate, or both, would be expected as well as low concentrations of serum 25-hydroxyvitamin D, and the abnormalities of bone structure and serum alkaline phosphatase concentration should disappear following several months of adequate vitamin D intake. Child abuse with traumatic periosteal separation is differentiated by the lack of metaphyseal fractures and the uniform nature of the lesion; in most instances of child abuse there will be evidences of both old healed or healing fractures and more recent injury.

The diagnosis of primary hyperphosphatasia is made on the basis of progressive and persistent periosteal separation with laying down of poorly mineralized subperiosteal bone, poor mineral density of the long bones, and persistent marked elevation of serum alkaline phosphatase concentrations despite normal serum calcium and phosphate and adequate intake of vitamin D. Because of the poor mineral density despite thickening of the bone, fractures, pain and deformity can result. Treatment with calcitonin has been suggested but long-term results are as yet unknown.

OSTEOSCLEROTIC DISORDERS

The osteosclerotic abnormalities of bone are the consequence of deficiency of the bone resorptive functions, so that normal remodeling with growth of bone does not occur and excessively dense bone results, which, however, may be structurally brittle because of excessive ratio of crystalline mineral to organic matrix. As mentioned in the section on renal osteodystrophy, osteosclerosis may exist along with secondary hyperparathyroidism in patients with renal insufficiency, possibly because of hyperphosphatemia and consequent precipitation of bone mineral, particularly at the plates of the vertebral bodies. In the specific genetic osteosclerotic abnormalities to be discussed, serum phosphate concentrations are normal unless vitamin D deficiency is superimposed.

OSTEOPETROSIS — AUTOSOMAL RECESSIVE

This severe form of defect of the normal resorptive functions of bone cells is transmitted by an autosomal recessive gene. A major clinical manifestation is loss of bone marrow cells owing to encroachment of bone crowding out the bone marrow cells. As a result, extramedullary hematopoiesis in liver and spleen is stimulated with enlargement of these organs. The extramedullary erythropoiesis, however, is inadequate to replace normal bone marrow function so that hypoplastic anemia ensues. The base of the skull is thickened and basilar obstruction with hydrocephalus can occur.

The dense bone seen on x-ray examination with disappearance of normal marrow cavity is characteristic. Despite the density of the bone it is brittle and subject to fracture because of abnormal mineral-to-matrix ratio. Secondary vitamin D deficiency rickets has also been observed with considerable frequency in these patients, which suggests that their vitamin D requirements are increased. There is no known abnormality of vitamin D metabolism unless the disturbance of liver function caused by extramedullary hematopoietic cell proliferation in the liver interferes with 25-hydroxylation in hepatocytes. No studies of this function have been made.

These infants are subject to infection because of the immunologic deficiency due to lack of bone marrow cells, and early death is common owing to severe anemia and infection. Dental infections with osteomyelitis of the mandible are particularly important. Blindness and deafness because of narrowing of skull foramina by bone encroachment are seen in the surviving children.

The underlying cellular abnormality is unknown, but it has been postulated that there is failure of differentiation of normally functioning osteoclasts. An excess of calcitonin inhibiting bone resorption was proposed as a mechanism but the evidence does not support this theory. The serum calcium concentrations are variable and can be low when these patients are placed on a low calcium diet. This is consistent with abnormality of the bone resorptive process and probable deficiency of PTH-responsive cells. Transplantation of bone marrow from normal donors has been tried with at least temporary improvement of bone structure and proliferation of osteoclasts. This treat-

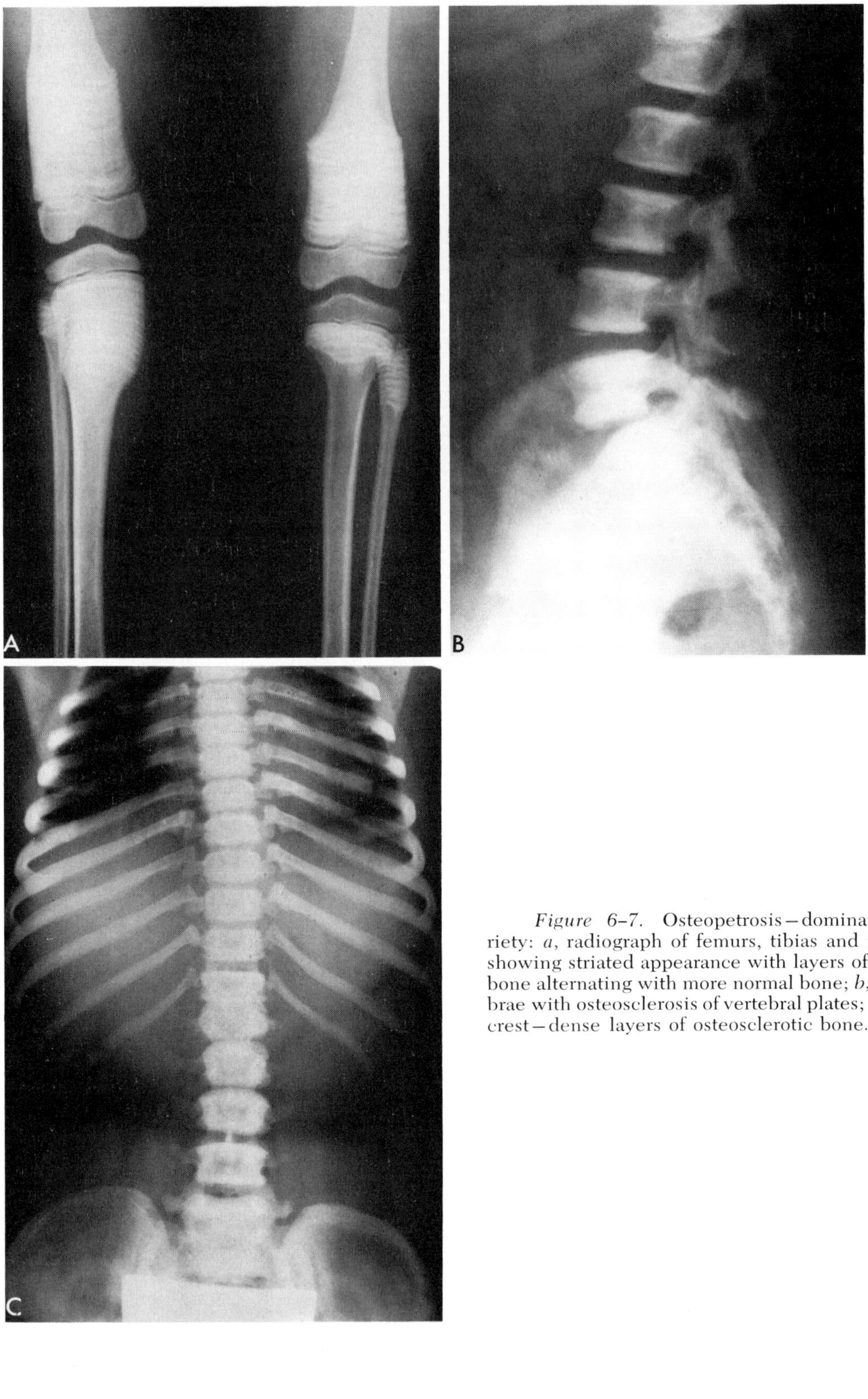

Figure 6–7. Osteopetrosis—dominant variety: *a*, radiograph of femurs, tibias and fibulas showing striated appearance with layers of dense bone alternating with more normal bone; *b*, vertebrae with osteosclerosis of vertebral plates; *c*, iliac crest—dense layers of osteosclerotic bone.

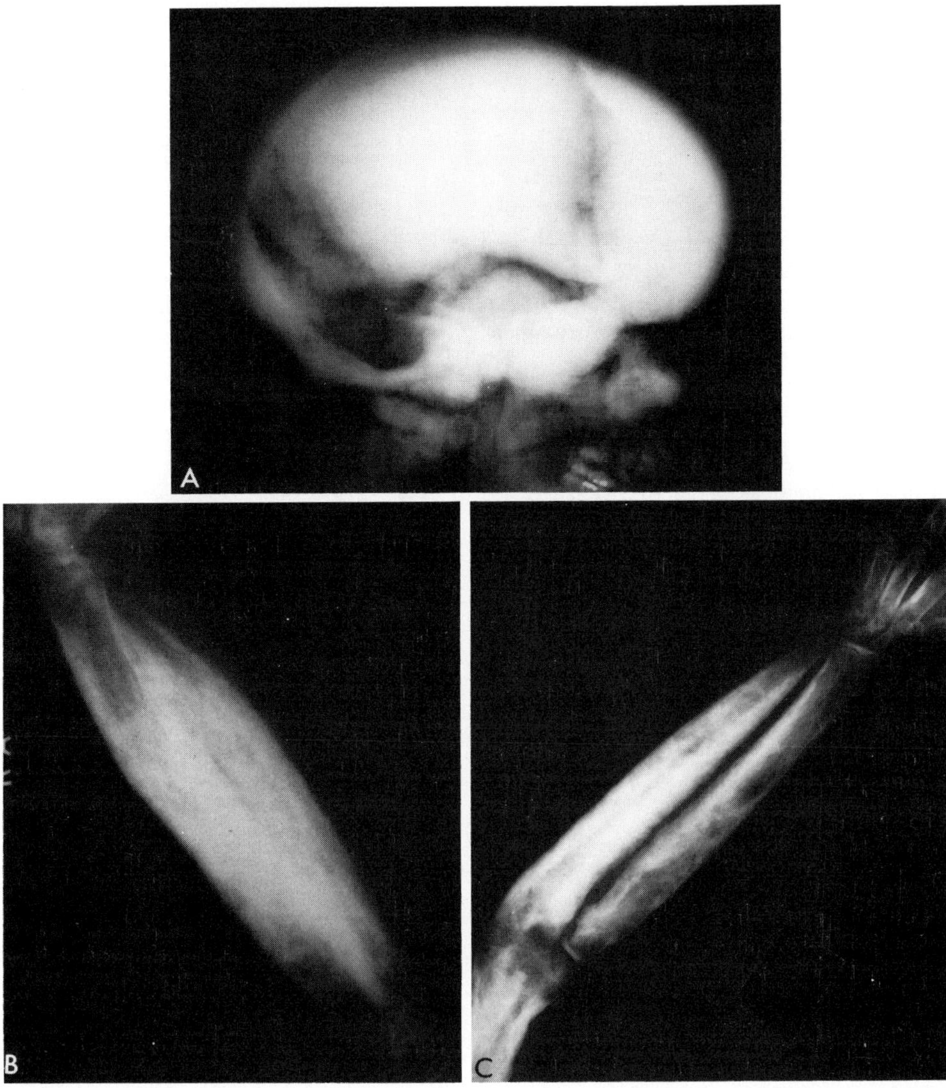

Figure 6–8. Engelmann's disease (diaphyseal dysplasia): *a*, skull radiograph showing extreme thickening and sclerosis of basilar bones and calvarium; *b*, tibia and fibula—marked cortical thickening and sclerosis of diaphyseal portions of long bones; *c*, radius and ulna—findings as in *b*.

ment was based on experimental studies of genetic osteopetrosis in two animal models, the rat and the mouse. In these models bone marrow or splenic cell transplants resulted in osteoclast formation from progenitor cells.

Osteopetrosis — Autosomal Dominant

This is a milder variant of osteosclerotic bone disease with autosomal dominant inheritance. Symptoms are mild during childhood, but severe back pains become progressively more marked and are the chief complaints in adult life. Skeletal x-rays are striking in these patients, showing alternate zones of osteopetrosis and more normal bone structure, providing an extraordinary striped effect which is shown in Figure 6–7. Bone marrow displacement does not occur in these patients so that hypoplastic anemia, hepatomegaly and splenomegaly are not seen. The mechanism is unknown. Serum calcium, phosphate and alkaline phosphatase concentrations are normal. Treatment of the bone pains, particularly the back pain, can only be symptomatic.

ENGELMANN'S DISEASE (DIAPHYSEAL DYSPLASIA)

This is included among the osteosclerotic disorders because of the marked thickening of the diaphyseal portion of the long bones, and particularly because of the extreme sclerosis and hyperostosis of the skull. It has been suggested that this is a genetic disorder with autosomal dominant transmission, but with variable and sometimes complete lack of penetrance, so that there may be no family history.

The onset is usually in childhood, with difficulty in walking owing to reduction of muscle mass of the legs with marked thickening of the shafts of the bones. Leg pain may be associated with the skeletal change and the muscle atrophy. In addition there is enlargement of the head owing to thickening of the calvaria, and neurologic complications such as optic atrophy and deafness resulting from bony encroachment on the foramina at the base of the skull. In severe cases the facies is grotesque because of the marked thickening of facial bones. There is shortness of stature that may be accentuated by severe kyphosis.

Radiographs of an unusually severe case are shown in Figure 6–8. In this instance there was no family history of skeletal abnormality.

The basis for this skeletal disorder is unknown. Treatment with corticosteroids has been helpful in improving muscle strength and reducing skeletal pain, but without effect on the underlying bony change. In mild cases there is apparent arrest of the disease at puberty and cessation of growth. Improvement of muscle strength and lessening of bone pain has been seen in such patients.

REFERENCES

OSTEOGENESIS IMPERFECTA

Abstracts of Symposium II on Osteogenesis Imperfecta. Clin. Orthop. *116*:258–263, 1976.

August, G. P., Shapiro, J. and Hung, W.: Calcitonin therapy of children with osteogenesis imperfecta. J. Pediatr. *91*:1001–1005, 1977.

Doty, S. B. and Matthews, R. S.: Electron microscopic and histochemical investigation of osteogenesis imperfecta tarda. Clin. Orthop. *80*:191–201, 1971.

Fujii, K. and Tanzer, M. L.: Osteogenesis imperfecta: biochemical studies of bone collagen. Clin. Orthop. *124*:271–277, 1977.

Ibsen, K. H.: Distinct varieties of osteogenesis imperfecta. Clin. Orthop. *50*:279–290, 1967.

McKusick, V. A.: Heritable Disorders of Connective Tissue. C. V. Mosby Co., St. Louis, 1972.

Rosenberg, E., Lang, R., Boisseau, V. et al: Effect of long-term calcitonin therapy on the clinical course of osteogenesis imperfecta. J. Clin. Endocrinol. Metab. *44*:346–355, 1977.

Sykes, B., Francis, M. J. O., Phil, F. D. et al.: Altered relation of two collagen types in osteogenesis imperfecta. N. Engl. J. Med. *296*:1200–1203, 1977.

CORTICOID OSTEOPOROSIS

Albright, F.: Cushing's syndrome. Harvey Lect. *38*:123–186, 1942–43.

Hahn, T. J.: Corticosteroid-induced osteopenia. Arch. Intern. Med. *138*:882–885, 1978.

Harrison, H. E. and Harrison, H. C.: Transfer of Ca^{45} across intestinal wall *in vitro* in relation to the action of vitamin D and cortisol. Am. J. Physiol. *199*:265–271, 1960.

Jee, W., Park, S. S., Roberts, W. E. et al: Corticosteroid and bone. Am. J. Anat. *129*:477–479, 1970.

Jowsey, J. and Riggs, B. L.: Bone formation in hypercortisonism. Acta Endocrinol. *63*:21–28, 1970.

Kimberg, D. V., Baerg, R. D., Gershon, E. et al: Effect of cortisone treatment on the active transport of calcium by the small intestine. J. Clin. Invest. *50*:1309–1321, 1971.

Ruder, H. J., Loriaux, D. L. and Lipsett, M. B.: Severe osteopenia in young adults associated with Cushing's syndrome due to micronodular adrenal disease. J. Clin. Endocrinol. Metab. *39*:1138–1147, 1974.

Sissons, H. A.: The osteoporosis of Cushing's syndrome. J. Bone Joint Surg. *38-B*:418–433, 1956.

COPPER DEFICIENCY

Baxter, J. H. and Van Wyk, J. J.: A bone disorder associated with copper deficiency. I. Gross morphological, roentgenological and chemical observations. Bull. Johns Hopkins Hosp. *93*:1–24, 1953.

Cordano, A. and Graham, G. G.: Copper deficiency complicating severe chronic intestinal malabsorption. Pediatrics *38*:596–604, 1966.

Danks, D. M., Campbell, P. E., Stevens, B. J. et al: Menkes' kinky hair syndrome: an inherited defect in copper malabsorption with widespread effects. Pediatrics *50*:188–201, 1972.

JUVENILE OSTEOPOROSIS

Davies, M., Mawer, E. B. and Adams, P. H.: Vitamin D metabolism and the response to 1,25-dihydroxycholecalciferol in osteoporosis. J. Clin. Endocrinol. Metab. *45*:199–208, 1977.

Dent, C. E.: Idiopathic osteoporosis. Proc. R. Soc. Med. (Section Med.) *48*:574–578, 1955.

Dent, C. E. and Friedman, M.: Idiopathic juvenile osteoporosis. Q. J. Med. *34*:177–210, 1965.

Lapatsanis, P., Kavadias, A. and Vretos, K.: Juvenile osteoporosis. Arch. Dis. Child. *46*:66–71, 1971.

Rich, C., Ensinck, J. and Ivanovich, P.: The effects of sodium fluoride on calcium metabolism of subjects with metabolic bone disease. J. Clin. Invest. *43*:545–556, 1964.

Rosenquist, J. B.: Fluoride in the prevention of osteopenia. Studies of the skeletal mass in calcium-deficient rats. Clin. Orthop. *125*:200–204, 1977.

CONGENITAL HYPERPHOSPHATASIA

Eyring, E. J., and Eisenberg, E.: Congenital hyperphosphatasia. J. Bone Joint Surg. *50-A*:1099–1117, 1968.

OSTEOPETROSIS

Johnston, C. C., Jr., Lavy, N., Lord, T. et al: Osteopetrosis. A clinical, genetic, metabolic, and morphologic study of the dominantly inherited benign form. Medicine *47*:149–167, 1968.

Loria-Cortés, R., Quesada-Calva, E. and Cordero-Chaverri, C.: Osteopetrosis in children. J. Pediatr. *91*:43–47, 1977.

Marks, S. C., Jr., and Walker, D. G.: Mammalian osteopetrosis—a model for studying cellular and humoral factors in bone resorption. *In* The Biochemistry and Physiology of Bone, 2nd ed., Vol. 4 (Ed.: G. H. Bourne). Academic Press, New York, 1976.

ENGELMANN'S DISEASE

Neuhauser, E. B. D., Schwachman, H., Wittenborg, M. et al: Progressive diaphyseal dysplasia. Radiology *51*:11–22, 1948.

Chapter Seven

URINARY TRACT STONES

Urinary lithiasis is relatively rare in childhood. Several inborn errors of metabolism associated with stone formation may be manifest in childhood, but most often not until adult life, despite the presumed genetic basis of the disorder. The diagnostic categories we use are given in Table 7–1.

HYPERCALCIURIA WITH STONE FORMATION

One of the factors in the precipitation of calcium phosphate in the urine is the product of the concentration of Ca^{++} ion and $HPO_4^{=}$ ion. At low urine pH most of the inorganic phosphate is present as the monovalent anion $H_2PO_4^{-}$, thus reducing the concentration of the divalent anion $HPO_4^{=}$. If the solubility product of $CaHPO_4$ is exceeded precipitation may occur, but there are constituents of the urine which protect against nucleation and growth of crystals of $CaHPO_4$ despite a high $Ca^{++} \times HPO_4^{=}$ product. There are also compounds such as citrate and amino acids that form un-ionized calcium complexes and reduce calcium ion concentration. Nevertheless, if urine calcium concentration is excessively high stone formation is likely to occur as the result of the formation of precipitates of both calcium phosphate and calcium oxalate. Organic matrices may act as nucleation sites on which these precipitates aggregate into calculi.

Table 7–1. Classification of Urinary Tract Calculi

I. Primary calcium stones
  A. Associated with hypercalciuria (urine calcium >4 mg/kg/day or > 0.25 mg/mg creatinine)
    1. Hyperparathyroidism
    2. Hypervitaminosis D
    3. Immobilization
    4. Renal tubular acidosis
    5. Idiopathic hypercalciuria
  B. Oxalate calculi
    1. Primary hyperoxaluria (oxalosis)
    2. Disorders of the intestinal tract
    3. Bladder stones of children
    4. Idiopathic stone formers
II. Stones associated with urinary stasis and infection
  A. Medullary sponge kidney
  B. Ammonium magnesium phosphate stones
III. Organic crystalluria with stones
  A. Cystinuria
  B. Uricosuria

HYPERPARATHYROIDISM WITH HYPERCALCIURIA

Most of the instances of hyperparathyroidism in childhood are familial parathyroid hyperplasia and, as discussed in Chapter 4, there are kindreds of familial parathyroid hyperplasia in which none of the affected individuals have hypercalciuria, whereas in other families hypercalciuria is consistently found. Renal stone as a presenting manifestation of hyperparathyroidism in childhood can occur but is uncommon (see Chapter 4, *case report*).

HYPERVITAMINOSIS D WITH HYPERCALCIURIA

Patients with hypoparathyroidism or primary hypophosphatemia who are receiving large doses of vitamin D or dihydrotachysterol are at risk for hypercalciuria, as has been discussed. Renal stone formation is unusual, however, and nephrocalcinosis with renal insufficiency is a more important complication of hypervitaminosis D. One of our adult patients with primary hypophosphatemic osteomalacia who was treated with oral phosphate salts and dihydrotachysterol had an episode of hypercalciuria and symptoms of ureteral calculus, namely, renal colic and hematuria. No stone was visualized by x-ray examination and there were no recurrences of this manifestation.

IMMOBILIZATION HYPERCALCIURIA

The imbalance between bone mineral deposition and resorption resulting from lack of stress on bone, either weight-bearing or muscle pull, has been discussed in the section on immobilization hypercalcemia in Chapter 4. The earliest manifestations of the net loss of bone mineral is increased urine excretion of calcium. As discussed in Chapter 4, acute immobilization of a previously active individual, particularly an adolescent male, in addition to hypercalciuria can also cause marked hypercalcemia. More chronic immobilization may result in hypercalciuria with renal stone formation and urinary tract infection. This has been seen, for example, in patients with muscular paralysis due to poliomyelitis or other neurologic disease. Treatment is directed at maintaining some degree of bone stress by weight bearing or passive motion. Physical therapy with encouragement of whatever motion is possible or a rocking bed can reduce hypercalciuria. A high phosphate intake to reduce urine calcium excretion, or a high magnesium intake to increase the solubility of calcium oxalate in urine *(see section on oxalosis)*, can also be tried.

RENAL TUBULAR ACIDOSIS

Untreated or relapsing distal renal tubular acidosis is associated with hypercalciuria. Symptoms of renal stone formation do occur but nephrocalcinosis is a more common complication. Why discrete stones are not passed more often when there is intratubular precipitation of calcium phosphate might be explained by the removal of calcium by precipitation in the tubules, so that its concentration in the more distal urinary tract is not high enough to cause aggregation of precipitate.

IDIOPATHIC HYPERCALCIURIA

This entity is an important cause of stone formation in adults but has received relatively little attention as a cause of symptoms in childhood. Since idiopathic hypercalciuria does seem to have a familial incidence and presumably a genetic basis, it should be detectable in childhood but few screening programs have been done. There have been a few reports of children with painful hematuria who were found to have excessively high urine outputs of calcium and were diagnosed as having idiopathic hypercalciuria. In these patients serum calcium and phosphate concentrations have been consistently normal, making hyperparathyroidism unlikely, and they have not had reduced serum bicarbonate concentrations, which rules out renal tubular acidosis.

There are two major theories of the physiologic basis of idiopathic hypercalciuria; these are based on studies in adults since very few children with this entity have been intensively investigated. The first theory postulates that the basic physiologic factor is hyperabsorption of calcium from the intestine with secondary increased urinary output of calcium. It has been further hypothesized that the hyperabsorption of calcium, in some patients at least, is the result of increased formation of 1,25-dihydroxyvitamin D as the consequence of slightly reduced concentrations of inorganic phosphate frequently found in adults with idiopathic hypercalciuria. Increased concentrations of serum 1,25-dihydroxyvitamin D have been found in some of these patients. If hyperabsorption of calcium is the basis of this disorder, it follows that serum IPTH concentrations should be low and that the specific treatment should be reduction of the calcium available for absorption, either by a very low calcium diet or by precipitation of calcium in the gut to make it unabsorbable.

The alternative theory is that hypercalciuria results from reduced renal tubular reabsorption of calcium, presumably as a primary tubular defect, and that intestinal hyperabsorption of calcium is a secondary adaptive response to the excessive calcium loss in the urine. This concept requires that serum IPTH concentrations be increased initially because of calcium loss, and that treatment should be directed at increasing tubular reabsorption of calcium. Both theories have their advocates and one compromise is that there are two varieties of idiopathic hypercalciuria: (1) primary intestinal hyperabsorption of calcium; and (2) reduced renal tubular reabsorption of calcium with secondary increase of intestinal absorption of calcium. These should not be separable by measurement of intestinal absorption of calcium since this will be increased in both types. Determination of serum IPTH should be helpful, however, particularly if determined before and after treatment. Proponents of the second theory have presented evidence that serum IPTH concentrations are increased in some patients with this disorder and are lowered when renal excretion of calcium is reduced by treatment with chlorothiazide or hydrochlorothiazide.

Clinical Presentation and Diagnosis

The presenting manifestation which arouses the suspicion of idiopathic hypercalciuria is that due to precipitation and aggregation of calcium phosphate or oxalate into calculi. The usual symptom, therefore, is painful hematuria. The type of pain depends on the location of the calculus in the urinary tract. It may be flank pain, characteristic renal colic of ureteral origin with transmission to groin or scrotum in the male, or suprapubic pain. If the calculus or small aggregates of calcium phosphate are in the bladder there is pain at the end of urination. In one of our patients there were recurrent episodes of hematuria, but pain was not a major complaint in most of these: it was not until an episode of renal colic occurred that determinations of urine calcium were made.

In any patient with unexplained gross or microscopic hematuria, particularly if associated with abdominal pain, whether or not the pain is typical of renal colic, measurement of 24-hour excretion of urine calcium should be made. If this value exceeds 150 mg/day in a child or 4 mg/kg/day or 0.25 mg/mg creatinine, depending on the standard of normality used, hypercalciuria is diagnosed. Serum calcium and phosphate concentrations should also be determined. If the former are consistently less than 10.8 mg/dl or if serum ionized calcium is less than 5.6 mg/dl, hypercalcemic hypercalciuria is eliminated. Although so-called normocalcemic hyperparathyroidism is seen in adults, we have not encountered children with hyperparathyroidism who have not had definite hypercalcemia. The differential diagnosis between hyperparathyroidism and hypervitaminosis D or idiopathic hypercalcemia has been presented in Chapter 4. A normal serum bicarbonate concentration eliminates renal tubular acidosis as a cause of normocalcemic hypercalciuria.

The differentiation between primary intestinal hyperabsorption of calcium and a defect of renal tubular reabsorption of calcium may be attempted by determination of serum IPTH concentrations before and after treatment with benzthiazide diuretics. This group of diuretics differs from others in that the benzthiazides increase renal tubular reabsorption of calcium and diminish urine calcium excretion despite sodium diuresis. Other diuretics such as furosemide and ethacrynic acid cause calcium as well as sodium diuresis. If serum IPTH concentration is increased secondary to calciuresis, treatment with the benzthiazide diuretics by reduction of urine calcium excretion should reduce the serum IPTH. If intestinal hyperabsorption of calcium is the primary factor the serum IPTH should be low initially, and presumable unaffected by chlorothiazide treatment.

Another diagnostic as well as therapeutic modality is reduction of calcium absorption to determine whether urine calcium excretion is decreased to normal by this means. It is usually difficult to reduce dietary calcium sufficiently to accomplish this. Addition of an excess of inorganic phosphate, or preferably cellulose phosphate, which complexes calcium and makes it nonabsorbable, can reduce calcium absorption and urine excretion of calcium if this is the major factor.

Although in theory the treatment of idiopathic hypercalciuria should differ depending on the pathogenesis, many cases have been treated with benzthiazide diuretics without classification as to type, and have shown reduction of urine calcium excretion if treatment is maintained. Transient hypercalcemia may result but serum calcium concentrations return to normal if treatment is maintained.

CASE REPORTS

B.U., a 13 year old boy, presented with "tea"-colored urine that had been noted for two days. He had not complained of abdominal pain or dysuria. Physical examination was unremarkable and blood pressure was normal. The urine had a trace of protein and numerous red blood cells. The blood count was normal including sedimentation rate. Serum electrolyte, urea nitrogen and creatinine concentrations were normal. Urine culture did not show a significant number of colonies and the intravenous pyelogram was interpreted as normal. The hematuria cleared rapidly and 36 hours after admission no RBCs were present. Cystoscopic examination revealed no bladder abnormalities. The past history was of interest in that he had had an episode of hematuria at 7 years of age which was diagnosed as nephritis. The hematuria at that time also cleared rapidly. There was no family history of urinary tract stones. The patient was one of dizygotic twins.

The serum calcium concentration was 11.1 mg/dl and serum phosphate 4.5 mg/dl. Urine calcium excretion was markedly increased, 485 mg/24 hrs. Subsequent serum calcium was below 10.7 mg/dl. The diagnosis of idiopathic hypercalciuria was considered and the patient was placed on a low calcium diet (150 mg per day) to determine the effect of reduced calcium intake on urine calcium excretion. The results are summarized in Table 7–2. Urine calcium excretion decreased following this diet but was still above normal. He was then allowed his usual calcium intake and treated with hydrochlorothiazide, 50 mg/day. On this regime urine calcium excretion decreased markedly without rise of serum calcium concentration.

COMMENTS: This is a patient who had had two episodes of painless hematuria, one of which had been diagnosed as nephritis, but the basis of that diagnosis is questionable. Urine calcium excretion was markedly increased. The initial serum calcium concentration was 11.1 mg/dl (method used was not atomic absorption spectrophotometry) and the possibility of hyperparathyroidism was considered. Serum phosphate was not low and subsequent serum calcium determinations were normal. Following four days of low calcium intake urine calcium excretion was reduced, but still excessive for the age of the patient. He was then placed on hydrochlorothiazide with marked reduction of urine calcium excretion and without increase of serum calcium concentration. It was thought that this patient had idiopathic hypercalciuria, possibly owing to impaired renal tubular reabsorption of calcium, with good response to hydrochlorothiazide treatment.

T.F., a white male age 15 years, was investigated because of a history of kidney

Table 7–2. Idiopathic Hypercalciuria (B.U.)

Serum Ca (mg/dl)	Serum P (mg/dl)	Urine Volume (ml/day)	Urine Ca (mg/day)	Urine P (mg/day)
		USUAL DIET		
11.1	4.5	1600	485	1408
		LOW CA DIET – 4 DAYS		
		1760	218	1020
		USUAL DIET		
		930	390	1116
		USUAL DIET. HYDROCHLOROTHIAZIDE–50 MG/DAY – 6 WEEKS		
		1450	89	1110
		USUAL DIET. HYDROCHLOROTHIAZIDE–50 MG/DAY – 5 MONTHS		
10.7	5.4	1070	85	940

stones. At age 13 he had a fall while playing ball. That night he complained of severe left flank pain, with nausea and vomiting, and voided grossly bloody urine. An emergency intravenous pyelogram showed delayed emptying of a dilated left renal pelvis. On bed rest with intravenous fluids the symptoms subsided and hematuria cleared. A second intravenous pyelogram showed a normal urinary tract. Two years later he again complained of left flank pain with gross hematuria. Abdominal radiographs showed calcific densities in the caliceal system and in the left ureter, thought to be renal stone.

The family history was of interest in that the mother had had a bladder stone at age 14 or 15 which was removed by cystoscopy. There were three siblings who were clinically well. The patient's past history was negative until the episode of pain and hematuria at age 13. Physical examination was unremarkable. He seemed to be a robust, well-developed adolescent boy.

Laboratory Data: serum Na 142 mEq/1, K 4.6 mEq/1, HCO_3^- 27.8 mEq/1, Cl 104 mEq/1, urea N 15 mg/dl, uric acid 4.6 mg/dl, creatinine 0.95 mg/dl, Ca 11.0 mg/dl, P 4.3 mg/dl, urine cystine — negative.

The patient's urine calcium excretion in relation to low calcium diet (300 mg per day) and to hydrochlorothiazide treatment is summarized in Table 7-3.

COMMENTS: This boy presented with renal colic and hematuria so that the possibility of a renal stone was immediately considered. However, definitive studies and a diagnosis were not made until a second episode of renal colic, hematuria and x-ray evidence of stones in the urinary tract. There was no HCO_3^- deficiency and serum calcium and phosphate concentrations were normal. However, urine calcium excretion was consistently increased while on the usual calcium intake. Urine calcium output was decreased when a low calcium diet (300 mg per day) was fed. He was given a low phosphate diet to determine the effect on the concentration of calcium in serum and urine. Serum calcium concentration did not rise but urine calcium excretion markedly increased, probably as the result of increased absorption of food calcium. Treatment with excess phosphate might well have been successful in this patient but it was elected to try the effect of hydrochlorothiazide, which reduced urine calcium to normal values. When treatment was interrupted hypercalciuria recurred so that continued treatment was recommended. No further episode of renal colic or hematuria occurred during the period of study. (We

Table 7–3. Idiopathic Hypercalciuria with Stone Formation (T.F.)

DATE	SERUM Ca (mg/dl)	SERUM P (mg/dl)	URINE Volume (ml/24 hr)	URINE Ca (mg/dl)	URINE Ca (mg/24 hr)	TREATMENT
5/8/68	11.0	4.3	830	23.7	197	Usual diet
5/13/68	10.4	4.6	1260	21.0	265	
						Low Ca diet–5 days
5/21/68	10.1	4.3	1275	11.4	145	
						Control diet
5/25/68			1320	17.2	227	
						Low P diet–4 days
5/29/68	9.8	4.6	2060	23	475	
						Hydrochlorothiazide 100 mg/day
6/6/68			920	6.1	56	
						Treatment stopped
8/14/69			790	44.2	341	
						Hydrochlorothiazide resumed
10/13/69			1260	10.3	128	

are indebted to Dr. Thomas Holcomb, Franklin Square Hospital, for the case histories on patients B.U. and T.F.)

R.P., a white male age 11 years, was investigated because of five episodes of hematuria during a period of three years. The initial episodes were not associated with pain or dysuria, and intravenous pyelograms, cystoscopic examination and hematologic studies for coagulation defect were all negative. The fifth episode of hematuria was accompanied by suprapubic pain radiating to the tip of the penis following micturition.

The family history was extraordinary in that the father, who had died at age 42, was known to have had several episodes of painful hematuria, and the father's brother was reported to have had hematuria. The boy's mother and her two sisters were reported to have passed stones. The mother's serum calcium concentration was 10.9 mg/dl and serum phosphate 2.7 mg/dl. Urine calcium excretion was less than 200 mg/24 hrs.

Laboratory studies on this patient were as follows: serum calcium 10.4 mg/dl, serum phosphate 4.5 mg/dl, urine calcium excretion 213 and 212 mg/24 hrs. Despite the mother's history of stones she did not have hypercalciuria and her normal serum calcium concentration made hyperparathyroidism unlikely. The possibility of idiopathic hypercalciuria was considered and the patient's response to chlorothiazide was determined. The results are summarized in Table 7-4.

Table 7–4. Idiopathic Hypercalciuria (R.P.)

DATE	SERUM Ca (mg/dl)	SERUM P (mg/dl)	URINE Volume (ml/24 hr)	URINE Ca (mg/dl)	URINE Ca (mg/24 hr)	TREATMENT
7/28/70	10.4	4.5	–	–	–	
7/31/70			725	29.4	213	
8/6/70	10.3	4.3	860	24.6	212	
						Chlorothiazide 400 mg/day
8/10/70	11.1	5.2	905	18.7	168	
8/14/70	11.9	4.7	1895	14.5	130	
8/25/70	11.0	5.0				
9/2/70	10.7	4.0				Hydrochlorothiazide 50 mg/day
9/16/70	9.6	4.7	710	18	128	
9/30/70	10.4	4.7	650	14.6	95	

COMMENTS: This boy with a strong family history of hematuria and renal stones had several episodes of painless hematuria which were self-limited and for which no etiology was found despite intensive urologic examination. When he had pain on urination radiating to the penile tip, along with hematuria, the possibility of crystal aggregates in the bladder was considered. Serum calcium concentrations were not elevated but urine calcium excretion of more than 200 mg/day was considered to be hypercalciuria. When he was given chlorothiazide serum calcium increased significantly, along with moderate decrease of urine calcium excretion. Hypercalcemia has been reported following chlorothiazide administration and has been thought to be due to stimulation of parathyroid hormone output. The hypercalcemia in this patient suggested the possibility of hyperparathyroidism, but serum calcium decreased to normal while he was still on chlorothiazide treatment, and serum phosphate did not decrease but actually rose somewhat. When treatment was changed to hydrochlorothiazide, serum calcium remained in the normal range while urine calcium excretion continued to be about 50 per cent of that before treatment. The further course has not suggested hyperparathyroidism. Compliance with treatment has been irregular. This patient is thought to have idiopathic hypercalciuria. The basis for the renal stones in mother and maternal aunts is undetermined.

All of the patients we have encountered who might be diagnosed as having idiopathic hypercalciuria have been males. This marked male predominance has also been seen in adult series. In a large group of 78 patients with kidney stones and definite hypercalciuria reported by Coe, 56 were males and in another series of 15 patients with idiopathic hypercalciuria and increased serum 1,25-dihydroxyvitamin D concentrations, 14 were males. Although reduction of serum phosphate concentration has been reported to be a feature of idiopathic hypercalciuria, particularly the group with primary intestinal hyperabsorption of calcium, none of the three patients with stones have had hypophosphatemia. However, we have seen two patients with the syndrome of growth retardation, hypercalciuria and hypophosphatemia but without stone formation or hematuria (*discussed in the following section*). Treatment with a thiazide diuretic has been effective in the children with hypercalciuria, as for the reported adults, and is probably the most suitable long-term treatment.

IDIOPATHIC HYPERCALCIURIA WITH GROWTH RETARDATION

A few cases have been reported of children with marked retardation of growth and excessively high urinary excretion of calcium. The basis of this metabolic disorder, and particularly of the growth retardation, is not clear but renal hypophosphatemia may be a factor.

Dent and Friedman reported two boys with hypophosphatemic rickets thought to be due to a defect of renal tubular reabsorption of phosphate, but with hypercalciuria rather than the low urine excretion of calcium characteristic of genetic primary hypophosphatemia. Proteinuria consisting mainly of proteins migrating as α_2 and β globulins was also present. This type of proteinuria is considered to be characteristic of nonspecific tubular injury. Retardation of growth was marked in both patients even when the rickets was healed by moderate doses of vitamin D. Hypercalciuria was marked on the vitamin D treatment but stone formation did not occur.

We have seen two boys under 10 years of age who presented for evaluation of growth below the third percentile. Serum calcium concentrations were normal but serum phosphate was consistently below 4 mg/dl, and as low as 3.0 mg/dl in one patient. Urine calcium excretions in both patients were increased. Serum HCO_3^- concentrations were normal and urine pH values were low, ruling out renal tubular acidosis, Proteinuria was not present and other renal function studies were normal. It was postulated that renal hypophosphatemia might be the primary disorder with excessive absorption of calcium and calciuria, as in the adult patients with low serum phosphate. The growth retardation was thought to be secondary to reduced extracellular concentrations of phosphate, as in primary hypophosphatemic rickets and osteomalacia. These patients are under treatment with oral phosphate providing 1.5 gm phosphorus per day, but it is too early to know whether growth will be stimulated.

HYPERCALCIURIA WITH NEPHROLITHIASIS AS A COMPLICATION OF TOTAL PARENTERAL NUTRITION

Hypercalciuria and stone formation has been noted to occur in children receiving prolonged total parenteral nutrition for disorders of intestinal function. In such patients, because of the added factors of immobilization and hypophosphatemia, the calcium given intravenously may not be deposited as bone mineral and is excreted chiefly in urine. In one patient in whom this complication occurred steroid treatment was also being given, which may have contributed to the failure of uptake of mineral by bone matrix. It has been suggested that urine calcium excretion be monitored in patients receiving total parenteral nutrition, and the calcium concentration of the infusion fluid reduced if hypercalciuria is found.

OXALATE CALCULI

PRIMARY HYPEROXALURIA (OXALOSIS)

Oxalosis is a genetically determined metabolic abnormality transmitted as an autosomal recessive characteristic. Oxalic acid is a nonmetabolizable organic acid, and all oxalic acid absorbed from the diet or endogenously produced is excreted in the urine. Calcium oxalate is a highly insoluble salt, so that accumulation of oxalate in body fluids or increased concentration in urine results in formation of calcium oxalate crystals which are deposited in tissues or form urinary calculi.

Two types of primary hyperoxaluria are now recognized of which type I is the more common. In this disorder the pathway of metabolism of glycine is altered so that an increased quantity of glyoxalate accumulates which is metabolized to oxalic acid. Excessive amounts of both glycolic acid and oxalic acid are excreted in the urine (Fig. 7–1), but only the excretion of oxalic acid is attended with toxic consequences because of the insolubility of calcium oxalate. The enzyme abnormality is deficiency of 2-oxoglutarate:glyoxalate carboligase.

Type II hyperoxaluria or l-glyceric aciduria is indistinguishable clinically from type I. Biochemically, however, urine glycolate excretion is normal and there is an excess of l-glyceric acid in urine. The enzymatic defect is that of d-glyceric dihydrogenase which also functions as a glyoxalate reductase. Increased conversion of glyoxalate to oxalic acid results from its absence. Genetically, type II hyperoxaluria is also transmitted as an autosomal recessive trait.

Clinical Presentation and Diagnosis

In both forms of hyperoxaluria the onset of symptoms may be in early childhood, although in many instances the serious mani-

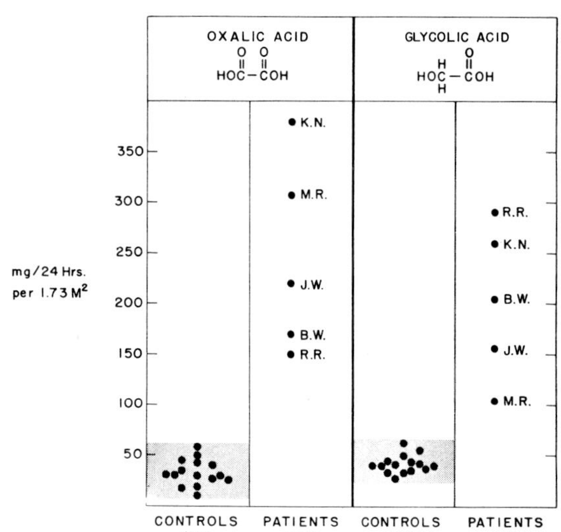

Figure 7–1. Excretion of glycolic acid and oxalic acid in patients with primary oxaluria type I in comparison with normal control subjects. (From Frederick, E. et al: Studies on primary hyperoxaluria. I. In vivo demonstration of a defect in glyoxylate metabolism. N. Engl. J. Med. *269*:824, 1963. Reprinted with permission.)

festations are not seen until later life. The major problem is renal injury due to precipitation of calcium oxalate in tubular contents and formation of calcium oxalate calculi, which break off and cause urinary tract obstruction with hydronephrosis and infection because of stasis. Although at postmortem examination calcium oxalate crystals are seen in tissues other than the kidney, the only important tissue injury is that of kidney.

The renal injury in the early stages may be confused with distal renal tubular acidosis since a defect of acidification of urine can be present.

The specific diagnostic determination is measurement of urine oxalate excretion. This is normally less than 50 mg/day and may be increased to several hundred mg per day in children with oxaluria. In type I primary hyperoxaluria the excretion of glycolic acid is also increased. On x-ray examination the radiopaque calculi are visualized (Fig. 7–2), and intravenous pyelography reveals the degree of obstruction produced in the upper urinary tract as well as the destructive effects of infection (Fig. 7–3).

The prognosis is quite variable. In the same family one child may have early severe renal calculus formation with obstruction and destruction of kidney substance, leading to early renal insufficiency, whereas a sibling may have only a small renal calculus that grows slowly and is relatively asymptomatic for a number of years. There is no specific treatment, but large loads of phosphate salts probably have been of benefit in reducing the rate of calculus formation. This treatment may operate both by reducing urine calcium output through interfering with intestinal calcium absorption, and by the formation of phosphate compounds that inhibit growth of calcium oxalate crystals. The dosage of phosphate is equivalent to 2 gm of phosphorus per day in divided doses diluted in a large volume of water. A secondary benefit may be the high water intake with excretion of a dilute urine that reduces calcium oxalate precipitate formation. Therapy with oral magnesium salts has also been proposed to increase urine magnesium concentration, since magnesium ion inhibits the quantitative precipitation of calcium salts. Dent and Stamp have reported the use of both magnesium hydroxide and sodium phosphate in the treatment of children with primary oxaluria, with suggestive reduction in stone formation.

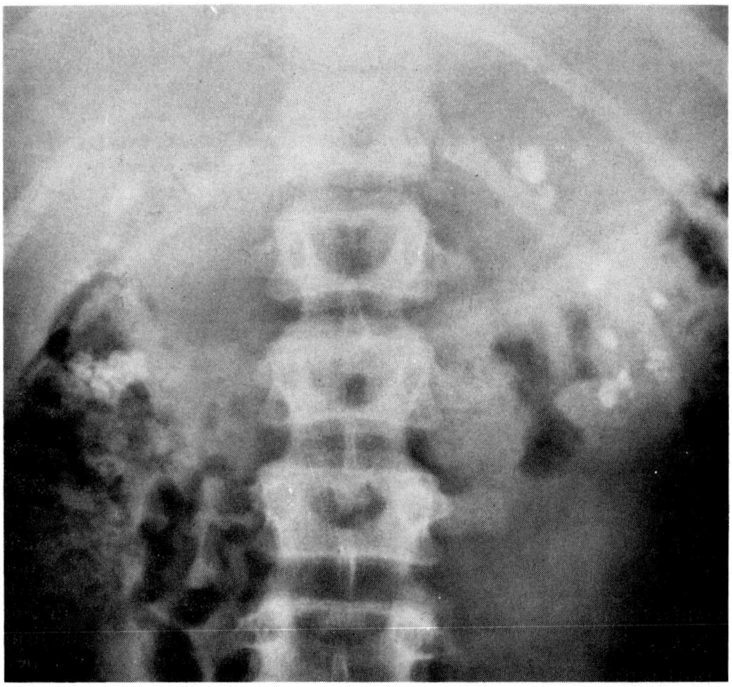

Figure 7–2. Radiopaque stones in kidneys of patient with primary oxaluria.

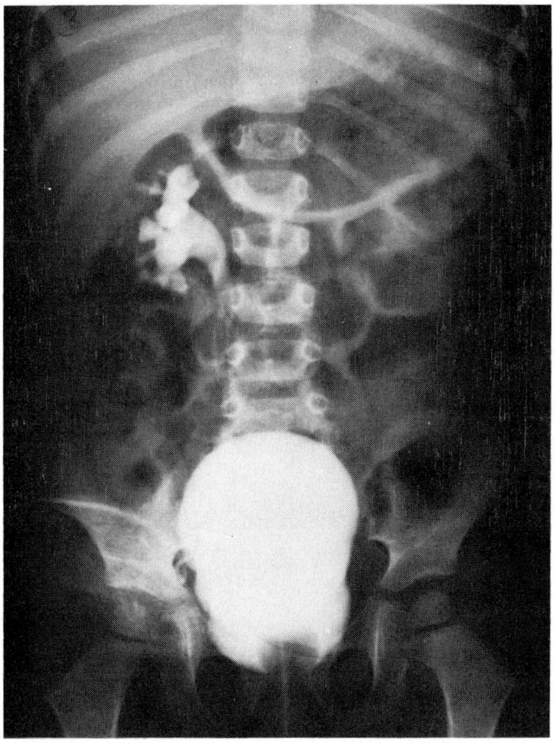

Figure 7–3. Caliceal dilatation resulting from obstruction by calcium oxalate stones in patient with oxaluria.

CASE REPORTS

M.R. was referred for consultation at 3½ years of age because of recurrent ureteral and renal calculus formation first noted at 2½ years. The initial presenting complaint was urinary tract infection and abdominal pain. X-ray examination of the abdomen revealed a large left ureteral calculus (Fig. 7–4), and on intravenous pyelography hydronephrosis of the left kidney was found. The calculus was removed by cystoscopy and ureteral catheterization. Serum calcium and phosphate determinations at this time were within normal limits. Because of recurrence of renal calculi (Fig. 7–5) she was referred elsewhere to determine whether she had hyperparathyroidism. This was ruled out on the basis of consistently normal concentrations of serum calcium and phosphate and low urinary excretion of calcium.

When we first saw M.R. she was an undersized thin child without any other specific physical findings. X-ray examination showed extensive renal calculus formation as before. Serum calcium and phosphate were again normal and urine calcium excretion was low. Serum bicarbonate concentration was also normal, ruling out renal tubular acidosis as a cause of the renal calcinosis. Urine oxalate measurements were then made and oxalate excretion was high, 200 to 300 mg per day. The diagnosis of oxaluria was made. Subsequent studies by Dr. H. Lloyd Smith at the Massachusetts General Hospital showed high excretion of glycolic as well as oxalic acid, classifying this as type I primary hyperoxaluria.

Since this disorder is genetic the entire family was studied. The parents showed normal urine oxalate output, but a sibling, R.R., had increased urine excretion of oxalate. This fits with the genetic pattern of an autosomal recessive gene. X-ray examination of the abdomen of R.R. showed an opaque mass in the region of the right kidney (Fig. 7–6), indicating a small renal calculus. He had been entirely asymptomatic, however.

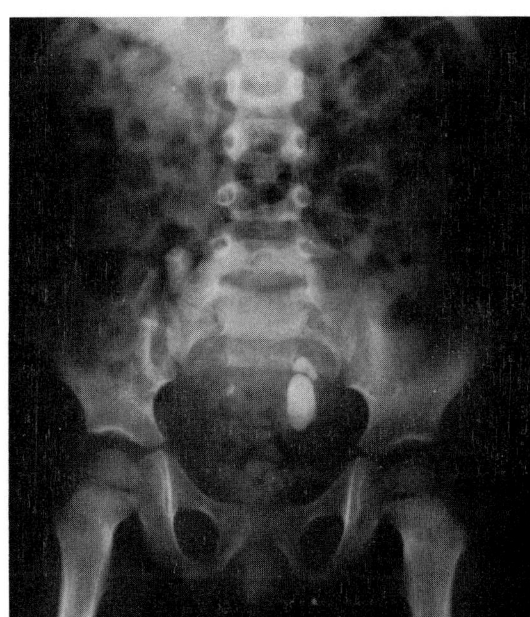

Figure 7–4. Ureteral calculus in patient (M.R.) with primary hyperoxaluria.

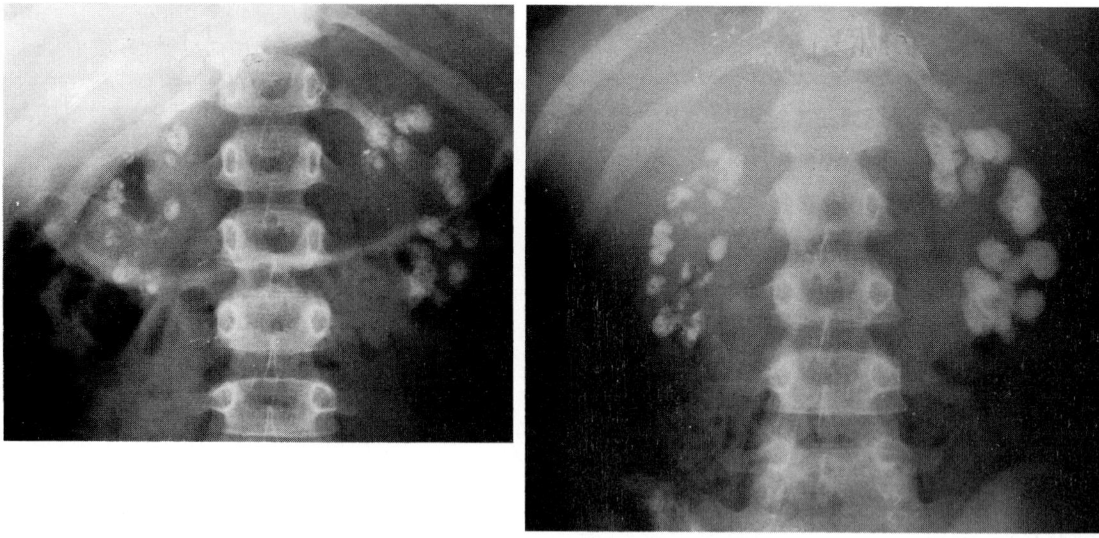

Figure 7–5. Multiple calculi and renal calcinosis (M.R.). (From Frederick, E. et al: Studies on primary hyperoxaluria. I. In vivo demonstration of a defect in glyoxylate metabolism. N. Engl. J. Med. *269*:824, 1963. Reprinted with permission.)

M.R. had a rapidly progressive course with severe renal insufficiency and death by 6 years of age. R.R. has had a mild course with only slow growth of the renal calculus and has been on phosphate treatment intermittently. Whether treatment has modified the course of his disease cannot be conclusively answered.

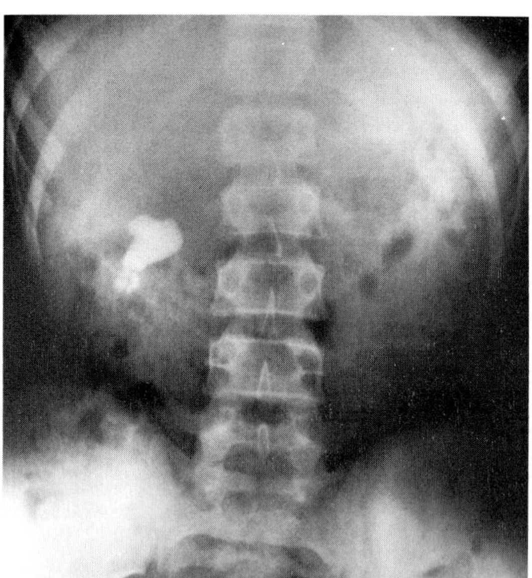

Figure 7–6. Renal calculus patient (R.R.) with primary oxaluria. Sibling of patient (M.R.) with severe disease.

COMMENTS: The diagnosis of primary hyperoxaluria in M.R. was suggested by the rapid and extensive formation of renal calculi with relative low urine output of calcium. Renal tubular acidosis was quickly eliminated by serum determinations, as was hypercalcemia. Urine oxalate measurements were diagnostic. The asymptomatic oxaluria of the sibling with an undiagnosed renal stone was of great interest. Why in the same family one sibling should have severe disease progressing to fatal renal insufficiency by 6 years of age, while another sibling should have an unusually mild course, is unknown. Although the inheritance is compatible with a single autosomal recessive gene, other genes possibly may influence the expression of the specific gene. In the instance of R.R. with a mild form of the disease, high intake of sodium and potassium phosphate (Neutra-Phos) may have helped to reduce the formation of stones during the period after the diagnosis was made.

OXALURIA SECONDARY TO INTESTINAL DISEASE WITH CALCULUS FORMATION

The incidence of renal calculus formation in patients with inflammatory bowel disease or following ileal bypass for obesity has attracted attention in recent years. It was found that these patients have increased urine oxalate excretion, and it has been proposed that the underlying abnormality is increased absorption of diet oxalate. The basis for this greater absorption is believed to be fat malabsorption with precipitation of calcium in the intestinal lumen as calcium soaps. Removal of calcium in this way prevents precipitation of oxalate by calcium and permits absorption of the soluble sodium and potassium oxalate. A test for fat malabsorption has been proposed utilizing this principle, measuring the urine excretion of oxalate following ingestion of a high oxalate meal. In patients with fat malabsorption urine oxalate is higher than in normals. It has also been suggested that the colon is the major site of oxalate hyperabsorption in enteric disease, and that this is secondary to the effects of unabsorbed bile acids on colonic mucosa. Since oxalic acid is not metabolized in the body, absorbed oxalate is excreted quantitatively in the urine.

The treatment of these patients is dietary, namely, elimination of foods high in oxalic acid such as certain of the green leafy vegetables, spinach, turnip greens, etc., as well as chocolate. Since one of the end-products of ascorbic acid metabolism is oxalic acid, excessively high intake of vitamin C should be avoided. A high water intake with output of a dilute urine can also be recommended.

BLADDER CALCULI OF INFANTS AND CHILDREN

This disorder, relatively common in Iran, Thailand and other Southeast Asian countries, is practically unheard of in the United States and Europe. In countries where it occurs large bladder calculi grow by accretion of calcium oxalate, and laminated stones with a diameter of 3 cm or more may arise (Fig. 7–7). One of the possible mechanisms is chronic dehydration in these hot climates with excretion of scanty, highly concentrated urine in which calcium salts can precipitate. Because of the high incidence of intestinal disorders, increased absorption of oxalate possibly may also play a role as in patients with chronic inflammatory bowel disease. The manifestations, as in other bladder stones, are hematuria, painful urination, and urinary tract infection. Treatment is removal of the stone via cystoscopic instrumentation or through suprapubic cystotomy if the stone is very large, and high fluid intake as an attempt to prevent recurrence.

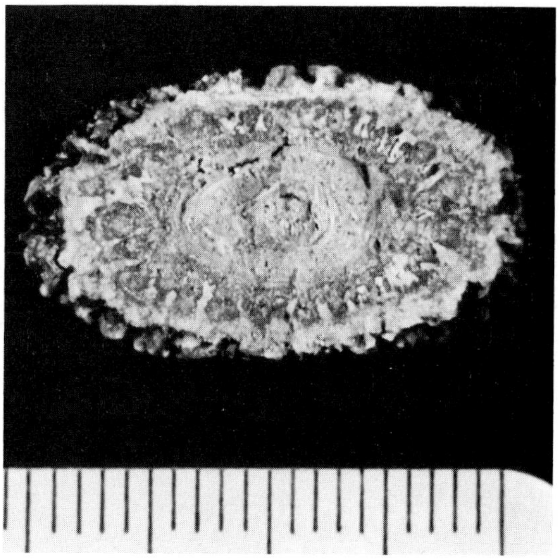

Figure 7–7. Bladder calculus removed from Iranian child. The layered structure is clearly shown.

IDIOPATHIC STONE FORMERS

This is a disorder primarily of adult life, possibly because of the time necessary for stones to grow to large enough dimensions to be clinically important. These subjects do not show excessive excretion of calcium in the urine, and although the stones often are chiefly calcium oxalate or mixed calcium phosphate and oxalate, urinary excretion of oxalate is not grossly excessive, differentiating these patients from those with primary hyperoxaluria. One of the theories of such stone formation is a deficiency in the urine of protective factors that are present in the urine of normal individuals and which prevent growth of crystals of calcium salts. The nature of these protective factors is still in doubt although pyrophosphate, which is present in minute amounts in urine has this property. Citrate in urine is a protective factor in that it reduces the concentration of ionized calcium, and a citrate compound, phosphocitrate, has been proposed as a specific protective factor in the prevention of growth of crystals of calcium salts.

These patients can be treated by oral intake of large amounts of inorganic phosphate, as in those with primary hyperoxaluria. This treatment reduces calcium excretion in urine, and possibly increases the excretion of pyrophosphate and other protective compounds.

CASE REPORT

D.B. was a 10 year old white male who had two episodes of right-sided abdominal pain with hematuria separated by a two-month interval. A year previously he had been hospitalized for right-sided abdominal pain suspected of being due to appendicitis, but it cleared spontaneously. No mention of hematuria was made at this time. At the time of the first episode of painful hematuria the urine had a Sp.Gr. of 1.014 and contained 50 RBC and 5 WBC per high power field. Serum calcium was 9.8 mg/dl, phosphate 7.0 mg/dl, urea nitrogen 13 mg/dl and urine calcium 80 mg/24 hrs. An intravenous pyelogram was normal and no stone was detected. When he was hospitalized for the second episode of painful hematuria the urine was loaded with RBC. Serum calcium was 10.4 mg/dl, phosphate 6.4 mg/dl, urea nitrogen 20 mg/dl, and creatinine 0.85 mg/dl. Urine calcium was 71 mg/24 hrs; urine cystine within normal limits; and urine oxalate 13 mg/24 hrs. The abdominal scout x-ray showed a calcific density in the right kidney region, and an intravenous pyelogram revealed a dilated ureter and pelvis on the right. He was given a high fluid intake and hematuria diminished. No stone was passed during hospitalization, but following discharge the child was asked to continue filtering his urine and a stone was passed. Unfortunately no analysis of the stone was made.

COMMENT: This is probably a case of idiopathic stone formation. Hypercalcemia and hypercalciuria were eliminated, as were cystinuria and oxaluria. The stone may well have been a calcium oxalate mass, since this is the common finding in idiopathic stone formation, but was not secondary to excessive urinary excretion of oxalate. Treatment of this patient was simply a high fluid intake to increase urine volume and reduce concentration of solutes in urine, including calcium. However, if there is recurrence of renal stone formation a high phosphate load in the form of Neutra-Phos, providing additional phosphate equivalent to 1 to 2 gm phosphorus per day, will be advised.

STONES ASSOCIATED WITH URINARY STASIS AND INFECTION

AMMONIUM-MAGNESIUM PHOSPHATE STONES

Obstructive lesions of the urinary tract commonly are complicated by infection with urea-splitting bacteria. The resulting conversion of urea to ammonia causes marked alkalinization of the urine and precipitation of phosphate as the ammonium–magnesium salt. This type of calculus is therefore a complication of failure of normal emptying of the urinary tract, whether due to neurologic lesions or organic obstruction and secondary infection. The diagnostic problem is that of the lesion causing urinary stasis, and treatment is directed at establishing normal emptying of the urinary tract and control of the infection.

MEDULLARY SPONGE KIDNEY WITH CALCULUS FORMATION

This anatomic lesion of the kidney consists of dilated collecting ducts and cysts lined with cuboidal epithelium, exclusively in the kidney medulla, mainly in the papillary portion. These cysts may communicate with the calyces, and frequently contain calcium precipitates and calculi. There is no known genetic transmission. The disorder does not cause progressive renal insufficiency, unlike medullary cystic disease, "familial juvenile nephronophthisis," and is ordinarily recognized only because of passage of stones with hematuria and renal or ureteral colic. For this reason the diagnosis is rarely made in childhood and sometimes may be diagnosed only incidentally in later life.

On intravenous urography, multiple cavities filled with contrast material are seen adjacent to the calyces, and opaque calculi are present. In the differential diagnosis, nephrocalcinosis secondary to renal tubular acidosis or to hypercalcemia must be considered, as well as primary hyperoxaluria. The diagnosis of these disorders has been discussed above. Treatment is directed only at the complications of passage of calculi. There is apparently no method of prevention of stone formation in the cavities. Ureteral calculi with obstruction and urinary tract infections are treated as they would be in the absence of medullary sponge kidney. Since the lesions are often bilateral, nephrectomy for recurrent stone formation would usually be contraindicated.

ORGANIC STONES

The two major organic compounds excreted in the urine which are sufficiently insoluble to form crystalline aggregates that grow into urinary tract calculi are cystine and uric acid. Xanthine can also form stones but this is a rare form of calculus.

CYSTINURIA

Cystine stones occur as the result of a genetic disorder transmitted as an autosomal recessive trait, cystinuria. This must be differentiated from cystinosis, which is a disorder of cystine metabolism resulting in intracellular cystine accumulation and toxic injury to kidney. Cystinuria is a disorder of tubular reabsorption of cystine and the basic amino acids, lysine, arginine and ornithine. There is, therefore, increased urinary excretion of these four amino acids. Since lysine, arginine and ornithine are highly soluble in urine, no clinical problems result from their loss in the urine; the amounts lost are not sufficient to produce a deficiency of amino acids in the body. Because of the relative insolubility of cystine at neutral or acid pH, it can crystallize out in urine and form calculi which can grow to large size and fill the calyces, with resultant hydronephrosis and destruction of kidney substance. Although cystine stones are less radio dense than calcium phosphate or oxalate stones, calcium salts can precipitate on the cystine nucleus, resulting in radio dense stones that can be mistaken for calcium calculi.

It is important to make this differentiation since treatment for cystine stones is quite different from that for calcium stones. It is true that for all kidney stones, including cystine, a major principle of treatment is the maintenance of a dilute urine at all times, reducing the possibility of precipitation of insoluble compounds. In the case of cystine, however, the urine should be maintained as alkaline as possible since cystine solubility is increased with high pH. The urine pH cannot be raised above 8, and it is at pH levels above this that the major cystine solubilizing effect of alkalinization of medium occurs; thus, some skepticism has been expressed about the value of alkalinization as compared to simple dilution of the urine by high water intake. Nevertheless, it seems worthwhile to increase urine pH as much as is feasible even if the increased solubility is only slight. For this purpose we prefer to use a solution of sodium and potassium citrate (Polycitra) because this is more acceptable in flavor than solutions of sodium bicarbonate and does not produce gastric discomfort. Sixty to 80 ml of this solution daily in divided doses is required to achieve maximum alkalinization of urine. In some cystinuric patients with large stones at the time of initial diagnosis, or who continue to form cystine stones despite dilution of urine and alkalinization, treatment with penicillamine is resorted to. This is a sulfhydryl compound that forms mixed disulfides with cysteine that are more soluble than cystine. By this mechanism the excretion of the less soluble cystine is reduced. Penicillamine treatment does have side-effects, including sensitivity reactions with skin rash and increased loss of trace metals in the urine because of its metal chelating action. Copper and zinc deficiency

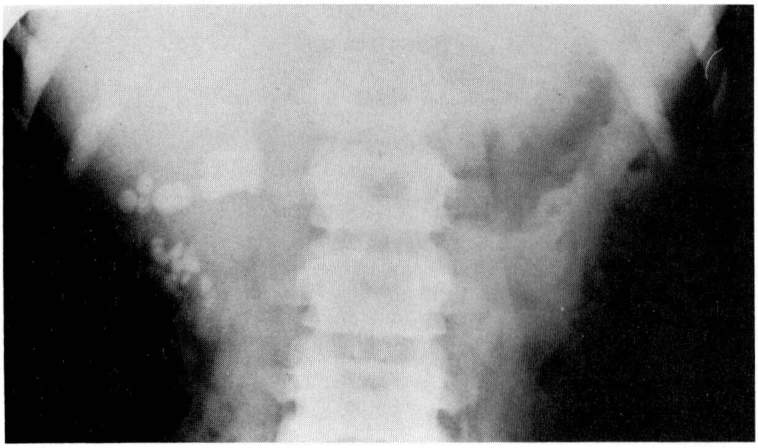

Figure 7–8. Radiopaque calculus in right hydronephrotic kidney of patient with cystinuria. The calculus on analysis had been classified as a calcium phosphate stone.

could be produced by long-term penicillamine treatment unless the supply of these trace metals in the diet is adequate.

CASE REPORT

A 12 year old boy was admitted to the hospital because of flank pain and hematuria. On x-ray examination a large radiopaque calculus was seen in the pelvis of the right kidney (Fig. 7–8). On retrograde pyelography this kidney was found to be hydronephrotic and infected, and thought to be too severely injured to be salvaged. A nephrectomy was done and the stone analyzed. The report indicated that it was a calcium phosphate stone and for this reason the patient was referred to us for study of calcium and phosphate metabolism. Serum calcium and phosphate concentrations were normal and urine calcium and oxalate excretions were not increased. Chromatography of the urine, however, showed excessive concentrations of cystine, lysine, arginine and ornithine, the typical pattern of cystinuria (Fig. 7–9). He was treated, therefore, by high fluid intake with the goal of maintaining a dilute urine at all times, including the overnight specimen, plus Polycitra, 60 ml per day in divided doses to increase urine pH. The remaining kidney had good function and the further course has been uneventful.

COMMENT: The initial diagnosis in this patient was a disorder of calcium metabolism because the stone was found on analysis to contain calcium phosphate. Urine amino acid chromatography should be done as part of the diagnostic study in all children with renal stones because cystine stones may form the nucleus for calcium salt deposition and thus be mistaken for calcium stones. A useful screening test is the qualitative nitroprusside reaction for sulfhydryl groups. A positive reaction indicates increased concentration of cystine in urine.

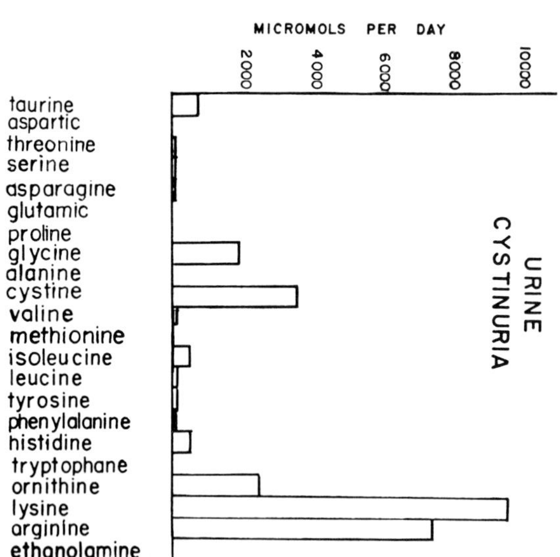

Figure 7–9. Pattern of amino acid excretion in urine of patient with cystinuria. (From Harrison, H. E. and Harrison, H. C.: Aminoaciduria in relation to deficiency diseases and kidney function. J.A.M.A. *164*: 1573, 1957. Reprinted with permission.)

URATE CALCULI

Sodium acid urate (Na H U), the salt of uric acid present in acid urine, is only slightly soluble, so that the combination of high urate excretion and acid urine can lead to formation of urate calculi or to urate nephropathy resulting from intratubular precipitation of urate with parenchymatous destruction of kidney. Urate stones are seen most frequently in childhood as the result of rapid cytolysis in the treatment of leukemia. It can also occur in children with disorders of uric acid metabolism, as in Lesch-Nyhan syndrome, and (more rarely) familial gout with onset in childhood. We have also seen several children who were being treated for seizure disorder with ketogenic diet who had painful hematuria, and the small stones recovered proved to be urate. Urate stones are not radiopaque, and it is unlikely that any of the conditions predisposing to urate calculi will be mistaken for disorders of calcium and phosphate metabolism.

REFERENCES

IDIOPATHIC HYPERCALCIURIA

Bordier, P., Ryckewart, A, Gueris, J. et al: On the pathogenesis of so-called idiopathic hypercalciuria. Am. J. Med. 63:398–409, 1977.

Bulusu, L., Hodgkinson, A., Nordin, B. E. C. et al: Urinary excretion of calcium and creatinine in relation to age and body weight in normal subjects and patients with renal calculus. Clin. Sci. 38:601–612, 1970.

Coe, F. L.: Treated and untreated recurrent calcium nephrolithiasis in patients with idiopathic hypercalciuria, hyperuricosuria, or no metabolic disorder. Ann. Intern. Med. 87:404–410, 1977.

Coe, F. L. and Kavalach, A. G.: Hypercalciuria and hyperuricosuria in patients with calcium nephrolithiasis. N. Engl. J. Med. *291*:1344–1350, 1974.

Coe, F. L., Canterbury, J. M., Firpo, J. J. et al: Evidence for secondary hyperparathyroidism in idiopathic hypercalciuria. J. Clin. Invest. 52:134–142, 1973.

Dent, C. E. and Friedman, M.: Hypercalciuric rickets associated with renal tubular damage. Arch. Dis. Child. 39:240–249, 1964.

Dent, C. E., Harper, C. M. and Parfitt, A. M.: The effect of cellulose phosphate on calcium metabolism in patients with hypercalciuria. Clin. Sci. 27:417–425, 1964.

Fanconi, A.: Idiopathische Hypercalciuria in Kindesalter. Helv. Paediatr. Acta *18*:306–322, 1963.

Gray, R. W., Wilz, D. R., Caldas, A. E. et al: The importance of phosphate in regulating 1, 25 $(OH)_2$ vitamin D levels in humans: studies in healthy subjects, in calcium stone formers and in patients with primary hyperparathyroidism. J. Clin. Endocrinol. Metab. 45:299–306, 1977.

Larbre, F., Freyon, M.-T., Hartemann, E. et al: Hypercalciurie idiopathique avec lithiase renale chez l'enfant. Pediatrie 24:421–434, 1969.

Nassim, J. R. and Higgins, B. A.: Control of idiopathic hypercalciuria. Br. Med. J. *1*:675–681, 1965.

Nordin, B. E. C.: Hypercalciuria. Editorial Review. Clin. Sci. Mol. Med. 52:1–8, 1977.

Pak, C. Y. C., Ohata, M., Lawrence, A. C. et al: The hypercalciurias: causes, parathyroid function and diagnostic criteria. J. Clin. Invest. 54:387–400, 1974.

Parfitt, A. M., Higgins, B. A., Nassim, J. R. et al: Metabolic studies in patients with hypercalciuria. Clin. Sci. 27:463–482, 1964.

Shen, F. H., Baylink, D. S., Nielsen, R. L. et al: Increased serum 1, 25-dihydroxyvitamin D in idiopathic hypercalciuria. J. Lab. Clin. Med. 90:955–962, 1977.

Sherwood, L. M.: Editorial. Idiopathic hypercalciuria: a mixed bag of stones. J. Lab. Clin. Med. 90:951–954, 1977.

Yendt, E. R., Gagné, R. J. and Cohanim, M.: The effects of thiazides in idiopathic hypercalciuria. Am. J. Med. Sci. *251*:449–460, 1966.

HYPERCALCIURIA ASSOCIATED WITH TOTAL PARENTERAL NUTRITION

Adelman, R. D., Abern, S. B., Merten, D. et al: Hypercalciuria with nephrolithiasis: a complication of total parenteral nutrition. Pediatrics 59:473–475, 1977.

IMMOBILIZATION HYPERCALCIURIA WITH STONE FORMATION

Bergstrom, W. H.: Hypercalciuria and hypercalcemia complicating immobilization. Am. J. Dis. Child. *132*:553–554, 1978.

Howard, J. E., Parson, W. and Bigham, R. S., Jr.: Studies on patients convalescent from fracture. III. The urinary excretion of calcium and phosphorus. Bull. Johns Hopkins Hosp. 77:291–313, 1945.

Rosen, J. F., Wolin, D. A. and Finberg, L.: Immobilization hypercalcemia after single limb fractures in children and adolescents. Am. J. Dis. Child. *132*:560–564, 1978.

Whedon, G. D. and Shorr, E.: Metabolic studies in paralytic acute anterior poliomyelitis. II. Alterations in calcium and phosphorus metabolism. J. Clin. Invest. 36:941–1033 (parts I-IV), 1957.

OXALURIA

Primary

Dent, C. E. and Stamp, T. C. B.: Treatment of primary hyperoxaluria. Arch. Dis. Child. 45:735–745, 1970.

Williams, H. E. and Smith, L. H., Jr.: Disorders of oxalate metabolism. Am. J. Med. 45:715–735, 1968.

Williams, H. E. and Smith, L. H., Jr.: l-glyceric aciduria. A new genetic variant of primary hyperoxaluria. N. Engl. J. Med. *278*:233–239, 1968.

Williams, H. E. and Smith, L. H., Jr.: Hyperoxaluria in l-glyceric aciduria: possible pathogenic mechanism. Science *171*:390–391, 1971.

Secondary to Intestinal Malabsorption

Anderson, H. and Gillberg, R.: Urinary oxalate on a high-oxalate diet as a clinical test of malabsorption. Lancet *2*:677–678, 1977.

Deren, J. J., Porush, J. G., Levitt, M. F. et al: Nephrolithiasis as a complication of ulcerative colitis and regional enteritis. Ann. Intern. Med. *56*:843–853, 1962.

Ogilvie, D., McCollum, J. P. K., Packer, S. et al: Urinary outputs of oxalate, calcium and magnesium in children with intestinal disorders. Potential cause of renal calculi. Arch. Dis. Child. *51*:790–795, 1976.

BLADDER STONES IN CHILDREN

Eckstein, H. B.: Endemic urinary lithiasis in Turkish children. A clinical study of 119 cases. Arch Dis. Child. *36*:137–145, 1961.

Halstead, S. B. and Valyasevi, A.: Studies of bladder stone disease in Thailand. I. Introduction and description of area studies. Am. J. Clin. Nutr. *20*:1312–1319, 1967.

URIC ACID STONES

Gutman, A. B. and Yü, T. F.: Uric acid nephrolithiasis. Am. J. Med. *45*:756–779, 1968.

Seegmiller, J. E.: Xanthine stone formation. Am. J. Med. *45*:780–783, 1968.

CYSTINURIA

Crawhall, J. C., Scowen, E. F. and Watts, R. W. E.: Effect of penicillamine on cystinuria. Br. Med. J. *1*:588–590, 1963.

Crawhall, J. C. and Watts, R. W. E.: Cystinuria. Am. J. Med. *45*:736–755, 1968.

Stephens, A. D. and Watts, R. W. E.: The treatment of cystinuria with N-acetyl-D penicillamine, a comparison with the results of D-penicillamine treatment. Q. J. Med. New Series *40*:355–370, 1971.

GENERAL

Muldowney, F. P., Freaney, R., McMullin, J. P. et al: Serum ionized calcium and parathyroid hormone in renal stone disease. Q. J. Med. New Series *45*:75–86, 1976.

Prien, E. L. and Prien, E. L., Jr.: Composition and structure of urinary stone. Am. J. Med. *45*:654–672, 1968.

Smith, L. H., Jr.: Symposium on Stones. Introduction. Am. J. Med. *45*:649–653, 1968.

Chapter Eight

ECTOPIC (PATHOLOGIC OR METASTATIC) CALCIFICATION

Ectopic calcification, namely, deposition of calcium phosphate in extraskeletal sites, will be discussed under two main categories: (1) pathologic calcification associated with abnormally high concentrations of either calcium, phosphate or both; and (2) ectopic mineralization with normal concentrations of both calcium and phosphate in extracellular fluid.

ECTOPIC CALCIFICATION ASSOCIATED WITH ABNORMAL CONCENTRATIONS OF CALCIUM OR PHOSPHATE IN PLASMA

HYPERCALCEMIA

Pathologic calcification occurs with hypercalcemia of any cause and the various categories of hypercalcemia have been discussed in Chapter 4. There is a definite predilection for certain tissues, and the pattern of distribution of the deposits of calcium phosphate is essentially the same irrespective of the mechanism of hypercalcemia.

The organ most commonly affected is the kidney and calcification occurs through two mechanisms. Because of hypercalciuria, calcium phosphate casts are formed in the tubule lumina that may rupture the tubule walls and extrude into peritubular tissue, or may form the nucleus for a renal calculus in the renal caliceal system or pelvis (see Chapter 4). In addition, calcium phosphate can precipitate in the peritubular tissue along with calcification of the cells and basement membranes of convoluted tubules, Henle's loop and collecting tubules. Another organ frequently involved is the lung with calcification of the elastic tissue of alveolar walls and the pulmonary venous system. The predilection for this site may be due to the change in pH resulting from diffusion of CO_2 into the pulmonary alveoli. Calcification also occurs in the media of large arteries and in the elastic tissue of the endocardium, particularly of the left auricle. Another site in which calcium phosphate deposits are sometimes found is in the connective tissue of the fundus of the stomach surrounding the acid-secreting glands. This also suggests that alkalinization of the tissue predisposes to calcium phosphate precipitation. In addition to these sites injured tissue, wherever it may occur, is prone to calcification. The calcification of necrotic tissue even in normocalcemic states is a well-known phenomenon, and the predisposing effect of tissue injury is even more striking in hypercalcemic states.

The composition of ectopic calcification is more variable than that of bone mineral. The major component is calcium phosphate, which may be amorphous $Ca_3(PO_4)_2$ or crystalline hydroxyapatite. Calcium carbonate may also be present, particularly in the deposits in the lung, and there usually is coprecipitation of iron salts which is revealed by iron stains of the affected tissues.

The treatment of ectopic calcification secondary to hypercalcemia is obviously directed toward the process causing the hypercalcemia. One of the objections to the treatment of acute hypercalcemia by intravenous or oral phosphate salts is the possibility that pathologic calcification of tissues may be intensified by this mode of therapy.

HYPERPHOSPHATEMIA

Ectopic calcification can also occur in hyperphosphatemic states even in the presence of reduced concentrations of extracellular calcium. The disease processes in which hyperphosphatemia-associated ectopic calcification can be seen are listed in Table 8–1.

Hypoparathyroidism and Pseudohypoparathyroidism

Hyperphosphatemia of considerable magnitude, up to 12 mg/dl in young children, is seen in both of these conditions, but for some reason ectopic calcification is more common in pseudohypoparathyroidism type I than in hypoparathyroidism despite comparable elevations of serum phosphate concentration. It must be presumed that other factors, possibly differences in matrix structure in connective tissue either of collagen or mucopolysaccharides, are significant in the sequence of events leading to precipitation of calcium phosphate in extraskeletal tissue. The location of ectopic calcification in hypoparathyroidism and pseudohypoparathyroidism is quite different from that in hypercalcemic states. One of the sites of predilection is the basal ganglion region of the brain, with stippled calcification in the perivascular tissue in this region. Why this portion of the brain is susceptible to calcium salt deposition is unknown. Another important site of ectopic calcification in these disease states is the subcutaneous tissue surrounding joints. The periarticular distribution is presumably related to the trauma of joint motion in making the connective tissue matrix calcifiable in the presence of high concentrations of inorganic phosphate in extracellular fluids.

Table 8–1. Hyperphosphatemic States Associated with Ectopic Calcification

Hypoparathyroidism
Pseudohypoparathyroidism
Renal Insufficiency
Acute Leukemia with Cell Lysis Following Chemotherapy

Renal Insufficiency

On the usual high phosphate intake of ordinary diets, serum phosphate concentration rises above the normal range when glomerular filtration rate is reduced below 50 percent of normal. As discussed in the section on renal disease and vitamin D metabolism (Chapter 5) this rise of serum phosphate reduces the formation of 1,25-dihydroxyvitamin D in the kidney, with resulting intestinal calcium malabsorption. The net result is reduced serum calcium concentration with consequent stimulation of parathyroid gland cell activity and growth and secondary hyperparathyroidism. If this is sufficiently marked solubilization of bone mineral results, but the effect of such excessive PTH concentration on renal tubule reabsorption of phosphate is inadequate to reduce serum phosphate to normal in the face of marked reduction of GFR, so that hyperphosphatemia and hypocalcemia persist.

Ectopic calcification in subcutaneous tissue may occur as the result of the hyperphosphatemia but, more importantly, ectopic calcification may be secondary to the attempted treatment of the hypocalcemia of these patients in the hope of suppressing the secondary hyperparathyroidism. Such treatment with large doses of vitamin D or the vitamin D-like compound, dihydrotachysterol, or vitamin D metabolites and analogues such as 25-hydroxyvitamin D, 1,25-dihydroxyvitamin D, or 1-α hydroxyvitamin D, can increase calcium absorption and extracellular calcium concentrations. If hyperphosphatemia persists the increased Ca × P product resulting from treatment causes increased extraskeletal calcium phosphate deposition in kidney and large blood vessels, as well as in lung and stomach mucosa, as in hypercalcemic states. The renal calcification can further reduce kidney function so that acute renal failure can be precipitated by injudicious treatment. It is important, as stressed in the section on treatment of renal hyperparathyroidism (Chapter 5), to reduce serum phosphate to 4 to 6 mg/dl in the child and 3 to 5 mg/dl in the adult by phosphate-binding compounds such as aluminum or calcium salts prior to the use of compounds with vitamin D activity.

Hyperphosphatemia Secondary to Massive Cell Destruction

This form of hyperphosphatemia occurs in patients with leukemia in whom rapid white blood cell lysis is occurring either spontaneously or, more commonly, after initiation of cytotoxic chemotherapy. The load of phosphate secondary to hydrolysis of cell phosphate esters is so great that phosphate concentration rises precipitously in extracellular fluid. This is particularly the case if renal function has been compromised by the coincident huge uric acid load with uric acid nephropathy. Under these conditions serum phosphate can rise to 12 to 15 mg/dl and serum ionized calcium is acutely lowered causing tetany. In order to prevent uric acid nephropathy with intratubular precipitation of urate, the patients are given alkalinizing agents such as sodium citrate, which may increase the severity of tetany. The treatment of these patients with intravenous calcium gluconate is cautiously approached to avoid raising the Ca × P product to such heights that renal calcification would occur, with further injury to renal function. The critical Ca × P product may not be a constant and possibly varies with age. For example, the serum phosphate of the young prematurely born infant is normally in the range of 7 to 8 mg/dl so that the Ca × P product exceeds 70. In the adult a Ca × P product of 70 is considered the maximum safe value above which renal calcification is likely to occur. It is likely that the young child can tolerate a product greater than that of the adult but there is no hard evidence on this point.

The treatment of these patients is maintenance of good renal function by high output of urine, and prevention of uric acid nephropathy through maintenance of an alkaline dilute urine and inhibition of uric acid production by the use of allopurinol. In addition, intestinal sequestration of phosphate by oral calcium carbonate or aluminum hydroxide is indicated. The intravenous injection of calcium gluconate for prevention or treatment of tetany should be cautiously approached and given at a rate that provides relief of tetany but does not raise serum calcium to normal. This is an acute, self-limited problem; after the initial massive cell cytolysis the patient comes back into equilibrium, and persistent hyperphosphatemia is not expected unless severe renal injury has resulted in the acute episode.

ECTOPIC CALCIFICATION WITH NORMAL CONCENTRATIONS OF CALCIUM AND PHOSPHATE IN PLASMA

Calcium phosphate deposition in extraskeletal tissues under conditions of normal Ca × P product must obviously depend on local tissue factors that in some way predispose to precipitation of calcium salts. Even under circumstances of raised Ca × P product local tissue factors play a role, as indicated by the selective calcification in certain organs and in specific sites in these organs. Since all connective tissue contains a collagen matrix suspended in mucopolysaccharide gel similar to that of calcifiable cartilage and bone, the question arises as to what makes this matrix noncalcifiable under normal conditions. Chapter 1 has dealt with the local factors that initiate cartilage mineralization, and the possible compounds such as the pyrophosphates and mucopolysaccharides that may prevent calcium phosphate crystal formation. The disorders of ectopic calcification with normal Ca × P product can be separated into two major groups: (1) those in which calcium phosphate aggregates are precipitated without organization into bone; and (2) those in which there is organized ectopic bone formation including, in some cases, marrow elements.

ECTOPIC CALCIFICATION WITHOUT ORGANIZED BONE STRUCTURE

Calcification of Necrotic Tissue

This is a well-recognized phenomenon, particularly the calcification of caseous lesions of tuberculosis, not only in lung but in lymph nodes, spleen and liver. The calcification of necrotic renal tubule cells, tumor cells, or necrotic brain tissue, as in toxoplasmosis or cytomegalic inclusion body disease, are further examples of such precipitation of calcium phosphate in dead tissue. Calcification may also occur at sites of inflammatory reaction even without necrosis. The mechanism of postnecrotic or inflammatory calcification is still debated. One possibility is local increase of inorganic phosphate concentration by hydrolysis of cell phosphoric acid esters by acid phosphatases released from lysosomes of injured cells. This ectopic calcification does not require increased concentration of calcium in the tissue, but if such concentration is increased calcification becomes more intense. An example of the latter is the calcification at sites of infiltration of tissues with calcium gluconate. There are many reports of patients who have had accidental infiltration of subcutaneous tissue owing to leakage from attempted intravenous infusion of calcium. There are also a few examples of intramuscular injection of 10 percent calcium gluconate before the injurious effect of such injections was recognized. These infiltrated subcutaneous or intramuscular lesions become necrotic and rapidly calcified, with slow healing of the calcified indurated masses. Some resorption of the calcified lesion occurs in time but residual scarring remains.

Calcinosis Universalis, Calcinosis Circumscripta, Tumoral Calcinosis

These disease entities of unknown pathogenesis are simply descriptive of the distribution of calcium phosphate deposits in subcutaneous tissue. In calcinosis universalis these precipitates are formed throughout the subcutaneous tissue although concentrated in periarticular regions or areas subject to trauma. In calcinosis circumscripta the precipitates are more localized and usually are periarticular. Tumoral calcinosis describes a form of calcium phosphate precipitation forming large masses which are predominantly in areas subjected to pressure, such as the gluteal region. This last disorder shows a racial and geographic distribution, but in none of these disease states is there a clear genetic transmission.

Calcinosis Universalis and Calcinosis Circumscripta. These syndromes often begin with manifestations of a connective tissue disorder resembling scleroderma or dermatomyositis. The initial symptoms are those of inflammatory changes in subcutaneous tissue, associated in some instances with muscle weakness. The muscle manifestation may predominate so that the diagnosis of polymyositis is entertained, or the joint pains and periarticular swellings may be sufficiently pronounced to suggest the diagnosis of rheumatoid arthritis. As the acute connective tissue disorder subsides, painful hard masses of calcium phosphate precipitate are felt in the subcutaneous tissue, often concentrated in areas subjected to continual trauma such as the periarticular regions or pressure zones. These masses vary in size, and form visible irregular nodules that sometimes erode through the overlying skin with discharge of a mass of chalklike white mineral precipitate. Histologically there is no necrosis but evidence of foreign body reaction to the unorganized calcium phosphate precipitate. Following ulceration of such a lesion secondary infection is an important complication, resulting in severe scarring. The end-result is a seriously disfiguring disorder with considerable limitation of motion in joints which are surrounded by masses of mineral precipitate.

The pathogenesis of the disorder is still a mystery. It is assumed that some alteration of matrix is present, either primarily or secondary to the connective tissue disease, which facilitates the precipitation of calcium phosphate. This abnormality may be a deficiency of so-called protective factors that ordinarily prevent such precipitation in connective tissue matrix but there is no direct evidence of absence of specific compounds. The process may become quiescent with only a limited portion of the body subcutaneous tissue affected (calcinosis circumscripta), or it may continue to progress with involvement of most of the subcutaneous tissue area (calcinosis universalis).

Treatment of these subcutaneous calcific deposits is unsatisfactory. During the initial manifestations of connective tissue inflammatory process the patient may have received steroid suppressive treatment or acetylsalicylic acid, but there is no evidence that such anti-inflammatory drug treatment has any influence on the development or course of the calcinosis. Medical treatment has been of three varieties: calcium chelating agents; diphosphonates which are analogues of pyrophosphate; and measures to reduce serum phosphate concentrations.

Treatment with intravenous courses of disodium ethylenediamine tetraacetic acid (Na_2EDTA), a calcium chelating agent, has been tried. There is no good evidence that the calcium deposits are resorbed under the influence of this treatment, and since EDTA has some nephrotoxic effect when given repeatedly or over a prolonged period its use seems inadvisable.

The diphosphonate treatment has been an experimental protocol. Because of the evidence that pyrophosphate and polyphosphates inhibit the growth of crystal formations of calcium phosphate, it has been thought that pyrophosphate may be one of the biologic factors in tissues that prevent formation of insoluble masses of calcium phosphate. However, administration of pyrophosphate does not increase tissue pyrophosphate concentrations because of the ubiquitous presence of pyrophosphatases in tissues, particularly the intestinal mucosa. Pyrophosphate analogues were prepared, therefore, which substitute a carbon bridge (-P-C-P) between the phosphate groups rather than the oxygen bridge (-P-O-P) of pyrophosphate. The diphosphonates are stable and not hydrolyzed by pyrophosphatase and in vitro, like pyrophosphates, have the property of inhibiting growth of calcium phosphate crystals. In vivo these compounds also affect mineralization of bone.

It was hoped that a particular diphosphonate, sodium etidronate, might be useful in the treatment of ectopic calcification, and this compound has been studied carefully in patients with various forms of calcinosis. Unfortunately, in doses that do not interfere with skeletal mineralization, no dramatic effect has been observed on the course of calcinosis of the subcutaneous tissues. One of the hopes for this treatment was that it might be used in conjunction with surgical removal of large masses of calcium precipitate to prevent reprecipitation of calcium phosphate at the site of the surgical incision. There has been no conclusive evidence that this can be accomplished, however. Prolonged treatment of children with sodium etidronate resulted in severe inhibition of bone mineralization with fractures, so that this investigational drug cannot be given in large doses with impunity, and it has now been withdrawn for this purpose. Further studies are under way in the hope that other diphosphonate derivatives might be clinically more useful but no compound of this type is now available for patient treatment.

The third form of treatment has been directed at reducing serum phosphate concentration by sequestration of phosphate in the intestinal lumen, with the aim of lowering the Ca × P product sufficiently to reduce the possibility of calcium phosphate precipitation in the altered connective tissue matrix. The deficiencies of this type of treatment are: (1) the need for large intake of aluminum hydroxide or carbonate to reduce serum phosphate sufficiently to impede calcium salt deposition; and (2) the impairment of mineralization of growing bone by hypophosphatemia. The hypophosphatemic therapy may be more useful in adults than in growing children because of this adverse effect on skeletal growth and mineralization. During periods of rapid progress of calcinosis, and particularly if surgical removal of calcified masses is contemplated, the induction of hypophosphatemia may be useful. Serum phosphate probably must be reduced to less than 3.5 mg/dl if significant inhibition of calcium salt deposition is to be obtained. In one young adult with calcinosis circumscripta, probenecid administration resulted in reduction of concentrations of phosphate and urate in serum with lessening of the calcinosis. This treatment has not been reported in children.

The surgical removal of large masses of subcutaneous calcification or the debridement of nodules of calcinosis that are ulcerating through overlying skin, particularly in periarticular regions, may be helpful cosmetically as well as functionally. Calcium salt deposition in the skin scar is a complication of this therapy. As mentioned above, it was hoped that diphosphonate treatment before and after surgery until the wound was completely healed would minimize this complication. Although we have used this expedient in several patients with possibly favorable results, the over-all experience has not been sufficiently good to warrant approval of this investigational drug, and it is not available at present. The possibility remains that hypophosphatemia may be helpful, and this is the only treatment that can be employed at present.

Tumoral Calcinosis. Whether this entity can be separated from other forms of calcinosis is unclear. The basis for considering it separately is the possible difference in incidence of this disorder in certain African peoples, suggesting genetic factors predisposing to its occurrence. In the tumoral calcinosis of the African variety the onset is usually in childhood, with the development of large masses of calcium phosphate deposition in sites where pressure may also play a role, i.e., in the gluteal regions. In this variety of calcinosis there is no previous evidence of a connective tissue disorder resembling scleroderma or dermatomyositis. This diagnosis has also been used for patients with calcinosis circumscripta who have large masses of calcium phosphate deposition causing visible and palpable tumors, so that in this sense it is simply a descriptive term rather than a special form of calcinosis.

Treatment of these patients has been primarily surgical, with removal of the large masses of calcium salt in subcutaneous tissue where they cause discomfort, interfere with normal function or are cosmetically unpleasant.

CASE REPORTS

Calcinosis Circumscripta Following Polymyositis

At the age of 5 years this girl, D.M.L., had generalized muscle weakness and tenderness without skin manifestations. Biopsy was compatible with polymyositis. She was treated with corticosteroids for two years with resultant improved muscle power. At the age of 10 there was a superficial tender mass adjacent to left elbow which was biopsied and reported to show calcification, focal necrosis and fibrosis, and chronic inflammation. Subsequently she developed similar lesions in the periarticular regions of knees and elbows (Fig. 8–1). Some of these lesions ulcerated and discharged a chalky material. The concentrations of serum calcium, phosphate and alkaline phosphatase were 9.8 mg/dl, 4.1 mg/dl and 3.9 Bodansky Units/dl respectively. The draining lesions became secondarily infected with staphylococcus. She was treated with the experimental diphosphonate, sodium etidronate, at a dose of 600 mg (12 mg/kg) increasing to 800 mg (15 mg/kg) per day. Following 18 months of treatment the calcified periarticular lesions seemed softer but did not regress. However, episodes of ulceration and

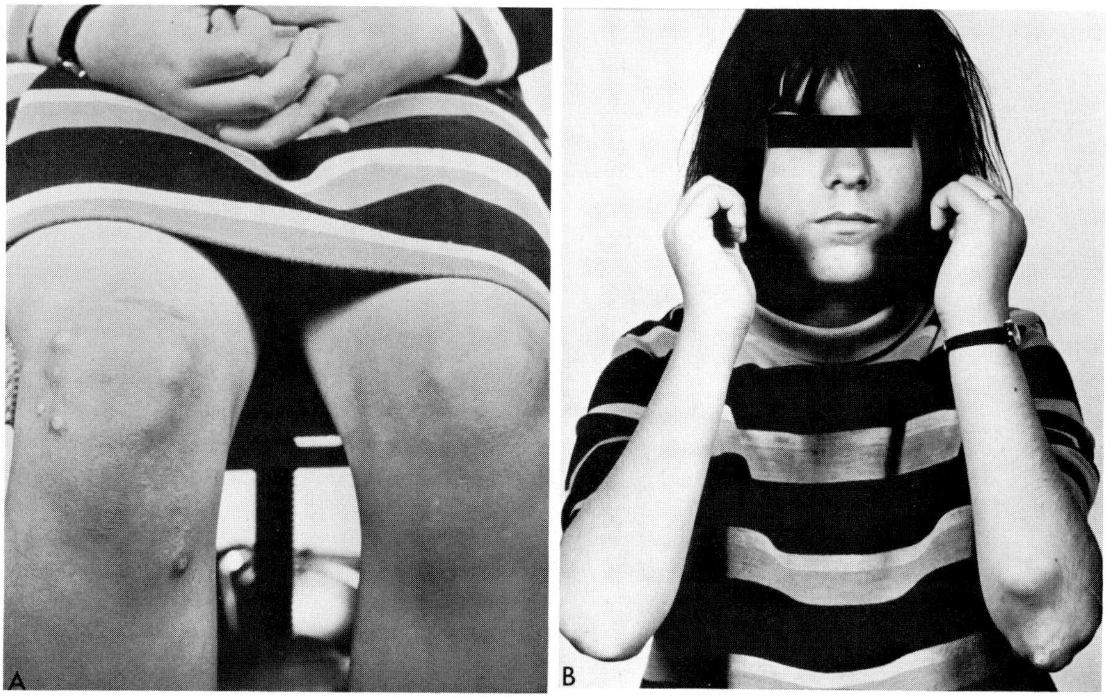

Figure 8–1. Calcinosis circumscripta following illness diagnosed as polymyositis (D.M.). Knees and elbows show nodular lesions of subcutaneous deposition of calcium phosphate. Some of these lesions eventually eroded through the skin, resulting in sinuses that drained chalky fluid.

drainage ceased. Treatment was discontinued because of pregnancy.

COMMENT: This is an instance of calcinosis in the periarticular subcutaneous tissues following an inflammatory disease of muscle, but without the characteristic skin lesions of dermatomyositis. The concentration of the calcified lesions in the periarticular subcutaneous tissue is characteristic. Treatment with diphosphonate did not result in complete solubilization and resorption of the masses of calcium phosphate in the tissue, but did seem to have some effect since the masses became softer and no longer perforated through the overlying stretched skin, so that draining sinuses and scarring ceased to develop. There were no obvious side-effects of treatment, but when the patient became pregnant administration of the drug was discontinued since possible teratogenic effects could not be excluded. The infant born of this pregnancy showed no abnormalities. Since D. M. L. was symptomatically improved it was elected not to give further treatment.

Tumoral Calcinosis

At the age of 4 years J. T., a black female, was hospitalized because of fever and pain and stiffness of knee joints of several days' duration. An infiltrate was found by chest x-ray and pneumonia was diagnosed and treated with penicillin. The pain in the lower extremities became worse and she had marked stiffness when she first began to walk in the morning. At the age of 5 years fusiform swellings of fingers were noted and the limitation of motion of hips, knees and ankles was more severe. A diagnosis of juvenile rheumatoid arthritis was suggested and salicylate treatment was given. The joint manifestations appeared to improve with such treatment but at age 7 a firm nodule was noted on the extensor surface of the left leg, followed by similar masses in the occipital region, right thigh and right calf. At age 10 she was first seen in our clinic. At this time there were large masses of calcium salts in the subcutaneous tissues of right upper arm and both thighs. The appearance of these lesions on x-ray examination is shown in Figure 8–2. She was given a six-month course of treatment with sodium etidronate without any evident change. After this period the lesion on the medial aspect of the left thigh became tender and ulcerated with drainage of white chalky fluid. The mass was excised and healed satisfactorily without calcification of the scar. Biopsy of the initial mass on the right thigh was reported as showing "chronic inflammation and cyst formation with a thick, fibrous wall, calcification and foreign body reaction." Pertinent laboratory data are summarized in Table 8–2.

COMMENTS: The initial manifestations in this patient suggested a connective tissue disorder thought to be rheumatoid arthritis. However, the subsequent course was not that of arthritis but of the

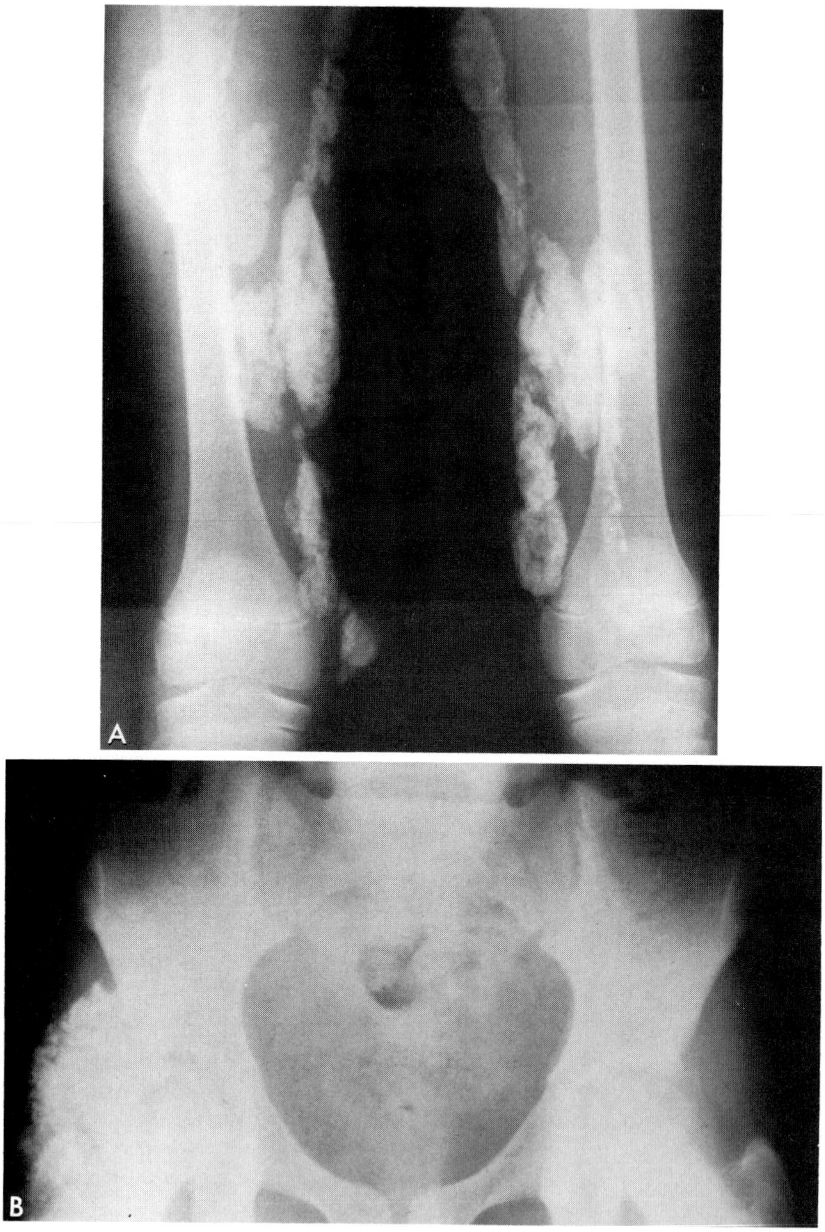

Figure 8–2. *a* and *b*, Tumoral calcinosis (J.T.). Large masses of subcutaneous calcification in thighs and gluteal region.

Table 8-2. J.T.

DATE	HCT %	WBC × 1000	SED. RATE *mm/hr*	SERUM *Ca* *mg/dl*	SERUM *P* *mg/dl*	*Alk. P'tase. Bodansky Units*
10-68	38	5.8	11			
10-70	39	6.4	2			
3-3-73				9.4	4.8	
12-12-73	Treatment with sodium etidronate begun					
1-2-74	36	10.1		10.2	5.4	12.8
1-30-74	37	4.7		9.9	5.6	15.6
2-27-74	37	5.1		9.5	6.3	19.8
3-27-74	36	4.6		8.9	6.3	20.1
4-10-74	35	6.0		8.9	7.6	16.2
5-7-74	36	7.0		9.7	6.8	22.0
7-10-74	34	6.7		10.6	7.0	26.5
	Treatment discontinued					
8-21-74	36	6.8		10.7	5.6	31.1

development of large masses of calcium phosphate deposition in the subcutaneous tissues, which fits the classification of tumoral calcinosis. The nature of the tissue insult is unknown. Treatment with diphosphonate (sodium etidronate) was not followed by any reduction of size of the calcified masses, and surgical excision of one of the masses was necessary to relieve discomfort and prevent a chronic fistula.

The laboratory findings are of interest in several respects. There were no hematologic abnormalities and sedimentation rate was never elevated. Serum calcium, phosphate and alkaline phosphatase concentrations were normal before diphosphonate treatment. However, during the six-month course of treatment serum phosphate and alkaline phosphatase both increased. Hyperphosphatemia following diphosphonate treatment has been seen frequently and presumably represents an effect of this compound on tubular reabsorption of phosphate. The increase of alkaline phosphatase may be a manifestation of the effect of diphosphonate on bone mineralization and may represent osteoblastic response to bone demineralization. In other patients osteopenia and fractures have resulted from prolonged treatment with diphosphonate. Because of the lack of striking therapeutic effect in calcinosis and the possible impairment of bone mineralization, this drug is no longer used. Treatment of calcinosis in children remains unsatisfactory since there is nothing that preferentially inhibits ectopic calcium salt precipitation without inhibition of bone mineral deposition. Surgical excision of large calcified masses remains the only treatment. This should be reserved for those masses which are causing pain or interfering with joint function. Calcium precipitation in the scar can be a complication of surgery.

ECTOPIC CALCIFICATION WITH BONE FORMATION

This variety of ectopic calcification is distinguished by the fact that the calcium salts are not deposited in unorganized masses in connective tissue but rather in an organized matrix which is true bone. This presumably is due to the stimulation of primordial cells in extraskeletal loci with conversion to osteoblasts, osteocytes and osteoclasts as well as concurrent transformation of primitive progenitor cells to marrow cells. The result is the development of tissue histologically identical with bone in extraskeletal locations. There are two major varieties of ectopic ossification: (1) post-traumatic myositis ossificans; and (2) ossifying fibrous dysplasia of congenital origin.

Myositis Ossificans

This is a localized conversion of connective tissue to bone following severe trauma to muscle. The trauma is often of the crushing variety as in athletic injuries, but can also be of surgical origin following resection of large masses of muscle. The healing response to muscle necrosis and hemorrhage includes calcification of the injured tissue and hematoma. There may also be induction of bone-forming cells which produce a calcifiable matrix, resulting in formation of ectopic bone that undergoes remodeling depending on local pressure forces. The process remains localized, however, the major manifestations being local pain and tenderness and a hard bony mass that may interfere with mobility if it is close to a joint. The treatment is primarily palliative. Sodium etidronate has been used experimentally in the attempt to stop the calcification of muscle injury, but there is no clear-cut evidence of success. Excision of the mass may be necessary if it interferes with normal function.

Ossifying Fibrous Dysplasia (Myositis Ossificans Progressiva)

This disorder is quite different from traumatic myositis ossificans, and the designation *ossifying fibrous dysplasia* is preferable to the older terminology of *myositis ossificans progressiva*. It is a congenital abnormality of connective tissue although the genetic transmission is uncertain. There is an unusual anatomic marker for this disease, namely, strikingly short big toes and thumbs. In one patient additional congenital malformations of the hands were present but this is unusual. These children begin to show evidence of abnormal connective tissue reaction in the first few years of life. Following mild trauma, or even without known trauma, tender indurated masses develop which slowly resolve and eventually form hard bony masses radiographically seen to penetrate deeply into muscle, replacing the muscle fascial structures with bone. Histologically the lesions show all the appearances of mature bone with haversian systems, i.e., concentric lamellar formations of mineralized tissue incorporating osteocytes and with osteoblasts and osteoclasts at the surfaces of the bony trabeculae. Marrow spaces are also present with true bone marrow cells. Not all of the indurated swellings progress to bony formation, but gradually the growth of the bony masses causes immobilization of joints, muscle contractures and severe deformity, both because of large nodular projecting masses of bony tissue and because of the severe contractures with limitation of joint motion. Eventually the patient becomes bedridden, and death results from respiratory complications of immobilization and impaired chest movement.

There are no detectable biochemical abnormalities. Serum calcium, phosphate and alkaline phosphatase concentrations are normal. There are no obvious abnormalities of the skeleton itself except as the result of treatment of the disease, or as the consequence of immobilization with resultant osteopenia.

Treatment has been discouraging. Corticoid therapy has been tried at the early stage of appearance of tender indurated masses, in the hope of hastening their resolution or preventing their development and progression into bony masses. There is no evidence that such treatment has altered the course of the process. Intensive investigation of the value of sodium etidronate in preventing this ectopic ossification or reducing the density of the mineralized tissue has shown that treatment with this particular diphosphonate has not been of clinical value. With prolonged high dose treatment there has been serious demineralization of the skeleton and abnormalities of mineralization of the growing bone which have necessitated discontinuance of treatment. It thus is not possible to inhibit mineralization of ectopic bone without marked inhibition of mineralization of the growing skeleton. Attempts at surgical manipulation, namely, fracture of the ectopic bone in order to mobilize a joint, has usually been of only temporary benefit with reossification, contracture and splinting of the joint again by the growth of bony tissue.

At the moment physical therapy with attempts to maintain joint mobility is all that can be offered. The severity of the disorder differs among patients, and some subjects may show apparent quiescence of the ossifying transformation of connective tissue so that they may not be completely immobilized.

CASE REPORTS

Ossifying Fibrous Dysplasia

N. P., a white male, was born at term weighing 11 lb, and showed absence of the third, fourth, and fifth digits of the left hand and short thumbs and big toes. He seemed otherwise normal, growing and developing without complication until at age 2 years a tender indurated swelling appeared in the parietal region of the skull. This resolved but other indurated swellings appeared on the back, anterior chest wall and abdomen. A diagnosis of connective tissue disorder was made and corticosteroid treatment was begun. This did not change the course of the disease. It was noted that the indurated swellings became hard and bony and these hard masses persisted, causing limitation of motion of neck and arms. At age 3½ he was admitted for study at the Naval Medical Center in Bethesda, where serum calcium, phosphate and alkaline phosphatase concentrations were found to be normal. X-ray examination showed dense masses with structure resembling bone at the sites of the swellings on back and neck, with bony trabeculae replacing biceps and triceps muscle and rectus abdominis. A diagnosis of ossifying fibrous dysplasia was made. There was no family history of similar disorder. The parents were entirely normal, as were two siblings.

Since we were investigating the possible benefit of diphosphonate treatment in this disorder the patient was transferred to our clinic at Baltimore City Hospitals. He was now 4 years of age and had large protuberant bony masses on back (Fig. 8–3) and limitation of extension of neck and back owing to bony transformation of muscles. There was also limited extension and flexion of arms, and hard bony masses were palpated in the muscles of the abdominal wall. On x-ray examination these masses had the appearance of true bone (Fig. 8–4). Initially the child was started on a low calcium diet in the hope that stimulation of PTH output by calcium deficiency would cause differential resorption of the ectopic bone. Serum calcium, phosphate and alkaline phosphatase concentration remained normal but x-ray studies indicated that skeletal demineralization was proceeding along with resorption of the ectopic bone.

Subsequently, treatment with sodium etidronate was initiated at a dose of 10 mg/kg in a single oral dose per day. After several months of this therapy there appeared to be regression of the large bony protuberances in the paraspinal region, and the indurated swellings that developed following mild trauma subsided without becoming ossified. However, it also became evident that the skeleton was progressively losing mineral, and following a fall while alighting from a school bus the patient sustained a fracture of the right femur. All treatment was discontinued. The fracture healed slowly but there was massive myositis ossificans of the injured tissues in the right upper thigh region, so that the right hip joint was held in flexion contracture.

Another course of diphosphonate was given, but again demineralization of the skeleton occurred with localized inhibition of calcification at the growing end of the long bones (Fig. 8–5). These lesions were associated with a rise in serum alkaline phosphatase concentration, presumably representing osteoblastic proliferation in the poorly mineralized bone. The boy fractured his tibia following another fall and all further diphosphonate treatment was discontinued, since inhibition of ectopic bone formation could only be attained at the cost of impairment of skeletal mineralization. The further course was one of progressive ectopic bone growth in the tissues surrounding the hip joints and the abdominal muscles, with additional ectopic bone at sites of pressure such as the pressure from the crutch he had to use. At 10 years of age he was markedly deformed, with severe kyphosis and contracture de-

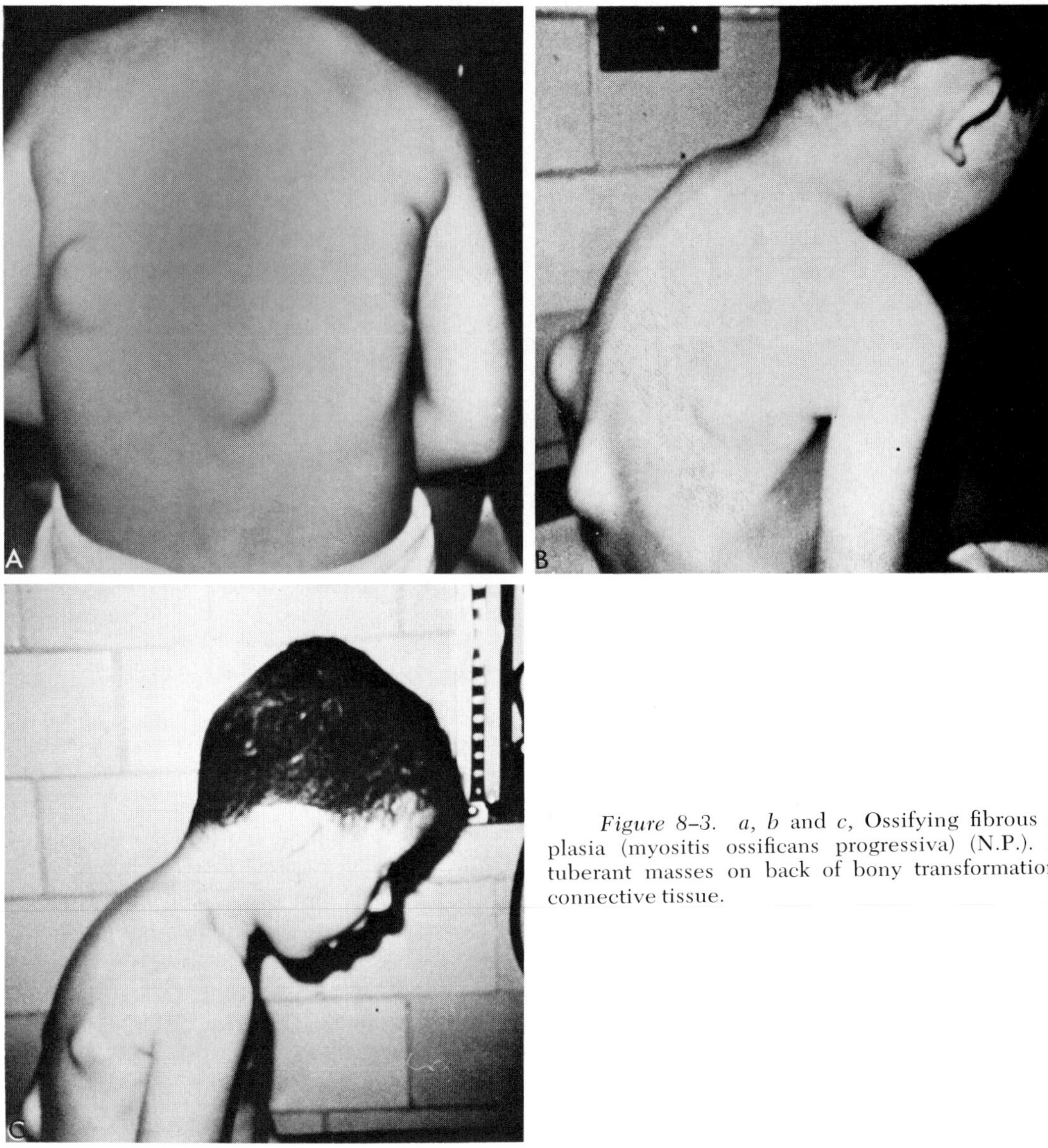

Figure 8–3. *a*, *b* and *c*, Ossifying fibrous dysplasia (myositis ossificans progressiva) (N.P.). Protuberant masses on back of bony transformation of connective tissue.

formity of the right hip and inability to abduct arms. He was able to walk with a crutch but had increasing difficulty in moving about.

COMMENTS: This is an example of severe ossifying fibrous dysplasia with the typical anatomic markers of short thumbs and big toes, but in addition an unusual absence of several digits of one hand. The onset suggested a connective tissue disorder, but it was likely that the indurated swellings which seemed to appear spontaneously were induced by mild trauma, and that lesions became the sites of ectopic bone formation that eventually so penetrated the large masses of skeletal muscle that severe joint contractures and immobility with deformity resulted. Although the density of ectopic bone could be altered by calcium deficiency or diphosphonate treatment, skeletal mineralization was similarly diminished, resulting in fractures and further deformity. Treatment of this disorder remains unsolved.

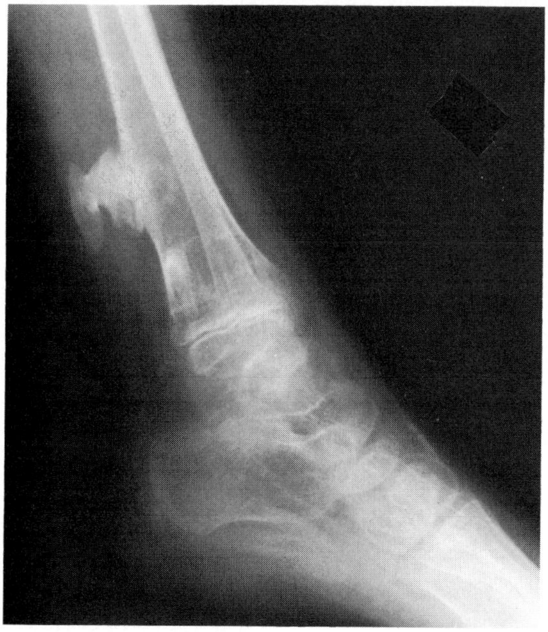

Figure 8–4. Radiograph of mass in leg showing appearance of bone (N.P.).

DISSEMINATED CONGENITAL OSTEOMAS OF SKIN

This is an extremely rare disorder of infancy. In one reported case the lesions were noted immediately after birth. These were "small bumps" on the left side of the scrotum, the left leg, and elsewhere on the body. During the first months of life the original lesions increased in size and new ones appeared. Most of the lesions were small, less than 1 cm in diameter, but others coalesced to form larger masses. They were firm, hard and gritty and on x-ray examination were radiopaque. Biopsy of the skin showed that the corium was replaced by bone spicules containing osteoblasts and marrow cells.

For the diagnosis of this disorder the lesions should be present at birth or soon after. There is no evidence of an inflammatory or traumatic lesion. It is suggested that the osteomas have their origin during embryogenesis and that they are chondro-osteomas. The lesions are characteristically within the skin and bony hard, and radiographs reveal obvious deposition of mineral. Biopsy shows true differentiation into bone. The patients do not have the short thumb and big toes of children with ossifying fibrous dysplasia, and the bony masses are more superficial than in the latter disorder and usually do not penetrate into muscle with loss of muscle function and joint contractures. Erosion of the skin and discharge of the osseous fragments can occur as in patients with calcinosis.

There is no known treatment of this disorder.

REFERENCES

METASTATIC CALCIFICATION

Bauer, J. M. and Freyberg, R. H.: Vitamin D intoxication with metastatic calcification. J.A.M.A. *130*:1208–1215, 1946.

Gabbiani, G., Badonnel, M. C. and Baud, C. A.: Relationship between iron, calcium and phosphate during experimental cutaneous calcinosis. Calcif. Tissue Res. *4*:224–230, 1969.

Holman, C. B.: Roentgenologic manifestations of vitamin D intoxication. Radiology *59*:805–815, 1952.

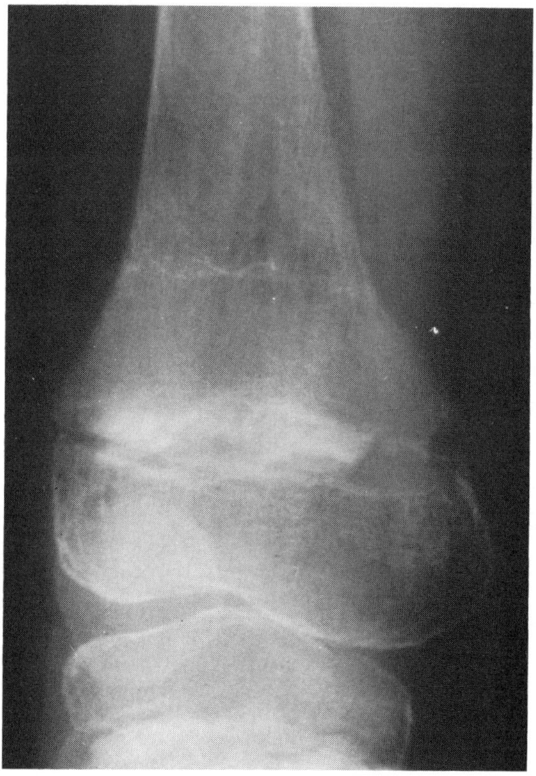

Figure 8–5. Demineralization of bone and irregular mineralization of metaphyseal zone resulting from inhibition of mineralization by diphosphonate treatment (N.P.).

Irnell, L., Werner, I. and Grimelius, L.: Soft tissue calcification in hyperparathyroidism. Acta Med. Scand. *187*:145–151, 1970.

Parfitt, H.A.M.: Soft-tissue calcification in uremia. Arch. Intern. Med. *124*:544–556, 1969.

Rasmurthy, R. S., Harris, V. and Pildes, R. S.: Subcutaneous calcium deposition in the neonate associated with intravenous administration of calcium gluconate. Pediatrics 55:802–806, 1975.

Seifert, G.: Morphologic and biochemical aspects of experimental extraosseous tissue calcification. Clin. Orthop. *69*:87–100, 1970.

CALCINOSIS UNIVERSALIS, CALCINOSIS CIRCUMSCRIPTA, TUMORAL CALCINOSIS

Cram, R. L., Barmada, R., Geho, W. B. et al: Diphosphonate treatment of calcinosis universalis. N. Engl. J. Med. *285*:1012–1013, 1971.

Dent, C. E. and Stamp, T. C. B.: Treatment of calcinosis circumscripta with probenecid. Br. Med. J. *1*:216–218, 1972.

Lafferty, F. W., Reynolds, E. S. and Pearson, O. H.: Tumoral calcinosis. A metabolic disease of obscure etiology. Am. J. Med. *38*:105–118, 1965.

Leonard, F., Boke, J. W., Ruderman, R. J. et al: Initiation and inhibition of subcutaneous calcification. Calcif. Tissue Res. *10*:269–279, 1972.

McClatchie, S. and Bremner, A. D.: Tumoral calcinosis—an unrecognized disease. Br. Med J. *1*:153–155, 1969.

Metzger, A. L., Singer, F. R., Bluestone, R. et al: Failure of disodium etidronate in calcinosis due to dermatomyositis and scleroderma. N. Engl. J. Med. *291*:1294–1296, 1974.

Nassim, J. R. and Connolly, C. K.: Treatment of calcinosis universalis with aluminum hydroxide. Arch. Dis. Child. *45*:118–121, 1970.

Wuthier, R. E., Bisaz, S., Russell, R. G. G. et al: Relationship between pyrophosphate, amorphous calcium phosphate, and other factors in the sequence of calcification *in vivo*. Calcif. Tissue Res. *10*:198–206, 1972.

FIBRODYSPLASIA OSSIFICANS PROGRESSIVA

Cabanela, M. E. and Jowsey, J.: The effect of diphosphonate on phosphate-induced bone loss and soft-tissue calcification. Mayo Clin. Proc. *46*:492–495, 1971.

McKusick, V. A.: Heritable Disorders of Connective Tissue, 4th Ed. C. V. Mosby Co., St. Louis, 1972, pp. 687–706.

Rogers, J. G., Dorst, J. P. and Geho, W. B.: Use and complications of high-dose disodium etidronate therapy in fibrodysplasia ossificans progressiva. J. Pediatr. *91*:1011–1014, 1977

Weiss, I. W., Fisher, L. and Phang, J. M.: Diphosphonate therapy in a patient with myositis ossificans progressiva. Ann. Intern. Med. *74*:933–936, 1971.

DISSEMINATED OSTEOMAS OF SKIN

Vero, F., Machacek, G. F. and Bartlett, F. H.: Disseminated congenital osteomas of the skin. J.A.M.A. *129*:728–734, 1945.

Index

Page numbers in *italics* refer to illustrations; page numbers followed by (t) refer to tables.